NARCOTIC DRUGS
Biochemical Pharmacology

NARCOTIC DRUGS
Biochemical Pharmacology

Edited by
Doris H. Clouet
New York State Research Institute for
Neurochemistry and Drug Addiction
and the
Narcotic Addiction Control Commission
New York, New York

ℚ PLENUM PRESS • NEW YORK–LONDON • 1971

First Printing — May 1971
Second Printing — May 1972

Library of Congress Catalog Card Number 76-128503

ISBN 0-306-30495-3

CONTRIBUTORS

Herbert L. Borison
  *Department of Pharmacology and
    Toxicology
  Dartmouth Medical School
  Hanover, New Hampshire*

Alan F. Casy
  *University of Alberta
  Edmonton, Alberta, Canada*

Doris H. Clouet
  *New York State Addiction Con-
    trol Commission
  Testing and Research Laboratory
  Brooklyn, New York*

Joseph Cochin
  *Department of Pharmacology
  Boston University School of Medi-
    cine
  Boston, Massachusetts*

Vincent P. Dole
  *Rockefeller University
  New York, New York*

Henry W. Elliott
  *Department of Medical Pharma-
    cology and Therapeutics
  California College of Medicine
  University of California
  Irvine, California*

Max Fink
  *Division of Biological Psychiatry
  Department of Psychiatry
  New York Medical College and
  Metropolitan   Hospital   Medical
    Center
  New York, New York*

Alfred M. Freedman
  *Department of Psychiatry
  New York Medical College and
    Metropolital Hospital Medical
    Center
  New York, New York*

James M. Fujimoto
  *Department of Pharmacology
  Marquette School of Medicine
  Milwaukee, Wisconsin*

Robert George
  *Department of Pharmacology and
    Brain Research Institute
  University of California
  The Center for the Health Science
  Los Angeles, California*

Louis S. Harris
  *University   of   North   Carolina
    School of Medicine
  Chapel Hill, North Carolina*

Carl C. Hug, Jr.
  Department of Pharmacology
  The University of Michigan
    Medical School
  Ann Arbor, Michigan

Hiroshi Kaneto
  Department of Pharmacology
  Faculty of Pharmaceutical Science
  Nagasaki University
  Nagasaki, Japan

Salvatore J. Mulé
  New York State Narcotic Addic-
    tion Control Commission
  New York, New York

Richard Resnick
  Department of Psychiatry
  New York Medical College and
    Metropolitan Hospital Medical
    Center
  New York, New York

Jiri Roubicek
  Division of Biological Psychiatry
  Department of Psychiatry
  New York Medical College and
    Metropolitan Hospital Medical
    Center
  New York, New York

Joseph T. Scrafani
  New York State Narcotic Addic-
    tion Control Commission
  Testing and Research Laboratory
  Brooklyn, New York

Fu-Hsiung Shen
  Department of Pharmacology
  University of California at San
    Franciso
  San Francisco, California

Louis Shuster
  Department of Biochemistry

Tufts University School of
    Medicine
  Boston, Massachusetts

Eric J Simon
  Department of Medicine
  New York University Medical
    School
  New York, New York

Jewell W. Sloan
  Addiction Research Center
  Lexington, Kentucky

Alfred A. Smith
  Departments of Anesthesiology
    and Pharmacology
  New York Medical College
  New York, New York

A. E. Takemori
  Department of Pharmacology
    College of Medical Sciences
  University of Minnesota
  Minneapolis, Minnesota

John F. Taylor
  New York State Research Institute
    for Neurochemistry and Drug
    Addiction
  Ward's Island, New York

Jan Volavka
  Division of Biological Psychiatry
  Department of Psychiatry
  New York Medical College and
    Metropolitan Hospital Medical
    Center
  New York, New York

E. Leong Way
  Department of Pharmacology
  University of California at San
    Francisco
  San Francisco, California

James R. Weeks
Experimental Biology Division
The Upjohn Company
Kalamazoo, Michigan

Marta Weinstock
Department of Pharmacology
St. Mary's Hospital School
London, W, 2, England

Arthur Zaks
Division of Biological Psychiatry
Department of Psychiatry
New York Medical College and
Metropolital Hospital Medical
Center
New York, New York

PREFACE

The riddle of the biochemical nature of drug dependence of the opiate type has stimulated many studies directed toward understanding the molecular basis of the action of opiates, and, particularly, the phenomena of tolerance, physical dependence, and drug-seeking behavior—phenomena exhibited by man and experimental animals exposed persistently to these drugs. The results of these studies provided a substantial body of information which has been published in the scientific and medical literature. The purely pharmacological responses in man and animals to the opiates have been described and evaluated in many monographs and text-books of pharmacology. However, there is no single source for specific and detailed information on the responses of the body and its tissues to narcotic analgesic drugs at the level of biochemical pharmacology; that is, the molecular history of the drug in the body and the biochemical consequences of its presence in tissue. This volume has been prepared in an effort to repair the deficiency.

Two factors have contributed a special urgency to making this information available in convenient form: (1) the current need for a better understanding of the biochemical mechanisms underlying addiction to narcotic drugs, and (2) the progress made in molecular biology which promises that significant advances in the elucidation of fundamental processes in the central nervous system and their drug-induced aberrations may soon be possible.

The question of which areas of research to include and which to omit has been difficult to decide, particularly for studies in disciplines other than biochemical pharmacology. A few such topics have been discussed with the hope that these chapters would be of special interest to scientists working on the basic aspects of drug dependence. Since most chapters describe biochemical events in opiate-tolerant and nontolerant man as well as in experimental animals, it is hoped that these chapters may provide a useful background for clinicians involved in therapy programs for addicts.

A discussion of the possible mechanisms for the action of narcotic drugs covers so many areas of scientific competence that a book of this kind could be produced only through the cooperation of a group of scientists active in the various areas who would be willing to contribute time and expertise to writing the appropriate chapters. I have been fortunate in obtaining the cooperation of such a group of distinguished scientists who have indeed contributed authoritative discussions of the status of research in their areas of specialization. Since each author has interpreted his assignment independently, there is some overlap of material in closely related chapters. I have not attempted to eliminate this minor duplication as I preferred to retain the flow of argument and clarity of discussion in each chapter.

I should like to express my sincere thanks to the contributors not only for their expenditure of valuable time and for the high caliber of their contributions, but also for their success in meeting the deadlines established to minimize the time interval between writing and publication.

Doris H. Clouet

New York, New York
February, 1971

CONTENTS

III. The Effects of Narcotic Analgesic Drugs on General Metabolic Systems

Chapter 7

Intermediary and Energy Metabolism 159
by A. E. Takemori

Chapter 8

Phospholipid Metabolism 190
by Salvatore J. Mulé

Chapter 9

Protein and Nucleic Acid Metabolism 216
by Doris H. Clouet

IV. The Effects of Narcotic Analgesic Drugs on Specific Systems

Chapter 10

Catecholamines and 5-Hydroxytryptamine 229
by E. Leong Way and Fu-Hsiung Shen

Chapter 11

Acetylcholine and Cholinesterase 254
by Marta Weinstock

Chapter 12

Corticosteroid Hormones 262
by Jewell W. Sloan

V. Sites of Action of Narcotic Analgesic Drugs

Chapter 18

VI. Tolerance and Dependence

Chapter 19

Chapter 20

Inhibitors of Tolerance Development...424
by Alfred A. Smith

Chapter 21

Role of Possible Immune Mechanisms in the Development of Tolerance...432
by Joseph Cochin

Chapter 22

Self-Administration Studies in Animals...449
by James R. Weeks

Chapter 1

THE STRUCTURE OF NARCOTIC ANALGESIC DRUGS

Alan F. Casy

University of Alberta
Edmonton, Alberta, Canada

I. INTRODUCTION

In this chapter a survey of the structure features of the chief groups of narcotic analgesics is presented. Emphasis is placed upon compounds in current clinical use or which have been used in the past to relieve pain in man, and only a few aspects of structure–activity relationships are discussed. In spite of the continual appearance of novel structures characterized as having morphine-like actions that are reversed by antagonists such as nalorphine, no significant analgesic has been identified that lacks either a basic center (of a pK_a permitting extensive protonation at physiological pH) or an aromatic moiety, and these two features appear to constitute fundamental structural requirements. The relative and absolute orientations of the various units within the molecule have a vital influence upon activity as is evident from the numerous examples of radical activity differences among the stereoisomers of asymmetric analgesics, and this point is emphasized in the following account.

The literature of analgesics is now extensive (a useful collection has been prepared recently by Ellis[1]), and key references are included in this chapter.

II. MORPHINE AND RELATED COMPOUNDS

Morphine [1], the chief active ingredient of opium, is a pentacyclic alkaloid that is regarded conveniently as a derivative of phenanthrene bridged across the 4,5 positions by oxygen and the 9,13 positions by an ethanamine chain. The latter link creates a 6-membered piperidine ring, the methylated nitrogen of which forms the basic center of the molecule. Ring A is fully aromatic and carries a phenolic hydroxyl group; ring C carries a

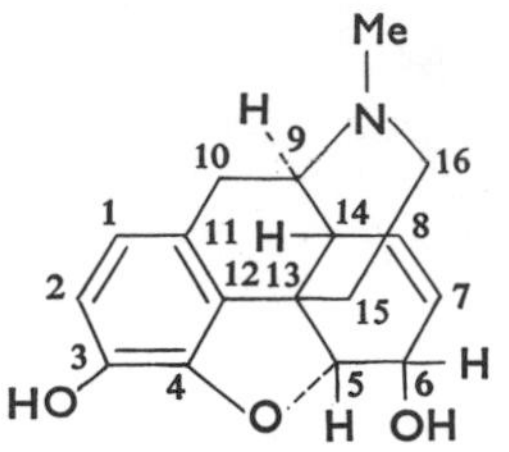

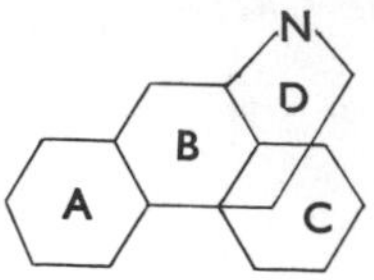

[1]

Ring designations

nonphenolic secondary hydroxyl group at position 6 and an isolated Δ^7-double bond. The structure [1], originally advanced in 1925 by Gullard and Robinson[2] and confirmed by synthesis in the 1950s[3], possesses asymmetric centers at positions 5, 6, 9, 13, and 14. The absolute configuration of the naturally occurring levo enantiomorph has been established by chemical degradation work and its relative configuration and solid-state conformation have been shown by the x-ray studies of Mackay and Hodgkins.[4] The dextro enantiomorph of morphine, available by synthesis,[3] is almost inactive as an analgesic.[5] The shape of the morphine molecule, shown in the conformational formula [2], may best be understood from the standpoint of the piperidine ring D that adopts the chair conformation. The aromatic ring A is fused to ring D by the axial C-13 bond and (via the C-10 methylene group) the axial C-9 bond. The C/D ring junction is *trans* while the ring C conformation is a boat, to allow formation of the ether linkage across positions 4 and 5, with an equatorial hydroxyl group at C-6. The overall shape of the molecule is rather like a letter T with rings C and D forming the horizontal and rings A and B the vertical parts of the capital.

Numerous modifications of morphine have been made and structure-activity relationships in this group have been extensively reviewed and clinical experience has been discussed by Eddy and his colleagues.[6] The potency of morphine is much reduced when its phenolic group is etherified and the best known ether, codeine [3a], is widely used for the relief of mild to moderate pain, and as an antitussive agent.[7] The ethyl (Dionine [3b]), benzyl (Peronine [3c]), and β-4-morpholinoethyl (pholcodine [3d]) ethers have a similar clinical utility. Acetylation of the 3- and 6-hydroxyl groups of morphine gives diacetylmorphine (diamorphine, heroin [4]) which, although more potent than the parent drug, is more toxic and has a higher addiction liability.[8] Addiction to heroin develops rapidly (it is the drug of

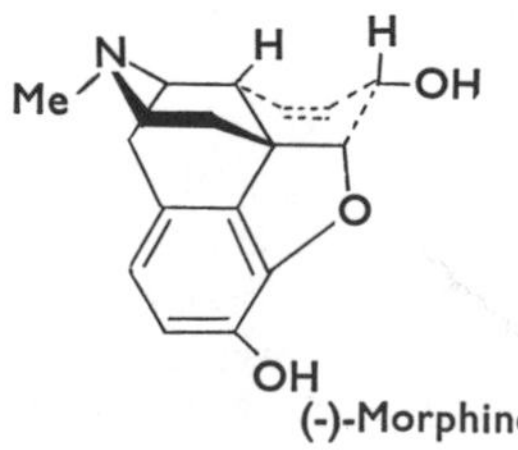

(-)-Morphine

Diagrammatic representation of the three-dimensional arrangement of morphine; surfaces in front of, behind, and in the plane of the paper are represented by —, ···, and — respectively.

[2]

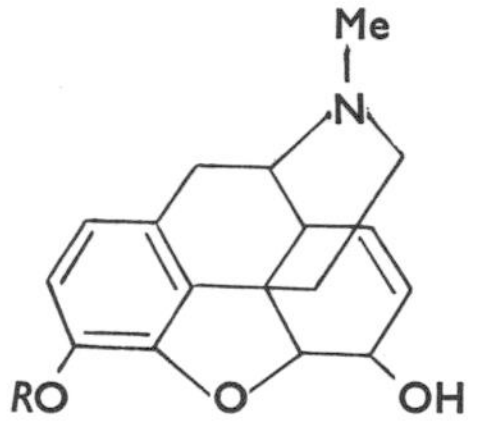

(a) R = Me
(b) R = Et
(c) R = CH$_2$Ph
(d) R = CH$_2$CH$_2$N⟨ ⟩O

[3]

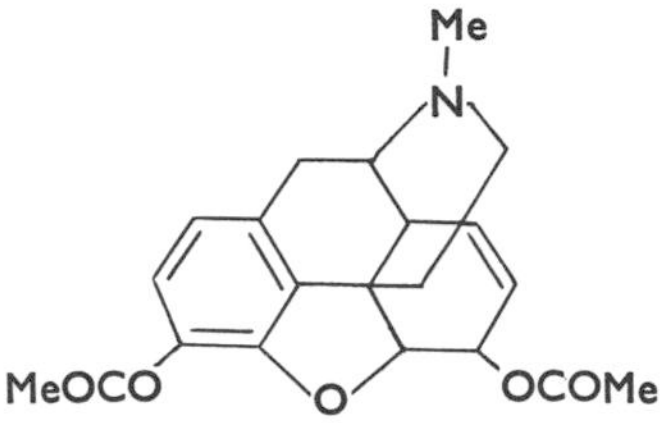

[4]

choice of many addicts), and the World Health Organization has recommended that it be no longer prescribed. The pharmacological effects of heroin are considered to be mediated primarily by its hydrolysis products, 6-monoacetyl morphine and morphine.[9]

The chemical transformation of ring C of morphine has led to several useful drugs that have morphine-like potencies. Clinically used examples are hydromorphone (Dilaudid [5]), dihydromorphine [6], and Metopon [7]. The 14-hydroxy derivative [8] (oxymorphone, Numorphan) is also marketed and is reported to be more potent than morphine with slightly fewer side effects.[10]

Removal of the N-methyl group of morphine gives normorphine; this compound, a probable metabolite of morphine, is less potent when given by normal routes of administration but is of equal potency intracisternally in mice.[11] Several alkylated derivatives of normorphine have been examined, the most significant being the N-allyl compound [9], nalorphine. This substance antagonizes a wide spectrum of morphine effects and is an antidote for morphine poisoning.[12] Although it lacks analgesic properties in laboratory animals, it is a potent, essentially nonaddicting analgesic in man, and this property has led to the development of several clinically useful analgesics based on morphine antagonists.[13] Nalorphine itself cannot be used clinically because it has undesirable psychotomimetic side effects.

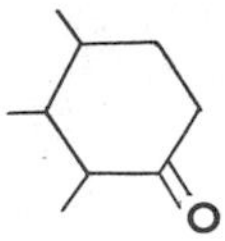

[5] 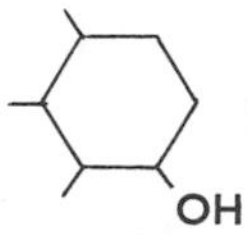 [6] 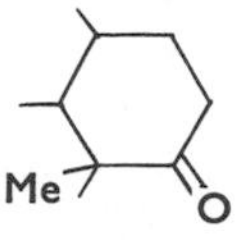 [7] 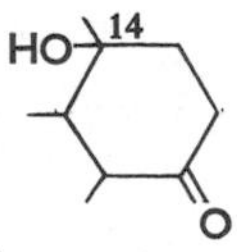 [8]

Partial formulae depicting ring C of morphine.

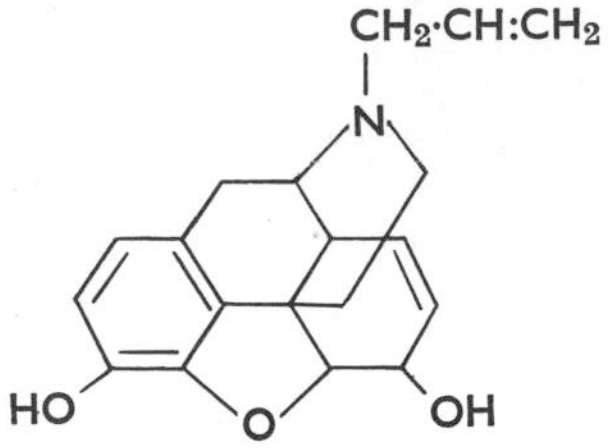

CH₂·CH:CH₂

[9]

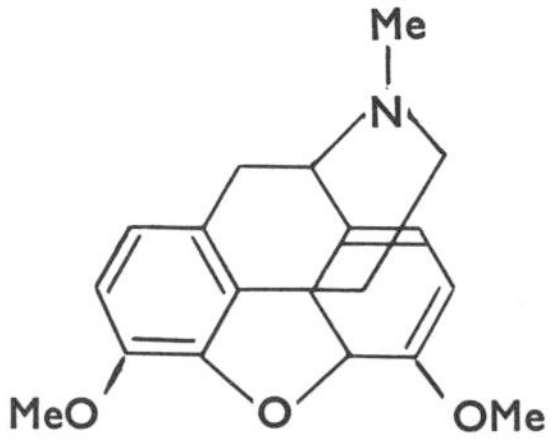

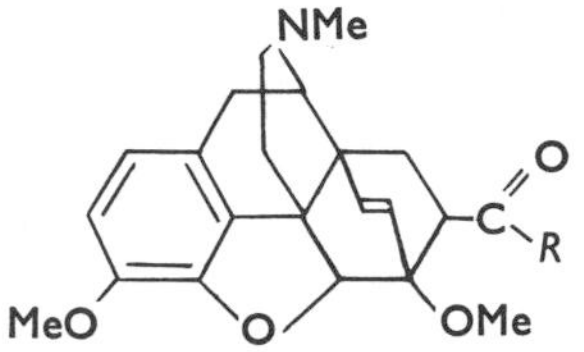

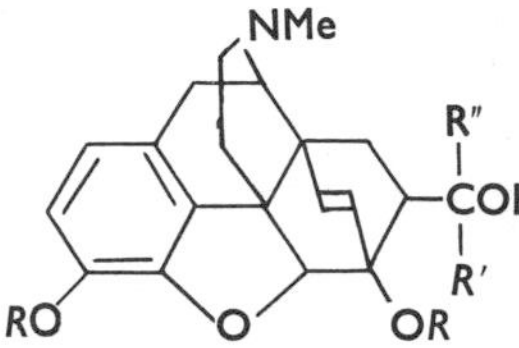

[10]   [11]   [12]

Naloxone, the *N*-allyl analog of oxymorphone [8], is one of the most potent morphine antagonists yet examined (19 × nalorphine in rats), and it gives no analgesic response in the phenylquinone writhing test, a procedure that detects analgesic properties in other morphine antagonists.[14] Harris[15] regards Naloxone as the most nearly pure antagonist so far tested.

There has recently been much interest in analgesics obtained by Diels Alder reactions between thebaine [10] (the diene component) and dienophiles such as vinyl methyl ketone.[16] The resultant ketonic adducts [11] have activities comparable to those of morphine, but derived tertiary alcohols [12] are much more potent and remarkably high potencies have been reported in several cases. Etorphine [12] ($R = H$, $R' = n\text{-Pr}$, $R'' = Me$), for example, is 850 (mice), 1700 (rats), and 8600 (guinea pigs) times as active in the animals specified after subcutaneous administration.[17] Structurally these alcohols may be regarded as morphine derivatives in which the 6 and 14 carbon atoms of ring C are bridged by a bimethylene chain substituted by an alcoholic function [13]. The ability of these derivatives to cause catatonia at very low dose levels has led to their use for immobilizing large

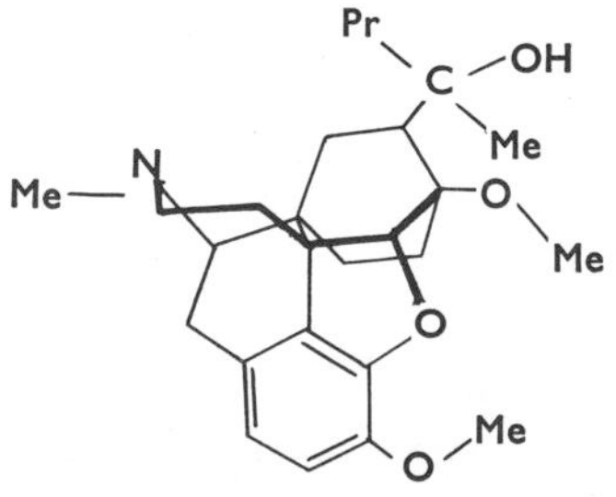

[13]

animals for game conservation and for veterinary purposes.[18] No derivative of this class is as yet marketed as an analgesic, but several clinical studies have been reported.[19]

III. MORPHINAN AND BENZOMORPHAN DERIVATIVES

Morphinan derivatives bear a very close resemblance to morphine; they possess the main skeleton of morphine, lacking only the ether bridge and the functionalities of ring C, while their C/D ring junctions are *trans* as in the natural alkaloid. The best known example is *l-N*-methyl-3-hydroxymorphinan (levorphanol, Dromoran [14]). Levorphanol is highly effective by oral administration and is more potent than morphine (2–3 mg of levorphanol are equal to 10 mg of morphine in man)[20]; its addiction liability, however, is at least as great as that of morphine.[21] The corresponding (+)-isomer (dextrorphan) is devoid of analgesic activity. Structure–activity relationships in the morphinans mirror those of morphine. Thus methylation of the phenolic group of (−)-[14] results in a large decrease in activity, while replacement of *N*-methyl by *N*-allyl gives a potent morphine antagonist, levallorphan.[22] Activity is retained in *l*-3-hydroxy-*N*-methyl isomorphinan [15], an isomer of levorphanol in which rings C and D are *cis* fused.[23]

6,7-Benzomorphan derivatives [16] represent a further simplification of the morphine skeleton, ring C in these compounds being replaced by methyl and other alkyl fragments at C-5 and C-9. Isomers having the 5,9-dimethyl groups *cis* (*α*-) and *trans* (*β*-) with respect to the hydroaromatic ring have been isolated, the former being related sterically to morphine and the morphinans and the latter to isomorphinan. The *α-N*-methyl analog [16a], metazocine, is a potent analgesic (the racemate is almost as active as

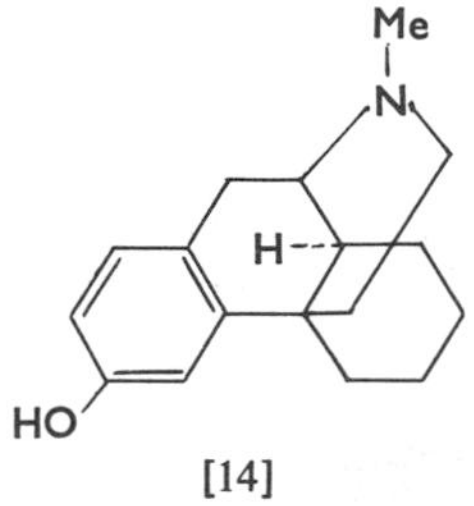

[14]

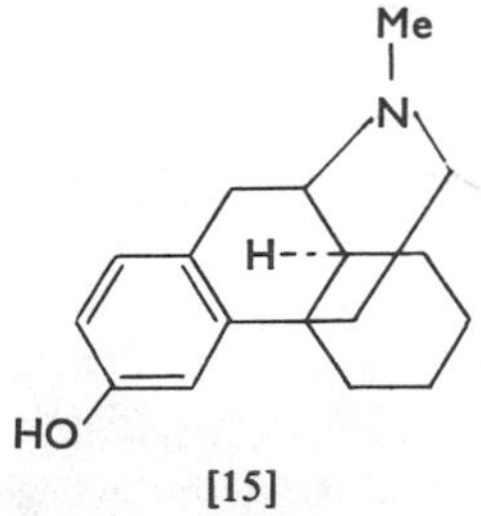

[15]

R
|
N

(a) R = Me
(b) R = CH₂CH₂Ph
(c) R = CH₂—CH=CMe₂
(d) R = CH₂—◁

9 — Me

5 Me

HO

[16]

morphine in mice),[24] but the clinically important benzomorphans are the
N-phenethyl (phenazocine, Prinadol [16b]) and N-3,3-dimethylallyl (pentazocine, Talwin [16c]) derivatives. Pentazocine is the most promising
clinical analgesic yet developed from analgesic antagonists. It is, itself, a
feeble antagonist of the effects of morphine on the tail-flick reaction,[25] but
in man it is an effective analgesic in a wide variety of pain situations as is
evident from the numerous clinical reports now available.[26] On the average
a 30–40 mg intramuscular dose is equal to 10 mg morphine, although higher
doses have been found necessary in cancer patients.[27] Most significantly,
the drug has a very low addiction liability[28]; it was released by the FDA
in 1967 and is not covered by the Harrison Narcotic Act.

The corresponding N-cyclopropylmethyl analog cyclazocine [16d] is
about twice as active a morphine antagonist as nalorphine and is a potent
analgesic in man.[29] It has addiction liability, but the abstinence syndrome
that follows its withdrawal from addicts is not so severe as that seen after
morphine.[30]

The analgesic potency of several 6,7-benzomorphan derivatives is
influenced both by the relative configurations of the 5,9-dialkyl substituents
(β-diastereoisomers are more potent than the α-forms) and by absolute
configuration within a particular enantiomorphic pair; in all cases examined
the activity of both α- and β-racemates resided largely in the levo antipode.
[31,32] In related derivatives that are analgesic antagonists, activity differences
between (±)-diastereoisomers are insignificant, but pronounced potency
variations among enantiomers are still found.[33]

Reviews on both morphinan[34] and benzomorphan analgesics[31] are
available, and Martin has reviewed related antagonists.[13]

IV. 4-PHENYLPIPERIDINES

The best-known example of this group is ethyl 1-methyl-4-phenylpi-
peridine-4-carboxylate (meperidine, pethidine, Dolantin) [17a]. It was first
introduced in 1939 by Eisleb and Schaumann,[35] and it remains the most
widely used synthetic analgesic. In potency, meperidine is graded between
codeine and morphine (50–100 mg equivalent to 10 mg morphine in man)[36]
and is useful for the management of mild to moderate pain, especially in
patients intolerant to opiates. Its toxicity is low, and its action is somewhat
shorter than that of morphine. At equivalent dosage, meperidine is at least

Ph CO$_2$Et

(a) $R = Me$
(b) $R = p\text{-}NH_2C_6H_4(CH_2)_2$
(c) $R = PhNH(CH_2)_3$
(d) $R = PhCH(OH)(CH_2)_2$

N

R

[17]

as depressant as morphine upon respiration and morphine-like side effects such as nausea and vomiting frequently occur. It has found extensive use in the relief of labor pain even though it increases the incidence of delay on the first breath and cry of the newborn infant.[6] Tolerance to the drug develops slowly, and its addictive liability is judged to be lower than that of morphine.[37] Full accounts of the clinical use of pethidine are available.[6,38]

The fact that both morphine and meperidine have a common 4-phenyl-piperidine unit allows a superficial correlation to be made between the two classes but should not be interpreted too narrowly in terms of molecular geometry. In morphine and related rigid analgesics, the 4-phenylpiperidine moiety is constrained to an axial-phenyl chair conformation with the aromatic plane parallel with the plane passing through a line joining C-2 and C-4 of the heterocyclic ring (see [2]). In meperidine and its congeners, however, equatorial 4-phenyl conformations will be preferred, as in [18], and these considerations, together with certain structure–activity differences between 4-phenylpiperidine derivatives and rigid analgesics typified by morphine, suggest that the two classes have different modes of association at the analgesic receptor.[39] The N-allyl analogs of some reversed esters of meperidine (see below), for example, display morphine-like properties in mice but fail to act as analgesic antagonists (like nalorphine) in rats.[39]

A vast number of 4-aryl (generally phenyl) piperidine derivatives have been screened as analgesics, and several reviews upon the structure-activity relationship data so accumulated have been published.[26,40−42] The structural variation most thoroughly investigated is that of replacement of N-methyl by other groups, notably phenalkyl. These studies probably stem from the observation made in 1956 by Perrine and Eddy[443]that N-phenethyl normeperidine is twice as active as meperidine in mice. Compounds in clinical use, developed in this way, are [17b] (anileridine, Leritine),[44] [17c] (piminodine, Alvodine),[45] and [17d] (phenoperidine). The last derivative has been used for neuroleptanalgesia,[46] an anesthetic technique in

R

Ph

Me

N

H

$R = CO_2Et$ or OCOEt

[18]

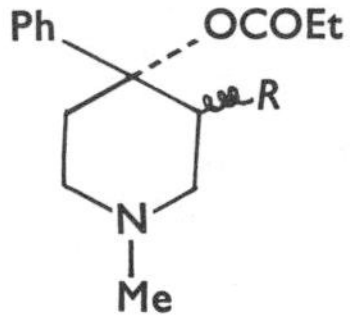

Ph　　　OCOEt

(a) R = H
(b) R = Me

[19]

which an analgesic-tranquilizer mixture is given intravenously usually as an adjunct to nitrous oxide.

Replacement of the ethoxycarbonyl group of meperidine [17a] by a propionyloxy function, giving the so-called reversed ester [19a], results in a potency rise.[47] This change, coupled with the introduction of a 3-methyl substituent into the piperidine ring, gives alphaprodine (Nisentyl [19b]), which has been used clinically.[6] It is about twice as potent as meperidine in man and its action is of very brief duration.[48] Alphaprodine is one of two possible racemic diastereoisomers (the C-3 and C-4 centers of [19b] are asymmetric) and is less potent than the second form, betaprodine.[49] The stereochemistry of the prodines has been investigated extensively,[50,51] and alphaprodine assigned a *trans*-3-Me/4-Ph, and the beta isomer a *cis* configuration. Stereoisomeric pairs of the same type, for example, the 3-methyl analogs of meperidine itself,[52] have been examined, and in all cases the *cis* member has the superior potency. The optical isomers of alpha- and betaprodine also show marked activity differences.[53]

Isomeric forms of the 1,2,5-trimethyl analog [20] also have been studied and activity variations observed.[54] The γ-isomer (2-Me *cis* to 4-Ph, and *trans* to 5-Me) (trimeperidine, Promedol) is used clinically in the U. S. S. R.[6,11]

Other meperidine congeners of importance are the 4-propionyl derivative [21] (ketobemidone), still used in Germany and Scandinavia[6]; the

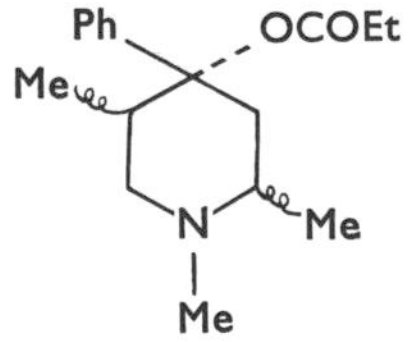

Ph　　　OCOEt
Me

[20]

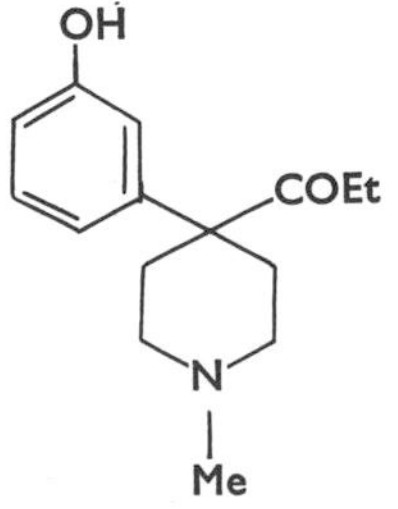

[21]

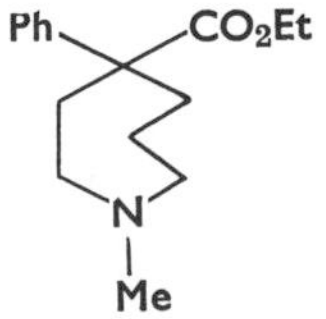

[22]

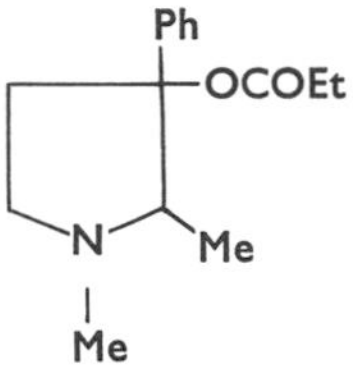

[23]

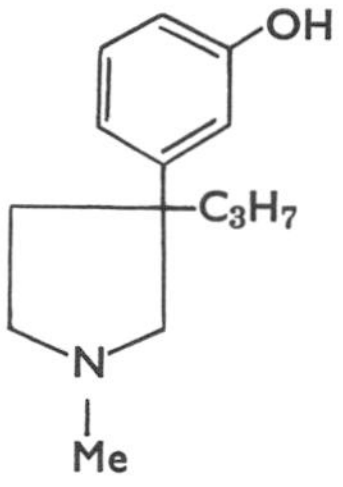

[24]

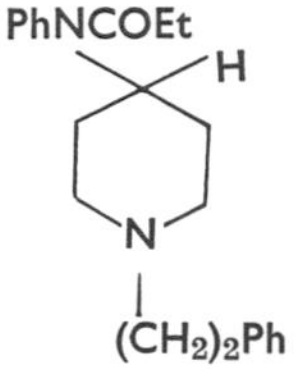

[25]

seven-membered ring analog [22] (ethoheptazine, Zactane); and the pyrrolidine [23] (Prodilidine). The last two derivatives both have potencies in the aspirin–codeine range,[55,56] but the derivative (Profadol) [24] is substantially more active (in postoperative patients 20–50 mg were equivalent to 10 mg of morphine)[57] and has a low addiction liability in monkeys.[58]

    The analgesic Fentanyl [25] is difficult to classify since it has structural features in common with both meperidine and the open-chain analgesic diampromid (described below). Although it is a piperidine derivative, it has an acylated anilino substituent in place of the usual 4-aryl substituent of this class. It is a very potent compound (200 times as active as morphine in mice)[59] and is in clinical use as an analgesic[60] and as a component of Innovan used in neuroleptanalgesia.[61] The N-phenethyl substituent of Fentanyl is essential for activity, the N-methyl analog, for example, being completely inactive in mice.[62] In contrast, the same N-substituted analogs of many 4-phenylpiperidines both have significant activity and the N-phenethyl member is more potent by a factor of only 3 to 4.[47]

V. DIPHENYLPROPYLAMINES AND RELATED ACYCLIC ANALGESICS

    A large number of derivatives of 3, 3-diphenylpropylamine [26] possess analgesic properties, the most active having an oxygenated function (X) at

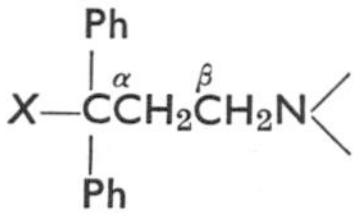

[26]

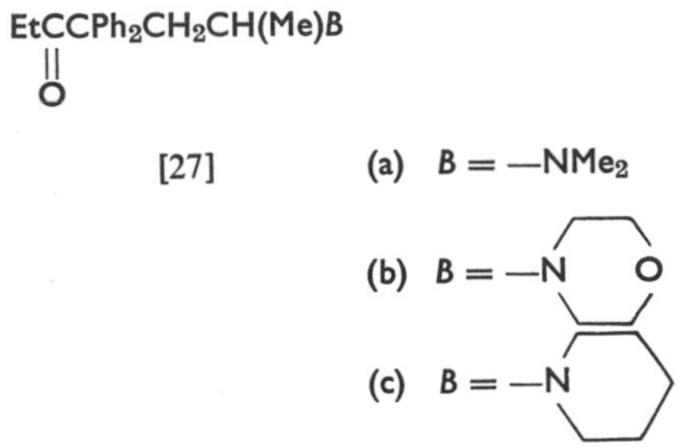

EtCCPh$_2$CH$_2$CH(Me)B
|
O

[27]   (a) B = —NMe$_2$

(b) B = —N⟩O

(c) B = —N⟩

EtCOCPh$_2$CH(R)CH$_2$NMe$_2$

[28]   (a) R = Me
                         (b) R = H

C-3 and a methyl substituent at C-2 (α) or C-1 (β); with few exceptions compounds of interest possess a tertiary dimethylamino or six-membered cyclic amino function. A comprehensive review of the group has been prepared by Janssen.[63] The best-known example is methadone [27a], which was introduced into medical practice on a wide scale in 1946. It is a strong analgesic in man, and its duration of effect is at least as long as that of morphine.[11] It is a powerful antitussive, and its spasmolytic properties make it useful against bladder spasms and renal colic.[6] Side effects are similar to those of morphine; tolerance to its therapeutic action develops after repeated administration, and it sustains addiction at one-quarter of the required dose of morphine, with a longer-lasting effect. After withdrawal, physical dependence signs are slow to develop and are less severe than those after morphine withdrawal; methadone thus may be used for the withdrawl of patients from morphine and other opiates.[64,65]

Variations in the basic group of methadone have led to the morphlino (phenadoxone, Heptalgin, Heptazone [27b]) and piperidino (dipipanone, Pipadone [27c]) analogs, both of which are employed clinically.[6,66] The α-methyl isomer Isomethadone [28a] and the straight-chain derivative nor-methadone [28b] are less potent than methadone.[11] The nor-compound is one of the ingredients of an antitussive preparation Ticarda, which is widely used in Germany and neighboring countries. The addiction liability of normethadone exceeds that of codeine,[37] and several cases have been recorded of the use of Ticarda by addicts instead of other narcotics.

A study of basic amides related to methadone led to the discovery of dextromoramide (R-875, Palfium [29]), which has a morpholino basic group, a pyrrolidino-amide oxygen function, and an α-methyl substituent (as in isomethadone). Clinically it is an analgesic of high potency (a dose of 5 mg is reported to be equivalent to 10 mg of morphine for the treatment of

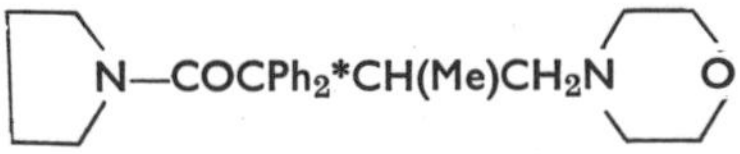

⟩N—COCPh$_2$*CH(Me)CH$_2$N⟩O

[29]

$$EtCH(OR)CPh_2CH_2CHMeNMe_2$$

[30]   (a) $R = H$
        (b) $R = COMe$

postoperative pain,[67] and it is effective by mouth as well as by injection. It is widely used in France and the Low Countries, and many reports of its clinical use are available.[68] Of special interest is the fact that dextromoramide is the dextrorotatory enantiomorph of structure [29] (the asymmetric center is starred), the corresponding levo-isomer being much less potent.[69] This is just one example of the many cases of stereochemical specificity found in diphenylpropylamine analgesics with α- or β-methyl substituents, analgesic potency and morphine-like side effects residing mainly in one isomer of each enantiomorphic pair.[40] The isomers of methadone itself have been tested in man, 4–6 mg of levo methadone being as effective as 7–9 mg of the racemate against postoperative pain.[70] Several stereochemical studies have been made of this group and the more active isomers of certain methadone and thiambutene-type compounds (see below) were shown to possess identical configurations related to R-(−)-alanine.[40] Configurational identity also obtains among the more active isomers of analgesics [26] with α-methyl substituents, for example, (−)-isomethadone and dextromoramide are both related to R-(−)-α-methyl-β-alanine.[71,72] Stereochemical results in the methadone series provided part of the basis for a hypothesis upon the nature of the analgesic receptor.[73]

Some clinical evaluation has been made of the diastereoisomeric secondary alcohols (α- and β-methadol [30a]) and corresponding acetates [30b] derived by reduction of methadone.[74] The methadols are of interest from two points of view: First, among stereoisomers, activity appears to governed by the C-3 rather than the C-6 center,[75] and, second, their monomethylamino analogs represent some of the rare examples of analgesics that have a secondary, rather than a tertiary, basic function.[76]

Isosteric replacement of phenyl in diphenylpropylamine analgesics is disadvantageous,[6] but the related dithienylbutenylamines (for example, dimethylthiambutene [31a]) are potent analgesics of a similar order of activity as morphine in animals.[77] The pharmacology of the diethylamino derivative [31b] has been investigated in the dog and the compound is marketed as Themalon for use in veterinary practice.[78]

In a further modification of the general formula [26] Pohland and Sullivan[79] prepared a compound [32] in which the two phenyl groups are on adjacent carbon atoms. The α-racemate, propoxyphene, is a weak analgesic in rats, the dextro isomer, dextropropoxyphene (Darvon, Doloxene),

[31]
        (a) $R = Me$
        (b) $R = Et$

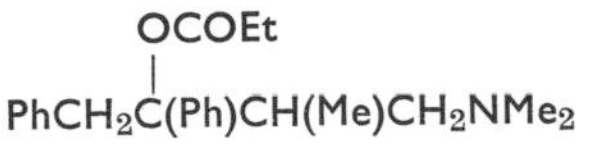

OCOEt
|
PhCH$_2$C(Ph)CH(Me)CH$_2$NMe$_2$

[32]

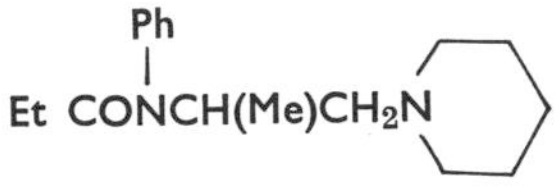

Ph
|
Et CONCH(Me)CH$_2$N⟨ ⟩

[33]

Ph   Me
| |
EtCONCH$_2$CH(Me)N
                     |
                  (CH$_2$)$_2$Ph

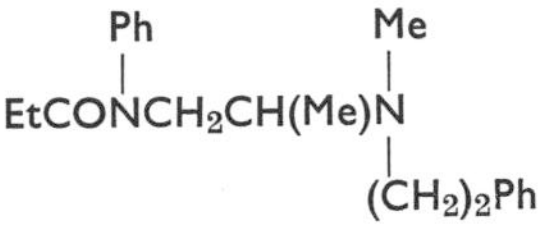

[34]

being twice as effective. Dextropropoxyphene has a potency in man that falls in the aspirin–codeine range and has been used extensively for the treatment of mild to moderate pain.[11] It has little, if any, addiction liability.[80]

Wright, Brabander, and Hardy[81] have modified the methadone structure [27a] by replacing one phenyl group and its attached quaternary carbon atom with nitrogen. Two of the resultant basic anilides, namely, phenampromide [33], and diampromide [34], are significantly active as analgesics (their actions are antagonized by nalorphine),[82] and some clinical studies of their use have been made.[83] The dextro base [34] is the more active of the two enantiomorphs, but its configuration is the reverse of that of the more active optical antipode of methadone[84]—an unexpected result in view of the fact that both compounds have asymmetric centers of the same chemical type. This result, together with differing basic group requirements for optimum activity in the two series [NMe$_2$ for methadone, NMe (CH$_2$CH$_2$Ph) for diampromid],[85] shows that the two analgesics probably differ in their modes of binding to the analgesic receptor.

Janssen has linked the cyanide precursor of normethadone to nor-meperidine to produce diphenoxylate [35]; this complex is not an analgesic but has the constipating action of morphine derivatives and is used as an antidiarrheal agent.[86] The related 4-aminocarboxamide [36], Pirinitramide,

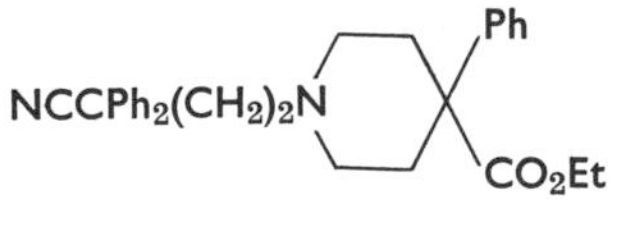

Ph
NCCPh$_2$(CH$_2$)$_2$N⟨ ⟩
                      CO$_2$Et

[35]

O
||
H$_2$NC⟨ ⟩N⟨ ⟩
      |
      N
      |
   (CH$_2$)$_2$CPh$_2$CN

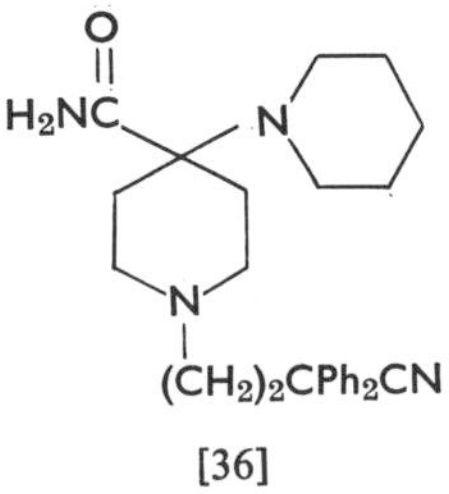

[36]

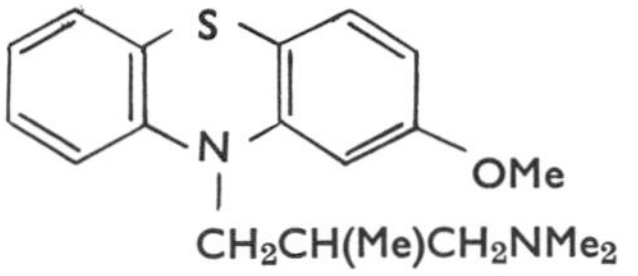

CH$_2$CH(Me)CH$_2$NMe$_2$

[37]

possesses analgesic properties (it is twice as active as morphine in mice)[87] and has been used in the clinic.[88]

VI. MISCELLANEOUS COMPOUNDS

Tranquilizing drugs based on the phenothiazine nucleus potentiate the action of narcotic analgesics,[89] and certain derivatives, notably methotrimeprazine (levo-mepromazine, Levoprome, levo-[37]), are claimed to have intrinsic analgesic activity. In mice (−)-[37] is more potent than morphine in a variety of tests, for example, hot plate ED$_{50}$ 1.02 mg/kg, vs morphine 2.1 mg/kg,[90] but studies with the racemic or dextro forms have not been reported; neither have studies of the antagonism of the effects of (−)-[37] by nalorphine. There is some doubt, therefore, as to its classification as a narcotic analgesic. It is, nevertheless, a valuable analgesic in man, and there are numerous reports of its clinical utility in this respect. The incidence of nausea and vomiting is less, there is no significant respiratory depression after its use,[91] and its addiction liability is of a low order.[92]

Before the advent of analgesics derived from thebaine, some derivatives of 2-benzylbenzimidazole [38] reported in 1957 led the field as the most potent analgesics available at that time.[93] The most active members of the group possess a 5-nitro and/or a 4′-alkoxy substituent plus a 1-β-diethylaminoethyl side chain (dimethylamino analogs are decidedly less potent). The derivative [38] (R = 5-NO$_2$, R' = 4′OEt, etonitazene) is stated to be about 1500 times as active as morphine in mice.[76] In man, Bromig[94] found the 4′-chloro derivative ([38], R=NO$_2$, R'=Cl) to be three to five times as potent as morphine against postoperative pain; in the same trial, the 4′-methoxy analog ([38], R=NO$_2$, R'=OMe) was ten times as potent as morphine but produced severe respiratory depression, and the clinical development of this class of analgesic has not been continued. Etonitazene has found application in experiments designed to condition drug-seeking behavior in rats, the benzimidazole derivative, which is effective at virtually homeopathic dose levels, being undetected by the animals in their drinking water.[11]

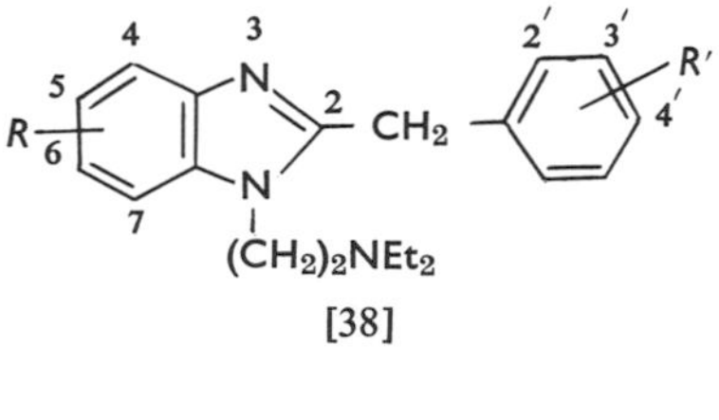

(CH$_2$)$_2$NEt$_2$

[38]

REFERENCES

1. G. P. Ellis, *in* "Progress in Medicinal Chemistry" (G. P. Ellis and G. B. West, eds.), Vol. 6, pp, 322, Butterworths, London (1969).
2. J. M. Gulland and R. Robinson, *Mem. Proc. Manchester Lit. Phil. Soc. 69*, 79 (1925).
3. M. Gates and G. Tschudi, *J. Amer. Chem. Soc. 74*, 1109–1110 (1952); *78*, 1380–1393 (1956).
4. J. Kalvoda, P. Buchschacher, and O. Jeger, *Helv. Chim. Acta 38*, 1847–1856 (1955); M. Mackay and D. C. Hodgkin, *J. Chem. Soc. 1955*, 3261–3267.
5. K. Goto, H. Yamasaki, I. Yamamoto, and R. Ohno, *Proc. Jap. Acad. 33*, 660 (1957).
6. L. F. Small, N. B. Eddy, E. Mosettig, and C. K. Himmelsbach, *Publ. Health Rep. Wash.*, Suppl. No. 138 (1938); O. J. Braenden, N. B. Eddy, and H. Halbach, *Bull. W. H. O. 13*, 937 (1955); N. B. Eddy, H. Halbach, and O. J. Braenden, *Bull. W. H. O. 17*, 569 (1957).
7. N. B. Eddy, H. Friebel, K-J. Hahn, and H. Halbach, *Bull. W. H. O. 38*, 673 (1968); N. B. Eddy, H. Friebel, K-J. Hahn, and H. Halbach, *Bull. W. H. O. 40*, 1 (1969).
8. N. B. Eddy, *Bull. Narcotics 5*, 39 (1953).
9. E. L. Way, J. W. Kemp, J. M. Young, and D. R. Grassetti, *J. Pharmacol. Exp. Ther. 129*, 144–154 (1960).
10. N. B. Eddy and L. E. Lee, *J. Pharmacol. Exp. Ther. 125*, 116–121 (1959).
11. A. H. Beckett and A. F. Casy, *in* "Progress in Medicinal Chemistry" (G. P. Ellis and G. B. West, eds.), Vol. 2, pp. 43, Butterworths, London (1962), and references there cited.
12. L. A. Woods, *Pharmacol. Rev. 8*, 175 (1956).
13. W. R. Martin, *Pharmacol. Rev. 19*, 463–521 (1967).
14. H. Blumberg, P. S. Wolf, and H. B. Dayton, *Proc. Soc. Exp. Biol. Med. 118*, 763 (1965).
15. L. S. Harris and W. L. Dewey, *in* "Annual Reports in Medicinal Chemistry" (C.K. Cain, ed.), p. 33 (1966)
16. K. W. Bentley and D. G. Hardy, *J. Amer. Chem. Soc. 89*, 3267–3273 (1967).
17. G. F. Blane, A. L. A. Boura, A. E. Fitzgerald, and R. E. Lister, *Brit. J. Pharmacol. 30*, 11 (1967).
18. A. M. Harthoorn, *Fed. Proc. 26*, 1251–1261 (1967).
19. D. Campbell, A. H. B. Masson, W. Norris, and J. M. Reid, *Brit. J. Anaesth. 38*, 603 (1966); D. Campbell, R. E. Lister, and G. W. McNicol, *Clin. Pharmacol. Ther. 5*, 193–200 (1964).
20. R. D. Hunt and F. F. Foldes, *New Engl. J. Med. 248*, 803 (1953).
21. H. Isbell and H. F. Fraser, *J. Pharmacol. Exp. Ther. 107*, 524–530 (1953).
22. K. Fromhertz and B. Pellmont, *Experientia 8*, 394–395 (1952).
23. M. Gates and W. G. Webb, *J. Amer. Chem. Soc. 80*, 1186–1194 (1958).
24. N. B. Eddy, J. G. Murphy, and E. L. May, *J. Org. Chem. 22*, 1370–1372 (1957).
25. L. S. Harris and A. K. Pierson, *J. Pharmacol. Exp. Ther. 143*, 141–148 (1964).
26. A. F. Casy, *in* "Progress in Medicinal Chemistry" (G. P. Ellis and G. B. West, eds.), Vol. 7, Butterworths, London, (1970), and references there cited.
27. W. T. Beaver, S. L. Wallenstein, R. W. Houde, and A. Rogers, *Clin. Pharmacol. Ther. 7*, 740 (1966).
28. H. F. Fraser and D. E. Rosenberg, *J. Pharmacol. Exp. Ther. 143*, 149–156 (1964).
29. L. S. Harris, A. K. Pierson, J. R. Dembinski, and W. L. Dewey, *Arch. Int. Pharmacodyn. 165*, 112 (1967).
30. W. R. Martin, C..W. Gorodetsky, and T. K. McClane, *Clin. Pharmacol. Ther. 7*, 455 (1966).
31. N. B. Eddy and E. L. May, "Synthetic Analgesics, Part II B, 6, 7-Benzomorphans," Pergamon, Oxford (1966).
32. J. Pearl and L. S. Harris, *J. Pharmacol. Exp. Ther. 154*, 319–323 (1966).

33. B. F. Tullar, L. S. Harris, R. L. Perry, A. K. Pierson, A. E. Soria, W. F. Wetterau, and N. F. Albertson, *J. Med. Chem. 10*, 383–386 (1967).
34. J. Hellerbach, O. Schnider, H. Besendorf, and B. Pellmont, "Synthetic Analgesics, Part II A, *Morphinans*," Pergamon, Oxford (1966).
35. O. Eisleb and O. Schaumann, *Dtsch. Med. Wschr. 65*, 967 (1939).
36. L. Lasagna and H. K. Beecher, *J. Pharmacol. Exp. Ther. 112*, 306–311 (1954).
37. N. B. Eddy, H. Halbach, and O. J. Braenden, *Bull. W. H. O. 14*, 353 (1956).
38. A. K. Reynolds and L. O. Randall, "Morphine and Allied Drugs," University of Toronto Press, Toronto (1957).
39. A. F. Casy, A. B. Simmonds, and D. Staniforth, *J. Pharm. Pharmacol. 20*, 768 (1968).
40. A. H. Beckett and A. F. Casy, *in* "Progress in Medicinal Chemistry" (G. P. Ellis and G. B. West, eds.), Vol. 4, pp. 171–218 Butterworths, London (1965).
41. A. H. Beckett and A. F. Casy, *Bull. Narcotics, 9*, 37–54 (1957).
42. G. de Stevens (ed.), "Analgetics," Academic Press, New York (1965).
43. T. D. Perrine and N. B. Eddy, *J. Org. Chem. 21*, 125–126 (1956).
44. N. B. Eddy, L. E. Lee, and L. S. Harris, *Bull. Narcotics 11*, 3 (1959).
45. T. J. DeKornfeld and L. Lasagna, *J. Chronic Dis. 12*, 252 (1960).
46. E. Nilsson and P. Janssen, *Acta Anasthesiol. Scand. 5*, 73 (1961).
47. P. A. J. Janssen and N. B. Eddy, *J. Med. Chem. 2*, 31 (1960).
48. E. H. Bachrach, A. D. Godholm, and A. M. Betcher, *Surgery 37*, 440 (1955).
49. A. Ziering and J. Lee, *J. Org. Chem. 12*, 911–914 (1947); A. H. Beckett, A. F. Casy, G. Kirk, and J. Walker, *J. Pharm. Pharmacol. 9*, 939 (1957).
50. A. H. Beckett, A. F. Casy, and N. J. Harper, *Chem. Ind. (London). 1959*, 19–20.
51. A. F. Casy, *Tetrahedron 22*, 2711–2719 (1966); A. F. Casy, *J. Med. Chem. 11*, 188–1891 (1968).
52. A. F. Casy, L. G. Chatten, and K. K. Khullar, *J. Chem. Soc.* (C), 2491–2495 (1969).
53. P. S. Portoghese and D. L. Larson, *J. Pharm. Sci. 57*, 711 (1968).
54. I. N. Nazarov, N. S. Prostakov, and N. I. Shvetsov, *J. Gen. Chem. USSR, 26*, 3117–3129 (1956); N. S. Prostakov, B. E. Zaitsev, N. M. Mikhailova, and N. N. Mikheeva, *J. Gen. Chem. USSR, 34*, 465–468 (1964).
55. A. J. Grossman, M. Golbey, W. C. Gittinger, and R. C. Batterman, *J. Amer. Geriatr. Soc. 4*, 187 (1956).
56. L. J. Cass and W. S. Frederik, *Curr. Ther. Res. 3*, 97 (1961).
57. R. W. Houde, personal communication.
58. C. V. Winder, M. Welford, J. Wax, and D. H. Kaump, *J. Pharmacol. Exp. Ther. 154*, 161–175 (1966).
59. J. F. Gardocki and J. Yelnosky, *Toxicol. Appl. Pharmacol. 6*, 48–62 (1964).
60. J. S. Finch and T. J. DeKornfeld, *J. Clin. Pharmacol. 7*, 46, (1967).
61. J. Yelnosky and J. F. Gardocki, *Toxicol. Appl. Pharmacol. 6*, 63–70 (1964).
62. A. F. Casy, M. M. A. Hassan, A. B. Simmonds, and D. Staniforth, *J. Pharm. Pharmacol. 21*, 434–440 (1969).
63. P. A. J. Janssen, "Synthetic Analgesics. Part I. Diphenylpropylamines," Pergamon, Oxford (1960)
64. H. Isbell, A. Wikler, N. B. Eddy, J. L. Wilson, and C. F. Moran, *J. Amer. Med. Ass. 135*, 888–894 (1947).
65. H. Isbell, A. Wikler, A. J. Eisenman, M. Daingerfield, and Frank, *Arch. Int. Med. 82*, 362 (1948).
66. A. S. Keats and H. K. Beecher, *J. Pharmacol. Exp. Ther. 105*, 109–129 (1952).
67. A. S. Keats, J. Telford, and Y. Kuroso, *J. Pharmacol. Exp. Ther. 130*, 212–217 (1960).
68. J. La Barre, *Bull. Narcotics 11*, 3 (1959).
69. P. A. J. Janssen and A. H. Jageneau, *J. Pharm. Pharmacol. 9*, 381 (1957).
70. J. E. Denton and H. K. Beecher, *J. Amer. Med. Ass. 141*, 1146–1153 (1949).
71. A. H. Beckett, G. Kirk, and R. Thomas, *J. Chem. Soc. 1962*, 1386–1388.
72. A. F. Casy and M. M. A. Hassan, *J. Chem. Soc. (C)*, 683–686 (1966).

73. A. H. Beckett and A. F. Casy, *J. Pharm. Pharmacol. 6*, 986–1001 (1954).
74. N. A. David, H. J. Semler and P. R. Burgner, *J. Amer. Med. Ass. 161*, 599–603 (1956); A. S. Keats and H. K. Beecher, *J. Pharmacol. Exp. Ther. 105*, 210–215 (1952).
75. A. F. Casy and M. M. A. Hassan, *J. Med. Chem. 11*, 601–603 (1968).
76. N. B. Eddy, *Chem. Ind. (London), 1959*, 1462–1469.
77. A. F. Green, *Brit. J. Pharmacol. 8*, 2 (1953).
78. J. Owen, *Vet. Rec. 1955*, 561.
79. A. Pohland and H. R. Sullivan, *J. Amer. Chem. Soc. 75*, 4458–4461 (1953); *77*, 3400–3401 (1955); E. B. Robbins, *J. Amer. Pharm. Ass. Sci. Ed. 44*, 497–500 (1955).
80. H. F. Fraser and H. Isbell, *Bull. Narcotics, 12,* 9 (1960).
81. W. B. Wright, H. J. Brabander, and R. A. Hardy, *J. Amer. Chem. Soc. 81*, 1518–1519 (1959); *J. Org. Chem. 26*, 476–485 (1961).
82. A. C. Osterberg and C. E. Rauh, *Pharmacologist 1*, 78 (1959).
83. T. J. DeKornfeld and L. Lasagna, *Anesthesiology 21*, 159 (1960).
84. P. S. Portoghese and D. L. Larson, *J. Pharm. Sci. 53*, 302 (1964).
85. A. F. Casy and M. M. A. Hassan, *J. Med. Chem. 11*, 599–601 (1968).
86. P. A. J. Janssen, *Brit. J. Anaesth. 34*, 260 (1962); P. A. J. Janssen, A. Jageneau, and J. Huygens, *J. Med. Pharm. Chem. 1*, 299 (1959).
87. C. van de Westeringh, P. V. Daele, B. Hermans, C. Van der Eychen, J. Boey, and P. A. J. Janssen, *J. Med. Chem. 7*, 619–623 (1964).
88. A. Saarne, *Acta Anaesthesiol. Scand. 13*, 11 (1969).
89. E. Schenker and H. Herbst, in *"Progress in Drug Research"* (E. Jucker, ed.), Vol. 5, pp. 274–550, Birkhauser, Basel (1963).
90. N. B. Eddy, quoted by L. B. Mellett and L. A. Woods, *in* "Progress in Drug Research" (E. Jucker, ed.), Vol. 5. p. 155, Birkhauser, Basel (1963).
91. W. T. Beaver, S. L. Wallenstein, R. W. Houde, and A. Rogers, *Clin. Pharmacol. Ther. 7*, 436 (1966); J. W. Pearson and T. J. DeKornfeld, *Anesthesiology 24*, 38 (1963).
92. H. F. Fraser and D. E. Rosenberg, *Clin. Pharmacol. Ther. 4*, 596 (1963).
93. A. Hunger, J. Kebrle, A. Rossi, and K. Hoffman, *Experientia 13*, 400–401 (1957); F. Gross and H. Turrian, *Experientia 13*, 401–402 (1957).
94. G. Bromig, *Klin. Wochenschr. 36*, 960 (1958).

Chapter 2

METHODS OF CHEMICAL ANALYSIS

John F. Taylor*

New York State Research Institute for
Neurochemistry and Drug Addiction
Ward's Island, New York

I. INTRODUCTION

The pharmacological properties of narcotic analgesics have made the identification of such compounds very important in drug control and in research in medicine, pharmacy, and criminology. In these fields the requirements of the analysis with respect to sensitivity and specificity vary greatly as do, consequently, the methods employed. A detailed description of all these methods is outside the scope of this review; therefore, brief mention and references will be given to those that are already well documented with fuller detail reserved for the more recently developed methods. The physicochemical methods for identification are obviously applicable to all compounds possessing the properties, or combination thereof, upon which the analysis is based. Therefore, evaluation of the specificity of the analysis for a given compound is of utmost importance.

II. SOLVENT EXTRACTION

All narcotic analgesics possess a tertiary aliphatic amino group except for some analogs of methadone[1] and an aromatic moiety that in some instances bears a phenolic hydroxyl group that may be substituted. They are mainly weak bases of pK_a 6.6–9.5,[2] and consequently their salts with organic acids and hydrochloric acid, besides being water soluble, are soluble to varying degrees in organic solvents depending on the tendency of the salt to form ion pairs. These solubility characteristics are utilized in routine forensic and toxicological extraction procedures that have been described in detail by Freimuth[3] and Curry,[4] with more recent developments being

---

* Present address: The Metropolitan Police Forensic Science Laboratory, 2 Richbell Place, London, W. C. 1, England.

reviewed by Jackson.[5] A general method for extraction of narcotic analgesics and basic organic compounds depends on their precipitation from aqueous alkaline solution and the ready solubility of the liberated free bases in organic solvents. Phenolic-substituted narcotics are soluble in strong aqueous alkali through formation of the phenate ion. For these compounds maximum base precipitation is achieved at about pH 8–10 by addition of (1) a strong alkali, such as sodium hydroxide solution, with pH monitoring; (2) a regulated amount of weak alkali, for example, ammonium hydroxide solution or the salt of a weak acid and a strong base such as sodium bicarbonate; or (3) an excess of strong pH 9–10 buffer solution for example, saturated borax. With certain samples protein precipitation or defatting procedures may be applied before alkalinization.[3,4] However, in many instances they may be circumvented by direct extraction of the aqueous alkaline sample with organic solvent followed by reextraction of basic compounds from the organic phase into dilute aqueous acid and then from the latter, after alkalinization, into fresh organic solvent.

The choice of extraction solvent is governed by many factors. These are as follows:

1. The solubility of the salt or basic form of the narcotic in various organic solvents. Such details are available[6] and may serve as a rough guide to which solvent would yield the most favorable recovery of drug.
2. Compatibility of the narcotic with solvent impurities. For example, solvents such as the ethers and chloroform may contain small amounts of highly reactive peroxides and phosgene, respectively. These impurities, being potential destroyers of the extracted drug, are particularly troublesome when the solvent extract is concentrated prior to analysis. Peroxides may be removed from ethers by *careful* redistillation, and procedures for stabilization of solvents against such impurities are available.[6]
3. The subsequent analysis to which the extract will be applied. For example, when analysis of narcotics by the indicator-dye method is to follow, the extraction solvents ethylene dichloride, chloroform, and benzene, in particular, are recommended because they yield a favorable partition of drug and low tissue-blank values.[7] Also, concentration of extracts prior to chromatography is an intermediate step most easily accomplished by choosing the most volatile solvent concomitant with good recovery of drug by extraction and low electron-capturing properties are a prerequisite for extraction solvents when electron capture detectors are used in gas chromatography.

Recoveries of compounds by extraction may have an important influence on the overall sensitivity of an analysis. Mulé,[8] emphasizing this fact recently, has developed and evaluated general procedures for the extraction of the main classes of narcotic analgesics (Table I). High recoveries of narcotics have also been obtained with (1) other aliphatic alcohol- or

TABLE I
Recoveries of Narcotic Analgesics by Solvent Extraction[8]

Sample	Class of Narcotic	pH of Sample	Solvent	Recovery (Percent±S.E.)
25–200 μg/ml	iminoethano phenanthrofuran	10.4	ethylene dichloride/ isobutanol (3:1)	88.5±2.6
of narcotic	iminoethano-	10.4	ethylene dichloride	87.2±6.2
in plasma	phenanthrene			
or urine (6 ml)	diarylalkoneamine	10.4	ethylene dichloride	92.5±7.1
saturated with	arylpiperidine	7.5	ethylene dichloride	81.3±5.1
sodium chloride	benzomorphan	10.4	ethylene dichloride	84.5±6.7

acetone-containing solvent mixtures for morphine,[9 11] nalorphine,[10] normorphine,[10,12] dihydromorphine,[11] and highly polar derivatives[11]; (2) benzene for codeine,[13] pethidine,[14] and pentazocine;[15,16] and (3) diethyl ether[16] for pethidine, propoxyphene, methadone, and methadone-type analgesics. In general it appears that the greater the polarity of the compound to be extracted the greater is the required polarity of the extracting solvent.

Other factors may influence the partition of narcotics into organic solvents. First, adsorption of drug onto biological material or glassware may reduce the recovery of small amounts of drug. For instance, Takemori,[17] using conventional high recovery extraction methods for morphine, only recovered about 30–50% of submicrogram quantities of the drug added to plasma. However, recoveries were restored to about 90% by TCA precipitation of proteins and siliconisation of glassware. Second, the presence of salts, such as ammonium sulfate or sodium chloride, may greatly increase the partition of basic compounds into organic solvents, for example, that of morphine into ether.[18,19]

III. QUALITATIVE IDENTIFICATION

A. Chromatographic Methods

Although chromatography serves primarily as a means of separation prior to physicochemical methods of identification, it is included here because of the intrinsic value that the chromatographic behavior of a compound yields to its identification. The underlying theory and practical applications of the methods in general have been reviewed by Consden[20] and their specific application to alkaloids and related bases by Curry[4] and Farmilo and Genest.[21]

1. Column Chromatography
a. Adsorption and Partition. This is the classical technique for separation of alkaloids.[22] In relation to narcotics it is mainly used for the

isolation of morphine from extracts of opium or poppy by chromatography on columns of alumina[23-26] or Florisil.[23] Morphine is eluted more readily than the secondary opium alkaloids or colored materials that are retained by the column. Büchi and Huber[27] extended the technique and have eluted the secondary alkaloids and separated them on Celite/pH 4.8 phosphate buffer into three fractions. Papaverine/narcotine and thebaine were eluted with ether:benzene (3:1) and codeine with chloroform containing 1% ethanol and saturated with ammonia. Such differential elution from Celite has been similarly applied to single-column separations of the major opium alkaloids,[28] to separation of heroin from the common diluents and degradation products found in illicit preparations,[29] and to quantitative partitions of codeine and its derivatives from ephedrine.[30,31] The practical application of column chromatography in toxicological studies also has been investigated: Fischer and Iwanoff,[32] using eight solid adsorbents and six solvents, examined the chromatographic behavior of 115 alkaloids and basic drugs; Ikekawa et al.[33] found that morphine and its conjugates may be adsorbed from urine onto activated charcoal and then eluted with glacial acetic acid, the procedure being quantitative for morphine; Stewart and co-workers[34,35] evaluated methods whereby alkaloids are adsorbed onto kaolin or Florisil and subsequently eluted with aqueous acidic or organic solvents. It was found that the optimum pH for adsorption onto Florisil is 7–8.5 for tissue extracts, 8–9 for blood, and 5–6 for urine and that the method is applicable to absorption analysis of morphine, codeine, and heroin in biological material.[35]

   b. Ion Exchange. Both cation and anion exchange chromatography are used for separation of alkaloids from unwanted impurities. The underlying theory of the methods is discussed by Büchi[36] and Saunders,[37] and practical aspects are reviewed by Jindra.[38]

   Cation exchange chromatography has been successfully applied to the isolation of morphine from opium[39] and of this narcotic and others from illicit narcotic mixtures,[40] pharmaceutical preparations,[41-43] and toxicological extracts.[44-47] Szerb[47] and Tompsett,[45] using columns of Amberlite IRC 50 (carboxylic) and Dowex 50 × 12 (sulfonic), respectively, have reported 80–100% recoveries of microgram to milligram quantities of morphine, codeine, and pethidine from biological samples. Recently Dole et al.[48] described a screening procedure in which paper containing Amberlite IR 120 (Na+ form) is used to isolate narcotics and other drugs of abuse from urine prior to solvent extraction and TLC. In this instance cation exchange enabled a more specific extraction compared with direct extraction methods, but concomitant low drug recoveries limit the overall sensitivity of the procedure.[49]

   Anion exchangers have been used in the same types of alkaloid analysis as cation exchangers. The main difference is that alkaloidal cations are not retained on the column by exchange but are converted to the free base that is adsorbed onto the resin and may be eluted quantitatively in this form by organic solvents. Such behavior of salts of narcotics and other alkaloids has

formed the basis for their rapid assay, eluted free bases being titrated poten-
tiometrically with dilute acid following chromatography on Amberlite
IR-4B[50,51] or Dowex-2.[52] Levi and Farmilo[51] report quantitative recoveries of 12 narcotics that were determined with a precision of $\pm 1.6\%$
by anion exchange chromatography. With strong anion exchangers it is not
possible to analyze phenolic bases in this manner as they are retained as
anions. On the other hand, this fact may be used to separate such bases from
the nonphenolic type, for example, in the separation of morphine from (1)
opium on Lewatit MN[53] and Amberlite IRA 411;[54] (2) plant and animal
tissues on Dowex-2,[55] and (3) codeine in mixtures of the two drugs.[56]
Nonphenolic bases are eluted in the normal way with methanol, and morphine is subsequently eluted with dilute acid. High purity and quantitative
elution of the narcotics facilitated assay by direct methods.[55,56]

Ion-exchange resins also have been used for the isolation of amphoteric
drug metabolites that cannot be extracted from aqueous solution by organic
solvents. Tompsett,[57] using the strong cation exchanger Dowex 50×8,
isolated pethidinic acid from urine. This metabolite was differentially eluted
with $6N$ aqueous ammonia, bases without weak acidic groupings being
retained and subsequently eluted with hydrochloric acid of the appropriate
concentration.[46] Recoveries of 2–8 mg of pethidinic acid, added to urine,
ranged from 49 to 54%. A similar method was used by Oka[58] to isolate
morphine-3-glucuronide from dog urine and by Yoshimura et al.,[59] who
incorporated preliminary charcoal adsorption and final anion exchange to
isolate both the 3- and 6-glucuronides from rabbit urine. However, the
isolation of such metabolites appears to depend not so much upon ion
exchange as it does upon adsorption because glucuronides of morphine
also have been isolated on activated charcoal[33] as have the ethereal sulfates
and glucuronides of morphine[60] and nalorphine[61] on the highly adsorbing
nonionic resin Amberlite XAD-2.

2. *Electrophoresis*

The theory, procedures, and applications of this method are discussed
by Bier,[62] and its general use for alkaloids is described by Kaye and Goldbaum,[63] who recommend $1N$ aqueous acetic acid as electrolyte with volt-
tages of 300–600 V. Mariani-Bettelo and Frugoni[64] investigated the
electrophoresis of 10–30 μg quantities of 70 alkaloids at seven pH values
(2.3–11.4) during a migration time of 3 hr. Although separation was possible
at the fixed pH, in the majority of instances complex mixtures were best
separated by the controlled variation of pH. The quantitative separation of
like alkaloids and the way it is affected by pH also was studied by Paris and
Faugeras.[65]

There have been two major investigations into the electrophoretic separation of narcotics and related bases. Wagner[66] reports mobilities for 12
narcotics at 12 pH values (2–14) and 210 V for a 6 hr running time (Table
II). At pH values well above 12 the phenolic narcotics, morphine, hydromorphone, and levorphanol, formed negatively charged phenate ions and

TABLE II
Electrophoretic Mobilities of Narcotics[66]
(Schleicher & Schull 2043 B, 50 × 5 cm, 210 V, d.c., 6 hr)

Distance (cm) (− = travel to anode)

Substance	0.1M Na-citrate–HCl				0.067M Na₂HPO₄–KH₂PO₄				0.1M glycine–NaOH 0.1N NaOH			
Buffer Solutions pH	2.0	3.0	4.0	5.0	6.0	7.0	8.0	9.0	10.0	11.0	12.0	NaOH
Morphine	4.6	4.6	4.9	6.3	6.3	3.9	1.2 2.8	1.3	0.8	0.7	0.6	−2.6
Codeine	4.6	4.7	4.9	6.4	6.0	4.0	2.9	2.2	2.1	2.1	1.7	1.9
Ethylmorphine	4.0	4.2	4.4	6.0	6.1	4.0	2.9	2.1	1.9	2.1	1.5	1.8
Hydromorphone	4.5	4.5	4.8	6.1	6.2	3.6	2.5	0.9	0.8	0.6	0.6	−2.6
Hydrocodone	4.8	4.8	5.0	6.5	6.4	4.3	3.4	2.3	2.2	2.2	1.6	1.8
Oxycodone	4.5	4.7	4.9	6.3	6.8	5.4	4.5	3.5	3.3	3.0	2.3	2.8
									partial travel; residue: starting point			
Diacetylmorphine	4.1	4.4	4.5	5.7	5.5	3.5	2.5	1.3	4.0	3.5	3.3	−0.8
l-Orphan	4.2	4.2	4.7	6.3	6.2	4.8	4.2	4.5	2.0	1.9	2.2	1.3
Pethidine	4.9	4.9	5.3	7.1	7.0	5.0	4.4	3.3	2.0			
Methadone	4.1	4.2	4.7	6.2	6.3	5.1	5.2	4.8 to	partial travel			
Papaverine	3.2	3.2	3.4	4.0	3.2 to partly	0.8				partly to 0.8		
Narcotine	2.8	3.2	3.4	4.4 to	4.0 to start					partly to 0.4		
				tailing								

migrated towards the anode to be completely separated from nonphenolic compounds. Willner[67] studied the electrophoretic properties of pure morphine, hydromorphone, codeine, oxycodone, hydrocodone, thebaine, meperidine, ketobemidone, atropine, and cocaine at 2 mA and 110 V. Glycine/NaOH buffers were used to separate several binary mixtures, and with a pH of 11.1 to 11.2 it was possible to separate a mixture of morphine, ketobemidone, oxycodone, codeine, cocaine, and atropine during the usual 15-hr running time. The mixture morphine/hydromorphone could not be separated.

Pharmaceuticals and galenicals have been analyzed electrophoretically: morphine has been isolated from opium,[68] extracts of poppy capsules,[69] and tincture of opium[70] and pholcodine, in tablets, syrups, or suppositories, can be identified and differentiated from codeine, morphine, and ethylmorphine.[71] Farmilo *et al.*[72] have applied electrophoresis to the separation and identification of the alkaloids of raw opium and established a semiquantitative method for determining its origin.

Application of electrophoresis also has been attempted in toxicological investigations. Morphine has been quantitatively isolated from urine,[73] and alkaloids have been separated from each other and from biological contaminants.[74,75] For identification purposes, Buff *et al.*[76] studied the effect of pH on the migration rate of nine alkaloids and narcotics that are of common occurrence in toxicological investigations. They concluded that electrophoresis showed greater reproducibility and, therefore, more reliability for identification than did chromatography. High voltage electrophoresis has enabled the rapid detection of narcotics in urine.[77-79] Sano and Kajita[77] separated microgram quantities of 17 drugs of addiction including methylamphetamine, codeine, morphine, ethylmorphine, and heroin. During a 20-min run 11 of the drugs did not differ much in migration, but after 35 min there was satisfactory separation. Williams *et al.*[80] described rapid electrophoresis of alkaloids upon agar gels. The running time was 25 min or less, and quantitative determination by spot size and quantitative recovery of alkaloids were possible. However, separation of the mixed alkaloids was poor and would have required longer running times.

3. Paper Chromatography (PC)

This technique has been excellently reviewed by Genest and Farmilo,[81] who described ascending, descending, and circular chromatography and discussed their practical application to systematic identification of narcotics and related compounds in natural products, pharmaceuticals, and biological materials. The same authors later developed methods for routine qualitative and quantitative identification of opium alkaloids and synthetic narcotics.[82] Detailed descriptions of the identification of narcotics by paper chromatography are additionally available in reports of the specific application of this technique in the systematic identification of alkaloids and related bases,[83-85] in the separation of opium alkaloids,[86-88] and in forensic,[89,90] toxicological,[85,91] and clinical analysis.[92-94] The more recent developments in

TABLE III

R_f Values of Narcotics in Six Solvent Systems[a]

Number	Name of Narcotic	R_f in System Number[b]					
		1	2	3	4	5	6
1	Oxymorphone	0.21	0.10	0.06	0.93	0.80	0.08
2	Morphine	0.34	0.13	0.05	0.92	0.78	0.01
3	Oxycodone	0.34	0.22	0.16	0.92	0.81	0.16
4	Hydrocodone	0.46	0.26	0.23	0.95	0.81	0.08
5	Codeine	0.49	0.24	0.12	0.95	0.79	0.10
6	Ethylmorphine	0.66	0.45	0.33	0.94	0.77	0.18
7	Diamorphine	0.73	0.50	0.52	0.91	0.75	0.06
8	Thebaine	0.85	0.67	0.41	0.95	0.61	0.23
9	Hydromorphone	0.87	0.64	0.10	0.93	0.77	0.03
10	Benzylmorphine	0.94	0.84	0.62	0.97	0.34	0.30
11	Myrophine	0.98	0.98	1.00	0.01	0.01	0.97
12	Cryptopine[c]	0.56	0.34	0.53	0.00[a]	0.55	0.10 (0.00)
13	Narceine[c]	0.80	0.75	0.62	0.82	0.78	0.00[a]
14	Papaverine[c]	0.88	0.78	0.64	0.98	0.02	0.04
15	Narcotine[c]	0.88	0.78	0.72	S	0.02	0.10(L)[e]
16	Cocaine	0.83	0.67	0.56	0.60(L)[e]	0.71	0.67
17	Pyrahexyl	1.00	1.00	1.00	0.01	0.00[a]	1.00
18	Anileridine	0.76	0.40	0.54	0.56	0.35	0.40
19	Pethidine	0.91	0.77	0.63	0.95	0.60	0.78

	System 1	System 2	System 3	System 4	System 5	System 6
20	Alphaprodine	0.91	0.81	0.77	0.51	0.62	0.77
21	Alphameprodine	0.94	0.90	0.84	0.85	0.33	0.84
22	Ethoheptazine[c]	0.92	0.82	0.66	0.87	0.70	0.80
23	Normethadone	0.97	0.95	0.90	0.11	0.30	0.87
24	l-Dipipanone	0.99	0.98	0.91	0.00	0.07	0.98
25	Phenadoxone	0.99	0.98	0.90	0.02	0.00	0.98
26	Methadone	1.00	0.96	0.88	0.15	0.27	0.94
27	l-Isomethadone	1.00	0.96	0.90	0.02	0.19	0.93
28	Alphaacetylmethadol	1.00	1.00	0.96	0.06	0.05	0.97
29	Propoxyphene	0.97	0.94	0.93	0.22	0.11	0.94
30	Diethylthiambutene	0.90	0.89	0.90	0.08	0.10	1.00
31	Levomoramide	0.97	0.95	0.87	0.0– 0.5S[f]	0.02	0.6S[f]
32	Levallorphan	0.95	0.92	0.66	0.88	0.60	0.58
33	dl-Methorphan	0.99	0.92	0.78	0.47(L)	0.36	0.91
34	Phenazocine	0.95	0.92	0.91	0.67	0.07	0.82

[a] From Genest and Farmilo.[99]

[b] System 1. IsoBuOH:AcOH:H_2O (100:10:24) on paper impregnated with 0.5M KH_2PO_4 (pH 4.2).
System 2. IsoBuOH:AcOH:H_2O (100:10:24) on paper impregnated with $(NH_4)_2SO_4$ (2%, pH 5.3).
System 3. Butylacetate:AcOH:H_2O (35:10:3) on paper impregnated with 0.5M KH_2PO_4 (pH 4.2).
System 4. PrOH:H_2O:diethylamine (1:8:1) on paper impregnated with 4% light paraffin (BP 1948) in hexane.
System 5. Ammonium formate (10%) in water, saturated with sec-octanol on paper impregnated with 20% sec-octanol in acetone.
System 6. Light paraffin (BP 1948):diethylamine (9:1) on paper impregnated with 20% formamide in acetone.

[c] Not narcotic under international law.

[d] (0.00), some material remains at start.

[e] (L), elongated spot.

[f] S, streaking.

these five fields of application of this technique to narcotics are elaborated in the following paragraphs.

Hilf et al.,[95] using systems at pH 3.0, 5.0, 6.5, and 7.5, have determined R_f values and behavior under UV light of 15 alkaloids of forensic interest, including five narcotics. Elution from paper and UV spectroscopy of the eluate were shown to be a further method of identification. Jackson and Moss[96] used four solvent systems in their study of 61 alkaloids and related bases (14 narcotics) by ascending paper chromatography. R_f values and reactions to UV light and six spray reagents were recorded.

Impurities in heroin seizures were investigated by Nakamura,[97] who separated morphine and 3- and 6-monoacetylmorphine from heroin and 6-acetylcodeine from codeine using the solvent systems: butanol:water:acetic acid (10:5:1); isoamyl alcohol, water, acetic acid (10:5:1); and isoamyl alcohol:water:ammonia (10:5:1). Good separations have been obtained for the opium alkaloids on succinyl cellulose paper[98] and for the most important natural and synthetic narcotics in six paper chromatographic systems.[99] The use of four systems with highly polar stationary phases and decreasingly polar mobile phases and two "reversed phase" systems provided a good distribution of the R_f values of narcotics over the full range of the R_f scale (Table III). Two reversed phase and one cellulose anion-exchange system also were used in the separation of morphine, nalorphine, and codeine.[100] Tewari[101] used two different buffered systems (pH 5.7 and pH 6.6) for separating opium alkaloids and determining the origin of a sample of opium by the pattern of alkaloid R_f values.

Much attention has been given to the use of paper chromatography for general qualitative screening in forensic, toxicological, and clinical analysis.

Separations of toxicologically important bases can be achieved in 5–15 min by centrifugally accelerated paper chromatography. Dal Cortivo et al.[102] used a Hi Speed Chromatograph at 750–1200 rpm to separate the basic constituents of narcotic seizures and extracts of urine and bile on paper impregnated with pH 5 phosphate buffer. R_f values were slightly higher than those obtained by the usual ascending technique. In addition, Street[103] has investigated the acceleration of chromatography by increased temperature.

The systematic separation and identification of toxicologically important poisons and drugs, including many narcotics, has been further investigated.[104,105] Vidic,[104] using specific extraction techniques, separated 74 basic drugs into six groups and identified them using chromatography in three solvent systems followed by color tests with spray reagents. Paulus et al.[106] investigated identification of 20 habit-forming drugs by circular chromatography; five solvent systems were used, and separation was possible even when the drugs, mostly narcotic analgesics, were extracted from urine.

Demonstration of the presence of drug metabolite(s) in biological samples may help to confirm the identification of an ingested drug. However, no general schemes have been developed toward this end and data on the paper-chromatographic separation of metabolites from the parent narcotics has resulted mainly from investigations of the fate of individual compounds. Milthers[10] separated normorphine from nalorphine and, with difficulty, from morphine. By using two solvent systems, Misra *et al.*[107] separated normorphine and morphine glucuronide from morphine more easily and Freundt[108] detected dextromoramide and its metabolite in urine. Pethidinic acid[57] and norpethidine[109] have been detected in human urine and norcyclazocine[110] in dog excreta following administration of meperidine and cyclazocine, respectively. Finally, Nakamura and Ukita[111] recently studied the *in vitro* and *in vivo* hydrolysis of heroin in blood by paper chromatography.

4. Thin-Layer Chromatography (TLC)

This technique, which is based upon adsorption rather than liquid-liquid partition as in paper chromatography, was originated in 1938.[112] The earlier developments[113] and applications[114] of the technique arose mainly from the work of Stahl, who recently has given a detailed account of methods and applications.[115] The most important advantages of TLC over PC are greater speed, sharper separations, higher sensitivity, and higher chemical and thermal stability of the layer that permit the use of methods of detection that normally destroy paper.

One of the original attempts to apply TLC to the analysis of narcotic analgesics was that of Borke and Kirch,[116] who using 1, 4-dioxane as a solvent separated some opium alkaloids on glass plates layered with a composite, buffered, fluorescent paste. Good separations of opium alkaloids and their semisynthetic derivatives were obtained with chloroform–ethanol (4:1) on Silica Gel G+0.5N KOH and benzene–heptane–chloroform–diethylamine (6:5:1:0.02) on formamide-impregnated cellulose.[117] Formamide was found to interfere with the detection of alkaloids by Dragendorff's reagent and could be removed by heating the plate in a vacuum-drying cabinet at 110°C and 20 mm pressure for 15 min or by conversion of formamide to formic acid with 0.25% sodium nitrite in 0.5% hydrochloric acid before applying Dragendorff's reagent.[118] The important opium alkaloids also have been successfully separated in nonbasic systems with benzene–methanol (4:1),[119] benzene–ethanol (4:1),[120] and chloroform–methanol (9:1)[120] on Silica Gel G. However, in the systematic analysis of 54 alkaloids by TLC, Waldi *et al.*[121] obtained minimum tailing by development of Silica Gel G layers with mobile phases containing a basic component. Basic layer material such as Aluminum Oxide G or Silica Gel G pretreated with sodium hydroxyde also can be used, but Silica Gel G plates in conjunction with basic mobile phases generally produce sharper separations. The following solvent systems were recommended:

1. Chloroform–acetone–diethylamine (5:4:1).
2. Chloroform–diethylamine (9:1) (for morphine, hydromorphone, codeine, dihydrocodeine, hydrocodeinone).
3. Cyclohexane–chloroform–diethylamine (5:4:1) (for cocaine).
4. Cyclohexane–diethylamine (9:1).

5.   Benzene–ethyl acetate–diethylamine (7:2:1).

On Aluminum Oxide G layers:

1.   Chloroform.
2.   Cyclohexane–chloroform (3:7)+0.5% diethylamine (three drops).

On Silica Gel G layers pretreated with 0.1N NaOH:

1.   Methanol

In addition, the system xylene–methylethylketone–methanol–diethylamine (20:20:3:1) is capable of separating all the principal alkaloids of opium in one development on Silica Gel G.[122] Complete separation of all opium alkaloids can be achieved with two-dimensional chromatography[123] or by the use of more than one solvent system on several plates, as reported by Pfeiffer,[124] who using five solvent systems investigated 50 alkaloids of known structure from genus *Papaver* on silica gel and alumina layers and buffered and unbuffered paper.

Modified methods for the general separation of natural and synthetic analgesics have been investigated recently. Emmerson and Anderson[125] report R_f values and development times for the TLC of ten analgesics and related compounds on Silica Gel G in 13 nonaqueous systems and an atmosphere of ammonia vapor. This procedure was more versatile than the existing method in which alkaline silica gel layers are used, especially in its adaptability to two-dimensional chromatography. Huang *et al.*[126] investigated the polyamide layer chromatography of microgram amounts of seven opium alkaloids and synthetic derivatives. Separations were achieved by using two solvent systems: cyclohexane–ethyl acetate–n-propanol–dimethylamine (30:2.5:0.9:0.1) and water–absolute ethanol–dimethylamine (88:12:0.1).

The rapid separations afforded by TLC have led to many investigations of its use as a screening procedure in forensic, toxicological, and clinical analysis.

Steele,[127] using eight solvent systems, systematically separated and identified 26 compounds, found in narcotic seizures, on Silica Gel G. Five opium alkaloids, morphine, codeine, thebaine, papaverine, and narcotine, were optimally separated by the solvents systems ethyl acetate–benzene–acetonitrile–ammonium hydroxide (50:30:15:5 and 25:30:40:5). Acetylcodeine recently has been separated from illicit heroin.[128]

Machata[129] has applied Stas–Otto extracts directly to TLC on Silica Gel G. Methanol was a suitable solvent for bases, providing 14 cm/hr development and good reproducibility of R_f. A case of suicide by overdose of opium tincture was reported where 108 mg of opium alkaloids were found in the stomach contents. Clear separation of morphine, codeine, and narcotine was achieved despite the high concentration of the alkaloids. Baümler and Rippenstein[130] recommended TLC for alkaline/ether extracts of stomach contents, stomach washes, blood, and urine. Development for a distance of 10 cm was obtained in 30 min on Silica Gel G plates with methanol or methanol–acetone (1:1), which were useful solvents for most alkaloids. Methanol–acetone–triethanolamine (1:1:0.03) was more generally useful for basic compounds. Vidic,[131] using the same adsorbent and $0.1N$ methanolic ammonia as developing solvent, separated dextromoramide (R_f 0.85) from its metabolite (0.57) in urine and from dextromethorphan (0.26), methadone (0.42), and normethadone (0.59). Sunshine[132] has described the use of TLC in the diagnosis of poisoning. R_f and $R_{propiomazine}$ values of 8 narcotic analgesics on Silica Gel G in three solvent systems were reported. Noirfalise[133] found that it was possible to characterize narcotics and other drugs of toxicological importance by their R_f and $R_{morphine}$ values in four solvent systems and the use of spray reagents. Methadone, dextromoramide, pethidine, morphine/codeine, ethylmorphine, heroin, and dihydrocodeinone were well separated. Morphine and codeine were differentiated by color reactions. In similar investigations, Schweda,[134] using three solvent systems, has compared TLC on Silica Gel G plates with that on polyester sheets. The main advantage of the latter was durability. Opium alkaloids and six narcotics were also among the basic compounds of toxicological importance investigated by Jackson and Moss.[96] Of the four systems used, n-butanol–5% wt/vol aqueous citric acid (9:1) on cellulose pretreated with sodium dihydrogen citrate was considered most useful in identification because it is essentially equivalent to the PC system of Curry and Powell[135] for which data on about 500 drugs has been published (see Clarke.)[136]

With respect to clinical analysis, Cochin and Daly[137] report R_f values of analgesic drugs after extraction from pH 9.0 urine with ethylene dichloride–isoamylalcohol (9:1) and chromatography on Silica Gel G and Aluminum Oxide G with various solvent mixtures. No interfering substances were found in control urines known to contain caffeine and nicotine. Eberhard and Norden,[138] using Silica Gel G and the solvent mixture dimethyl formamide–ethylacetate (1:3), reported R_f values and color reactions with iodoplatinate for 12 narcotics and nicotine extracted from urine with ethyl acetate after addition of ammonium hydroxide. They compared the detection of analgesics in urine after direct addition of the compounds with that after their therapeutic administration. It was possible to demonstrate the presence of metabolites that can be used to identify the analgesic in question. Probably the most comprehensive investigation of identification of microgram quantities of narcotic analgesics in biological materials is that of Mulé.[8] R_f values were reported for 31 narcotics and related compounds on Silica Gel G with seven solvent systems (see Table IV). Solvent S3 was most effective in separating the commonly abused narcotics: morphine, codeine, dihydrohydrocodeinone, and pethidine. Successful separations of mixtures of narcotic analgesics in extracts of biological material were obtained by selection of the appropriate solvent system(s) or two-dimensional chromatography with solvent S4 followed by solvent, S1, S2, S3, or S5.

Davidow et al.[139] improved the resolution of morphine from other narcotics and commonly prescribed sedatives, tranquilizers, and antihista-

TABLE IV

Thin-Layer Chromatographic Data for the Various Narcotic Analgesics Arranged According to Chemical Families[a]

Compound (Free Base)	R_f Values ($\times$100) in Various Solvent Systems[b]						
	S1	S2	S3	S4	S5	S6	S7
Iminoethanophenanthrofurans							
Morphine	29	27	11	21 86[c]	7 85[c]	54	34
Normorphine	8	48	4	7 48	S[a] 25	66	62
Codeine	30	29	39	25 86	8 91	53	30
Norcodeine	12	50	13	9 59	6 56	63	49
Heroin	37	35	76	35 90	15 95	61	32
Nalorphine	71	55	35	67 88	25 96	59	41
Methyldihydromorphinone	16	24	25	15 76	S 92	45	26
Dihydromorphine	11	21	17	13 65	S 85	41	25
Ethylmorphine	33	25	46	27 84	8 96	53	33
Dihydrohydroxymorphinone	46	29	34	24 63	10 81	45	28
Dihydromorphinone	15	21	10	10 67	S 73	43	29
Dihydrocodeinone	17	25	41	19 76	S 94	42	23
Dihydrohydroxycodeinone	46	24	87	29	16	32	34
6-Monoacetylmorphine	38	40	64	29	19	37	37
Iminoethanophenanthrenes							
l-3-Hydroxy-N-methylmorphinan	11	47	80	10	7	51	60
l-3-Hydroxymorphinan	5	68	19	10	8	72	80
l-3-Methoxy-N-methylmorphinan	13	43	91	8	7	55	59
l-3-Methoxymorphinan	7	65	38	S	S	66	81
l-3-Hydroxy-N-allylmorphinan	65	70	98	41	44	64	73

	S1	S2	S3	S4	S5	S6	S7
Diarylalkoneamines							
dl-Methadone	34	59	99	17	17	55	62
l-Acetylmethadol	64	60	99	40	38	52	62
d-Propoxyphene	73	68	97	54	56	53	61
Arylpiperidines							
Pethidine	42	41	97	36	20	46	44
Norpethidine	12	65	51	10	11	58	63
Ketobemidone	31	39	47	24	12	42	40
dl-Alphaprodine	39	40	93	34	20	42	40
Piminodine	88	73	99	85	76	69	58
Benzomorphans							
dl-2'-Hydroxy-5,9-dimethyl-2-phenethyl-6,7-benzomorphan	88	87	97	82	70	76	77
l-2'-Hydroxy-2,5,9-trimethyl-6,7-benzomorphan	12	36	56	8	5	43	51
2'-Hydroxy-5,9-dimethyl-2-(3,3-dimethylallyl)-6,7-benzomorphan	73	81	96	25	34	65	77
2'-Hydroxy-5,9-dimethyl-2-cyclopropylmethyl-6,7-benzomorphan	45	71	92	15	16	55	67

[a] From Mulé[8] by courtesy of *Analytical Chemistry*.
[b] S1 Ethanol:pyridine:dioxane:water (50:20:25:5).
S2 Ethanol:glacial acetic acid:water (60:30:10).
S3 Ethanol:dioxane:benzene:ammonium hydroxide (5:40:50:5).
S4 Methanol:n-butanol:benzene:water (60:15:10:15).
S5 tert-Amyl alcohol:n-butyl ether:water (80:7:13).
S6 n-Butanol:glacial acetic acid:water (4:1:2).
S7 n-Butanol:concentrated HCl, saturated with water (90:10).
[c] R_f values obtained on chromatoplates prepared by dissolving 15 g of cellulose powder in 90 ml of $0.1M$ phosphate buffer, pH 8.0 (250 μ of adsorbent layer).
[d] Compound streaked.

mines by using a new developing solvent: ethylacetate–methanol–ammonia (85:10:5). Pure drugs added to urine did not interfere with detection, but substances with R_f values close to that of morphine were detected in the urine of patients taking phenothiazines. Fortunately these metabolites were distinguishable from morphine by the colors or UV fluorescence produced with 5% sulfuric acid and potassium iodoplatinate spray reagents. This TLC system was subsequently used by Dole et al.[48] for the general detection of drugs of abuse. Information concerning this subject is also available in a collection of reports from different laboratories.[140,141] TLC identification of alkaloids and basic drugs in the saliva, plasma, and urine of horses has been investigated in connection with doping.[142] Cocaine, codeine, heroin, morphine, and papaverine were among the compounds examined. Best separations of 12 alkaloids were obtained with hexane–acetone–diethylamine (6:3:1) used in one development or as a second solvent in two-dimensional chromatography with chloroform–methanol–diethylamine (95:5:0.05) for initial development. On Silica Gel GF nearly all the compounds could be detected as absorbing spots under UV light (254 mμ). Recent investigations in the screening of human urine for narcotics have been aimed mainly at increasing the specificity and sensitivity of the methods.

Goenechea and Bernard[143] have drawn attention to the possibility of obtaining erroneous results with extracts of smokers' urine when screening for basic drugs such as morphine. Methods are mentioned for differentiating between morphine and nicotine. Ono et al.[144] described the TLC of common basic drugs of abuse and confirm identification by microcrystal tests. Mulé,[49] using 13 solvent systems, described separation, R_f values and sequential color tests for drugs of abuse in general. Heaton and Blumberg[145] suggested methods for differentiating narcotics, amphetamine, and psychotropic drug metabolites when using the general screening procedure of Dole et al.[48] Fujimoto and Wang,[146] using columns of Amberlite XAD-2 resin, have separated commonly used narcotic drugs and their metabolites from interfering urinary constituents prior to chromatography on silica gel. Comparison of the pattern of R_f values for drug and metabolites with that of a urine standard allows definitive identification, especially of drugs with several well-separated metabolites. The authors only reported 68% success in the identification of samples but proposed methods for improvement.

TLC is useful in drug metabolism studies, and the following narcotics and their metabolites have been separated and identified: morphine and normorphine[8,137,147]; morphine and its 3- and 6-glucuronides[59]; nalorphine and normorphine[8,137,147]; nalorphine, its 3-ethereal sulfate, and its 3-glucuronide;[61] codeine and norcodeine;[8] heroin, 6-monoacetyl morphine and morphine;[8] l-methorphan, levorphanol, levallorphan, and their N-dealkylated derivatives;[8] methadone and the product formed by mono-N-demethylation;[148] acetylmethadol, noracetylmethadol, methadol, and normethadol;[149] propoxyphene and norpropoxyphene;[150] pethidine and norpethidine;[8] and pentazocine and the two isomeric alcohols and one of the corresponding carboxylic acids produced by metabolic oxidation of the terminal methyl groups of the dimethylallyl side chain.[151]

5. Gas-Liquid Chromatography (GLC)

In this technique separation of a mixture of compounds depends upon their relative partitioning between a mobile gaseous phase and a stationary liquid phase. Theory, instrumentation, and applications of the technique to analysis of drugs and pesticides are reviewed by Gudzinowicz.[152]

Attempts to use GLC for separation and identification of opium alkaloids were made by Lloyd et al.[153] and later by Brochmann-Hanssen and Furuya.[120] Columns, conditions, and results (Table V) were similar in both investigations. In comparison with PC and TLC the method gave better separations and was faster, more reproducible, and equally as sensitive, sharp peaks being easily obtained with as little as 5 μg alkaloid. A more comprehensive study was undertaken by Yamaguchi et al.,[154] who chromatographed 43 compounds including morphine alkaloids, sinomenine, and its derivatives on SE-30 to determine whether a correlation existed between retention times and structures. Increases or decreases in retention time could be correlated specifically with several types of structural changes. Anders and Mannering[155] described the on-column preparation

TABLE V
Relative Retention Times of Opium Alkaloids[a]

Codeine	0.38
Neopine	0.41
Morphine	0.48
Thebaine	0.59
10-Hydroxycodeine	0.66
Laudanosine	1.00
Laudanine	1.12
Laudanidine	1.12
($\pm$)-Reticuline	1.18
Papaverine	1.65
Protopine	1.95
Cryptopine	2.41
Narcotine (noscapine)	4.40
Carrier gas	argon
Inlet pressure, psi	19
Column dimensions	6 ft $\times$ 1/8 in.
Liquid phase	2% SE-30
Column temperature	207°
Injection port temperature	302°
Cell bath temperature	230°
Laudanosine time, min	16.4

[a] From Brochmann-Hanssen and Furuya,[120] courtesy of *Journal of Pharmaceutical Sciences.*

of acetate and propionate derivatives of a number of alkaloids and steroids by following injection of the parent compound with one of acetic and propionic anhydride, respectively. Derivatives may be formed with compounds containing alcoholic or phenolic hydroxyl groups and with primary and secondary amines, the observed change in retention time (peak shift) being useful for characterization of the parent compound. In this respect results for a number of compounds including 11 morphine and morphinan derivatives were reported. Alkaloids also have been identified by peak shift upon methylation and trimethylsilylation with, for example, trimethylanilinium hydroxide[156] and a mixture of hexamethyldisilazane and trimethylchlorosilane.[155] In contrast, Kingston and Kirk[157] investigated the identification of alkaloids by pyrolysis GLC. Specificity was good, and by using the four most prominent peaks it was possible to identify each of ten morphine alkaloids studied. However, the results were too variable for general practical use and required handling by computer. Pyrolysis products of narcotine and papaverine have been characterized by GLC, and schemes of degradation have been outlined.[158]

In addition to SE-30, previously used for the separation of alkaloids, Brochmann-Hanssen and Fontan[159] have used liquid phases of progressively greater polarity: XE-60 (a cyanosilicone polymer), EGGS-Y (a polyester methylsilicone copolymer), and HI-EFF-8B (cyclohexanedimethanol succinate polyester). Retention times relative to that of codeine for 18 alkaloids, including six from opium, were plotted against the stationary phases arranged in order of increasing polarity. From such a plot one can determine the best phase for gas chromatographic separation of mixtures. For example, the principal alkaloids of opium (morphine, codeine, and thebaine) require a very efficient column for separation on SE-30, but the same mixture is readily separated on the more polar stationary phases.

The application of GLC in screening for microgram quantities of toxicologically important compounds has been investigated.[11,160,161] Retention data were reported for a large number of alkaloids, basic drugs, and barbiturates on SE-30 at several temperatures. Of 11 narcotics examined all were separated on 5-ft columns at 210°C.[160] In many instances compounds separated with difficulty on SE-30 were readily separated on the more polar stationary phases HI-EFF-8B[11] and QF-1-0065 (fluorosiloxane polymer).[161]

The detection of narcotic analgesics in human biological materials was especially investigated by Mulé.[8] A chromatograph equipped with an argon ionization detector and a 6 ft by 3 mm i.d. borosilicate glass coiled column packed with 2% SE-30 on 80–100 mesh Gas Chrom S was used under the following conditions: detector and column temperature, 215°C; flash heater, 260°C; inlet pressure, 20 lb/in.^{2}; outlet pressure, atmospheric; argon carrier gas flow rate, 47 ml/min; detector voltage, 1000 V; relative gain, 10^{-8} A. The procedure involved injection into the chromatograph of 2–25 μg free base dissolved in methanol or ethanol. Acetate and propionate derivatives of the drugs were formed on column by the method[248] already described herein. The retention data for the free base of the narcotic and its derivatives are given in Table VI. All the iminoethanophenanthrofurans had relative retention times (RRT) greater than that of codeine and norcodeine, whereas RRT of members of the other chemical families, except for piminodine and phenazocine, were less than that of codeine. When injection of heroin, dihydrohydroxycodeinone, *l*-3-methoxy-*N*-methylmorphinan, *dl*-methadone, *d*-propoxyphene, pethidine, or *dl*-alphaprodine was followed by injection of acetic or propionic anhydride no change in retention time was observed since esterification was theoretically impossible. Incomplete esterification was observed with compounds such as codeine, ethylmorphine, dihydro-codeinone, *l*-3-hydroxy-N-methylmorphinan, *l*-3-hydroxy-*N*-allylmorphinan, ketobemidone, *l*-acetylmethadol, piminodine, and the benzomorphans; all other narcotic analgesics reacted completely. The retention times of the ester derivatives were always greater than that of the corresponding free base. The first peak of acetylated morphine appeared at a retention time similar but not identical to that of 6-monoacetylmorphine and may represent either 3-monoacetylmorphine or a mixture of 3- and 6-monoacetyl-morphine. However, the second peak of acetylated morphine did correspond to that of diacetylmorphine (heroin). When morphine was subjected to a homologous series of anhydrides the retention times of the first and second peaks increased as the series was ascended in the following manner: acetic, 8.06, 11.34; propionic, 10.50, 18.37; *n*-butyric, 12.75, 29.06; valeric, 16.87, 48.19; and hexanoic 22.31, 79.50. Attempts to esterify morphine with *n*-heptanoic, glutaric, succinic, and trifluoroacetic anhydrides were unsuccessful.

Typical chromatograms of extracts of biological fluids containing analgesic drugs are illustrated in Fig. 1. Interference by normal plasma

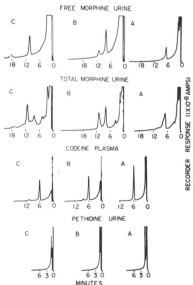

FREE MORPHINE URINE

TOTAL MORPHINE URINE

CODEINE PLASMA

PETHIDINE URINE

MINUTES

RECORDER RESPONSE (1 X 10⁻⁸ AMPS)

Fig. 1. Gas-chromatographic analysis of analgesic drugs extracted from human fluids. Free morphine urine. A. 5 μl of un-hydrolyzed extract; B. 5 μl of extract followed by 5 μl of acetic anhydride; and C. 5 μl of extract followed by 5 μl of propionic anhydride (concentration about 4 μg/5 μl). Total morphine urine. A. 1 μl of hydrolyzed extract; B. 0.8 μl of extract followed by 5 μl of acetic anhydride; and C. 0.8 μl of extract followed by 5 μl of propionic anhydride (concentration about 7 μg/μl). Codeine plasma. A. 1 μl of extract; B. 1 μl of extract followed by 5 μl of acetic anhydride; and C. 1 μl of extract followed by 5 μl of propionic anhydride (concentration about 2 μg/μl). Pethidine urine. A. 0.5 μl of extract; B. 0.5 μl of extract followed by 5 μl of acetic anhydride; and C. 0.5 μl of extract followed by 5 μl of propionic anhydride (concentration about 4 μg/μl). From Mulé,[8] courtesy of *Analytical Chemistry*.

TABLE VI

Retention Data on the Various Narcotic Analgesics Arranged According to Chemical Families[a]

Compound (Free Base)	Free[b]	RRT[c]	With 5 μl Acetic Anhydride		With 5 μl Propionic Anhydride	
			Acetic	Anhydride	Propionic	Anhydride
Iminoethanophenanthrofurans						
Morphine	6.94	1.16	8.06	11.34	10.50	18.37
Normorphine	6.94	1.16	23.81	29.34	31.70	35.62 57.37
Codeine	6.00	1.00	6.00	8.40	6.00	10.54
Norcodeine	6.00	1.00	16.12	22.50	19.78	34.87
Heroin	11.34	1.89		11.34		11.34
Nalorphine	9.84	1.64	11.62	15.94	14.81	26.34
Methyldihydromorphinone	7.50	1.25		10.12		12.56
Dihydromorphinone	7.31	1.22	9.94	11.44	12.37	18.94
Ethylmorphine	6.71	1.12	6.71	9.28	6.71	11.68
Dihydrohydroxymorphinone	8.81	1.47	12.56	15.94	15.62	26.63
Dihydromorphine	6.94	1.16	8.06	10.31	10.31	16.31
Dihydrocodeinone	6.94	1.16		6.94	6.94	10.87
Dihydrohydroxycodeinone	8.62	1.44		8.62		8.62
6-Monoacetylmorphine	8.62	1.44		11.34		14.67
Iminoethanophenanthrenes						
l-3-Hydroxy-N-methylmorphinan	4.12	0.69	4.12	4.59	4.12	5.81
l-3-Hydroxymorphinan	4.12	0.69		12.97		20.44
l-3-Methoxy-N-methylmorphinan	3.19	0.53		3.19		3.19
l-3-Methoxymorphinan	3.19	0.53		9.28		11.06
l-3-Hydroxy-N-allylmorphinan	5.72	0.95	5.72	6.56	5.72	8.25

	I	II	III	IV	V	VI	VII
Diarylalkoneamines							
dl-Methadone	3.32	0.55	3.32			3.32	
l-Acetylmethadol	3.65	0.61	3.65		3.65	3.32	
d-Propoxyphene	3.56	0.59	3.56			3.56	4.50
Arylpiperidines							
Pethidine	1.22	0.20	1.22			1.22	
Norpethidine	1.22	0.20	3.01			3.71	
Ketobemidone	2.53	0.42	2.53		2.53		
dl-Alphaprodine	1.22	0.20	1.22			1.22	3.28
Piminodine	21.94	3.66	36.38		21.94	40.69	
Benzomorphans							
dl-2'-Hydroxy-5,9-dimethyl-2-phenethyl-benzomorphan	12.66	2.11	12.66	14.62	19.12		
l-2'-Hydroxy-2,5,9-trimethyl-6,7-benzomorphan	1.87	0.31	1.87	2.06	1.87	2.66	
2'-Hydroxy-5,9-dimethyl-2-(3,3-dimethylallyl)-6,7-benzomorphan	4.27	0.71	4.27	4.97	4.27	6.34	
2'-Hydroxy-5,9-dimethyl-2-cyclopropylmethyl-6,7-benzomorphan	3.78	0.63	3.78	4.50	3.78	5.81	

[a] From Mulé[8] by courtesy of *Analytical Chemistry*.
[b] Retention time of the free unreacted drug.
[c] RRT, retention time relative to free codeine.

TABLE VII
Chromogenic Reagents for Narcotic Analgesics

Reagent and Preparation	Spot Color	Sensitivity and Specificity
Dragendorff's—Solution A: bismuth subnitrate (2.13 g), water (100 ml) and glacial acetic acid (25 ml). Solution B: potassium iodide (50 g) and water (125 ml). Spray solution: (10 ml A), (10 ml B), acetic acid (20 ml), and water (100 ml).	orange-brown on light yellow background	General reagent for alkaloids, stable for several weeks. Limit of detectability 3–10 μg compound.[21]
Potassium iodoplatinate—10% platinic chloride (1 ml) mixed with 4% potassium iodide (25 ml) and water (to 50 ml).	blue-violet on pink background	Similar to Dragendorff's. Some selectivity in that color not observed with compounds containing only primary or secondary amino groups. Minimum level of detectability 0.1–1.0 μg compound.[49]
Iodine in potassium iodide (Mandel's)—Stock solution: iodine (2 g), potassium iodide (4 g), and water (94 ml). Spray solution: mix stock solution. (10 ml) with aqueous ethanol (90 ml).	yellow-brown fading rapidly	Nonspecific. Sensitivity less than Dragendorff's. Narceine becomes blue.
Bromocresol green—(0.5 g) in ethanol (100 ml).	green-blue on yellow-green background	Nonspecific. Reaction with nearly all amines either immediately or within half an hour.[96]
Potassium permanganate—0.1–0.5% in water.	yellow on pink background	About 50% narcotics give positive reaction.[96] Reagent highly unspecific and best used for differentiating pure alkaloids.
Cobalt thiocyanate—2% wt/vol in acetone.	blue on pale pink background	Nearly all bases give positive reaction either immediately or after a few hours. Some compounds differentiated by green color and others by speed of color development. Limited value.

Reagent	Color	Comments
Marquis reagent—Formalin 5% vol/vol in concentrated sulfuric acid (used on thoroughly dried chromatogram, which is then baked at 110°C for 10 min).	purple	Good for differentiating morphine and its derivatives from other types of narcotics, but certain stimulants and tranquilizers may interfere. Other classical color tests involving concentrated H_2SO_4 have been adopted for detection of opiates[162] and are described later in the text.
Ammoniacal silver nitrate—50% silver nitrate (15 ml) + 5N ammonium hydroxide (15 ml) then dropwise until solution is clear). Heat chromatogram on hot plate for 1–2 min.	black	General reagent. Good for phenolics and a less pronounced reaction with other opium alkaloids. Used to confirm morphine in urine screening.[48]
Potassium ferricyanide (Kiefer's reagent)—0.5% potassium ferricyanide (5 ml), ethanol (10 ml), and 50% ferric chloride (three drops).	blue	Specific for phenolic alkaloids, e.g., morphine, cotarncline, laudanine, and narcotoline.[162]
Diazo-reagents.—(1) Sulfanilic acid (5 g) dissolved in water (700 ml) without heating. 25% HCl (50 ml) and water to 1 liter were added. For spraying, 5 ml of solution are mixed with 0.5% aqueous sodium nitrite (5 ml) followed by N NaOH (20 ml) (2) p-Nitro-aniline (0.25 g) dissolved by gentle heating in N HCl (25 ml) and the solution made up to 50 ml with ethanol. Sodium nitrite (0.1 g) is added to this solution (10 ml) before spraying. Chromatograms are air-dried for 3–5 min and then sprayed with 0.5N ethanolic NaOH.	orange-yellow / red-purple	As above. Cryptopine and thebaine yield faint pink colors—probably due to decomposition products.[162] / Phenolic group reagent, differentiating between free and substituted phenolic groups. Primary amines also react but give different colors, e.g., amphetamine pink.
Phosphomolybdic acid—10% aqueous solution is sprayed, and then chromatogram is exposed to ammonia vapor.	blue	Coloration specific for phenolic compounds. This reagent and a similar one, Folin–Ciocalteau, used for morphine.[81]

and urinary constituents was negligible. However, acid hydrolysis of urine in the manner prescribed to obtain free morphine from its conjugates did present some difficulties in chromatography because hydrolysis products of normal urinary constituents are extracted and appear as extraneous peaks. Such interference does not prevent the use of GLC during such experiments provided that adequate controls consisting of known normal biological samples are analyzed concurrently with the unknown samples. In addition, it has been found[16] that interference by hydrolysis products may be greatly reduced by washing the hydrolyzed urine with organic solvent prior to alkalinization and drug extraction.

Normorphine, norcodeine, *l*-3-hydroxymorphinan, and norpethidine are metabolites of their corresponding *N*-alkylated analgesics. Demonstration of the presence of such metabolites in biological fluids by retention data (Table VI) would aid identification of the administered analgesic. Supplementary to the metabolite data in Table VI are the recently reported retention times for 2'-hydroxy-5,9-dimethyl-6,7 benzomorphan (norcyclazocine)[110] and the cyclized metabolite produced upon mono-*N*-demethylation of methadone.[148]

B. Color Reactions

1. Location of Compounds on Chromatograms

Most of the reagents used to detect alkaloids and basic compounds are of high sensitivity but low specificity. Their composition, modifications, and typical reactions with and sensitivities to various compounds or classes of compounds have been described in detail[4,21,81,96] and are summarized with respect to narcotic analgesics in Table VII. Curry[4] described a scheme for identification of toxicologically important alkaloids based upon their R_f values on paper and reactions with Dragendorff's, cobalt thiocyanate, potassium permanganate and *p*-dimethylaminobenzaldehyde spray reagents. A similar scheme based upon R_f value on silica gel and sequential color reactions with iodoplatinate, ammoniacal silver nitrate, and potassium permanganate is used in urinary screening for narcotics in addict treatment programs.[48]

2. Direct Tests on Solids

This subject has been reviewed for opium alkaloids,[163] morphine,[164] and its derivatives,[165-167] and newer synthetic analgesics.[168-170] The most widely used color reactions are those in which a compound in concentrated sulfuric acid is added to a minute portion of the solid alkaloid and the resulting color changes are noted. The most useful compounds for this purpose include formaldehyde (5% vol/vol Formalin, Marquis), ammonium molybdate (0.5%, Fröhde), ammonium vanadate (0.5%, Mandelin), selenious acid (0.5%, Mecke), *p*-dimethylaminobenzaldehyde (10% in glacial acetic acid, Wasicky), sodium tungstate (1%, Reichard), titanium dioxide (0.5%, Flueckiger), and sucrose (10%, Schneider–Weppen). Original references to these tests,[163] various modifications of reagents, and results for numerous alkaloids are available.[21] The sensitivity of such tests can be increased by using the microtechnique of Clarke and Williams.[163] For example, with Fröhde's test a microdrop each of the test solution and a 0.5% aqueous solution of ammonium molybdate are mixed on an opal glass plate. After evaporation, a microdrop of concentrated sulfuric acid is applied to the residue and the color changes are noted. This procedure is not practicable for Marquis' or Flueckiger's tests and here the evaporated test solution is treated with a trace of the original sulfuric acid reagent.

Carried out in the above manner, these tests have sensitivities in the range of 1.0–0.025 μg. With minimal quantities, the whole range of colors normally observed with larger quantities may not be detected. The colors obtained for 1 μg quantities of a number of narcotic analgesics together with the absolute sensitivities are given in Table VIII, and other color tests for alkaloids are given in Table IX. Marked specificity may be exhibited for certain chemical groups; for instance, in the microdiazo test phenolic and primary aryl amino groups give rise to red or purple colors. In most tests of this type the reaction mechanisms and structures of colored derivatives have been established. In contrast, there has been little research on the classical tests involving strong acids and, except for the more recent electron paramagnetic resonance studies of colored free radicals,[177] the available information and theories concerning reaction mechanism are contained in a previous review.[21]

Interpretation of the results is very important in these tests because, although there is generally no question as to whether or not a positive reaction has occurred, there may be a subjective difference between observers in the way that shades and tints of primary colors are described. Such differences may be partly overcome if the description of colors follows the spectral arrangement: red, orange, yellow, brown, green, blue, violet (magenta), and colorless or negative. In addition, the colors themselves are influenced by certain variable factors such as (1) contaminants or tissue extractives, (2) the arrangement of chromophores in a given structure, (3) the solvent medium, and (4) the concentration of the solute. The effect of solvent is especially noteworthy with strong acids such as sulfuric, which itself may yield brown and orange colors with aromatics. Therefore, with such compounds the reaction to sulfuric acid reagents is only considered positive when additional colors are observed; that is, the behavior of an unknown compound in sulfuric acid should be determined in advance.

When color reactions are used for identification of an unknown compound, it is advisable to try, if possible, the effect of the reagent on a known sample of the suspected substance. Moreover, it is imperative to use more than one type of test in such an identification because color reactions are relatively nonspecific[21] and would serve only as preliminary tests.

Color tests, besides being used extensively in toxicological investigations of narcotics and other alkaloids, also have been applied as official pharmaceutical tests for such compounds. For instance, Marquis' test was used for morphine (USP XVII) and Fröhde's and the ferric chloride/potassium

TABLE VIII

Results of Marquis', Fröhde's, Mecke's, and Reichard's Color Tests[a]

Alkaloid	Formaldehyde	Ammonium Molybdate	Selenious Acid	Sodium Tungstate
Iminoethanophenathrofurans				
Morphine	violet (0.05)[b]	violet→blue→light green (0.05)	blue-green→grey-green (0.1)	violet (0.25)
Pseudomorphine	rose red (0.1)	blue→violet→green (0.1)	purple→brown (0.1)	deep violet (0.1)
Dihydromorphine	red purple (0.025)	blue-violet (0.025)	brown→green (0.25)	violet (0.25)
Dihydromorphinone	yellow→red→purple (0.25).	purple→blue→green (0.05)	orange→brown (0.25)	dark brown (0.25)
Methyldihydromorphine	purple (0.1)	violet→blue→green (0.1)	green (0.25)	gray (1.0)
Metopon	purple (0.5)	black-violet→blue→green (0.1)	yellow-brown (0.5)	—[c]
Morphine-N-oxide	purple (0.25)	deep violet→blue→green (0.25)	green (0.25)	gray-purple (0.5)
Nalorphine	purple (0.25)	deep blue→green (0.25)	brown (0.25)	black-purple (0.5)
Normorphine	purple (0.1)	purple→blue (0.1)	green (0.25)	gray (1.0)
Oxymorphone	purple (0.5)	bright blue→green (0.25)	yellow (1.0)	deep violet (1.0)
Codeine	violet (0.05)	blue, slowly fading (0.1)	blue-green→yellow→brown (0.5)	pale violet (1.0)
Dihydrocodeine	purple (0.1)	green→blue (0.1)	green→yellow-brown (0.1)	—
Dihydrocodeinone	yellow→brown→purple (0.25)	green→blue (0.1)	yellow-green (0.25)	—
Dihydrohydroxycodeinone	yellow→brown→purple (0.5)	yellow→green→blue (0.1)	orange→olive green (0.25)	deep violet (0.5)
Heroin	violet (0.05)	red-violet→blue→lt. green (0.05)	as codeine	violet (0.25)
Benzylmorphine	red→purple (0.025)	violet→green (0.05)	green (0.1)	brownish green (0.25)
Thebaine	red→orange (0.05)	greenish brown→red-brown (0.1)	green→brown→orange (0.25)	gray (0.5)
Norcodeine	purple (0.1)	blue-green→blue→green (0.1)	green (0.1)	—
Acetyldihydrocodeine	purple (0.25)	green→blue (0.25)	green (0.5)	—
Levallorphan	—	blue-green (0.25)	yellow-brown (0.5)	purple (0.25)
Racemorphan	—	blue-green (0.25)	yellow-brown (0.5)	purple (0.25)
Racemethorphan	—	blue-green (0.25)	yellow-brown (0.5)	faint brown (1.0)
Levorphanol	—	blue-green (0.25)	yellow-brown (0.5)	purple (0.25)
Phenomorphan	—	blue-green (0.25)	yellow-brown (0.5)	purple (0.25)
l-Phenacylmorphan	faint purple→grey→green (1.0)	bright blue (0.5)	pale yellow (1.0)	black-purple (0.25)
Norlevorphanol	olive (1.0)	blue→yellow green (0.1)	yellow-brown (1.0)	black-purple (0.25)
Diphenylpropylamines and Related Analgesics				
Acetylmethadol	purple-brown→gray-green (0.25)	brown-purple→green (0.25)	purple-brown→brown (0.25)	faint green-gray (1.0)
Methadol	purple-brown→gray green (0.25)	brown-purple→green (0.25)	purple-brown→brown (0.25)	faint green-gray (1.0)
Dipipanone	These compounds are nearly all colorless in these four tests. However, colors with ammonium vanadate (Mandelin's) test are:	deep green-blue (0.25)	light brown (0.5)	—
Isomethadone		brown-purple→violet-blue (0.1)	—	—
Methadone		faint green-blue (0.5)	—	—
Normethadone		yellow-green (1.0)	—	gray (1.0)
Phenadoxone		deep green-blue (0.25)	faint brown (0.5)	—
Propoxyphene	black-violet→dull green (0.5)	black-green (0.5)	—	—

Diampromide	orange (10)	—	—	—
Ethylmethylthiambutene	purple-brown (0.1)	orange-brown→pale green (0.1)	violet-blue (0.1)	orange (0.25)
Di(m)ethylthiambutene	purple-brown (0.1)	orange-brown→pale green (0.1)	violet-blue (0.1)	orange (0.25)
Dimenoxadol	orange greenish-blue (0.1)	orange-brown→dull purple (0.1)	orange (0.1)	orange→purple (0.1)
Arylpiperidines and Related Analgesics				
Anileridine	dull orange (1.0)	—	—	—
Etoxeridine	dull orange (1.0)	—	—	—
Morpheridine	dull orange (1.0)	—	—	—
Pethidine	dull orange (1.0)	—	—	—
Properidine	dull orange (1.0)	—	—	—
Hydroxypethidine	dull orange (1.0)	bright blue, fading (0.25)	gray-blue→brown (0.5)	reddish purple (0.5)
Ketobemidone	dull orange (1.0)	bright blue, fading (0.25)	blue-green (0.25)	reddish purple (0.5)
Benzethidine	orange (1.0)	orange-brown (1.0)	orange-brown (1.0)	faint gray (1.0)
Furethidine	faint orange (1.0)	—	—	—
Phenoperidine	—	faint gray-purple (1.0)	faint brown (1.0)	—
Alphameprodine	brownish red (0.5)	blue-gray→green with blue rim (0.5)	orange-brown (1.0)	—
Alphaprodine	brownish red (0.5)	blue-gray→green with blue rim (0.5)	orange-brown (1.0)	—
Betameprodine	red-purple (0.5)	blue-gray→green with blue rim (0.5)	orange-brown (1.0)	—
Betaprodine	red-purple (0.5)	blue-gray→green with blue rim (0.5)	orange-brown (1.0)	—
Trimeperidine	red-purple (0.5)	—	yellow (1.0)	—
Allylprodine	blue-black (0.1)	gray→green (1.0)	pale olive (1.0)	blue-gray (1.0)
Ethoheptazine	dull orange (0.5)	—	—	—
Proheptazine	dull purple (0.5)	blue-gray→green (0.25)	yellow-brown→orange (0.5)	yellow-brown→green-gray (1.0)
Benzomorphans				
dl-Phenazocine	brown (1.0)	bright blue→green (1.0)	brown (1.0)	black-purple (0.1)
Metazocine	brown (1.0)	bright blue→green (1.0)	brown (1.0)	gray-purple (1.0)
Pentazocine	dull red→olive green (0.25)	bright blue (0.1)	not tested	not tested
2-Benzylbenzimidazoles				
Etonitazine	faint orange (1.0)	pale green (1.0)	yellow (1.0)	yellow-brown (1.0)
Clonitazine	—	grey-blue (1.0)	—	—

a From Clarke [163,169,178]
b Sensitivities (μg) in parentheses.
c Dashes indicate colorless or negative reaction.

TABLE IX
Additional Color Tests for Narcotic Analgesics

Reagent and Preparation	Color	Sensitivity (μg)	Specificity	Reference
Zernik—Concentrated nitric acid	red to orange red	0.1–1	all phenolic and etheric opiates and many other alkaloids give positive reaction	168, 21
Vitali—One drop of test solution is evaporated to dryness, a drop of fuming HNO_3 added, and color noted. HNO_3 evaporated to dryness. Color noted again and yet again after addition of a drop of 4% ethanolic potassium hydroxide solution to the cold residue.	a variety of three color sequences (see reference)	0.1–1	the sequence: yellow/yellow orange is observed predominantly with morphines and morphinans; some degree of specificity with this and other classes of narcotic	21 169
Microdiazo—Microdrop of saturated p-nitroaniline placed on opal glass. Then, in order, microdrop of 0.1% $NaNO_2$, the test solution, and 2N NaOH	red or purple	0.25	phenolic group reagent; masked groups do not react	171
Modified nitroso—Test substance + 1–2 drops of sulfuric acid + a drop of nitrobenzene; heat to boiling. After cooling dilute with 3–4 drops of water and shake with 1–2 drops of butanol.	violet (max 552 mμ) in organic layer	2	phenolic group reagent; masked groups do not react	172
1-Nitroso-2-naphthol—in dilute nitric acid	red	10	phenolic group reagent	172
Phosphomolybdic[173]	blue	0.5	phenolic group reagent	164

Reagent / Procedure	Color	Value	Notes	Ref.
Silicomolybdic[173]	blue	0.5	phenolic group reagent	164
Folin–Ciocalteau—Sodium tungstate, sodium molybdate, H_3PO_4, HCl, $LiSO_4$ (see reference for details)	blue	0.5	phenolic group reagent	81
Chloramine T—in dilute HCl + 4 aminopyrimidine (see reference)	red	5	related alkaloids give the same reaction	172
Ceric ammonium nitrate followed by 2,4-dinitrophenylhydrazine (see reference for details)	orange red	1.5	codeine, thebaine, and papaverine interfere	174
Ferric chloride (one drop) + test solution in H_2SO_4. Heat on boiling water bath for 2 min. Note color. Add one drop nitric acid and note color again.	blue/changing to dark red-brown		codeine and morphine give same but dihydromorphinone and papaverine do not produce this color change	175
One drop of test solution evaporated to dryness in microtest tube. A drop of syrupy zinc chloride is added and tube is kept at 180°C for 5 min. After cooling contents of the tube are dropped on filter paper impregnated with a saturated benzene solution of 2,6-dichloroquinone-4-chloroimine. Paper then fumed over concentrated ammonia.	blue	2–4	reaction specific for morphine and codeine in the absence of other phenols with a free para position; in same test with 0.1 g oxalic acid (anhyd) instead of zinc chloride, codeine is positive but morphine is negative (see reference)	176

ferricyanide tests for morphine (USP XVII), nalorphine (USP XVII), and levorphanol (NF XI). Similar tests have been useful for routine detection of commonly abused narcotics in illicit drug preparations. In addition, a fairly specific test for heroin[178] has been formulated as follows: Ten drops of a concentrated nitric–85% phosphoric acid mixture (12:38) are placed in a 5 ml glass-stoppered centrifuge tube, and 3.25 ml chloroform are added. A little chloroform is used to wash heroin into the tube, which is then shaken vigorously for 30 sec. The bottom layer acquires a color that depends on the amount of heroin present. For example, after 10 min the blank sample produces light green; 10 μg heroin, light yellow; 1 mg heroin, yellow-brown; and 10 mg heroin, dark red-brown. Of 25 other narcotics and possible diluents or adulterants only antipyrine had a positive reaction, and this compound can be distinguished from heroin by a negative Marquis reaction.

C. Microcrystal Tests

The microcrystal test is one of the oldest and simplest tests used in toxicology. It is most commonly applied to basic nitrogenous drugs or alkaloids, but it also may be adapted to neutral and acidic compounds. Historical development and various specific adaptations of this test have been excellently reviewed by Fulton.[179] In this and other reviews it is generally agreed that the microcrystal test, although more specific and sensitive than color tests when working with suitable reagents, is unsuitable as a primary method of identification of unknown compounds because it cannot form the basis of an identification scheme. Its real value is that it allows simple, rapid, and specific confirmation of a provisional identification made from chromatographic or spectrophotometric evidence.

In its simplest form the test consists of mixing a drop of test solution with one of reagent on a microscope slide and observing crystal formation through a microscope. This method works if sufficient test material is available and the crystalline derivative forms immediately. However, if crystals form very slowly or if one attempts to increase the sensitivity of the test by decreasing the size of the drop, loss of water by evaporation will cause the drop to become saturated with reagent which will crystallize and mask the crystalline derivative. These difficulties were overcome by the hanging microdrop technique,[180] which is sensitive enough to be used with microgram quantities of material eluted from chromatograms. The procedure is as follows: A microdrop (about 0.1 μl) of the test material in 1% acetic or hydrochloric acid is transferred to a cover slip by means of a glass rod 1 mm in diameter at its tapered end. A similar drop of reagent is added, and the two are stirred together, the glass being scratched slightly to promote crystal formation. A cavity slide is ringed with gum arabic solution, inverted

A   B

Fig. 2. Microdrop technique. A. sealed with gum arabic, B. elevated cover slip for faster results.

over the cover slip, and then reinverted to leave a "hanging drop" (Fig. 2A). Evaporation is greatly retarded by the gum arabic seal, and crystals form in regular, reproducible shapes. The drop is examined at once at low magnification under ordinary and polarized light, and the appearance of any precipitate (amorphous, oily droplets, or crystals) is noted. In the absence of crystals it is reexamined every few hours until crystallization is complete. This may take 24 hr or more. This delay may be an intrinsic characteristic in

TABLE X
Descriptive Terms for Microcrystals[a]

Appearance	Descriptive Term	Definition of Term
	needle	long thin crystal with pointed ends
	rod	similar to a needle but with ends square cut
	blade	broad needle
	plate	blade when length and breadth are of the same order of magnitude
	tablet	plate with appreciable thickness
	prism	much thicker tablet
	splinters	small, irregular rods and needles
	grains	very small lenticular crystals
	rosette	collection of crystals radiating from a single point
	burr or hedgehog	rosette so dense that only the tips of the component crystals are seen
	tuft	sector of a rosette
	fan	sector of a rosette
	sheaf	double tuft
	star	rosette with only four to six components
	cross	single cruciform crystal
	cluster	loose complex of crystals
	bundle or bunch	cluster with majority of crystals lying in one direction
	dendrites	multibrachiate branching crystals
	wedges, dumbells and combs	self-explanatory—denote similarity with these objects
	smudge rosettes and gelatinous rosettes	reproducible assemblies that possibly are not truly crystalline

[a] More-detailed illustrations and information on descriptive terms are given in Farmilo and Genest[21] and in Fulton.[179]

TABLE XI

Microcrystal Tests for Common Narcotic Analgesics and Related Compounds[a]

Narcotic	Reagent (Compound and/or Quantity Dissolved in 100 ml Water, Unless Otherwise Indicated)	Type of Crystal	Sensitivity[b] (μg)
Morphine	potassium cadmium iodide (1 g cadmium iodide + 2 g potassium iodide)	sheaves of fine needles	0.01
	potassium tri-iodide (Type 1:2 g iodine + 4 g potassium iodide)	plates	0.1
Codeine	potassium cadmium iodide	gelatinous rosettes—aggregates of small tablets	0.01
	potassium tri-iodide (Type 1)	feathery rosettes forming overnight	0.05
Ethylmorphine	potassium mercuric iodide (1.5 g mercuric chloride + 5 g potassium iodide)	smudge rosettes	0.05
	potassium tri-iodide (Type 3:1 g iodine + 50 g potassium iodide)	bunches of feathery needles	0.05
Nalorphine	potassium chromate (5 g)	four-pointed stars	0.5
	potassium mercuric iodide	smudge rosettes or fibers	0.25
Dihydromorphinone	sodium nitroprusside (1 g)	rods	0.5
	potassium bismuth iodide (5 g bismuth subnitrate + 25 g potassium iodide + 2% wt/wt sulfuric acid to produce 100 ml)	grainlike crystals forming overnight	0.05
Dihydrocodeinone	mercuric chloride (5 g)	dense rosettes	0.05
	potassium bismuth iodide	rhomboidal plates	0.01
Dihydrocodeine	potassium bismuth iodide	bunches of hexagonal plates	0.005
Methyldihydromorphinone (metapon)	picrolonic acid (saturated solution)	serrated needles	0.5
	potassium bismuth iodide	very small rods	0.1
Diacetyl-morphine (heroin)	mercuric chloride	fine dendrites	0.1
	platinic chloride (5 g)	rosettes	0.25
Levorphanol	platinic bromide (5 g platinic chloride + 10 g sodium bromide)	bunches of small, irregular plates	0.25
	platinic chloride	masses of small, irregular crystals	0.25

Levallorphan	ammonium thiocyanate (5 g)	bunches of rods	1.0
	potassium iodide (5 g)	needles and plates forming overnight	0.5
Methadone	mercuric chloride	rosettes of branching rods	0.05
	platinic bromide	bunches of prisms	0.25
Propoxyphene	gold bromide-hydrochloric acid solution (5 g gold chloride + 5 g sodium bromide + concentrated HCl to 100 ml)	curving blades forming overnight	0.25
	potassium tri-iodide (Type 1)	small plates in fernlike patterns best viewed under polarized light	0.025
Dextromoramide	gold chloride (5 g)	oily needles, some serrated, forming overnight	0.25
	picrolonic acid	rosettes of branching rods, sometimes dense	0.25
Phenadoxone	picrolonic acid	dense, oily rosettes	0.1
	potassium tri-iodide (Type 1)	rods and plates, overnight	0.1
Dipipanone	platinic bromide	curving blades, mostly serrated	0.25
	platinic chloride	bunches of serrated blades	0.1
Pethidine	picric acid (saturated solution) or picrolonic acid	feathery rosettes	0.1
Anileridine	potassium iodide	large plates and small needles from edge	0.25
Alphaprodine	picric acid	small, serrated plates often in bunches	0.25
	styphnic acid (5 g)	very small crystals best viewed under polarized light	0.1
Phenazocine	potassium iodide	dense, oily rosettes	0.1
	sodium carbonate (5 g)	bunches of irregular prisms	0.1
Metazocine	picrolonic acid	curving needles	0.25
	platinic chloride	bunches of small plates	0.25

a Compiled from the results of Clarke.[163,169,170]
b Using method of Clarke and Williams.[180]

the formation of the particular crystals or may result from very slow evaporation and concentration of the drop. To promote evaporation and obtain quicker results in urgent cases the method described above is used, but distance pieces (small pieces of broken cover slip) are cemented on opposite sides of the depression of the cavity slide (Figure 2B).

When definite crystals have been formed, their form (that is, needles, rods, and the like) and habit (that is, the positions they occupy relative to each other—rosettes, sheaves, and the like) are noted or recorded by sketch or photograph fairly promptly because certain crystals are unstable and may disappear within an hour or two. Tentative identification then may be made by comparing the descriptions, sketches, or photographs of the crystals formed with those listed for known compounds. However, final identification must depend on comparison of the crystals formed from the unknown with those from an authentic sample of the drug that has been treated identically.

The descriptive terms (Table X) used in making tentative identifications are not absolutely specific because various crystal forms do not differ distinctly but gradually merge into one another. Additionally, irregular forms may prevent an exact match between crystals of test and control. The source of such trouble may be impurities in the test solution or, occasionally, polymorphism where (under different conditions of temperature, pH, and particularly concentration) a compound may crystallize in entirely different forms. Such interference can be overcome by using material eluted from chromatograms and by ensuring that test and control crystals are formed under identical conditions. This is difficult because the concentration of the test solution usually will be obtained by guesswork. However, should the control drop appear to be more crowded with crystals than the test drop, the control solution may be diluted and the test repeated.

Solutions of test material should be in the range of concentration of 0.1–1%. One of the best solvents is $2N$ acetic acid. Crystals form more readily from dilute hydrochloric acid, but this may hydrolyze unstable compounds. They form even more readily in more concentrated mineral acids, and this observation has given rise to a series of reagents made up in moderately concentrated sulfuric and phosphoric acids.[179] Several hundred reagents have been used for precipitation of alkaloids in microcrystal tests.[21] These include halides of heavy metals; double halides of cadmium, bismuth, and mercury; various solutions of iodine; alkalis; oxygen acids of elements occurring in groups 4, 5B, 6A, and 7 of the periodic table; and sundry organic acids or their salts. Details of the classification and preparation of reagents and their reaction with alkaloids are given in other reviews.[21,179]

Specific application of crystal tests to toxicological detection of microgram quantities of narcotics has been thoroughly investigated.[163,168–170] The results for commonly used compounds in their most appropriate tests are summarized in Table XI. The tests were selected according to sensitivity, rapidity and certainty of crystal production, and reagent stability.

Illicit narcotic seizures also have been analyzed by microcrystal tests. Certain diluents or adulterants, for example, quinine,[181] can interfere in the test for heroin. However, specificity has been increased and reagents have been formulated that can differentiate between such closely related compounds as morphine, 3-monoacetylmorphine, 6-monoacetylmorphine, and diacetyl-morphine (heroin).[182] The reagents iodine–potassium iodide in concentrated hydrochloric and syrupy phosphoric acids (1:1) and bromauric acid in syrupy phosphoric acid were found to be most sensitive for such tests.

Microcrystal tests also afford the only simple and relatively cheap method for distinguishing between microgram quantities of optical isomers. The method depends on the fact that, with certain reagents, racemic forms of compounds give crystals whereas the optical isomers each give an amorphous precipitate. For instance, a solution of racemorphan in $2N$ hydrochloric acid gives bunches of small plates within half an hour when treated with sodium carbonate solution by the microdrop technique, while under similar conditions levorphanol and dextromorphan only give amorphous precipitates even after standing for 48 hr. To distinguish between these optical isomers in an unknown, a microdrop each of test solution, authentic levorphanol solution, and sodium carbonate solution are placed, in that order, onto a cover slip and examined in the usual way. If the test solution contains levorphanol, then the addition of levorphanol will not affect it and an amorphous precipitate will be formed. However, if it contains dextromorphan, the test drop will contain both isomers and crystals typical of racemic compound will be formed.[183] This method is of general application provided that a reagent can be found that will distinguish between the racemic compound and its optical isomers in the described fashion and that one of the isomers is available for cross testing. Examples of other reagents successfully applied to narcotics are trinitrobenzoic acid, which gives rosettes of crystals with racemethorphan but oily amorphous precipitates with the individual isomers; ammonium thiocyanate, which gives large plates (best seen under polarized light) with racemoramide but oily droplets with the isomers; and gold chloride, which gives small, curved, irregular needles almost at once with racemic propoxyphene but no crystals with the isomers for several hours after which large straight needles may form.[183]

D. Physicochemical Methods

The methods suitable for identification of narcotics and related compounds were generally reviewed in 1953.[184] Since then more versatile and automated instrumentation such as recording spectrophotometers and spectropolarimeters have appeared and complex analytical techniques such as nuclear magnetic resonance and mass spectroscopy have become more readily available. But for these, the methods remain basically the same and the success of their application depends, as always, upon the use of compounds in either the pure state or when isolated as such by chromatography.

TABLE XII
Physical Data for Narcotic Analgesics and Related Compounds[a]

Compound	Molecular Weight	Water Analysis		Melting Point (°C)	pKa
		%	Moles		
Codeine phosphate	424.38	5.26	1.22	233–236	8.22
Dihydrocodeinone	299.36	0.00	0.00	197–200	6.61
Dihydrohydroxycodeinone	315.36	0.00	0.00	224–225.5	8.05
Morphine hydrochloride	375.84	14.83	3.00	265–275	8.02
Morphine sulfate	758.82	12.16	5.14	235–250	8.55
Dihydromorphine	305.37	6.19	1.05	125–128	8.20
Ethyl Morphine, HCl	385.88	9.40	2.01	123	7.83
Diacetylmorphine (Heroin) HCl	423.88	4.46	1.05	228–232	4.82
Morphine-N-oxide	301.33	0.57	0.10	245–270	8.15
Dihydromorphinone HCl	312.79	0.34	0.06	280–295	8.08
Methyldihydromorphinone HCl	335.82	0.19	0.04	above 310	7.83
Nalorphine HCl	347.83	0.29	0.06	242–243	8.15
Thebaine	311.37	0.05	0.01	191–193	5.30
Pholcodine	416.53	4.77	1.10	69.7	8.18
Levorphanol tartrate	443.48	8.30	2.05	112–117	8.97
Racemorphan HBr	347.29	2.64	0.51	186–192	8.83
Racemethorphan HBr	388.34	9.57	2.07	112–115	8.30
Levallorphan tartrate	433.49	0.35	0.08	178–179	7.30
Phenomorphan HBr	428.40	1.09	0.26	289–292	8.25
dl-Methadone HCl	345.90	0.10	0.02	230–249 / 235[6]	8.99[16]

dl-Phenadoxone	387.94	0.14	0.03	204–213	6.89
dl-Isomethadone HCl	363.91	4.70	0.95	224–225[6] 130–135 ca 175[6]	6.75[16] 8.07 8.21[16]
dl-Pipidone HCl	386.00	0.18	0.04	186–196	6.80
dl-Dipipanone HCl	403.98	4.87	1.09	112–116	8.70 9.08[16]
Normethadone HCl	331.88	1.21	0.22	175–177	6.00 8.14[16]
Levomoramide	392.52	0.09	0.02	190	6.60
Dextromoramide	392.52			180–184[6]	
Phenampromid HCl	310.90			201–202[6]	
Diampromid H$_2$SO$_4$	422.50			110–111[6]	
Propoxyphene HCl	375.93	0.50	0.10	169–170	6.30
dl-Alphaprodine HCl	297.82	0.13	0.02	223–225	8.73
Pethidine HCl	283.79	0.17	0.03	186.5–188	8.72
Ketobemidone HCl	283.79	0.21	0.03	202–203	8.67
Anileridine 2 HCl	425.46			280–287 (decomposition)	
Ethoheptazine citrate	453.46	0.24	0.06	138.5	8.45
Piminodine ethane sulfonate	476.63	0.59	0.16	135.5	6.90
Phenazocine	402.37	1.04	0.235	164.5	8.50
Cyclazocine	271.39			201–204[6]	
Pentazocine	285.41			145.4–147.2[6]	8.76[16]
Naloxone HCl	363.87			200–205[6]	

[a] Compiled from the data of Farmilo and co-workers[2,185] unless otherwise indicated.

1. Common Physical Constants

Physical constants are determined mainly for authentication of pure chemicals and pharmaceuticals in official tests. Physical data for narcotics was reviewed by Farmilo et al.,[2] was updated for modern analgesics by Martin et al.,[185] and is also available in reviews of certain classes of analgesics such as the phenylpropylamines[186] and in monographs concerning individual analgesics.[6] Unless otherwise indicated the data for common narcotic analgesics and related compounds (Table XII) was obtained as follows: *water analysis* by Karl Fischer titration,[187] *melting points* in a Fisher–Johns microfusion apparatus by the offical procedure for class 1 compounds,[187] and *acid dissociation constants* (pK_a) by fractional neutralization generally using aqueous ethanol (50%) as solvent for the narcotic salt or base that was titrated electrometrically with $0.01N$ NaOH or $0.01N$ HCl, respectively.[188] Methods for calculating the pKa of narcotics titrated as their monobasic acid salts[188] were less complex than those for narcotics titrated as their polybasic acid salts or for narcotics with more than one titratable basic function.[185] Results of pK_a determination were variable because of nonreproducible laboratory conditions, especially temperature. Methods that eliminate such variation and results of their application to

TABLE
Ultraviolet Absorption Spectra

Compound (Free Base)	Maximum Wavelength (mμ)				Absorptivity Molar (ε_{max})				Minimum Wavelength (mμ)			
	0.1N HCl	0.2N NaOH	Abs. ETOH	Ib-ETCl$_2$[c]	0.1N HCl	0.2N NaOH	Abs. ETOH	Ib-ETCl$_2$	0.1N HCl	0.2N NaOH	Abs ETOH	Ib-ETCl2
Morphine	285	297	287	287	1583	2421	1644	1514	260	278	262	263
Normorphine	285	297	287	287	1345	2233	1346	1245	260	278	262	263
Codeine[d]	285	285	286	287	1587	1479	1491	1494	262	262	263	263
Norcodeine[d]	285	285	285	286	1555	1478	1732	1478	262	262	263	263
Diacetylmorphine (heroin)	285	298	281	281	1950	3294	2190	1740	261	279	255	257
N-Allylnormorphine (nalorphine)	281	298	287	287	1466	2437	1635	1551	257	279	263	263
Methyldihydromorphinone (metapon)	281	291	285	285	1164	1982	1138	1121	262	275	266	267
Dihydromorphinone (dilaudid)	281	292	282	283	1331	2253	1250	1187	262	272	265	267
Ethylmorphine (dionine)	285	285	285	285	1720	1501	1564	1551	262	262	263	263
Dihydrohydroxymorphinone (oxymorphone)	281	292	284	286	1270	2280	1392	1097	262	274	265	267
Dihydromorphine	284	297	285	287	1600	2556	1765	1582	256	271	256	257
Dihydrocodeinone[d] (hydrocodone)	280	280	282	283	1389	1302	1302	1344	261	262	264	265
Dihydrohydroxycodeinone[d] (oxycodone)	281	281	284	285	1375	1280	1309	1305	262	264	265	265
6-Monoacetylmorphine	285	297	287	287	1418	2498	1559	1513	260	287	263	263

a From Mulé[8] by courtesy of *Analytical Chemistry*.
b Difference between the maximum and minimum wavelengths.
c The free base was dissolved in ethylene dichloride containing 25% isobutanol (vol/vol).
d The bathochromic shift was not observed in the 0.2N NaOH sample.

organic bases, including narcotic analgesics, in aqueous solution are listed by Perrin.[189]

2. Ultraviolet Spectroscopy

The theory, instrumentation, and application of ultraviolet spectroscopy to qualitative and quantitative analysis of narcotics have been thoroughly reviewed by Farmilo.[190] Oestreicher et al.[191] subsequently applied the method to 90 different narcotics and published their absorption spectra and tables of maxima, minima, and molecular extinction coefficients. Data on anileridine was presented individually.[192] Bradford and Brackett[193] used this method to identify 166 compounds of forensic and toxicological interest, including many narcotics, and Martin et al.[185] extended the work of Oestreicher et al. by presenting data for 12 modern narcotics. Substances with similar absorption spectra were often identified by spectroscopy at different pHs. For instance, the wavelength of the absorption maximum (λ_{max}) for morphine sulfate changes from 285 mμ at acidic pH to 297 mμ at alkaline pH. However, λ_{max} of codeine, which has a UV spectrum similar to that of morphine at acidic pH, does not change upon alkalinization. This increase in λ_{max} or "bathochromic shift" has not been completely explained

XIII

Data for the Iminoethanophenanthrofurans[a]

Absorptivity Molar, (ε_{min})				Peak Ratio ($\varepsilon_{max}/\varepsilon_{min}$)				$\Delta m\mu$[b]			
0.1N HCl	0.2N NaOH	Abs ETOH	Ib-ETCl$_2$	0.1N HCl	0.2N NaOH	abs. ETOH	Ib-ETCl$_2$	0.1N HCl	0.1N NaOH	Abs ETOH	Ib-ETCl$_2$
540	1439	534	534	2.93	1.68	3.12	2.83	25	19	25	24
388	1361	469	385	3.47	1.64	2.87	3.23	25	19	25	24
594	554	512	512	2.74	2.67	2.91	2.91	23	23	23	24
599	722	719	562	2.59	2.05	2.41	2.63	23	23	22	23
654	2037	716	462	2.98	1.62	3.06	3.77	24	19	26	24
430	1407	588	489	3.41	1.03	2.80	3.17	24	19	24	24
602	1607	682	707	1.93	1.23	1.67	1.58	19	16	19	18
756	1663	802	816	1.76	1.35	1.56	1.45	19	20	17	16
596	590	570	524	2.88	2.54	2.74	2.96	23	23	22	22
779	1731	973	790	1.63	1.32	1.43	1.39	19	18	19	19
415	1154	324	327	3.85	2.21	5.45	4.84	28	26	29	30
781	919	760	770	1.78	1.42	1.71	1.74	19	18	18	18
807	902	750	725	1.70	1.42	1.74	1.80	19	17	19	20
450	1530	534	505	3.15	1.63	2.92	2.99	25	19	24	22

TABLE XIV
Ultraviolet Absorption Spectra Data for Narcotic Analgesics Arranged According to Chemical Families[a]

Compound (Free Base)	Maximum Wavelength, (mμ)		Absorptivity Molar, (ϵ_{max})		Minimum Wavelength, (mμ)		Absorptivity Molar (ϵ_{min})		Peak Ratio, ($\epsilon_{max}/\epsilon_{min}$)		Δ mμ[b]	
	0.1N HCl	0.2N NaOH	0.1N HCl	0.2N NaOH	0.1N HCl	0.2N NaOH	0.1N HCl	0.2N NaOH	0.1N HCl	0.2N NaOH	0.1N HCl	0.2N NaOH
Iminoethanophenanthrenes												
l-3-Hydroxy-N-methyl-morphinan (levorphanol)	279	299	1621	2375	245	269	77	551	21.05	4.31	34	30
l-3-Hydroxymorphinan	279	298	1947	2927	245	269	102	652	19.09	4.49	34	29
l-3-Methoxy-N-methyl morphinan (levomethorphan)	278	279	1970	2092	245	249	108	214	18.24	9.77	33	30
l-3-Methoxymorphinan	278	279	2113	2265	245	249	139	226	15.20	10.02	33	30
l-3-Hydroxy-N-allylmorphinan (levallorphan)	279	299	2120	3157	245	269	130	740	16.30	4.27	34	30
Diarylalkoneamines[c]												
dl-Methadone	292		554		275		372		1.49		17	
l-Acetylmethadol	258		423		238		145		2.92		20	
d-Propoxyphene (darvon)	258		397		235		112		3.54		23	
Arylpiperidines												
Pethidine (meperidine)	257	256	210	210	254	255	158	166	1.33	1.26	3	1
Norpethidine	257	257	198	198	254	254	150	157	1.32	1.26	3	3
Ketobemidone	280	299	2075	2941	250	271	336	1308	6.17	2.25	30	28
dl-Alphaprodine (nisentil)	257	257	195	201	254	254	137	147	1.42	1.37	3	3
Piminodine (alvodine)	257		357		250		328		1.09		7	

Benzomorphans

dl-2'-Hydroxy-5,9-dimethyl-2-phenethyl-6,7-benzomorphan (phenazocine)	278	299	1944	2894	245	271	254	845	7.65	3.43	33	28
l-2'-Hydroxy-2,5,9-trimethyl-6,7-benzomorphan	278	299	1816	2684	245	270	211	726	8.60	3.70	33	29
2'-Hydroxy-5,9-dimethyl-2-(3,3-dimethylallyl)-6,7-benzomorphan	278	299	2071	3035	245	270	279	867	7.42	3.50	33	29
2'-Hydroxy-5,9-dimethyl-2-cyclopropylmethyl-6,7-benzomorphan	278	299	2061	3097	245	270	201	858	10.25	3.61	33	29

[a] From Mulé[8] by courtesy of *Analytical Chemistry.*

[b] Difference between the maximum and minimum wavelength.

[c] Ultraviolet spectra data were not obtained for this group in $0.2N$ NaOH due to the apparent precipitation of the free base upon the addition of $19N$ NaOH to the $0.1N$ HCl sample. Data obtained in ethylene dichloride containing 25% isobutanol (vol/vol) as follows:

	$m\mu_{max}$	$m\mu_{min}$	ε_{max}	ε_{min}
dl-Methadone	295	280	433	390
d-Propoxyphene	258	245	447	302
l-Acetylmethadol	259	256	550	494

but is attributable to the presence of a free phenolic hydroxyl in compounds exhibiting the effect. Consequently, it has been used to demonstrate the position of conjugation in compounds such as nalorphine,[61] which may be conjugated at positions other than the phenolic hydroxyl. Narcotics have also been identified by spectroscopy after conversion to derivatives such as reineckates[194,195] and tetraphenylborides.[196]

For toxicological investigations, UV spectroscopy has been adapted to paper[95] and thin-layer chromatography[132] for the identification of narcotics and other alkaloids. It was similarly adapted to gel electrophoresis, where the differential spectrum obtained by spectroscopy under acid and alkaline conditions served as a qualitative aid to identification.[80] Mulé[8] investigated the application of UV analysis to the microgram amounts of narcotics likely to be recovered from chromatograms. He tabulated wavelengths, wavelength differences, molecular extinctions, and molecular extinction ratios for absorption maxima and minima (Tables XIII and XIV). More recent information on the UV absorption of narcotics and related compounds may be obtained from atlases of spectra for organic compounds.[197]

3. Fluorometry

Measurement of the fluorescence of drugs in UV light has been used for a long time as a method of identification and assay.[198-200] Numai[201] published fluorescent color reactions of a number of narcotics, sympathomimetics, antihistamines, and alkaloids. The native fluorescence of some of the compounds such as morphine was low but could be greatly increased by heating with acids, for example, H_2SO_4, H_3PO_4, or $HCHO/H_2SO_4$.[201] Reaction with sulfuric acid was subsequently used in a method for detecting and assaying morphine in urine and opium.[202] This method recently has been modified for the simultaneous detection of morphine and quinine in urine and bile. Sensitivity to as little as 5 μg of compound was reported.[203] However, Mulé,[204] in attempting to use the method in urinary screening, has been unable to differentiate between urine containing no drug and urine containing 10 μg morphine.

The fluorescent moiety produced by reaction of morphine with sulfuric acid and other oxidizing agents[165] is considered to be pseudomorphine that has been shown to be 2:2'-bimorphine.[205] This compound has been produced quantitatively by oxidation with potassium ferri- and ferrocyanide mixture, which has been used in aqueous solution (57 and 7.8 mg%) as a selective spray reagent for morphine and similar compounds capable of forming fluorescent dimers, for example, normorphine, dihydromorphine, 6-monoacetylmorphine, and nalorphine.[206]

4. Optical-Rotation—Rotatory Dispersion (ORD) and Circular Dichroism (CD)

The interrelation of optical rotation and circular dichroism and their theory, instrumentation, and applications are well documented.[207–210] The measurement of optical rotation at a single wavelength, usually the sodium D-line (*ca.* 589 mμ), has been used to correlate the configurations of homologous series of optically active compounds. With regard to identification, this measurement also has been used in pharmacopeial tests for confirming identification of many optically active drugs including narcotics. For example, in the test for propoxyphene hydrochloride in capsules (USP 17) the specific rotation of evaporated chloroform extracts of capsule contents, after being taken up in a specified volume of 0.1N hydrochloric acid, should not be less than $+52°$. Details of D-line rotations of most other optically active narcotics are available in individual monographs.[6]

ORD and CD spectra are used mainly as an aid to structure determination or conformational and configurational correlations in complex, polycyclic organic molecules including those of narcotics such as morphine and related compounds,[211–213] ketonic morphines,[214] morphinans,[212,213] and methadone-type analgesics.[16,215]

5. Infrared Spectroscopy (IR)

The IR spectroscopic method and its application to various salts and bases of 45 narcotics and related compounds have been described in detail.[216–218] Three atlases, respectively containing spectra of compounds in mineral oil (Nujol), chloroform, and potassium bromide disks, were published. This major work has since been updated for (a) 12 newer narcotics in potassium bromide disks and carbon tetrachloride;[185] (b) pharmaceuticals,[219] more than 200 USP and NF reference compounds[220] and new and nonofficial drugs[221] including many narcotics; and (c) new derivatives of 14-hydroxymorphine[222] and 14-hydroxydihydrocodeinone.[223] The general features of the IR spectra of narcotics and the specific relationships of their functional groups to the absorption bands are discussed elsewhere.[185,224]

The IR method also has been successfully applied to qualitative and quantitative analysis of narcotics such as pethidine, codeine, and methadone[225] and in multicomponent systems: papaverine and oxycodone,[226] papaverine and narcotine,[227] papaverine, narcotine, and thebaine,[228] and morphine, codeine, and porphyroxine in opium.[229] However, similar analysis was not feasible with mixtures of the closely related narcotics: heroin, 3- and 6-monoacetylmorphines, and morphine and color tests were applied instead.[230]

The high degree of specificity of IR analysis compared to UV and fluorometric methods recommends its application to forensic and clinical toxicology as witnessed by its increasing usage in these fields and in pharmacopeial identification tests. For example, in the USP 17 one method of identification of capsules of propoxyphene hydrochloride is by comparison of the IR absorption of chloroform solution of capsule contents with that of a reference standard. The main factor limiting the application of conventional IR analysis to toxicology has been the large sample requirements (mg). This is gradually being overcome by the introduction of infrared microtechniques such as those applied to gas chromatographic fractions.[231,232]

In specific investigations of drug metabolism the sample requirements are less limiting and metabolites produced by N-dealkylation of pethidine,[14] levallorphan,[234] propoxyphene,[233] and methadone[148] and by ethereal sulfate and glucuronide conjugation of morphine[59] and nalorphine[61] have been identified by comparison of their IR spectra with those of authentic reference compounds. In instances where no particular route of metabolism is suspected or synthetic reference metabolites are unavailable the IR spectrum of the metabolite itself or its comparison with that of the parent compound may yield important information. Such was the case with levallorphan, whose major metabolite in the rat appeared to differ only slightly from the parent compound as confirmed by elemental analysis that showed the addition of one oxygen atom.[234]

6. X-Ray Diffraction

Of the x-ray diffraction methods available,[235] that involving measurements on powdered crystalline materials[236] is considered to be most convenient and superior to IR analysis for identification purposes.[184,237] x-ray diffraction patterns of narcotics and related compounds have been published,[238-242] and data for 95 narcotics is available in two articles.[185,237]

Reviews concerning the forensic application of x-ray diffraction have been written.[243,244] The principal limitations of the method are that it is restricted to crystalline materials, that structurally similar compounds give rise to very similar patterns, and that sensitivity toward small quantities of one constituent in the presence of a large quantity of another is not very high. Important advantages of the method are that it is nondestructive and gives a specific characterization of a crystalline phase.

The main application of x-ray diffraction is in the determination of the structure of crystalline compounds. In this respect the opium alkaloids cryptocavine and cryptopine have been characterized and absolute configurations of the enantiomorphs of methadone[245] and alphaprodine[246] have been confirmed.

7. Nuclear Magnetic Resonance Spectroscopy (NMR)

NMR, like x-ray diffraction, is mainly used for structure determination.[247] With respect to narcotics, the method has been used (1) to obtain information on the conformation of the B[248] and C[249,250] rings of a number of morphine alkaloids, (2) in conformational studies of synthetic morphines and morphinans,[251] and (3) on codeine isomers and their derivatives where the chemical shifts of the protons in different isomers was explained by the anisotropy of the double bond or aromatic ring.[252] In addition, the applications of NMR spectroscopy in medicinal and pharmaceutical chemistry have been the subject of a review[253] in which examples of conformational applications have been derived from studies on compounds from nearly all classes of narcotic analgesics.

The application of NMR to identification of compounds has not been investigated very much. Alexander and Koch,[254] in investigating the role of NMR in the analysis of pharmaceuticals, consider the possibility of fingerprint identification of a sample with the aid of spectral catalogues. Specificity was good, but the method was poor with regard to the detection of impurities where it was only sensitive enough in certain cases. Quantitative identification also was investigated, and two methods for determination of the relative amounts of drugs in multicomponent mixtures were described.

The main factor, other than high instrument costs, which limits the use of NMR in toxicology is the large sample requirement (about 10–50 mg compound in 0.3 ml solvent). However, the method has been successfully applied in drug metabolism studies, for example, in confirming the structure of the metabolite of methadone produced by mono-N-demethylation.[148]

8. Mass Spectrometry (MS)

This method of analysis[255,256] is used mainly for structure elucidation of compounds and has been extensively applied in the chemistry of natural products including alkaloids.[257] The main investigations with narcotics have involved determination of fragmentation patterns for a large series of morphine and morphinan derivatives. Audier et al.,[258] in investigations of morphine and 27 related compounds, found that one electron is initially removed from the nitrogen atom, followed by fragmentation of the resultant ion, and that this process is more rapid for the tetracyclic morphinans than the pentacyclic morphine derivatives. Nakata et al.[259] discussed fragmentation by various routes for 14 morphinans, tabulating the most important fragmentation ions and indicating where some of the illustrated fragmentations differ markedly from those of the pentacyclic morphine system described by Audier et al. Other investigators studied the steric direction of fragmentation in morphine derivatives with cis or trans fused B:C rings[260] and the effect of extent and position of unsaturation in ring C upon the fragmentation pathways of morphine and ten related alkaloids.[261]

Mass spectroscopy is also useful as a method of identification since the spectrum of a compound obtained under a given set of conditions is a highly specific characteristic. Tatematsu and co-workers investigated this application of MS in the pharmaceutical analysis of opium alkaloids. They obtained mass spectra[262] and examined the relationship between intensity of the molecular ion peak and ionization voltage for codeine, dihydrocodeine, and papaverine.[263] Their results indicated that quantitative determination of these compounds based upon intensity of the molecular ion peak relative to a reference standard was possible at ionization voltages of 10–80 eV and was optimal at 70 eV.[263] Since the molecular ion peak of codeine was not disturbed by that of other opium alkaloids and dihydrocodeine, the latter could be used as a reference substance for quantitative analysis of codeine (and vice versa).[264] Further investigations with various salts of codeine were carried out to confirm specificity. The mass spectrum of codeine appeared consistently with molecular ion peak at 299 m/e and constant relative intensity for all salts. The temperature at which the molecular ion appeared was related to the volatility or sublimability of the acid radical and in an inverse sense to the strength of the acid.[265]

Microgram or submicrogram amounts of drugs encountered in toxicology and clinical chemistry also may be identified by mass spectrometry. Preliminary separation of the drug from contaminating biological material by solvent extraction and chromatography is usually necessary. Gas chromatography is most suitable for this purpose because the effluent can be introduced directly into a rapidly scanning spectrometer (GC-MS). Coupled in this way, these techniques offer greater combined specificity and sensitivity than all other physicochemical methods for the identification of compounds. The instrumentation applications and data interpretation for such systems have been described in detail by McFadden.[266,267] In addition, Crawford and Morrison fed results from GC-MS into a digital computer and have shown, by using the six strongest peaks for 3200 catalogued compounds, that an unknown can be identified in 1.7 μsec.[268] Examples of the application of GC-MS to toxicological investigations of narcotics are not readily available in the literature. However, other drugs of abuse have been studied. For instance, the technique has been used for confirming the presence of amphetamines in urine[269] and for detecting metabolites of barbiturates and other drugs in forensic analysis.[270] Recently systems also have been described for combined TLC-MS and capillary GC-MS of various derivatives of amines.[271]

E. Biological Methods

The biological detection of narcotic analgesics depends on the catatonic rigidity of the tail, which is carried in an S-shaped curve when such compounds are administered to mice. This phenomenon was first observed upon subcutaneous injection of morphine[272] and found to be roughly quantitative, as little as 5 μg of the compound being detectable.[273] Quantitative estimation by finding the threshold of the "tail reaction," found to be more accurate than by measuring duration of the effect,[274] was further developed by Munch for the investigation of "doping" in racehorses.[275,276] He found that in a 20-g mouse the threshold amounts of narcotics were 1 μg for heroin, 12 μg for dilaudid, and 60 μg for codeine and morphine. However, the method is nonspecific, and although Munch[275] states that there is a qualitative difference between the effect of morphine and that of heroin, it is not usually possible to distinguish between closely related narcotics. The great advantage of the method is that results can be obtained rapidly by direct injection of body fluids such as urine and saliva into the mouse without the tedious extraction and purification required by chemical and physical methods. In general, the tail reaction method is less sensitive than chemical methods[277] with which it has been used mainly for identification of narcotics in cases of poisoning or suicide by overdosage of, for example, methadone,[278] ketobemidone,[279] and dipipanone.[280]

A method for detecting previous consumption of narcotics in man is based upon the pupillary dilation (mydriasis) induced by the narcotic antagonist, nalorphine, which reverses the meiotic effect produced by morphine-like drugs. The nalorphine test has correlated well with chemical tests for the detection of narcotics abuse and is widely used for screening purposes with urinalysis reserved for confirmation.[281]

IV. QUANTITATIVE DETERMINATION

A. Photometric Methods

1. Ultraviolet Absorptiometry

Quantitative application of this method and the factors affecting it are described by Farmilo.[190] Absorption is most conveniently measured at wavelengths of absorption maxima (Tables XIII and XIV) whereby sensitivity is also rendered optimal. However, of the photometric methods available, UV absorptiometry has the lowest sensitivity (usually milligram quantities) and is therefore seldom applied to biological samples obtained in toxicological and drug distribution studies. The most common usage of the method is in the analysis of pharmaceutical and illicit drug preparations where it is applied differently according to the particular preparation.

Injectable solutions or solutions of pure compounds may be assayed by direct measurements upon the preparation or a suitable dilution thereof, as demonstrated previously with phenadoxone,[282] codeine,[283] and nalorphine.[284] This kind of measurement, although least specific, has been applied with some success to narcotics in multicomponent mixtures such as narcotic seizures. For instance, in the UV assay of pethidine, a common constituent of such mixtures, only one compound (phenacetin) out of 20 normally used adulterants was found to interfere.[285] Furthermore, simultaneous determination of heroin and quinine was achieved by measurements at 297.5 and 330 mμ in alkaline methanolic solutions of the compounds, as their absorptivities are similar at 297.5 mμ but that of heroin is almost zero at 330 mμ.[286] Benzocaine, brucine, procaine, and papaverine were the only compounds out of a large number tested that were found to interfere.[287] In the UV assay of methadone in tablets interference with complex mixtures could be removed by separating the compound by extraction with chloroform or steam distillation.[288]

Differential spectrophotometry ($\Delta\varepsilon$ analysis) is applicable to certain compounds which exhibit large spectral shifts with changes in pH. This method has been used, for example, in the determination of morphine in opium,[289] various opium preparations,[290,291] paregoric,[289] tincture of opium, camphorated tincture of opium, morphine tablets,[292] and alkaline solutions of the drug.[293] The procedure was to determine spectra of the extracted alkaloids in acidic and alkaline solutions and then to subtract absorption in acid from that in alkali to obtain the differential spectrum which, for morphine, exhibits a maximum at 298–300 mμ of suitable magnitude for quantitative determination in a manner analogous to the classical method but with added specificity.

The various forms of chromatography followed by elution and UV absorptiometry yield the most specific quantitation. Examples are column partition chromatography on Celite 545 in the assay for heroin in illicit preparations,[29] cation exchange in assays for common drugs of abuse[41] and of dextromethorphan,[43] codeine[42] and trimeperidine[294] in pharmaceuticals, paper chromatography in assays for opium alkaloids,[295] and TLC in assays for heroin in pharmaceutical preparations.[296] Vidic[297] developed a microprocedure in which the sensitivity of the UV assay of morphine derivatives after elution from paper chromatograms is increased three- to ninefold over the UV assay of the free bases by treatment with 72% sulfuric acid at 85°C for 1 hr. After cooling the UV absorbance was measured against a similarly treated blank and compared with standard curves. Yields averaging 92% were reported. The acid treatment caused a dehydration reaction to form apomorphine-type compounds in the case of morphine, codeine, and dihydrocodeine, whereas compounds containing a ketone group and levorphanol were changed to a lesser degree. Absorption maxima were reported as follows: morphine, 276 mμ; hydromorphone, 280 mμ; levorphanol, 288 mμ; codeine, 245 and 272 mμ; dihydrocodeine, 251 mμ; and hydrocodone and oxycodone, 280 mμ. Also the ratios between the maxima and minima in the case of morphine (276/252 mμ) and dihydrocodeine (251/232 mμ) served to control background absorbance.

2. Spectrofluorometry

The instrumentation,[298] techniques,[298,299] and application[298,300] of this method of assay have been described recently. Two types of assay are available: one utilizes the native fluorescence of the compound and the other is dependent upon the fluorescence induced by previous chemical treatment of the compound. Use of the first type of assay for narcotics was described by Udenfriend et al.,[301] who reported that dromoran (levorphanol) could be extracted from basic solution, pH 11, into benzene and then reextracted into 0.1N hydrochloric acid. With excitation at the 275 mμ maximum and reading at the 320 mμ fluorescence maximum, 0.2 μg of compound could be determined, but the procedure was not used with biological material. Fluorescence assay of nalorphine was also investigated and could be accomplished using the excitation maximum at 285 mμ and the fluorescence maximum at 355 mμ.[301] The optimum pH for assay was 1.0 for both narcotics. It was considered that morphine would probably fluoresce in the same way as nalorphine, but this was not confirmed until 1961 when Brandt et al.[302] reported excitation maxima at 285 mμ and fluorescence maxima at 350 mμ for morphine and codeine in 0.1N sulfuric acid. A secondary excitation maximum was observed at 245 mμ. Both narcotics fluoresced with similar intensity except at pH values of 10–12, where fluorescence of morphine solutions was negligible but that of codeine solutions was unchanged. This difference between the two compounds served as the basis of a method for distinguishing and determining the two alkaloids in the presence of each other. Quantitative studies showed that the optimal range of determination was between 0.1 and 50 μg/ml and that the lower limit of detection was 10 ng/ml.[302] At concentrations above 100 μg/ml quenching was quite noticeable and was later ascribed to absorption effects that prevented excitation of the whole system.[303] The application of fluorometry to the determination of pentazocine in biological material was investigated in human distribution studies of the drug.[15] Plasma was subjected to a procedure similar to that described for solutions of dromoran[301] but with extraction at pH 8.5 to 9.5, excitation at 278 mμ, and fluorescence measurement at 310 mμ. It was found that after correction for reagent blank the plasma blank was low, being equivalent to less than 0.006 μg drug/ml. However, countercurrent distribution studies with plasma from drug-treated patients indicated that pentazocine metabolite(s) with similar solubility characteristics to those of the unchanged drug were present and caused interference in the assay.

The second type of assay, as applied to narcotics, depends upon the fluorescence that can be produced by oxidation.[201] For example, sulfuric acid originally was used for this purpose in the assay of morphine in opium.[202] The procedure was to evaporate the unknown solution of morphine to dryness with precautions to remove the last traces of water. The residue obtained was then heated with concentrated sulfuric acid (0.5 ml) on a water bath at 50°C for 8 min after which water (5 ml) and concentrated ammonia (6 ml) were added. The mixture was then heated at 50°C for a further 2 hr after which it was allowed to cool and then shaken with isobutanol (10 ml) for 4 min. An aliquot (6 ml) of the alcoholic phase was taken for fluorescence measurements. Excitation was at the 365 mμ maximum and fluorescence readings at about the 430 mμ maximum. The method was reported to be specific for morphine and sensitive to 0.02 μg. A similar method was used by Balatre et al.[304] for the assay of codeine and ethylmorphine in pharmaceuticals. These compounds also were excited maximally at 365 mμ but fluoresced maximally at 478 and 499 mμ, respectively.

Sulfuric acid oxidation of morphine has been reported to yield at least three main products, all of which fluoresce blue. Two of these are excited maximally at 366 mμ, and the other at 254 mμ.[202] The latter compound may be pseudomorphine, which has been produced specifically by oxidation of morphine with aqueous potassium ferri- and ferrocyanides (58 and 5 mg%) and found to possess the same excitation maximum.[17,285] This reaction also has formed the basis for the assay of morphine. Amounts as low as 0.1 μg were measurable in plasma and brain,[305] and recently the assay was modified to provide a tenfold increase in sensitivity by eliminating ferrocyanide and reducing reaction volumes.[17] The method is as follows:
1. Precipitation of protein in plasma (1 ml) with 7% trichloracetic acid (3 ml) in a siliconized centrifuge tube:
2. An aliquot (3 ml) of supernatant is adjusted to pH 9.0 and extracted

with 10% n-butanol in chloroform (10 ml) 9 ml of which is reextracted with 0.01N HCl (1.2 ml):

3. One milliliter of the acid layer is carefully evaporated to dryness under vacuum at 40°C:

4. To the residue is added 0.5 M Tris-HCl buffer, pH 8.5 (40 μl), followed, after mixing, by 0.4 mM potassium ferricyanide (2 μl) and then, after mixing for 15 min, by distilled water (3 ml):

5. The resulting solution is read in a fluorometer at 436 mμ with excitation at 254 mμ.

The above procedure may be modified to measure normorphine, nalorphine, dihydromorphine, and 6-monoacetylmorphine in biological fluids. Compounds that do not interfere with this method are the morphinans, codeine and its derivatives, dihydromorphinone, diacetylmorphine (heroin), apomorphine, pethidine, anileridine, and methadone.[305]

3. Colorimetry

a. Acidic Dye Technique. This is based on the fact that organic bases react with acidic dyes to form addition complexes which, unlike the unreacted dye, are soluble in organic solvents. The basic procedure is alkaline extraction of the sample with organic solvent (ethylene dichloride, chloroform, or benzene), an aliquot of which is then mixed with an aqueous buffered solution of the dye that quantitatively complexes with basic compounds in the organic phase and dissolves therein. An aliquot of the organic phase is then added to aqueous mineral acid (*ca.* 1 ml), which disrupts the base-dye complex, the dye entering the acidic phase in an amount proportional to that of the base-dye complex, which in turn reflects the amount of organic base in the original sample. The concentration of dye in the acidic phase is determined by measuring its absorbance in a spectrophotometer.

The application of the acidic dye technique to the assay of narcotics in biological materials was excellently reviewed by Way and Adler.[7] The dye, methyl orange, originally proposed in a general assay procedure for organic bases[306,307] has been most widely and successfully employed with narcotics.[7] Some of the other dyes used include bromothymol blue,[308,150] bromocresol purple,[309] bromocresol green,[91] and tropaeolin 00.[310] Since many organic bases are capable of forming a solvent-soluble dye complex and of being mistakenly measured for the drug under estimation, the dye procedure must be evaluated for specificity. In many instances modifications such as washing the base-containing organic phase with the appropriate buffer solution have been found to remove contaminants. However, nicotine may interfere in several of these modified versions of the methyl orange assay for basic drugs as shown, for example, with amphetamine.[311] A high degree of specificity can be conveyed to the dye method by using it in conjunction with a separatory technique such as countercurrent distribution[7] and the various forms of chromatography. For instance, it has been combined with (1) column chromatography in codeine assay[312] and in an automated assay of microgram quantities of basic drugs such as

TABLE XV
Quantitative Determination of Phenolic Narcotics

Reagent	Color	λ_{max} (mμ)	Application	Range (μg/ml)	Reference
Iodic acid, ammonium carbonate, nickel salt complex	green	670	Morphine in: opium	300	(315)
			poppy		(316)
			galenicals		
Iodic acid at pH 7 (KH$_2$PO$_4$–NaOH buffer)	blue-green	597	morphine HCl	10–500	(317)
Folin–Ciocalteau	blue	690	morphine in blood plasma	0.6–3.0	(47)
Silico-molybdic acid	blue	675	morphine, nalorphine, and levorphanol in biological materials,		(318)
			morphine in biological materials and pharmaceuticals	6	(319)
Phosphomolybdic acid	blue	750	morphine in pharmaceuticals	0.02	(320)
Nitrous acid (NaNO$_2$/HCl)	yellow	450	morphine in poppy	1000	(321)
			morphine in opium	1–500	(322, 323)
Modified nitrous acid		590	morphine in concentrated H$_2$SO$_4$	50–500	(68, 324)
p-Nitrobenzoyl chloride (esterification)	complexing with methyl orange	515	morphine in plasma	3	(325)
			morphine in urine	10	(326)
			morphine in tissue	5	
p-Nitrobenzoyl chloride (esterification)	coupling with N-1-naphthyl-ethylenediamine	555	morphine in aqueous solution	0.5	(326)
Diazonium compounds			morphine in opium		(322)
			ketobemidone in injection solutions		(294)

pethidine and morphine in urine[313] and (2) TLC for quantitation of similar compounds in pharmaceuticals.[314]

   b. Phenolic Reagent Technique. This is based upon the ability of certain compounds to react specifically and quantitatively with phenolic groups to form colored products that can then be estimated spectrophotometrically. Various applications of the technique to phenolic narcotics are given in Table XV. Specificity of assay has been increased by preliminary column chromatography of extracts on alumina[316] and ion-exchange resins.[47,294]

   Other types of reagent have been less frequently used for colorimetric assay of narcotics. Methadone, 5–50 μg/ml in urine, has been extracted with ether and assayed by nitration of its phenyl radicals and subsequent development of color with ethylmethylketone.[327] Since any phenyl-substituted base extracted was measured as drug, the method was considered nonspecific. Another method makes use of the violet color (λ_{max}, 490 mμ) that codeine develops with ferric chloride in the presence of a mixture of sulfuric and acetic acids after heating. The reaction is also positive for morphine, heroin, and ethylmorphine. Ethylmorphine can be analyzed in the presence of codeine by adding sodium arsenate and measuring absorbance at 530 mμ.[166] Oxidation with $(NH_4)_2Ce(NO_3)_6$ and reaction with 2,4-dinitrophenyl hydrazine to give an orange color (λ_{max}, 500 mμ) has been used in the assay of morphine (30–240 μg/ml) with codeine in opium preparations, but thebaine and papaverine had to be removed first.[174] Colorimetric determination of morphine utilizing chloromine T/4-aminopyrimidine[172] and 1-nitroso-2-naphthol[328] also has been described.

B. Tracer Techniques

   Since these are mainly based on measurement of radioactivity in samples after the administration of known amounts of radioactive compounds to an organism, they cannot be considered quantitative in the same way as the generally applied methods described herein and will be discussed only briefly.

   Radioactively labeled drugs are mainly utilized to facilitate their determination in biological materials in studies of drug disposition in animals. Such studies with narcotics were reviewed previously by Way and Adler in 1962.[7] Since then tracer techniques also have been employed for assays of N-^{14}C-methyl-codeine,[329] N-^{14}C-dl-acetylmethadol,[149] N-^{14}C-d-propoxyphene[330] and tritiated-pethidine,[331] tritiated dihydromorphine,[332,333] tritiated cyclazocine,[110] tritiated pentazocine,[151] and tritiated etorphine.[333]

   The radio tracer method is highly sensitive, but when used alone in drug disposition studies it has been found to be relatively nonspecific because of interference by radioactive drug metabolites. Specificity has been greatly increased by using techniques such as isotope dilution[7] and chromatography[110,149,151] in conjunction with tracers.

   In the last decade increasing use has been made of the isotope derivative technique, which permits the use of tracer techniques in assays of non-radioactive compounds. For example, paper chromatograms upon which alkaloids had been separated were treated with ^{32}P-labeled phosphomolybdic acid and then washed to remove excess reagent and the individual spots assayed radiometrically.[334] More recently, secondary amines such as desimipramine[335] have been reacted with tritiated acetic anhydride and, upon removal of excess reagent, estimated radiometrically. This method presumably also could be used with secondary amine derivatives of narcotics produced by N-dealkylation.

C. Chromatographic Methods

1. Paper and Thin-Layer Chromatography

The most commonly used methods utilize the relationship between the amount of a compound and the area, length, or color intensity of the spot that it produces. The reproducibility of such methods depends on the precise application of a given volume of solution to the origin of the chromatogram, the use of solvent systems that give round spots, and a quantitative or reproducible conversion of compound to the form in which it is detected (for example, colored derivative).

a. Spot Measurement. It has been shown that the size of a spot on a paper chromatogram is directly proportional to the amount of material present.[336] A similar relationship was proposed for thin-layer chromatograms.[337] However, Brenner and Niederwieser[338] found that a linear relationship between spot area and amount of compound only existed for very small amounts. Purdy and Truter[339] recently demonstrated that in TLC the square root of the area of the spot is a linear function of the logarithm of the weight of material in the spot and that this relationship is valid over a wide range of weight. For quantitative analysis one graphical[339] and two algebraic methods[339,340] have been described that are based on determination of spot size. This is measured with a planimeter for greatest accuracy. However, precision of this type of assay is low and there are only isolated instances of its application to narcotics, for example, the determination of codeine in pharmaceutical preparations.[341]

b. Densitometry. This method involves *in situ* measurements of the color intensity of spots following separation of compounds by chromatography and treatment with suitable spray reagents. Both paper and thin-layer chromatograms may be evaluated using either transmitted[342] or reflected[343] light. It is advantageous to make paper more transparent by drawing it through paraffin oil[344] and, with TLC, to scan a photographic negative,[345] print,[346] or autoradiograph[347] of a chromatoplate. In addition to Dragendorff's[348] and potassium iodoplatinate reagents,[349] the phosphomolybdic acid detection method may be recommended, in which the precipitate is reduced with stannous chloride.[358] Genest and Farmilo[349] used densitometry for determination of opium alkaloids on paper chromatograms and obtained linear calibration graphs for codeine, morphine, papaverine, and thebaine. Quantities as low as 5–50 μg could be determined in pharmaceutical preparations with a standard deviation 3.94%. In the estimation of morphine in opium, various phenolic group reagents[351] including nitrous acid[68] have been adapted for the colorimetry of the obtained spots. In the range of 100–200 μg of material 90% recovery was obtained.[351] Densitometry of opium alkaloids on thin-layer chromatograms with a Zeiss extinction recording apparatus also has been described.[352]

The main advantages of direct quantitative assay of compounds on chromatograms are greater simplicity and speed. However, methods applied to chromatographic eluates such as colorimetry and fluorometry give more accurate results.

2. Gas-Liquid Chromatography (GLC)

Quantitation of chromatogram recordings is based on the fact that

TABLE

Gas Chromatographic Data for the

Analgesic (or Metabolite)	Retention Time (min)	Internal Marker	Retention Time (min)	Column Specifications		
				Material	Length	Diameter
Morphine-TMS	10.8	tetraphenyl-ethylene (TPE)	8.1	glass	4 ft	4 mm, i.d.[a]
		landanosine	17.2			
Morphine-TMS		cholesterol		glass	6 ft	1/4 in., o.d.
Codeine-TMS						
Morphine-TMS	14.2	nalorphine-TMS	21.0	glass	1.8 m	4 mm, i.d.
Morphine-TMS	3.5	nalorphine-TMS	5.3	glass	1 m	3 mm, i.d.
Morphine-TMS	6.5	TPE	5.0	stainless steel	5 ft	1/8 in., o.d.
Codeine	2.0	cholestane	5.6	glass	6 ft	4 mm, i.d.
Norcodeine	2.0	cholesterol	16.0			
Codeine-acetate	2.6	acetate				
Morphine-acetate	3.4					
Codeine	2.2	cholestane	1.6	glass	4 ft	4 mm, i.d.
Norcodeine	2.8	cholesterol	6.7			
Codeine-acetate	2.8	acetate				
Morphine-acetate	5.1					
Propoxyphene	4.6	d-pyrroliphene	10	glass	1 ft	4 mm, i.d.
Pethidine	7.5	4-benzylpyridine	5.8	stainless steel	1 m	1/8 in., o.d.
Methadone	8.3	tripelennamine	6.5	stainless steel	1 m	1/8 in., o.d.
Metabolite by N-Demethylation	5.2					
Pentazocine	5.3	a-methadol	3.7	glass	2 m	1/4 in., o.d.
Phenadoxone	11.5	marcain	5.3	glass	2 m	1/4 in., o.d.
Pipadone	9.8					
Pyrolidino-Methadone	7.7					

[a]i.d. = Inside diameter; o.d. = outside diameter.
[b]Argon-ionization detector.

under constant conditions the area under a chromatographic peak is proportional to the amount of compound producing the peak. Furthermore, if the peak is sharp and symmetrical, the height of the peak is directly proportional to its area.

GLC assays of narcotics are based mainly upon the internal marker method in which a compound (internal marker) is chosen that has a retention time close to, but not overlapping, that of the drug to be assayed. A known constant amount of internal marker is added to a solution containing the unknown amount of drug, and (after concentration) a small volume (1–5 μl) of the mixture is chromatographed. The amount of drug is then determined by measuring the peak-height ratio of drug:internal marker and relating this to a previously constructed calibration graph of this same ratio obtained with, and plotted against, known amounts of the drug. If cali-

XVI
Determination of Narcotic Analgesics

Column Packing Material	Temperature (°C)			Carrier Gas Flow Rate (ml/min)	Range of Calibration	Reference
	Injection Port	Oven	Detector			
4% SE-30 on 60/80 mesh gas Chrom P- AW-HMDS treated and coated with 0.1% polyethylene glycol 9000	280	183	240[b]	argon-250	10–25 mg	(353)
3.8% SE-30 on 60/80 mesh Diataport S		207		helium-60	0.25–1.0 mg	(354)
1.5% OV-17 on 80/100 mesh Shimalite W-AW-HMDS treated		220		nitrogen-45	1–10 μg	(33)
2.9% OV-17 on 100/120 mesh Chromosorb W-AW-DMCS treated	250	205		nitrogen-53	2.5–50 μg	(19)
3% OV-1 on 100/120 mesh gas-Chrom Q	250	215	250	nitrogen-30	0.025–0.05 μg	(358)
5% SE-30 on 80/100 mesh gas-Chrom P-AW	260	240	260	nitrogen-70		(360)
4% XE-60 on 90/100 mesh Anakrom A, acid washed	260	240	260	nitrogen-70		(360)
3.8% W 98 silicone rubber on 80/100 mesh Diataport S-silanized	205	172	195	helium-60	0.25–1.5 μg	(361)
2% Carbowax 20M on 80/100 mesh Chromosorb G, AW, DMCS-coated with 5% KOH	250	170		nitrogen-28	1–10 μg	(16)
as for pethidine	250	200		nitrogen-28	1–10 μg	(16)
2.5% SE-30 on 80/100 mesh Chromosorb G-AW-DMCS treated	250	200		nitrogen-60	1–10 μg	(362)
as for pentazocine	250	200		nitrogen-65	1–5 μmol	(16)

bration is linear a calibration factor may be calculated from the slope of the graph. The product of this factor and the peak-height ratio from an unknown sample gives the amount of drug in that sample.

The internal marker method has certain advantages over other assay methods: First, if the internal marker compound has similar chemical and physical properties to those of the drug under determination, it (IM) may be added initially and carried through various extraction steps with the drug, thereby eliminating errors due to spillage; second, this method eliminates the need to accurately measure the volume of (1) the concentrated solution or extract and (2) the aliquot that is injected into the chromatograph.

Certain highly polar phenolic narcotics such as morphine can be chromatographed directly but are adsorbed onto the column and give rise to excessively tailing peaks, making quantitative analysis impossible.[353,354] In assays of opium[353,354] this difficulty has been overcome by converting morphine to its trimethylsilyl (TMS) derivative by reaction with hexamethyldisilazane (HMDS) in pyridine[355] under *anhydrous* conditions for 24 hr at room temperature.[353] The derivative, which could be chromatographed at a lower temperature (183°C) than that for the free alkaloid, produced a sharp symmetrical peak and was eluted just after the internal marker, tetraphenylethylene (TPE, see Table XVI). The most important step in the procedure was the extraction of the total alkaloids from opium by hot water infusion and chromatography on strongly acidic cation-exchange resin (Dowex 50-X2) followed by separation of phenolic and nonphenolic alkaloids on strong anion-exchange resin (Dowex 1-X1). Morphine was eluted (after washing) with $0.5N$ acetic acid, the eluate evaporated to *dryness*, internal marker added, and the mixture silylated as described above.[353] A similar gas chromatographic procedure was developed for the determination of morphine and codeine in urine.[356] The internal marker was the same (TPE), but the derivatives used were the acetates or propionates that were formed on column according to the method of Anders and Mannering[155] (see GLC identification).

More recently Klebe *et al.*[357] introduced the silylating reagent bis(trimethylsilyl) acetamide (BSA) which, because of its increased silylating power, is replacing HMDS. This has been used in two recently published methods for the determination of morphine in urine.[33,19] Both methods were the same except for the extraction procedures employed. In the first,[33] morphine, after adsorption from urine onto activated charcoal and washing with water, was eluted with glacial acetic acid that was then adjusted to pH 9 with NaOH and extracted with chloroform/isopropanol (9:1). After drying over anhydrous Na_2SO_4, the organic extract was evaporated to dryness. In the second procedure[19] urine was acidified with concentrated HCl, saturated with ammonium sulfate, and extracted with diethyl ether that was subsequently rejected. Upon adjustment to pH 9 with strong ammonia, the sample was extracted three times with ether that was bulked and evaporated to dryness. In both methods, nalorphine was added initially as internal marker and the residues of extracts were silylated by dissolving them in 0.1–0.2 ml BSA and standing the solutions at room temperature for 10 min before the injection of 2 μl into the chromatograph. Both methods also may be adapted for measurement of free plus conjugated morphine in urine by including an acid-autoclaving step in the procedure. Analytical data is given in Table XVI.

GLC also has been applied to the estimation of submicrogram quantities of morphine in plasma and cerebrospinal fluid.[358] At such low concentrations, precautions had to be taken to ensure complete or constantly reproducible extraction of the drug and its quantitative silylation. The latter was achieved by using the new trifluoro analog of BSA, bis(trimethylsilyl) trifluoroacetamide (BSTFA)[359] which, in addition to the potent silylating properties of BSA, has the following advantages over that reagent: (a) production of a much lower response with the flame ionization detector (FID), thereby reducing the possibility that small amounts of TMS-morphine will be eluted during the off-scale response due to excess silylating reagent; (b) less crystallization when in solution with morphine; and (c) lower deposition of silicon dioxide in the FID.

Briefly the general procedure involved extraction of morphine from biological fluid saturated with $NaHCO_3$ into ethylacetate/isopropanol (9:1), back extraction into $1N$ HCl, evaporation of the latter to dryness under vacuum at 35°C, addition of a methanol solution of internal marker (TPE), evaporation to dryness and reaction of the residue with 25 μl of BSTFA/trimethylchlorosilane (TMCS) reagent at room temperature for 20–30 min after which 1–2 μl of the reaction mixture was injected into the chromatograph. The limit of sensitivity was found to be about 25 ng morphine per sample (Table XVI), and recovery, although not quantitative, was constantly reproducible at about 59%.[358]

By selection of the appropriate stationary phases, narcotics that are less polar than morphine may be chromatographed directly to give sharp, symmetrical peaks suitable for use in quantitative analysis. Such has been the case, for example, in the determination of codeine and its metabolites[360] in serum and urine; propoxyphene[361] in plasma; pethidine[16] and pentazocine[362] in blood and urine; methadone and its mono-N-demethylated metabolite in urine[16]; and phenadoxone, pipadone, and pyrrolidinomethadone[16] in liver homogenate supernatant (Table XVI).

The advantages of gas chromatography over other methods for assaying narcotics are greater rapidity and intrinsic specificity of the technique and the high sensitivity of the detection systems that are available.[152] Unless otherwise indicated, the data in Table XVI was obtained using flame ionization detection.

D. Polarography

This basically involves the electrolysis of solutions of electro-oxidizable of electroreducible materials between a dropping mercury electrode (DME) and a reference electrode. The potential applied between electrodes is varied, and the resulting changes in current flow are recorded against

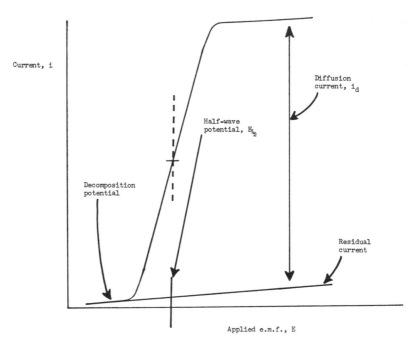

Fig. 3. Main features of a polarographic current-voltage curve, or "polarogram".

voltage. The current-voltage curve obtained (Fig. 3) is known as a "polarogram" or "polarographic wave." The potential at the midpoint of the wave, where the current is exactly half the limiting value, is known as the half-wave potential ($E_{\frac{1}{2}}$). This quantity is characteristic of a particular species under fixed solution conditions and may be used for identification purposes. Furthermore, the limiting diffusion current is directly proportional to the concentration of the species being reduced, so that both qualitative and quantitative information may be obtained from the same polarogram. This fact greatly enhances the specificity of this technique as an assay procedure whose wide application, including that to narcotics, has been reviewed in detail.[364]

1. Morphine Normorphine and Nalorphine

Original attempts to polarograph morphine were unsuccessful because it was not reduced at the DME in the available range of potential and the catalytic evolution of hydrogen in its solution was unsuitable for specific determination of this compound. However, morphine[365] and other phenolic compounds were able to be accurately determined using the wave of the readily reducible derivatives that they form after reaction with nitrite in acidic solution. It was originally supposed that the morphine derivative formed was the 2-nitroso compound,[365] but it has been recognized more recently that the reducible compound showing a wave at −0.9 V is, in fact, 2-nitromorphine.[366]

Polarographic determination as described above was subsequently applied to milligram quantities of morphine in poppy seeds[367] and capsules,[368,369] opium,[370] and pharmaceutical preparations.[371] By combination with preliminary paper chromatographic and elution techniques the method also has been successfully applied to microgram quantities of (1) morphine in blood,[372,373] urine,[374] blood plasma[375] and tissues;[376] (2) normorphine, nalorphine (which also form 2-nitro derivatives) and morphine together in blood and tissues;[10] and (3) normorphine and morphine in urine and feces.[377]

2. Codeine and Methadone

At normal temperatures (0–20°C) for reaction with nitrous acid, codeine does not form a 2-nitro derivative and will not, therefore, interfere with the morphine assay. However, at higher temperatures the derivative was formed[378] but assays based on this type of reaction proved unsatisfactory when applied to codeine in mixtures. Thus the possibility of determination after nitration also was studied.[379]

In polarographic investigations of methadone, no reduction waves have been found up to −2.0 V, but a *catalytic wave*[363] was present at −2.0 V.[380] This wave could only be used for assay of methadone in pure solutions. Polarography of the drug after nitration also has been investigated and could be used in assays of methadone in tablets, blood, urine and liver.[381]

3. Heroin (Diacetylmorphine)

Pure solutions of the drug can be determined using the catalytic wave[382] in borate buffers of pH 8–9. This method could not be used for heroin in mixtures or pharmaceuticals since the wave was sensitive to other variables. These complications can be avoided by using the method involving reaction with nitrous acid as developed for morphine.[383] First, however, it is necessary to hydrolyze the 3-acetyl group by heating with 1N HCl for the reaction to proceed in the 2-position to form 2-nitro-6-monoacetylmorphine. The nitro group of this compound is easily reduced polarographically, and the half-wave potential is practically the same as that for 2-nitromorphine. The optimum concentrations for analysis of heroin are the same as those for morphine (5–100 mg %). Alkaloids with phenolic groups such as morphine, apomorphine, and dihydromorphine interfere.

To assay heroin in the presence of morphine, the nitrous acid reaction is carried out in normal acetic acid in which heroin is sparsely hydrolyzed and 2-nitromorphine can be obtained for morphine assay. Then, using a new sample of the mixture, assay by hydrolysis with 1N HCl gives the combined concentrations of morphine and hydrolyzed heroin.

4. Keto Derivatives of Morphine and Related Compounds

Detailed studies have been carried out on morphine-type compounds that have a ketone group in place of the hydroxyl group in ring C.[348] All the compounds examined produced polarographic waves except dihydrothebainone, which lacks an oxygen bridge and has an isolated ketone group.

Thebainone, metathebainone, and hydroxythebainone, which do not have the 4,5-oxygen bridge, each give a wave in pH 7.5 buffer at -1.2 to -1.3 V corresponding to a two-electron reduction of the ketone group conjugated with the ring. At pH values less than 5, the wave clearly divides into two waves of the same height.

Codeinone, pseudocodeinone/hydroxycodeinone, and acetoxycodeinone, which have the oxygen bridge, are reduced mainly in two waves at -0.94 and -1.58 V; -1.01 and -1.60 V; and -0.9 and -1.3 V, respectively, at pH 7.5. The results do not show which wave corresponds to the reduction of the keto group and which one is for the reduction of the system involving the oxygen bridge and conjugated double bond. At lower pH, the second wave splits, and in acidic solutions, the second portion is obscured by the reduction of hydrogen ions.

With dihydromorphinone, dihydrocodeinone, and dihydroxycodeinone, which have an oxygen bridge and carbonyl group nonconjugated with ring C, the polarographic wave is equivalent to a two-electron reduction with half-wave potentials at -1.45, -1.37, and -1.38 V, respectively, at pH 7.5. The concentration of dihydromorphinone in Dilaudid and dihydrocodeine in Multacodin have been determined polarographically after solution in a Britton-Robinson buffer of pH 8. The polarographic behavior of dihydroketo derivatives of morphine also have been studied in relation to the analysis of other pharmaceutical preparations such as Dicodid, Cardiazol D, Coretonin D, Optikor D, Eukodal, Scophedral, Benacros, and Dinarkon.[383,385,386] Dihydrohydroxycodeine preparations were analyzed in pH 12.4 buffer ($E_{\frac{1}{2}} = -1.5$ V), whereas dihydromorphinone ($E_{\frac{1}{2}} = -1.5$ V) and dihydrocodeinone ($E_{\frac{1}{2}} = -1.5$ V) were assayed in $0.1N$ lithium hydroxide solution. Acetyldimethyldihydrothebainone was assayed after deacetylation to dihydrocodeinone by using a pH greater than 10, the sample (Acedicon) being mixed or dissolved in $0.2N$ LiOH containing 0.005% gelatin and polarographed after 10 min.

In some instances it is possible to determine simultaneously two or three ketoderivatives in a mixture. For instance, in pH 5 acetate buffer oxycodeinone has two polarographic waves of identical height at -0.79 and -1.18 V. Dihydrooxycodeinone forms one wave at -1.2 V. In mixtures of both compounds, oxycodeinone is determined from the wave height at -0.79 while dihydro-oxycodeinone is determined from the difference in wave heights at -1.2 and -0.79 V.[387]

5. Oscillographic Polarography

This is the study of electrolysis using polarizable mercury electrodes connected to an oscillograph on whose screen various relations between voltage, current intensity, and time may be recorded. The application of this method to the determination of narcotics was reviewed by Kalvoda and Zyka.[388] Morphine was determined by conversion into pseudomorphine through the addition of $0.5N$ potassium ferricyanide to an alkaline solution of the drug. $2 \times 10^{-4}M$ was the limit of detection for this rapid and precise method that has been investigated recently.[389]

E. Titrimetric and Gravimetric Methods

These are mainly restricted to macrodeterminations of milligram quantities of narcotics in the pure state or in pharmaceutical or illicit drug preparations after preliminary extraction or separation. Application of the methods to alkaloids in general[390] and to morphine[164] have been reviewed, and the titration of narcotics separated by ion-exchange chromatography[38,51] or with amperometric end-point detection[391] has been described.

Aqueous and nonaqueous titrations[392,393] are the main types of assay adopted for official preparations. For instance, in the *United States Pharmacopeia* (Vol. XVII) and the *National Formulary* (Vol. XI) morphine, codeine, dihydrocodeine, and methadone, after extraction from the preparation, are determined by the addition of $0.02N$ sulfuric acid and back titration of the excess acid with $0.02N$ NaOH using methyl red as indicator. The purity of the powdered compounds such as methadone HCl, propoxyphene HCl, and levorphanol tartrate is officially determined by nonaqueous titration in glacial acetic acid or anhydrous alcohol with perchloric acid as titrant and crystal violet or methyl red as indicator. It should be noted that nalorphine preparations that can be assayed by methods similar to those for morphine are assayed by UV absorptiometry in *United States Pharmacopeia* (Vol. XVII).

Gravimetric assays of dextromoramide,[394] morphine,[395,396] and codeine[397] after precipitation respectively as free base, dinitrophenyl derivative, and tetraphenylboride salt have been described recently, and nonaqueous titration in glacial acetic acid with perchloric acid has been reported for trimeperidine (crystal violet indicator),[398] dextromoramide in injections, tablets and suppositories (α-naphthol benzein indicator),[394] and codeine in terpin hydrate and codeine elixir after cation-exchange separation (methyl violet indicator).[399] Methadone, hydroxycodeine, and dihydrocodeine were likewise titrated with *p*-toluenesulfonic acid using chloroform as the solvent and methyl yellow indicator.[400] Recently, amperometric,[401] conductometric,[402] coulometric[403] and redox[404] titrations of morphine and codeine also have been reported.

V. REFERENCES

1. N. B. Eddy, *Chem. Ind. (London)*, 1462–1469 (1959).
2. C. G. Farmilo, P. M. Oestreicher, and L. Levi, *Bull. Narcotics, U. N. Dept. Social Affairs* 6 (1), 7–19, (1954).
3. H. C. Freimuth, *in* "Toxicology, Mechanisms and Analytical Methods" (C. P.

Stewart and A. Stolman, eds.), Vol. 1, pp. 285–302, Academic Press, New York (1960).
4. A. S. Curry *in* "Methods of Biochemical Analysis" (D. Glick, ed.), 7, pp. 39–76, Interscience, New York (1959).
5. J. V. Jackson *in* "Identification of Drugs and Poisons," pp. 15–26, Pharmaceutical Society of Great Britain (1965).
6. P. G. Stecher (ed.), "The Merck Index of Chemicals and Drugs," 8th ed., Merck & Company, Rahway, New Jersey (1968).
7. E. L. Way and T. K. Adler, *Bull. W. H. O. 27*, 359–394 (1962).
8. S. J. Mulé, *Anal. Chem. 36*, 1907–1914 (1964).
9. S. J. Mulé and L. A. Woods, *J. Pharmacol. Exp. Ther. 136*, 232–241 (1962).
10. K. Milthers, *Acta Pharmacol. Toxicol. 18*, 119–206 (1961).
11. N. C: Jain and P. L. Kirk, *Microchem. J. 12*, 229–241 (1967).
12. A. L. Misra, H. I. Jacoby, and L. A. Woods, *J. Pharmacol. Exp. Ther. 132*, 311–316 (1961).
13. M. E. Latham and H. W. Elliott, *J. Pharmacol. Exp. Ther. 101*, 259–267 (1951).
14. J. J. Burns, B. L. Berger, P. A. Lief, A. Wollack, E. M. Papper, and B. B. Brodie, *J. Pharmacol. Exp. Ther. 114*, 289–298 (1955).
15. B. A. Berkowitz, J. H. Asling, S. M. Shnider, and E. L. Way, *Clin. Pharmacol. Ther. 10*, 320–328 (1969).
16. J. F. Taylor, Ph.D. thesis, University of London (1968).
17. A. E. Takemori, *Biochem. Pharmacol. 17*, 1627–1635 (1968).
18. A. S. Curry, *J. Forensic Sci. 6*, 373–399 (1961).
19. F. Fish and W. D. C. Wilson, *J. Chromatog. 40*, 164–168 (1969).
20. R. Consden, *in* "Toxicology, Mechanisms and Analytical Methods" (C. P. Stewart and A. Stolman, eds.), Vol. 1, pp. 303–352, Academic Press, New York (1960).
21. C. G. Farmilo and K. Genest, *in* "Toxicology, Mechanisms and Analytical Methods" (C. P. Stewart and A. Stolman, eds.), Vol. 2, pp. 209–595, Academic Press, New York (1960).
22. E. Heftmann, "Chromatography," Reinhold, New York (1961).
23. F. C. Klee and E. R. Kirch, *J. Am. Pharm. Ass. Sci. Ed. 42*, 146–150 (1953).
24. G. Pruner, *R. C. Inst. Sup. Sanit. 22*, 710–716 (1959), through *Anal. Abstr.* 2422 (1960).
25. C. A. Teijgler, *Pharm. Weekbl. 94*, 201–209 (1959).
26. D. C. Garratt, C. A. Johnson, and C. J. Lloyd, *J. Pharm. Pharmacol. 9*, 914–928 (1957).
27. J. Büchi and R. Huber, *Pharm. Acta Helv. 36*, 571–586 (1961), through *Chem. Abstr. 56*, 7425 i (1962).
28. C. G. Farmilo, A. Almond, and K. Genest, Third Conference, Canadian Conference on *Pharmacological Research*, Ottowa, August (1956).
29. G. R. Nakamura and H. J. Meuron, *Anal. Chem. 41*, 1124–1126 (1969).
30. S. Kori and M. Kono, *Yakugaku Zasshi 80*, 728–732 (1961), through *Chem. Abstr. 55*, 23932 g (1961).
31. S. Kori and M. Kono, *Yakugaku Zasshi 81*, 776–778 (1961), through *Chem. Abstr. 55*, 26368 d (1961).
32. R. Fischer and W. Iwanoff, *Arch. Pharm. 281*, 361–367 (1943).
33. N. Ikekawa, K. Takayama, E. Hosoya, and T. Oka, *Anal. Biochem. 28*, 156–163 (1969).
34. C. P. Stewart, S. K. Chatterji, and S. Smith, *Brit. Med. J.* 790–792 (1937).
35. A. Stolman and C. P. Stewart, *Analyst 74*, 536–546 (1949).
36. J. Büchi, *J. Pharm. Pharmacol. 8*, 369–381 (1956).
37. L. Saunders, *in* "Toxicology, Mechanisms and Analytical Methods" (C. P. Stewart and A. Stolman, eds.), Vol. 1, pp. 353–372, Academic Press, New York (1960).
38. A. Jindra, *Bull. Narcotics, U. N. Dept. Social Affairs 7*, (2), 20–27 (1955).

39. C. H. Van Etten, F. R. Earle, T. A. McGuire, and F. R. Senti, *Anal. Chem. 28*, 867–870 (1956).
40. J. R. Broich, M. M. DeMayo, and L. A. DalCortivo, *J. Chromatog. 33*, 526–529 (1968).
41. J. S. Foster and J. W. Murfin, *Analyst 86*, 32–36 (1961).
42. K. O. Montgomery, P. V. Jennings, and M. H. Weinswig, *J. Pharm. Sci. 56*, 141–143 (1967).
43. K. O. Montgomery and M. H. Weinswig, *J. Pharm. Sci. 55*, 1141–1142 (1966).
44. F. W. Oberst, *J. Lab. Clin. Med. 24*, 318–329 (1938).
45. S. L. Tompsett, *Acta Pharmacol. Toxicol. 17*, 295–303 (1960).
46. S. L. Tompsett, *Acta Pharmacol. Toxicol. 18*, 414–418 (1961).
47. J. C. Szerb, D. P. MacLeod, F. Moya, and D. H. McCurdy, *Arch. Intern. Pharmacodyn. 109*, 99–107 (1957).
48. V. P. Dole, W. K. Kim, and I. Eglitis, *J. Amer. Med. Ass. 198*, 115–118 (1966).
49. S. J. Mulé, *J. Chromatog. 39*, 302–311 (1969).
50. A. Jindra and J. Pohorsky, *J. Pharm. Pharmacol. 2*, 361–363 (1950); *3*, 344–350 (1951).
51. L. Levi and C. G. Farmilo, *Can. J. Chem. 30*, 793–799 (1952).
52. F. O. Gundersen, R. Heiz, and R. Klevstrand, *J. Pharm. Pharmacol. 5*, 608–614 (1953).
53. J. Bosvart and A. Jindra, *Cesk. Farm. 6*, 82–87 (1957).
54. M. Asahina and M. Ono, *Eisei Shikenjo Hokoku 77*, 139–152 (1959), through *Chem. Abstr. 55*, 9781 g–i (1961).
55. L. B. Achor and E. M. K. Geiling, *Anal. Chem. 26*, 1061–1062 (1954).
56. E. W. Grant and W. W. Hilty, *J. Amer. Ass. Sci. Ed. 42*, 150–152 (1953).
57. S. L. Tompsett, *Acta Pharmacol. Toxicol. 19*, 368–370 (1962).
58. T. Oka, *Keio J. Med. 16*, 31–36 (1967), through *Chem. Abstr. 67*, 105850 p (1967).
59. H. Yoshimura, K. Oguri, and H. Tsukamoto, *Biochem. Pharmacol. 18*, 279–286 (1969).
60. J. M. Fujimoto and V. B. Haarstad, *J. Pharmacol. Exp. Ther. 165*, 45–51 (1968).
61. J. M. Fujimoto, W. M. Watrous, and V. B. Haarstad, *Proc. Soc. Exp. Biol. Med. 130*, 546–549 (1969).
62. M. Bier (ed.), "Electrophoresis: Theory, Methods and Applications," Academic Press, New York (1959).
63. S. Kaye and L. R. Goldbaum, *in* "Legal Medicine" (R. B. H. Gradwohl, ed.), pp. 663–674, Mosby (1954).
64. G. B. Marini-Bettolo and J. A. C. Frugoni, *Gazz. Chim. Ital. 86*, 1324–1331 (1956), through *Chem. Abstr. 52*, 653 (1958).
65. R. Paris and G. Faugeras, *Ann. Pharm. Fr. 16*, 305–310 (1958).
66. G. Wagner, *Wiss. Z. Ernst Moritz Ardnt–Univ. Greifswald, 5* (1/2) (1955).
67. K. Willner, *Arch. Toxikol. 17*, 347–356 (1959).
68. A. B. Svendsen and K. Bergane, *Bull. Narcotics, U. N. Dept. Social Affairs 10* (4), 17 (1958).
69. P. Horak, J. Holubek, V. Bumba, and A. Cekau, *Collect. Czech. Chem. Commun. 27*, 1037–1042 (1962), through *Anal. Abstr.* 4874 (1962).
70. E. Graf and D. P. H. List, *Arzneim.–Forsch. 4*, 450–453 (1954), through *Chem. Abstr. 48*, 13168 f (1954).
70. A. Roux and J. Roux-Matignon, *Ann. Pharm. Franc. 18*, 135–138 (1960).
72. C. G. Farmilo, W. P. McKinley, J. C. Bartlet, P. Oestreicher, and A. Almond, U. N. Document ST/SOA/Ser K/61 (1957).
73. L. R. Goldbaum and L. Kazyak, *J. Pharmacol. Exp. Ther. 106*, 388–389 (1952).
74. C. L. Brown and P. L. Kirk, *Mikrochim. Acta* 720–723 (1957).
75. S. K. Niyogi, S. L. Tompsett, and C. P. Stewart, *Clin. Chim. Acta 6*, 741–743 (1961).
76. G. Buff, J. Orantis, and P. L. Kirk, *Microchem. J. 3*, 13–18 (1959).

77. I. Sano and H. Kajita, *Klin. Wochschr. 33*, 956–958 (1955).
78. H. Kajita, *Seikagaku 27*, 461–463 (1955).
79. G. A. Spengler, *Helv. Med. Acta 25*, 430–436 (1958).
80. L. A. Williams, Y. M. Brusock, and B. Zak, *Anal. Chem. 32*, 1883–1885 (1960).
81. K. Genest and C. G. Farmilo, *Bull. Narcotics, U. N. Dept. Social Affairs 11*, (4), 20–37 (1959).
82. K. Genest and C. G. Farmilo, *Bull. Narcotics, U.N. Dept. Social Affairs, 12* (1), 15–24 (1960).
83. K. Macek, J. Hacaperkova', and B. Kakáč, *Pharmazie 11*, 533–538 (1956).
84. D. Waldi, *Arch. Pharm. 292*, 206–220 (1959).
85. L. R. Goldbaum and L. Kazyak, *Anal. Chem. 28*, 1289–1290 (1956).
86. J. Büchi and H. Schumacher, *Pharm. Acta Helv. 32*, 273–288 (1957).
87. A. A. A. Rahman, *Arch. Pharm. 290*, 321–325 (1957).
88. G. Nadeau, G. Sobolewski, L. Fisef, and C. G. Farmilo, *J. Chromatog. 1*, 327–337 (1958).
89. F. Vorel, *Soudni Lek. 6*, 43–95 (1958), through *Anal. Abstr.* 696 (1959).
90. E. Vidic, *Arzneim.–Forsch. 5*, 291–295 (1955).
91. E. Vidic, *Arzneim.–Forsch. 7*, 314–319 (1957).
92. G. J. Mannering, A. C. Dixon, N. V. Carroll, and O. B. Cope, *J. Lab. Clin. Med. 44*, 292–300 (1954).
93. H. Kajita, *Seikagaku 27*, 456–461 (1955).
94. P. J. Morgan, *Analyst 84*, 418–422 (1959).
95. R. Hilf, F. F. Castano, and G. A. Lichtbourn, *J. Lab. Clin. Med. 54*, 634–639 (1959).
96. J. V. Jackson and M. S. Moss, *in* "Chromatographic and Electrophoretic Techniques" (I. Smith, ed.), Vol. 1, pp. 516–530, Interscience, New York (1969).
97. G. R. Nakamura, *J. Forensic Sci. 5*, 259–265 (1960).
98. J. Michael and W. Leifels, *Mikrochim. Acta 3*, 444–448 (1961).
99. K. Genest and C. G. Farmilo, *J. Chromatog. 6*, 343–349 (1961).
100. H. V. Street, *J. Pharm. Pharmacol. 14*, 56–57 (1962).
101. S. N. Tewari, *Mikrochim. Acta* (4), 704–707 (1967), through *Chem. Abstr., 67*, 67626 k (1967).
102. L. A. Dal Cortivo, C. H. Willumsen, S. B. Weinberg, and W. Matusiak, *Anal. Chem. 33*, 1218–1220 (1961).
103. H. V. Street, *Acta Pharmacol. Toxicol. 19*, 325–329 (1962).
104. E. Vidic and J. Schuette, *Arch. Pharm. 295*, 342–360 (1962).
105. M. Yoshimura, M. Deki, and H. Tsukamoto, *Yakugaku Zasshi 83*, 223–226 (1963).
106. W. Paulus, U. Janitzki and W. Hoch, *Arzneim.–Forsch 12*, 1086–1087 (1962).
107. A. L. Misra, S. J. Mulé, and L. A. Woods, *J. Pharmacol. Exp. Ther. 132*, 317–322 (1961).
108. K. J. Freundt, *Arzneim.–Forsch 12*, 614–617 (1962).
109. A. M. Asatoor, D. R. London, M. D. Milne, and M. L. Simenhoff, *Brit. J. Pharmacol. 20*, 285–298 (1963).
110. S. J. Mulé, T. H. Clements, and C. W. Gorodetzky, *J. Pharmacol. Exp. Ther. 160*, 387–396 (1968).
111. G. Nakamura and T. Ukita, *J. Pharm. Sci. 56*, 294–295 (1967).
112. N. A. Izmailov and M. S. Schraiber, *Farmatsiya (Sofia)* (3) 1 (1938), through *Anal. Chem. 33*, 1138–1142 (1961).
113. E. Stahl, *Chem.–Ztg. 82*, 323–329 (1958).
114. E. Stahl, *Angew. Chem. 73*, 646–654 (1961).
115. E. Stahl (ed.), "Thin-Layer Chromatography," Springer-Verlag, New York (1969).
116. M. E. Borke and E. R. Kirch, *J. Amer. Pharm. Ass. Sci. Ed., 42*, 627–629 (1953).
117. K. Teichert, E. Mutschler, and H. Rochelmeyer, *Deut. Apoth.-Ztg. 100*, 477–480 (1960).
118. K. Teichert, E. Mutschler, and H. Rochelmeyer, *Z. Anal. Chem. 181*, 325–331 (1961).

119. D. Neubauer and K. Mothes, *Planta Med. 9*, 466–470 (1961), through *Chem. Abstr. 57*, 2332 c (1962).
120. E. Brochmann-Hanssen and T. Furuya, *J. Pharm. Sci. 53*, 1549–1550 (1964).
121. D. Waldi, K. Schnackerz, and F. Munter, *J. Chromatog. 6*, 61–73 (1961).
122. I. Bayer, *J. Chromatog. 16*, 237–238 (1964).
123. W. J. Kamp, W. J. M. Onderberg, and W. A. Van Seters, *Pharm. Weekbl. 98*, 993–1007 (1963).
124. S. Pfeiffer, *J. Chromatog. 24*, 364–371 (1966).
125. J. L. Emmerson and R. C. Anderson, *J. Chromatog. 17*, 495–500 (1965).
126. J-T. Huang, H-C. Hsiu, and K-J. Wang, *J. Chromatog. 29*, 391–392 (1967).
127. J. A. Steele, *J. Chromatog. 19*, 300–303 (1965).
128. G. R. Nakamura, *J. Ass. Offic. Anal. Chem. 49*, 1086–1090 (1966), through *Chem. Abstr. 66*, 22154 t (1967).
129. G. Machata, *Mikrochim. Acta 47*, 79–86 (1960).
130. J. Baümler and S. Rippenstein, *Pharm. Acta Helv. 36*, 382–383 (1961).
131. E. Vidic, *Arch. Toxikol. 19*, 254–268 (1961).
132. I. Sunshine, *Amer. J. Clin. Pathol. 40*, 576–582 (1963).
133. A. Noirfalise, *J. Chromatog. 20*, 61–77 (1965).
134. P. Schweda, *Anal. Chem. 39*, 1019–1022 (1967).
135. A. S. Curry and H. Powell, *Nature 173*, 1143–1144 (1954).
136. E. G. C. Clarke *in* "Methods of Forensic Science" (F. Lundquist, ed.), Vol. 1, Interscience London (1962).
137. J. Cochin and J. W. Daly, *Experientia 18*, 294–295 (1962).
138. H. Eberhardt and O. Norden, *Arzneim.-Forsch. 14*, 1354–1355 (1964).
139. B. Davidow, N. Li Petri, B. Quame, B. Searle, E. Fastlich, and J. Savitzky, *Amer. J. Clin. Pathol. 46*, 58–62 (1966).
140. K. D. Parker and C. H. Hine, *Psychopharmacol. Bull. 3* (3), 18–42 (1966).
141. *Psychopharmacol. Bull. 3* (4), 27–62 (1966).
142. M. Debackere and L. Laruelle, *J. Chromatog. 35*, 234–247 (1968).
143. S. Goenechea and W. Bernard, *Fresenius' Z. Anal. Chem. 246*, 130–132 (1969), through *Chem. Abstr. 71*, 79349 t (1969).
144. M. Ono, B. F. Engelke, and C. Fulton, *Bull. Narcotics, U.N. Dept. Social Affairs 21* (2), 31–40 (1969), through *Chem. Abstr. 71*, 10130 x (1969).
145. A. M. Heaton and A. G. Blumberg. *J. Chromatog. 41*, 367–370 (1969).
146. J. M. Fujimoto and R. I. H. Wang, *Toxicol. Appl. Pharmacol. 16*, (1970) (in press).
147. A. Penna-Herreros, *J. Chromatog. 14*, 536 (1964).
148. A. H. Beckett, J. F. Taylor, A. F. Casy, and M. M. A. Hassan, *J. Pharm. Pharmacol. 20*, 754–762 (1968).
149. R. E. McMahon, H. W. Culp, and F. J. Marshall, *J. Pharmacol. Exp. Ther. 149*, 436–445 (1965).
150. M. E. Amundson, M. L. Johnson, and J. A. Manthey, *J. Pharm. Sci. 54*, 684–686 (1965).
151. K. A. Pittman, D. Rosi, R. Cherniak, A. J. Merola, and W. D. Conway, *Biochem. Pharmacol. 18*, 1673–1678 (1969).
152. B. J. Gudzinowicz, "Gas Chromatographic Analysis of Drugs and Pesticides," Arnold, London (1967).
153. H. A. Lloyd, H. M. Fales, P. F. Highet, W. J. A. Vandenheuvel, and W. C. Wildman, *J. Amer. Chem. Soc. 82*, 3791 (1960).
154. S. Yamaguchi, I. Seki, S. Okuda, and K. Tsuda, *Chem. Pharm. Bull. (Tokyo) 10*, 755–757 (1962).
155. M. W. Anders and G. J. Mannering, *Anal. Chem. 30*, 730–733 (1962).
156. E. Brochmann-Hanssen and T. O. Oke, *J. Pharm. Sci. 58*, 370–371 (1969).
157. C. R. Kingston and P. L. Kirk, *Bull. Narcotics, U.N. Dept. Social Affairs 17*, (2) 19–25 (1965).
158. M. Yoshimura and M. Deki, *Bunseki Kagaku 12*, 941 (1963).
159. E. Brochmann-Hanssen and C. R. Fontan, *J. Chromatog. 19*, 296–299 (1965).

160. K. D. Parker, C. R. Fontan, and P. L. Kirk, *Anal. Chem. 35*, 356–359 (1963).
161. L. Kazyak and E. C. Knoblock, *Anal. Chem. 35*, 1448–1452 (1963).
162. R. Miram and S. Pfeifer, *Sci. Pharm. 26*, 22–40 (1958).
163. E. G. C. Clarke and M. Williams, *Bull. Narcotics, U.N. Dept. Social Affairs 7* (3–4), 33–42 (1955).
164. S. Ehrlich-Rogozinsky and N. D. Cheronis, *Microchem. J. 7*, 336–356 (1963).
165. K. W. Bentley, "The Chemistry of the Morphine Alkaloids," Oxford University Press (Clarendon), London and New York (1954).
166. H. Wachsmuth and L. Van Koeckhoven, *J. Pharm. Belg. 14*, 215–220 (1959), through *Chem. Abstr. 54*, 3474 b (1960).
167. R. G. Splies and J. M. Shellow, *J. Chem. Eng. Data 11* (1), 123–124 (1966).
168. C. G. Farmilo, L. Levi, P. M. L. Ostreicher, and R. J. Ross, *Bull. Narcotics, U.N. Dept. Social Affairs 4* (4), 16–42 (1952).
169. E. G. C. Clarke, *Bull. Narcotics, U.N. Dept. Social Affairs 11*(1), 27–44 (1959).
170. E. G. C. Clarke, *Bull. Narcotics, U.N. Dept. Social Affairs 13* (4), 17–20 (1961).
171. Y.-H. Yu, *Yao Hsueh Hsueh Pao 6* (2), 101–106 (1958), through *Chem. Abstr. 53*, 11765 d (1959).
172. H. Sakurai, *Yakugaku Zasshi 81*, 155–159 (1961), through *Chem. Abstr. 55*, 14824 d (1961).
173. F. D. Snell and C. T. Snell, "Colorimetric Methods of Analysis," Van Nostrand, Princeton, New Jersey (1954).
174. H. Sakurai, *Yakugaku Zasshi 81* 869–872 (1961), through *Chem. Abstr. 55*, 22715 e (1961).
175. "United States Pharmacopeia," 17th rev., p. 401, Mack Publishing, Easton, Pennsylvania (1965).
176. F. Feigl, "Spot Tests in Organic Analysis," pp. 635–637, Elsevier, Amsterdam (1966).
177. D. W. Schieser, *J. Pharm. Sci. 53*, 909–913 (1964).
178. M. Lerner, *Anal. Chem. 32*, 198 (1960).
179. C. C. Fulton, "*Modern Microcrystal Tests for Drugs*," Interscience, New York and London (1969).
180. E. G. C. Clarke and M. Williams, *J. Pharm. Pharmacol. 7*, 255–262 (1955).
181. C. C. Fulton, *Bull. Narcotics, U.N. Dept. Social Affairs 5*(2), 27 (1953).
182. C. C. Fulton, *Microchem. J. 6*, 51–65 (1962).
183. E. G. C. Clarke, *J. Pharm. Pharmacol. 10*, 642–644 (1958).
184. C. G. Farmilo and L. Levi, *Bull. Narcotics, U.N. Dept. Social Affairs 5*(4), 20–27 (1953).
185. L. Martin, K. Genest, J. A. R. Cloutier, and C. G. Farmilo, *Bull. Narcotics, U.N. Dept. Social Affairs 15*(3–4), 17–38 (1963).
186. P. Janssen, "Synthetic Analgesics, Part I: Diphenylpropylamines," International Series of Monographs on Organic Chemistry, Vol. 3, Pergamon, Oxford (1960).
187. "United States Pharmacopeia," 14th rev., pp. 795 and 734, Mack Publishing Easton, Pennsylvania (1950).
188. L. Saunders and R. S. Srivastava, *J. Pharm. Pharmacol. 3*, 78–86 (1951).
189. D. D. Perrin, "Dissociation Constants of Organic Bases in Aqueous Solution", Butterworths, London (1965).
190. C. G. Farmilo, *Bull. Narcotics, U.N. Dept. Social Affairs 6*(3), 18–41 (1954).
191. P. M. Oestreicher, C. G. Farmilo, and L. Levi, *Bull. Narcotics, U.N. Dept. Social Affairs 6* (3), 42–70 (1954).
192. J. Weijlard, P. D. Orahovats, A. P. Sullivan, G. Purdue, F. K. Heath, and K. Pfister, *J. Amer. Chem. Soc. 78*, 2342 (1956).
193. L. W. Bradford and J. W. Brackett, *Mikrochim. Acta* (3), 353–382 (1958).
194. L. Levi and C. G. Farmilo, *Can. J. Chem. 30*, 783–792 (1952).
195. K.-T. Lee, *J. Pharm. Pharmacol. 12*, 666–676 (1960).
196. J. A. Gautier, J. Renault, and J. Rabiant, *Ann. Pharm. Fr. 17*, 401–408 (1959), through *Anal. Abstr.*, 2917 (1960).

197. L. Lang (ed.), "Absorption Spectra," Vols. 1–5, Academic Press, New York (1961–1965).
198. J. A. Radley and J. Grant, "Fluorescence Analysis in Ultraviolet Light," p. 70, Chapman and Hall, London (1933).
199. S. Udenfriend, "Fluorescence Assay in Biology and Medicine," Academic Press New York (1962).
200. S. Fleury, *Chim. Anal. (Paris) 48*, 321–325 (1966).
201. T. Numai, *Kagaku To Sosa 8* (2), 5–9 (1955).
202. G. Nadeau and G. Sobolewski, *Can. J. Biochem. Physiol. 36*, 625–631 (1958).
203. J. R. Broich, *Fluorescence News 4* (4), 7–8 (1969), through *Chem. Abstr. 71*, 121972 t (1969).
204. S. J. Mulé, private communication.
205. K. W. Bentley and S. F. Dyke, *J. Chem. Soc.* 2574–2577 (1959).
206. H. J. Kupferberg, A. Burkhalter, and E. L. Way, *J. Chromatog. 16*, 558–559 (1964).
207. C. Djerassi, "Optical Rotatory Dispersion," McGraw-Hill, New York (1960).
208. P. Crabbé, "Optical Rotatory Dispersion and Circular Dichroism in Organic Chemistry," Holden-Day, San Francisco (1965).
209. L. Velluz, M. Legrand, and M. Grosjean, "Optical Circular Dichroism," Academic Press New York (1965).
210. P. Crabbé and W. Klyne, *Tetrahedron 23*, 3449–3503 (1967).
211. J. M. Bobbit, U. Weiss, and D. D. Hanessian, *J. Org. Chem. 24*, 1582–1584 (1959).
212. P. Crabbé, "Optical Rotatory Dispersion and Circular Dichroism in Organic Chemistry," pp. 249–251 and 286–292, Holden-Day, San Francisco (1965).
213. U. Weiss and T. Rüll, *Bull. Soc. Chim. Fr.* 3707–3714 (1965).
214. T. Rüll, *Bull. Soc. Chim. Fr.* 3715–3718 (1965).
215. P. Crabbé, P. Demoen, and P. Janssen, *Bull. Soc. Chim. Fr.* 2855–2860 (1965).
216. C. E. Hubley and L. Levi, *Bull. Narcotics, U.N. Dept. Social Affairs, 7* (1), 20–41 (1955).
217. L. Levi, C. E. Hubley, and R. A. Hinge, *Bull. Narcotics, U.N. Dept. Social Affairs 7* (1), 42–84 (1955).
218. J. J. Manning, *Bull. Narcotics, U.N. Dept. Social Affairs 7* (1), 85–100 (1955).
219. J. Carol, *Ann. N.Y. Acad. Sci. 69*, 190–193 (1957).
220. A. L. Hayden, O. R. Sammul, G. B. Selzer, and J. Carol, *J. Ass. Offic. Agr. Chem. 45*, 797–900 (1962), through *Chem. Abstr. 58*, 3823 g (1963).
221. O. R. Sammul, W. L. Brannon, and A. L. Hayden, *J. Ass. Offic. Agr. Chem. 47*, 918–991 (1964), through *Chem. Abstr. 61*, 14467 c (1964).
222. I. Seki, *Chem. Pharm. Bull. (Tokyo) 14*, 445–453 (1966).
223. M. G. Lester, V. Petrow, and O. Stephenson, *Tetrahedron 20*, 1407–1417 (1964).
224. L. J. Bellamy, "Infrared Spectra of Complex Molecules," 2nd ed. p. 102, Methuen, New York (1958).
225. J. Carol, *J. Ass. Offic. Agr. Chem. 37*, 692–697 (1954), through *Chem. Abstr. 48*, 11726 g (1954).
226. B. Salvesen, L. Domange, and J. Guy, *Ann. Pharm. Fr. 13*, 354–359 (1955), through *Chem. Abstr. 49*, 16345 h (1955).
227. K.-T. Lee, R. A. Rockerbie, and L. Levi, *J. Pharm. Pharmacol. 10*, 621–624 (1958).
228. V. J. Bakre, Z. Karaata, J. C. Bartlet and C. G. Farmilo, *J. Pharm. Pharmacol. 11*, 234–243 (1959).
229. K. Genest and C. G. Farmilo, *Anal. Chem. 34*, 1464–1468 (1962).
230. M. Lerner and A. Mills, *Bull. Narcotics, U.N. Dept. Social Affairs 15* (1), 37–42 (1963).
231. A. S. Curry, J. F. Read, C. Brown, and R. W. Jenkins, *J. Chromatog. 38*, 200–208 (1968).
232. A. B. Littlewood, *J. Gas Chromatog. 6*, 65–68 (1968).
233. H. M. Lee, E. G. Scott, and A. Pohland, *J. Pharmacol. Exp. Ther. 125*, 14–18 (1959).

234. G. J. Mannering and L. S. Schanker, *J. Pharmacol. Exp. Ther. 124*, 296–304 (1958).
235. E. W. Nuffield, "X-ray Diffraction Methods," John Wiley, New York (1966).
236. W. H. Barnes, *Bull. Narcotics, U.N. Dept. Social Affairs 6* (1), 20–31 (1954).
237. W. H. Barnes and H. M. Sheppard, *Bull. Narcotics, U.N. Dept. Social Affairs 6* (2), 27–68 (1954).
238. W. H. Barnes and W. J. Forsyth, *Can. J. Chem. 32*, 984, 988, 991, 993 (1954).
239. W. H. Barnes, *Can. J. Chem. 32*, 994–995 (1954); *33*, 444–446 (1955).
240. W. H. Barnes and J. M. Lindsey, *Can. J. Chem. 33*, 565–570 (1955).
241. C. E. Hubach and F. T. Jones, *Anal. Chem. 22*, 595–598 (1950).
242. E. Pedley, *J. Pharm. Pharmacol. 7*, 527–532 (1955).
243. M. Bradford and J. J. Barbarino, *J. Criminal Law, Criminal Police Sci. 44*, 525 (1953).
244. L. Levi, *Roy. Can. Mounted Police Quart., 20*, 1–6 (1954).
245. A. W. Hanson and F. R. Ahmed, *Acta Crystallogr., 11*, 724–728 (1958).
246. F. R. Ahmed, W. H. Barnes, and G. Kartha, *Chem. Ind. (London)* 485 (1959).
247. L. M. Jackman and S. Sternhell, "Applications of Nuclear Magnetic Resonance Spectroscopy in Organic Chemistry," Pergamon Press, Oxford (1969).
248. T. Rüll and D. Gagnaire, *Bull. Soc. Chim. France* (10), 2189–2192 (1963), through *Chem. Abstr. 60*, 3640 c (1964).
249. T. Rüll, *Bull. Soc. Chim. Fr.* (3), 586–593 (1963).
250. S. Okuda, S. Yamaguchi, Y. Kawazoe, and K. Tsuda, *Chem. Pharm. Bull. (Tokyo) 12*, 104–112 (1964).
251. I. Seki, *Yakugaku Zasshi 84*, 631–637 (1964), through *Chem. Abstr. 61*, 9545 g (1964).
252. T. J. Batterham, K. H. Bell, and U. Weiss, *Aust. J. Chem. 18*, 1799–1806 (1965), through *Chem. Abstr. 64*, 8254 h (1966).
253. A. F. Casy, *J. Pharm. Sci. 56*, 1049–1063 (1967).
254. T. G. Alexander and S. A. Koch, *J. Ass. Offic. Agr. Chem. 48*, 618–621 (1965), through *Chem. Abstr. 63*, 6783 a (1965).
255. J. H. Beynon, R. A. Saunders, and A. E. Williams, "The Mass Spectra of Organic Molecules," Elsevier, Amsterdam (1968).
256. J. Roboz, "Introduction to Mass Spectrometry: Instrumentation and Techniques," Interscience, New York (1968).
257. H. Budzikiewicz, C. Djerassi, and D. H. Williams, "Structure Elucidation of Natural Products by Mass Spectrometry," Vol. 1, "The Alkaloids," Holden-Day, San Francisco (1964).
258. H. Audier, M. Fetizon, D. Ginsburg, A. Mandelbaum, and T. Rüll, *Tetrahedron Lett.* (1), 13–22 (1965).
259. H. Nakata, Y. Hirata, A. Tatematsu, H. Tada, and Y. K. Sawa, *Tetrahedron Lett.* (13), 829–836 (1965).
260. A. Mandelbaum and D. Ginsburg, *Tetrahedron Lett.* (29), 2479–2489 (1965).
261. D. M. S. Wheeler, T. H. Kinstle, and K. L. Rinehart, *J. Amer. Chem. Soc. 89*, 4494–4501 (1967).
262. A. Tatematsu and T. Goto, *Yakugaku Zasshi 85*, 152–157 (1965).
263. A. Tatematsu and T. Goto, *Yakugaku Zasshi 85*, 778–785 (1965).
264. A. Tatematsu and T. Goto, *Yakugaku Zasshi 85*, 786–790 (1965), through *Chem. Abstr. 63*, 17798 h (1965).
265. A. Tatematsu, T. Goto, T. Nakamura, and S. Yamaguchi, *Yakugaku Zasshi 86*, 195–199 (1966), through *Chem. Abstr. 64*, 17356 a (1966).
266. W. H. McFadden, *Separ. Sci. 1*, 723–746 (1966).
267. W. H. McFadden, *Advan. Chromatog. 4*, 265–332 (1967), through *Chem. Abstr. 69*, 55259 w (1968).
268. L. R. Crawford and J. D. Morrison, *Anal. Chem. 40*, 1464–1469 (1968).
269. A. H. Beckett, G. T. Tucker, and A. C. Moffat, *J. Pharm. Pharmacol. 19*, 273–294 (1967).

270. W. Arnold and H. F. Grützmacher, *Fresenius' Z. Anal. Chem. 247*, 179–188 (1969).
271. A. Zeman and I. P. G. Wirotama, *Fresenius' Z. Anal. Chem. 247*, 155–157 and 158–163 (1969).
272. W. Straub, *Deut. Med. Wochenschr. 37*, 1462 (1911).
273. O. Herman, *Biochem. Z. 39*, 216–231 (1912).
274. L. Maier, *Arch. Exp. Pathol. Pharmakol. 161*, 163–172 (1931).
275. J. C. Munch, *J. Amer. Pharm. Ass. Sci. Ed. 23*, 766, 1185 (1934); *24*, 557 (1935).
276. J. C. Munch, A. B. Sloane, and A. R. Latven, *Bull. Narcotics, U. N. Dept. Social Affairs 4* (3), 23–27 (1952).
277. C. E. Morgan and A. Gellhorn, *Ind. Eng. Chem. Anal. Ed. 19*, 806 (1947).
278. A. R. Alha and K. Ohela, *Acta Pharmacol. Toxicol. 11*, 156–162 (1955).
279. J. Schmidlin-Mészáros and H. Hartmann, *Arch. Toxikol. 18*, 259–262 (1960).
280. J. J. Taylor, *J. Forensic Sci. Soc. 5*, 188–191 (1965).
281. H. W. Elliott, N. Nomof, K. D. Parker, and G. R. Turgeon, *Calif. Med. 109*, 121–125 (1968).
282. W. H. C. Shaw and J. P. Jeffries, *J. Pharm. Pharmacol. 3*, 823–827 (1951).
283. S. Chiba and T. Kawai, *Ann. Rep. Takamine Lab. 6*, 75–79 (1954), through *Chem. Abstr. 49*, 16346 e (1955).
284. W. J. Seagers, J. D. Neuss, and W. J. Mader, *J. Amer. Pharm. Ass. Sci. Ed. 41*, 640–642 (1952).
285. M. J. Pro and R. A. Nelson, *J. Ass. Offic. Agr. Chem. 40*, 1103–1108 (1957).
286. M. J. Pro, W. P. Butler, and A. P. Mathers, *J. Ass. Offic. Agr. Chem. 38*, 849–857 (1955).
287. M. J. Pro, *J. Ass. Offic. Agr. Chem. 42*, 458–459 (1959).
288. M. J. Pro, *J. Ass. Offic. Agr. Chem. 42*, 177–180 (1959).
289. C. Milos, *J. Pharm. Sci. 50*, 837–839 (1961).
290. W. A. Clark and A. J. McBay, *J. Amer. Pharm. Ass., Sci. Ed. 43*, 39–42 (1954).
291. H. J. van der Pol, *J. Pharm. Belg. 40*, 426–436 (1958).
292. J. L. Casinelli and J. E. Sisheimer, *J. Pharm. Sci. 51*, 336–338 (1962).
293. R. C. Gupta, *J. Forensic Sci. 11*, 95–100 (1966).
294. J. Vacek and L. Tyrolova, *Mitt. Deut. Pharm. Ges. 28*, 176 (in *Arch. Pharm. 291*) (1958).
295. G. R. Nakamura, *Bull. Nacrotics, U.N. Dept. Social Affairs 12* (4), 17–20 (1960).
296. E. A. Davey, J. B. Murray, and A. R. Rogers, *J. Pharm. Pharmacol. 20*, 51s–53s (1968).
297. E. Vidic, *Arzneim.–Forsch. 11*, 408–413 (1961).
298. S. Udenfriend, "Fluorescence Assay in Biology and Medicine," Vol. 2, Academic Press, New York (1969).
299. R. F. Chen, *Anal. Biochem. 20*, 339–357 (1967).
300. S. Udenfriend, "Fluorescence Assay in Biology and Medicine," pp. 418–423, Academic Press, New York (1962).
301. S. Udenfriend, D. E. Guggan, B. M. Vasta, and B. B. Brodie, *J. Pharmacol. Exp. Ther. 120*, 26–37 (1957).
302. R. Brandt, S. Ehrlich-Rogozinsky, and N. D. Cheronis, *Microchem. J. 5*, 215–223 (1961).
303. R. Brandt, M. J. Olsen, and N. D. Cheronis, *Science 139*, 1063–1064 (1963).
304. P. Balatre, M. Fraisnel, and J. P. Delcambre, *Ann. Pharm. Fr. 19*, 171–174 (1961), through *Chem. Abstr. 55*, 27779 e (1961).
305. H. J. Kupferberg, A. Burkhalter, and E. L. Way, *J. Pharmacol. Exp. Ther. 145*, 247–251 (1964).
306. B. B. Brodie and S. Udenfriend, *J. Biol. Chem. 158*, 705–714 (1945).
307. B. B. Brodie, S. Udenfriend, and W. Dill, *J. Biol. Chem. 168*, 335–339 (1947).
308. R. A. Lehman and T. Aitken, *J. Lab. Clin. Med. 28*, 787–793 (1943).
309. G. Cronheim and P. A. Ware, *J. Pharmacol. Exp. Ther. 92*, 98–102 (1948).
310. G. Fiese and J. H. Perrin, *J. Pharm. Pharmacol. 20*, 98–101 (1968).
311. A. H. Beckett, M. Rowland, and E. J. Triggs, *Nature 207*, 200–201 (1965).

312. D. J. Smith, *J. Ass. Offic. Anal. Chem.* *49*, 536–541 (1966).
313. C. McMartin, P. Simpson, and N. Thorpe, *J. Chromatog.* *43*, 72–83 (1969).
314. F. Matsui, J. R. Watson, and W. N. French, *J. Chromatog.* *44*, 109–115 (1969).
315. R. R. Pride and E. S. Stern, *J. Pharm. Pharmacol.* *6*, 590–606 (1954).
316. C. A. Johnson and C. J. Lloyd, *J. Pharm. Pharmacol.* *10*, 60 T–71 T (1958).
317. Y.-M. Chu and Y.-H. Yu, *Yao Hsueh Hsueh Pao 6*, 179–183 (1958), through *Chem. Abstr. 53*, 6526 i (1959).
318. J. Axelrod and J. K. Inscoe, *Proc. Soc. Exp. Biol. Med. 103*, 675–676 (1960).
319. J. M. Fujimoto, E. L. Way, and C. H. Hine, *J. Lab. Clin. Med. 44*, 627–635 (1954).
320. P. G. Smolyanskaya, *Tr. Lenigrad. Khim. Farm. Inst. 13*, 268–278 (1961), through *Anal. Abstr. 10*, 1558 (1963).
321. J. Celechovsky and D. Svoboda, *Cesk. Farm. 8*, 380–384 (1959), through *Anal. Abstr. 1888* (1960).
322. S. Pfeifer, *Bull. Narcotics, U.N. Dept. Social Affairs 10* (1), 18–33 (1958).
323. D. C. M. Adamson and F. P. Handsyde, *Analyst 70*, 305–306 (1945).
324. E. G. Clair, *Analyst 87*, 499–500 (1962).
325. T. Duktiewicz and I. Lisikowa, *Acta Pol. Pharm. 19*, 149 (1962).
326. L. A. Woods, J. Cochin, E. G. Fornefield, and M. H. Seevers, *J. Pharmacol. Exp. Ther. 111*, 64–73 (1954).
327. I. C. Rickards, G. E. Boxer, and C. C. Smith, *J. Pharmacol. Exp. Ther. 98*, 380–391 (1950).
328. M. Ono and K. Takashashi, *Bull. Nat. Inst. Hyg. Sci. 82*, 50–53 (1964).
329. S. Y. Yeh and L. A. Woods, *J. Pharmacol. Exp. Ther. 166*, 86–95 (1969).
330. J. L. Emmerson, J. S. Welles, and R. C. Anderson, *Toxicol. Appl. Pharmacol. 11*, 482–488 (1967).
331. J. Shapira, W. H. Perkins, and M. Hara, *Clin. Res. 8*, 92 (1960).
332. J. H. Sanner and L. A. Woods, *J. Pharmacol. Exp. Ther. 148*, 176–184 (1965).
333. G. F. Blane and H. E. Dobbs, *Brit. J. Pharmacol. 30*, 166–172 (1967).
334. M. Sarsunova, J. Toelgyessy, and J. Majer, *Pharm. Acta Helv. 35*, 271–275 (1960).
335. W. M. Hammer and B. B. Brodie, *J. Pharmacol. Exp. Ther. 157*, 503–508 (1967).
336. R. B. Fisher, D. S. Parsons, and G. A. Morrison, *Nature 161*, 764–765 (1948).
337. A. Seher, *Mikrochim. Acta* 308–313 (1961).
338. M. Brenner and A. Niederwieser, *Experientia 16*, 378–383 (1960).
339. S. J. Purdy and E. V. Truter, *Chem. Ind.* (*London*) 506–507 (1962).
340. S. J. Purdy and E. V. Truter, *Analyst 87*, 802–809 (1962).
341. G. Vitte and E. Boussmart, *Bull. Trav. Soc. Pharm. Bordeaux 89*, 83 (1951).
342. R. Klaus, *J. Chromatog. 16*, 311–326 (1964).
343. C. B. Barrett, M. S. J. Dallas, and F. B. Padley, *J. Amer. Oil Chem. Soc. 40*, 580–584 (1963).
344. R. B. Ingle and E. Minshall, *J. Chromatog. 8*, 369–385 and 386–392 (1962).
345. E. Vioque and A. Vioque, *Grasas Aceites 15*, 125–128 (1964), through *Chem. Abstr. 62*, 7 c (1965).
346. F. W. Hefendehl, *Planta Med. 8*, 65–70 (1960).
347. H. K. Mangold, R. Kammereck, and D. C. Malins, *Microchem. J. Symp. Ser. 2*, 697–714 (1962), through *Chem. Abstr. 58*, 12850 c (1963).
348. T. J. Quinn, J. G. Jeffrey, and W. C. MacAulay, *J. Amer. Pharm. Ass. Sci. Ed. 46*, 384–386 (1957).
349. K. Genest and C. G. Farmilo, *J. Amer. Pharm. Ass. Sci. Ed. 48*, 286–289 (1959).
350. P. H. List, *Naturwissenschaften 41*, 454 (1954).
351. A. B. Svendsen, *Pharmazie 10*, 550–552 (1960).
352. W. Poethke and W. Kinze, *Pharm. Zentralh. 103*, 577–583 (1964), through *Chem. Abstr. 62*, 2669 g (1965).
353. E. Brochmann-Hanssen and A. B. Svendsen, *J. Pharm. Sci. 52*, 1134–1136 (1963).
354. G. E. Martin and J. S. Swinehart, *Anal. Chem. 38*, 1789–1790 (1966).
355. C. C. Sweeley, R. Bentley, M. Makita, and W. W. Wells, *J. Am. Chem. Soc. 85*, 2497–2507 (1963).

356. H. W. Elliott, N. Nomof, K. Parker, M. L. Dewey, and E. L. Way, *Clin. Pharmacol. Ther. 5*, 405–413 (1964).
357. J. F. Klebe, H. Finkbeiner, and D. M. White, *J. Amer. Chem. Soc. 88*, 3390–3395 (1966).
358. G. R. Wilkinson and E. L. Way, *Biochem. Pharmacol. 18*, 1435–1439 (1969).
359. R. W. Zumwalt, D. L. Stalling, and C. W. Gehrke, c 159, *154th, Meeting American Chemical Society, September* (1967).
360. E. Schmerzler, W. Yu, M. I. Hewitt, and I. J. Greenblatt, *J. Pharm. Sci. 55*, 155–157 (1966).
361. R. L. Wolen and C. M. Gruber, *Anal. Chem. 40*, 1243–1246 (1968).
362. A. H. Beckett, J. F. Taylor, and P. Kourounakis, *J. Pharm. Pharmacol. 22*, 123–128 (1970).
363. J. Heyrovsky and P. Zuman, "Practical Polarography," Academic Press, New York (1968).
364. M. Brezina and P. Zuman, "Polarography in Medicine, Biochemistry and Pharmacy," Interscience New York (1958).
365. H. Baggesgaard-Rasmussen, C. Hahn, and K. Ilver, *Dan. Tidsskr. Farm. 19*, 41–67 (1945).
366. H. Lund, *Acta Chem. Scand. 12*, 1444–1450 (1958).
367. J. Nosek and O. Krestynova, *Cas. Lek. Cesk. 63*, 49–51 (1950), through *Chem. Abstr. 46*, 4173 h (1952).
368. E. Schulek and K. Burger, *Ann. Univ. Sci. Budapest Rolando Eotvos Nominatae, Sect. Chim. 2*, 531–536 (1960), through *Chem. Abstr. 57*, 2337 f (1962).
369. J. Holubeck, *Pharm. Zentralh. 95*, 435–436 (1956).
370. B. Jambor, *Agrokem. Talajtan 1*, 201 (1952).
371. K. Matsumoto, *Yakugaku Zasshi 72*, 1393, 1396, and 1398 (1952); *77*, 367–370 (1957).
372. F. Santavy, *Collect. Czech. Chem. Commun. 12*, 422–428 (1947).
373. M. Brezina and P. Zuman, "Polarography in Medicine, Biochemistry and Pharmacy," p. 363, Interscience, New York (1958).
374. P. Paerregaard, *Acta Pharmacol. Toxicol. 14*, 38–52 (1957).
375. K. Milthers, *Acta Pharmacol. Toxicol. 15*, 21–28 (1958).
376. K. Milthers, *Acta Pharmacol. Toxicol. 16*, 10–12 (1959).
377. K. Milthers, *Acta Pharmacol. Toxicol. 19*, 149–155 (1962).
378. M. Brezina and P. Zuman, "Polarography in Medicine, Biochemistry, and Pharmacy," p. 364, Interscience, New York, (1958).
379. B. Novotny, *Cesk. Farm. 3*, 199–206 (1954).
380. B. Jambor and E. Bajusz, *Pharmazie 14*, 447–452 (1959), through *Chem. Abstr. 54*, 6363 g (1960).
381. M. Skora, *Diss. Pharm. 15*, 433–437 (1963), through *Chem. Abstr. 61*, 2910 b (1964).
382. H. F. W. Kirkpatrick, *Quart. J. Pharm. Pharmacol. 18*, 338–350 (1945).
383. J. Volke and V. Fortova, *Symposium on Practical Polarography, Bratislava*, p. 104 (1952).
384. F. Santavy and M. Cernoch, *Chem. Listy 46*, 81–85 (1952), through *Chem. Abstr. 46*, 4172 b and 11582 c (1952).
385. V. Fortova, PhD Thesis, Charles University, Prague (1952).
386. V. Fortova and J. Volke, *Symposium on Practical Polarography, Bratislava*, p. 92 (1952).
387. D. R. Dzhalilov, N. G. Ermachenkova, and M. J. Gorjaev, *Aptech. Delo 15*, 49–51 (1966), through *Chem. Abstr. 64*, 19318 g (1966).
388. R. Kalvoda and J. Zyka, *Bull. Narcotics, U.N. Dept. Social Affairs 9* (2), 41–45 (1957).
389. P. Balatre, J. C. Guyot, and M. Traisnel, *Ann. Pharm. Fr. 24*, 425–428 (1966), through *Chem. Abstr. 65*, 5733 n (1967).
390. T. Higuchi and J. I. Bodin, *in* "Pharmaceutical Analysis" (T. Higuchi and E. Brochmann-Hanssen, eds.), pp. 313–543, Interscience, New York (1961).

391. J. Zyka, *Bull. Narcotics, U.N. Dept. Social Affairs* 10 (1), 35–38 (1958).
392. J. Kucharsky and L. Safarik, "Titrations in Non-Aqueous Solvents," pp. 144–169, Elsevier, New York (1965).
393. A. H. Beckett and E. H. Tinley, "Titration in Non-Aqueous Solvents," British Drug Houses, Poole, England.
394. P. J. A. Demoen, *J. Pharm. Sci.* 50, 79–82 (1961).
395. R. R. Paris, G. Faugeras, L. Balard, and M. Capmal, *Ann. Pharm. Fr.* 24, 411–417 (1966), through *Chem. Abstr.* 66, 22240 e (1967).
396. I. R. Juniper, *J. Pharm. Pharmacol.* 21, 632 (1969).
397. J. Bonnard, *J. Pharm. Belg.* 21, 363–374 (1966), through *Chem. Abstr.* 66, 14060 u (1967).
398. C-Y Chen, *Yao Hsueh Hsueh Pao* 5, 249 (1957), through *Chem. Abstr.* 55, 23929 (1961).
399. M. I. Blake and B. Carlstedt, *J. Pharm. Sci.* 55, 1462 (1966).
400. L. Safarik, *Cesk. Farm.* 15, 360–364 (1966), through *Chem. Abstr.* 66, 31974 n (1967).
401. V. Y. Vengerova, *Izv. Vyssh. Ucheb. Zaved. Khim. Khim. Tekhnol.* 8, 1030 (1965), through *Anal. Abstr.* 14, 4232 (1967).
402. W. Wisniewski, H. Piasecka, and S. Jablonski, *Farm. Polska* 23, 211–216 (1967), through *Chem. Abstr.* 68, 24573 u (1968).
403. Z. Blagojevic, R. Popovic and K. Nikolic, *Acta Pharm. Jugoslav.*, 16, 145–150 (1966), through *Chem. Abstr.* 66, 108277 x (1967).
404. S. S. Chausokovskii and M. Brombergiene, *Farmatsiya* (Moscow) 16, 59–61 (1967), through *Chem. Abstr.* 67, 102817 x (1967).

Chapter 3

STRUCTURE–ACTIVITY RELATIONSHIPS

Louis S. Harris

University of North Carolina School of Medicine
Chapel Hill, North Carolina

I. INTRODUCTION

One of the most intriguing developments in the field of narcotic analgesics has been the discovery of the narcotic antagonists.[1-3] These compounds, which reverse or prevent most of the pharmacological actions of the narcotic analgesics, may have analgesic activity themselves. This finding was uncovered serendipitously in the clinic with nalorphine[4] and was unexpected since this compound was essentially devoid of analgesic activity in the common animal test procedures then in use.[5-7] The importance of this finding lies in the fact that the narcotic-antagonist analgesics have a much lessened abuse potential. They do not produce a typical morphine-like physical dependence.[2,3]

A. Agonists, Antagonists, and Partial Agonists

Before discussing structure–activity relationships among the narcotic antagonists, it is important, as Cavallito has pointed out,[8] to define what one means by "structure" and "activity." For this review, structure is considered only as the three-dimensional spatial arrangement of the molecule. Physical-chemical properties such as linear free energy charge and partition characteristics will not be generally considered.

When describing antagonistic activity, the classical scheme of Gaddum[9] and Ariens[10] will be used. Thus we will broadly speak of agonists where the compound has both an affinity for the receptor and a high intrinsic activity. Antagonists would have a variable affinity and little or no intrinsic activity. Partial agonists, then, are compounds having an intermediate intrinsic activity and would thus not produce a maximal effect yet could act as an antagonist against a pure agonist.

With the narcotic antagonists we will, in the main, depend on their ability to interfere with the effects of morphine in the tail-flick or a similar test as a measure of antagonistic activity. For the agonistic activity in this chapter, either their analgesic effects in man or their activity in the hot-plate or tail-flick test in rodents will be used as criteria.

II. MORPHINE ANALOGS

The first narcotic antagonist, N-allylnorcodeine (IA), was discovered in 1915.[11,12] This finding lay dormant until the properties of N-allylnor-morphine (IB), nalorphine, were reported.[13-15]

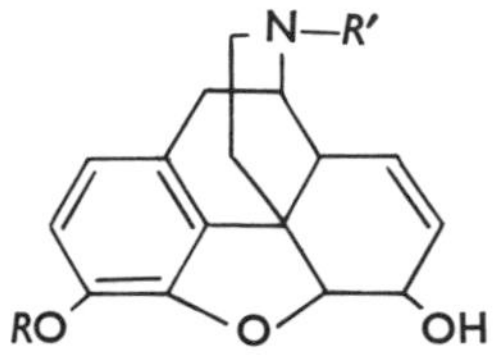

A. $R = CH_3$, $R' = CH_2CH{=}CH_2$

B. $R = H$, $R' = CH_2CH{=}CH_2$

C. $R = CH_3$, $R' = CH_2$◁

D. $R = CH_3$, $R' = CH_2$⟋▱

E. $R = H$, $R' = CH_2$◁

[I]

The key to antagonistic activity appears to reside in the alkylation of the piperidine nitrogen. In the homologous N-alkyl series of normorphine, compounds containing a three-carbon side-chain with or without branching, such as propyl, allyl, isopropyl, or 2-methylallyl, have antagonistic activity. The N-ethyl and N-butyl compounds have little agonistic or antagonistic activity. As the chain length increases to amyl and hexyl, agonistic activity is restored and these compounds are nearly as potent analgesics as morphine.

Within the five-ring structure of morphine, many modifications have been made while keeping the pentannular ring system intact and the alkylation of the nitrogen constant. For instance, we have the morphine series, IIA, IIB, IIC, IID, IIE, and the dihydromorphine series, IIIA, IIIB, IIIC, IIID, IIIE. When one examines the consequence of replacing the N-methyl

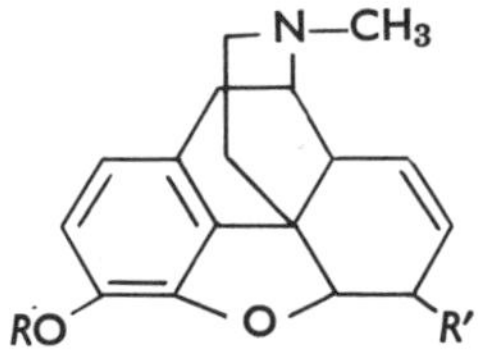

A. $R = H$, $R' = OH$
B. $R = CH_3$, $R' = OH$
C. $R = CH_3CO$, $R' = OH$
D. $R = CH_3CO$, $R' = OCOCH_3$
E. $R = R' = H$

[II]

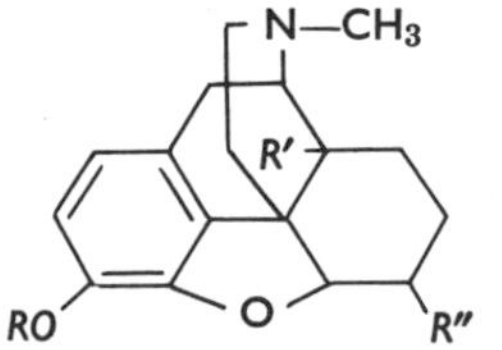

A. $R = R' = R'' = H$
B. $R = CH_3CO$, $R' = H$, $R'' = OCOCH_3$
C. $R = CH_3$, $R' = R'' = H$
D. $R = R' = H$, $R'' = {=}O$
E. $R = H$, $R' = OH$, $R'' = {=}O$

[III]

group in either series by an *N*-allyl constituent, in every case an antagonist is produced. Furthermore, there appears to be a good correlation between the agonistic potency of the *N*-methyl derivative and antagonistic potency of their respective *N*-allyl congeners.[1] That is, the more potent the analgesic activity of the *N*-methyl compound, the more potent the antagonistic potency of its *N*-allyl derivative. Such a relationship breaks down in the six-ring oripavine series, which is discussed in a later section.

In addition to replacement of the *N*-methyl with alkyl radicals having a length of about three carbons, it was found that cycloalkylmethyl groups also conferred antagonistic activity. Indeed the first of these, *N*-cyclopropyl-methylnorcodeine (IC) and *N*-cyclobutylmethylnorcodeine (ID), were described in 1926,[16] some 20 years prior to the introduction of nalorphine. Unfortunately, the original material was amorphous and did not represent truly homogenous compounds. Later work revealed *N*-cyclopropylmethyl-normorphine (IE) to be a potent antagonist. This also is true for the oxymor-phone analog (IVA).[17] The oxymorphone series is of great interest since the *N*-allyl compound naloxone (IVB) is closest to being a pure antagonist of any of the compounds reported to date. Thus naloxone is a potent antagonist in many systems[18-20] and is essentially inactive as an analgesic in animals[21,22] or man.[23] Recently the *N*-cyclobutylmethyl derivative (IVC) has been reported to have analgesic activity in man.[24]

A.   R = —CH₂△

B.   R = CH₂CH=CH₂

C.   R = CH₂◁

[IV]

Since all of the compounds mentioned in this section are derived from natural morphine, they are all levorotatory isomers. The classical synthesis of morphine by Gates and Tschudi[25] did provide a racemic compound. The racemic mixture has been resolved, and the dextrorotatory isomer has been shown to be inactive as an analgesic, but no other dextrorotatory *N*-sub-stituted derivatives have been reported. The importance of both optical isomerism and conformation become more evident in later sections.

III. MORPHINAN ANALOGS

The morphinan ring system (V) was first synthesized by Grewe and his colleagues[26,27] and exploited by Schnider and Grussner.[28] Removal of the furan ring from morphine provides a 4-ring structure with no loss of potency. Indeed the direct analog of morphine, levorphanol (VA), is about four times more potent. Replacement of the *N*-methyl group by an allyl residue gives the potent antagonist levallorphan (VB).[29] The dimethylallyl

A. R = CH$_3$
B. R = CH$_2$CH=CH$_2$
C. R = CH$_2$CH=C(CH$_3$)$_2$
D. R = CH$_2$C≡CH

E. R = CH$_2$—△

[V]

derivative (VC) was first considered to be a weak antagonist, but subsequent study showed it to be a morphine-like analgesic both in animals and man.[30]

The propargyl compound (VD) is a potent antagonist in animals and man and did relieve severe clinical pain. However, it also produced nalorphinelike side effects.[30] Gates and Montzka[31] have prepared a series of N-cyclopropylmethylmorphinan derivatives. One of these, cyclorphan (VE), is a potent antagonist in animals[32] and a potent analgesic in man,[33] being 40 times more active than morphine. Unfortunately, the compound produces a high incidence of psychotomimetic activity, which precludes its use as an analgesic.

IV. BENZOMORPHAN ANALOGS

The triannular benzomorphan structure, VI and VII, has been the most extensively explored in regard to analgesic antagonistic activity. This is probably due to the facile syntheses devised by May and his associates[34-36] and to the more easily modified basic structure. Thus both the *cis* and *trans* forms can be selectively prepared and their racemates resolved. In this way both configuration and optical isomerism can be examined for their influence on biological activity.

The first reported antagonists in this series were described simultaneously.[37,38] Actually, Ager and May[39] had prepared the N-propyl derivative (VIC) some time prior to this, but it had not been evaluated for antagonistic activity. This compound was later shown to be a potent antagonist.[40] In the homologous N-alkyl series the N-methyl (VIA) is an analgesic more potent than morphine. The N-ethyl derivative (VIB) has little or no analgesic

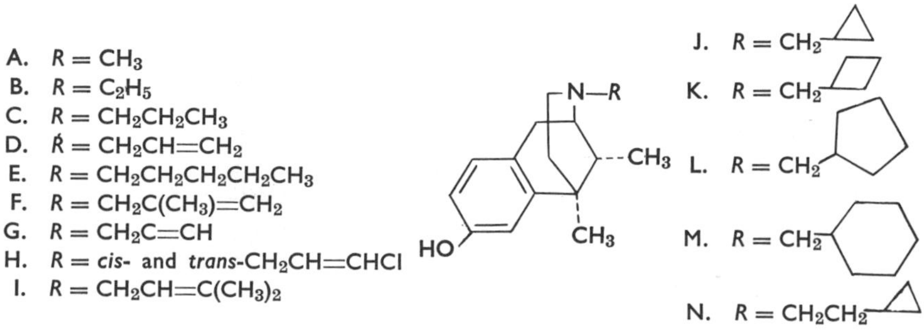

A. R = CH$_3$
B. R = C$_2$H$_5$
C. R = CH$_2$CH$_2$CH$_3$
D. R = CH$_2$CH=CH$_2$
E. R = CH$_2$CH$_2$CH$_2$CH$_2$CH$_3$
F. R = CH$_2$C(CH$_3$)=CH$_2$
G. R = CH$_2$C≡CH
H. R = *cis*- and *trans*-CH$_2$CH=CHCl
I. R = CH$_2$CH=C(CH$_3$)$_2$

J. R = CH$_2$—△
K. R = CH$_2$—▱
L. R = CH$_2$—⬠
M. R = CH$_2$—⬡
N. R = CH$_2$CH$_2$—△

[VI]

or analgesic-antagonist activity, while the *N*-allyl (VID) and *N*-propyl (VIC) were found to be three and seven times more potent than nalorphine as antagonists. When alkylation is increased to *N*-amyl (VIE), morphine-like analgesia returns. The *N*-β-methylallyl (VIF) and *N*-propargyl (VIG) are somewhat less active than the allyl, while the *N*-β-chloroallyl derivative is a very weak antagonist. This is quite analgous to the findings in the morphine series. On the other hand, the *N*-γ-chloroallyl compounds (VIH) are potent antagonists. The *N*-dimethylallyl derivative (VII, pentazocine) is a weak antagonist that is active in the writing test and may be considered as a partial agonist. The corresponding compound in the morphinan series was originally thought to be an antagonist but was later shown to be a potent agonist both in laboratory animals and man.[30] Pentazocine is an active analgesic in man, has little nalorphine-like psychotomimetic activity, and has a much lessened addiction liability. Thus pentazocine represents the first successful translation into clinical utility of the original finding by Lasagna and Beecher[4] with nalorphine.

Substitution on the nitrogen of cycloalkylmethyl groups has produced a highly interesting series of compounds. The *N*-cyclopropylmethyl derivative (VIJ, cyclazocine) is a potent antagonist, and in addition it has a high degree of internuncial blocking and tranquilizing activity. Cyclazocine is a potent analgesic in man but produces too high a level of psychotomimetic activity to be clinically useful.[41] This agent has recently aroused a great deal of interest in the therapy of narcotic addiction.[42,43] The *N*-cyclobutylmethyl compound (VIK) has a curious mixture of agonist and antagonist activity in laboratory animals.[1] This compound also has been evaluated in man but produces such intense psychotomimetic activity that its analgesic activity could not be assessed accurately.[23] The *N*-cyclopentylmethyl (VIL) derivative is still a potent antagonist but has no agonistic activity and has a lessened degree of tranquilizing activity. Increasing the ring size to six carbons (VIM) causes a marked dimunition of antagonistic activity, and the internuncial blocking activity essentially disappears. The *N*-cyclopropylethyl (VIN) compound has about one-fifth of the antagonistic activity of cyclazocine and is devoid of the internuncial blocking activity.

All of the compounds discussed in this section are racemates. In cases where resolution was carried out the major portion of the antagonistic activity was found in the *l*-isomer.[44,45] Thus with pentazocine the *l*-isomer is 20 times more potent than the *d*-isomer while the ratio with cyclazocine is 500. It is of interest to note that in the several cases examined the *l*-isomer proved to be four times more potent than the racemic mixture. Theoretically, a ratio of two would .be expected. No experimental evidence exists to explain this behavior.

In addition to optical isomerism, geometrical isomerism exists in this series as it does in the morphine and morphinan series, and the stereochemical interrelationships have been detailed.[46] With the *N*-methyl compounds, the *trans* derivatives (VIIA and B) were found to be more potent agonists than the analogous *cis* compounds. With the antagonists in the

A. $R = R' = CH_3$
B. $R = C_2H_5, R' = CH_3$
C. $R = CH_3, R' = CH_2CH{=}CH_2$
D. $R = CH_3, R' = CH_2CH{=}C(CH_3)_2$
E. $R = CH_3, R' = CH_2{-}\triangle$
F. $R = C_2H_5, R' = CH_2CH{=}C(CH_3)_2$
G. $R = C_6H_5, R' = CH_3$

[VII]

5,9-dimethyl series (VIIC–E) there was little difference in antagonistic potency between the *cis* and *trans* isomers of either the racemic or resolved compounds.[45] The first inkling of a separation of activity came with the comparison of the *cis* and *trans* isomers in the 5-ethyl-9-methyl series. For instance, the *cis* N-dimethylallyl compound (VIIIA) is a weak antagonist, while the *trans* analog (VIIF) is a potent antagonist that also has high agonistic activity in the tail-flick test.[1] The picture becomes even more complex when the optical isomers are examined. The *l-cis* compound is a moderate antagonist some six times more potent than the *d*-isomer. In the *trans* series both isomers had equivocal antagonistic activity. That is, some antagonism was seen at low doses but there was no increase in activity as the dose was increased. On the other hand, the *l-trans* compound was a potent agonist (five times morphine) in the rat tail-flick test while the *d-trans* derivative was devoid of activity.[44]

Recent work from May's laboratory where the alkylation at both the 5- and 9-position has been altered has provided some theoretically important compounds.[47] Thus in the N-methyl-5,9-diethyl series both the *d*- and *l-cis* compounds (VIIIB) were found to be potent agonists in the mouse hot plate test with the *l*-isomer being six times more potent than the *d*-isomer. However, in the substitution test in morphine-dependent monkeys, the more potent *l*-isomer behaved as an antagonist while the *d*-isomer behaved as an agonist.[48] A similar situation was seen with the *cis*-5-propyl-9-methyl (VIIIC) and the *trans*-5-phenyl-9-methyl (VIIG) derivatives.[49] Also of great interest is N-methyl-5-ethyl derivative (VIIID) with no alkylation at 9-position. This compound does not exhibit *cis-trans* isomerism. Again, the *l*-isomer was nearly ten times more potent than morphine as an agonist in the hot plate test, yet it behaved as an antagonist in morphine-dependent monkeys. The *d*-isomer was only one-fortieth as potent as an agonist, yet it suppressed abstinence in the addicted monkey. It should be noted that

A. $R = C_2H_5, R' = CH_3, R'' = CH_2CH{=}C(CH_3)_2$
B. $R = R' = C_2H_5, R'' = CH_3$
C. $R = C_3H_7, R' = R'' = CH_3$
D. $R = C_2H_5, R' = H, R'' = CH_3$

[VIII]

these compounds are all *N*-methyl derivatives and thus represent the first antagonists that do not have an *N*-alkyl group approximating a three-carbon chain. Further developments in this area will be of great interest.

V. MEPERIDINE ANALOGS

In the meperidine series, which may be considered as 2-ring analogs of morphine, replacement of the *N*-methyl group by an allyl radical (IXA) did not lead to an antagonist. Indeed, the compound behaved like a typical narcotic analgesic.[50,51] A similar situation was seen with the *N*-cyclopropyl-methyl derivative (IXB). We postulated[1] that this might be related to the lack of a phenylethylamine fragment in the structure. Some evidence for this view comes from the work of Kugita,[52,53] who has prepared the *N*-allyl-3-phenyl piperidine derivative (X). This compound is a phenylethylamine and proved to be devoid of analgesic activity and to be about one-fifth as potent as nalorphine as an antagonist. Also of interest in this regard is the pyrrolidine derivative profadol (XI). This compound has a phenylethylamine structure and behaves much like the *N*-methyl benzomorphans described previously that have a mixture of agonist and antagonistic activity. For instance, profadol is active in the mouse hot plate and rat tail-pressure tests[54] yet precipitates abstinence in morphine-dependent monkeys[48] and human subjects.[55]

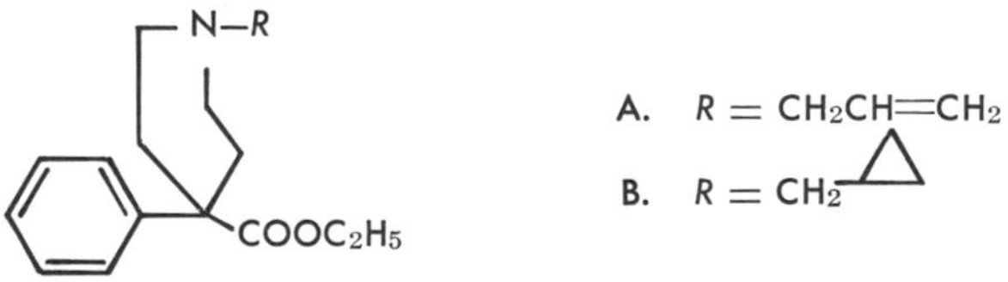

A.   R = CH₂CH=CH₂

B.   R = CH₂—△

[IX]

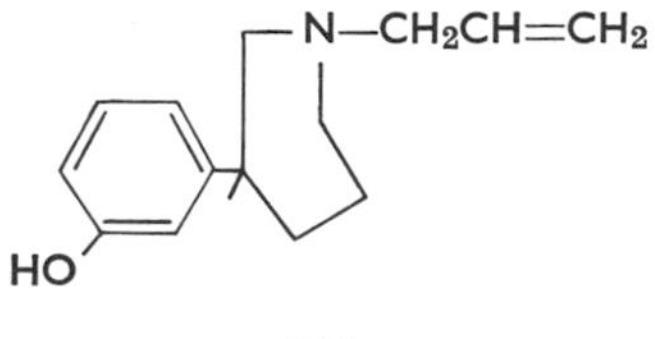

[X]

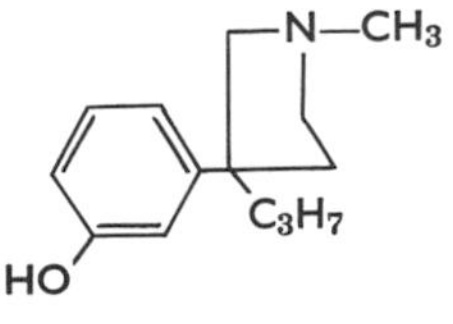

[XI]

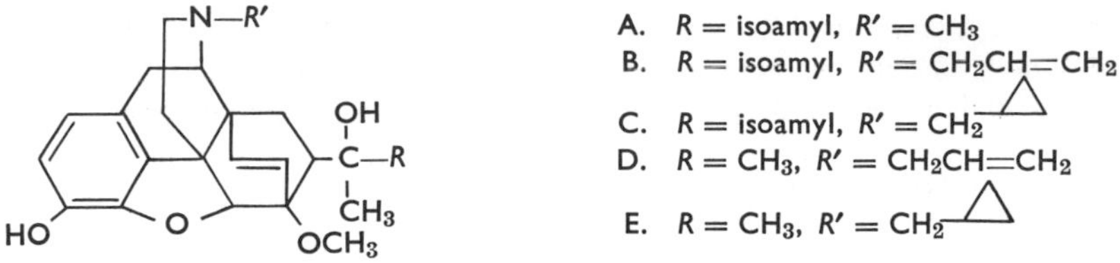

A. $R =$ isoamyl, $R' = CH_3$
B. $R =$ isoamyl, $R' = CH_2CH{=}CH_2$

C. $R =$ isoamyl, $R' = CH_2{-}\triangle$
D. $R = CH_3$, $R' = CH_2CH{=}CH_2$

E. $R = CH_3$, $R' = CH_2{-}\triangle$

[XII]

VI. ORIPAVINE DERIVATIVES

In addition to ring contraction, attention has recently been given to ring expansion. Bentley and his group[56] have reported the preparation of some 6,14-*endo*-ethenotetrahydrothebaine derivatives, the *N*-methyl compounds having a surpassing potency (i.e., XIIA is 5000–10,000 times morphine). The *N*-allyl and *N*-cyclopropylmethyl derivatives in this series have been prepared.

In the most potent analgesic series where R' is isoamyl, the *N*-allyl (XIIB) and *N*-cyclopropyl (XIC) derivatives are not antagonists. They have atypical morphine-like analgesic activity. That is, although they do not antagonize the antinociceptive activity of morphine they do not add to it.[57] When a shorter C side chain is employed, potent nalorphine-like antagonists (XIID and E) can be generated.[58] The *N*-cyclopropylmethyl compound XIIE is about 35 times more potent than nalorphine. It has been evaluated in man as an analgesic but produces a high incidence of psychotomimetic side effects.[23]

VII. MISCELLANEOUS

To date there is no report of narcotic antagonist activity in the diphenylmethane or nitrobenzimidazole series of analgesics. Of some interest is the finding[59] of antagonistic activity among a series of benzodiazepines (XIII). No agonistic activity was seen with these compounds.

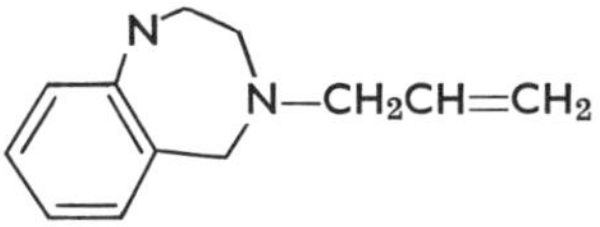

[XIII]

VIII. SUMMARY AND CONCLUSIONS

Since the majority of the classical type of narcotic antagonists have little or no activity in the mouse hot plate and rat tail-flick tests we have chosen to use these an indication of agonistic activity in this review. This has been supplemented by data from man where this was available. It should be noted, however, that many narcotic antagonists have agonistic activity in such procedures as the phenylquinone[21,22,60,61] or bradykinin[62] writhing test or the coaxially stimulated guinea pig ileum.[19,20] In these test procedures there is a good correlation between agonistic activity and analgesia in man. These and other observations have led us to conceive of compounds such as nalorphine, pentazocine, cyclazocine, and cyclorphan as partial agonists while compounds such as naloxone can be considered as pure antagonists. Many of the agonistic effects of the antagonists differ qualitatively from morphine. Even when the effects are similar they may be attributed to interactions with different populations of receptors.[2] Explanations like this must certainly be invoked to explain the findings with the N-methylbenzomorphan isomers (that is, VIIIB and C) as well as the qualitatively different type of abstinence seen after prolonged medication with nalorphine[63] and cyclazocine[64] in man.

This field is one that can still be characterized as actively germinal. Important practical and theoretical findings are continuing to emerge that are helping us to gain a greater insight into the problems of pain, analgesia, and physical dependence.

REFERENCES

1. S. Archer, and L. S. Harris, *Progr. Drug Res. 8*, 261–320 (1965).
2. W. R. Martin, *Pharmacol. Rev. 19*, 463–521 (1967).
3. H. F. Fraser, and L. S. Harris, *Ann. Rev. Pharmacol. 7*, 277–300 (1967).
4. L. Lasagna, and H. K. Beecher, *J. Pharmacol. Exp. Ther. 112*, 356–363 (1954).
5. L. A. Woods, *Pharmacol. Rev. 8*, 175–198 (1956).
6. C. A. Winter, P. D. Orahovats, L. Flataker, E. G. Lehman, and J. T. Lehman *J. Pharmacol. Exp. Ther. 111*, 152–160 (1954).
7. L. S. Harris, and A. K. Pierson, *J. Pharmacol. Exp. Ther. 143*, 141–148 (1964).
8. C. A. Cavallito, *Ann. Rev. Pharmacol. 8*, 39–66 (1968).
9. J. H. Gaddum, *Pharmacol. Rev. 9*, 211–218 (1957).
10. E. J. Ariens, J. M. Van Rossum, and A. M. Simonis, *Pharmacol. Rev. 9*, 218–236 (1957).
11. J. Pohl, *Z. Exp. Pathol. Ther. 17*, 370–382 (1915).
12. J. von Braun, *Ber. Deut. Chem. Ges. 49*, 977–989 (1916).
13. E. R. Hart, *J. Pharmacol. Exp. Ther. 72*, 19 (1941).
14. J. Weislard, and A. E. Erickson, *J. Amer. Chem. Soc. 64*, 869–870 (1942).
15. E. R. Hart, and E. L. McCawley, *J. Pharmacol. Exp. Ther. 82*, 339–348 (1944).
16. J. von Braun, M. Kuhn, and S. Siddiqui, *Ber. Deut. Chem. Ges. 59*, 1081–1089 (1926).
17. G. A. Deneau, and M. H. Seevers, *Addendum 1, Comm. Drug. Addiction Narcotics* (1967).
18. H. Blumberg, H. B. Dayton, M. George, and D. N. Rapaport, *Fed. Proc. 20*, 311 (1961).
19. H. W. Kosterlitz, and A. J. Watt, *Brit. J. Pharmacol. 33*, 266–276 (1968).
20. L. S. Harris, W. L. Dewey, J. F. Howes, J. S. Kennedy, and H. Pars, *J. Pharmacol. Exp. Ther. 169*, 17–22 (1969).
21. H. Blumberg, P. S. Wolf, and H. B. Dayton, *Proc. Soc. Exp. Biol. Med. 118*, 763–766 (1965).
22. J. Pearl, and L. S. Harris, *J. Pharmacol. Exp. Ther. 154*, 319–323 (1966).
23. A. S. Keats, and J. Telford, personal communication.
24. H. W. Elliot, G. Navarro, and N. Nomof, personal communication.

25. M. Gates, and G. Tschudi, *J. Amer. Chem. Soc. 74*, 1109–1110 (1952).
26. R. Grewe, and A. Mondon, *Ber. Deut. Chem. Ges. 81*, 279–286 (1948).
27. R. Grewe, A. Mondon, and E. Nolte, *Ann. Chem. 564*, 161–198 (1949).
28. O. Schnider, and A. Grüssner, *Helv. Chim. Acta. 32*, 821–828 (1949).
29. J. Hellerbach, A. Grüssner, and O. Schnider, *Helv. Chim. Acta. 39*, 429–440 (1956).
30. J. Telford, C. N. Papadopoulos, and A. S. Keats, *J. Pharmacol. Exp. Ther. 133*, 106–116 (1961).
31. M. D. Gates, and T. Montzka, *J. Med. Chem. 7*, 127–131 (1964).
32. L. S. Harris, A. K. Pierson, J. R. Dembinski, and W. L. Dewey, *Arch. Int. Pharmacodyn. 165*, 112–126 (1967).
33. L. Lasagna, J. W. Pearson, and T. DeKornfeld, personal communication.
34. E. L. May and E. M. Fry, *J. Org. Chem. 22*, 1366–1369 (1957).
35. E. L. May and H. Kugita, *J. Org. Chem. 26*, 188–193 (1961).
36. J. H. Ager, and E. L. May, *J. Org. Chem. 27*, 245–247 (1962).
37. M. Gordon, J. J. Lafferty, D. H. Tedeschi, N. B. Eddy, and E. L. May, *Nature 192*, 1089 (1961).
38. S. Archer, N. F. Albertson, L. S. Harris, A. K. Pierson, J. G. Bird, A. S. Keats, J. Telford, and C. Papadopoulos, *Science 137*, 541–543 (1962).
39. J. H. Ager, and E. L. May, *J. Org. Chem. 25*, 984–986 (1960).
40. S. Archer, M. F. Albertson, L. S. Harris, A. K. Pierson, and J. G. Bird, *J. Med. Chem. 7*, 123–127 (1964).
41. L. Lasagna, T. J. DeKornfeld, and J. W. Pearson, *J. Pharmacol. Exp. Ther. 144*, 12–16 (1964).
42. W. R. Martin, C. W. Gorodetzky, and T. K. McClane, *Clin. Pharmacol. Ther. 7*, 455–465 (1966).
43. J. H. Jaffe, and L. Brill, Int. J. Addictions *1*, 99–123 (1966).
44. S. Archer, L. S. Harris, N. F. Albertson, B. F. Tullar, and A. K. Pierson, *Advan. Chem. Ser. 45*, 162–169 (1964).
45. B. F. Tullar, L. S. Harris, R. L. Perry, A. K. Pierson, A. E. Soria, W. F. Wetterau, and N. F. Albertson, *J. Med. Chem. 10*, 383–386 (1967).
46. S. E. Fullerton, E. L. May, and E. D. Becker, *J. Org. Chem. 27*, 2144–2147 (1962).
47. E. L. May, and N. B. Eddy, *J. Med. Chem. 9*, 851–852 (1966).
48. J. E. Villarreal, and M. A. Seevers, personal communication.
49. F. B. Block, and F. H. Clarke, *J. Med. Chem. 12*, 845–847 (1969).
50. P. J. Costa, and D. D. Bonnycastle, *J. Pharmacol. Exp. Ther. 113*, 310–318 (1955).
51. C. A. Winter, P. D. Orahovats, and E. G. Lehman, *Arch. Int. Pharmacodyn. 110*, 186–202 (1957).
52. H. Kugita, H. Inoue, T. Oine, G. Hayashi, and S. Nurimoto, *J. Med. Chem. 7*, 298–301 (1964).
53. H. Kugita, T. Oine, H. Inoue, and G. Hayashi, *J. Med. Chem. 8*, 313–316 (1965).
54. C. V. Winder, M. Welford, J. Wax, and D. H. Kaump, *J. Pharmacol. Exp. Ther. 154*, 161–175 (1966).
55. D. R. Jaskinski, W. R. Martin, and J. D. Sapira, personal communication.
56. K. W. Bentley, and D. G. Hardy, *J. Amer. Chem. Soc. 89*, 3267–3273 (1967).
57. L. S. Harris, unpublished observations.
58. K. W. Bentley, A. L. A. Boura, A. E. Fitzgerald, D. G. Hardy, A. McCoubrey, M. L. Aikman, and R. E. Lister, *Nature 206*, 102–103 (1965).
59. P. M. Carabeteas, and L. S. Harris, *J. Med. Chem. 9*, 6–9 (1966).
60. R. I. Taber, D. D. Greenhouse, and S. Irwin, *Nature 204*, 189–190 (1964).
61. J. Pearl, M. D. Aceto, and L. S. Harris, *J. Pharmacol. Exp. Ther. 160*, 217–230 (1968).
62. G. F. Blane, *J. Pharm. Pharmacol. 19*, 367–373 (1967).
63. W. R. Martin, and C. W. Gorodetzky, *J. Pharmacol. Exp. Ther. 150*, 437–442 (1965).
64. W. R. Martin, H. F. Fraser, C. W. Gorodetzky, and D. E. Rosenberg, *J. Pharmacol. Exp. Ther. 150*, 426–436 (1965).

Chapter 4

PHYSIOLOGICAL DISPOSITION OF NARCOTIC AGONISTS AND ANTAGONISTS

Salvatore J. Mulé

New York State Narcotic Addiction Control Commission
New York, New York

I. INTRODUCTION

The disposition and metabolism of narcotic drugs have been an extremely active area of research during the past few decades. Much effort has been extended in this area due to two factors: (1) a search for the mechanism of action of analgesic drugs and/or the mechanism of tolerance and physical dependence and (2) the need for basic data concerning the disposition and metabolism of a new drug prior to release for clinical experimentation and subsequent filing of a new drug application.

Although the literature on this topic is extensive in animals, there is a paucity of information in man. This is primarily due to the lack of sensitivity in detecting small levels of drug in human biological materials as well as the inherent difficulty in conducting clinical drug research. The extensive use of tracer techniques should help to provide much needed information and data in this area. It is also obvious that very little data was obtained until a few years ago on the CNS levels of narcotic drugs. The principal reason for the lack of experiments appears to be the sensitivity of the available chemical analytical techniques.

General accounts of the disposition and metabolism of narcotics and their surrogates are given in excellent reviews by Way and Adler.[1-3] It is, therefore, the purpose of this paper to focus attention on the distribution of the narcotic analgesics and antagonists within the CNS. Furthermore, an assessment of the kinetics of absorption, disposition, and excretion of these drugs are provided only in reference to the CNS action of these drugs.

II. DYNAMICS OF DRUG DISPOSITION

The plasma is the physiological medium of exchange between the drug

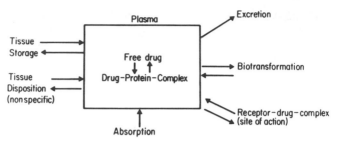

Fig. 1. Dynamics of drug disposition.

and the tissue where pharmacological activity is exerted. The overall dynamics of drug activity are depicted schematically in Fig. 1. While in the plasma compartment the drug exists in both the free and bound form. When bound to plasma proteins the drug is not freely diffusible, and in fact only a small fraction of the free drug actually complexes with the specific drug receptor; most of the drug is associated with nonspecific binding sites. All the drug eventually leaves the circulatory system to be metabolized, excreted, or deposited in tissue depots. The rate at which a drug leaves the plasma and gains access to the site of action and is eventually eliminated from the body is highly dependent upon the chemical and physical properties of the drug.

It is obvious that many factors are involved in the pharmacological action of drugs, and there are several excellent monographs[4−8] that provide information on the absorption, disposition, and excretion of drugs. It is the purpose of this report to consider these factors as they relate to the central nervous system and in particular to the action of narcotic drugs.

A. Absorption

The absorption of any drug is dependent upon the preparation, the physical and chemical properties of the compound, and the route of administration. However, most drug absorption follows first-order kinetics. The rate of absorption may be determined by the following equation:

$$\log D = \log D_0 - \frac{k_a t}{2.30}$$

where D is the quantity of drug remaining at the absorption site at time t and k_a is the rate constant for absorption and D_0 is the dose of drug at the site of absorption.

Drugs that are highly lipid soluble, predominently un-ionized, and primarily free (unbound) gain ready access to the CNS. The narcotic analgesics in general have pK_a values ranging from 7.5 to 9,[9] which provides ionization from 84 to 99% at physiological pH. The concentration of the narcotic drugs in the CNS does appear to correlate with lipid solubility as measured by organic solvent:water partition ratios. The percentage of narcotic drugs bound to plasma proteins may vary from 18 to 68%.[10] In essence a very small percentage of the administered narcotic drug gains access to the CNS (usually less than 0.1%), and that appears to be primarily through a mechanism of simple diffusion.

B. Disposition

In order to effectively understand the data obtained on the disposition of narcotic analgesics and antagonists within the CNS, some knowledge of the brain is required. The brain comprises about 2% of the body weight, with a blood flow of about 0.5 ml/g/min.[11] A drug may gain access to the CNS through two routes: (1) capillary circulation a (2) cerebrospinal fluid (CSF). Blood flow rates that are extremely important for disposition of the drug vary considerably between the white and gray cerebral cortex, as does the total capillary density (three times higher in gray cortex). The blood flow (in milliliters per gram per minute) in visual cortex (gray) of the cat was 1.25 vs 0.23 in the cerebral cortex white matter.[11]

Penetration of the CNS and CSF by drugs from the circulation have been the subject of intensive study.[11-16] The factors directly associated with the penetration of drugs into the CNS are as follows:

1. *Ionization.* The degree of ionization is dependent upon the pK_a of the drug at the physiological pH of the cell. Permeability of the neuronal cell appears to occur primarily with the unionized lipid soluble form of the drug:
2. *Protein binding.* Only the free drug passes across the many biological membranes to the neuronal cell, because a portion of the drug molecules are bound to plasma proteins and/or cellular sites. It is important, therefore, to determine the free drug in the plasma, since the rate of diffusion is dependent upon free drug levels. At an equilibrium phase the concentration in the plasma is equal to the level of drug in the CSF. Since the CSF is practically protein free, all the drug present in this compartment may be considered in the free form:
3. *Lipid solubility.* This lipid solubility determines to a great extent the ease or difficulty with which a drug passes into the brain. The lipid solubility is often determined by partition coefficients between an organic solvent and water. This approach to lipid solubility for a specific drug does not necessarily directly resemble the cell membrane chemical characteristics. However, an order of correlation may be achieved by using this approach.

Drugs that penetrate the CNS therefore must display characteristics that allow ready access to the CNS, namely, low ionization at physiological pH, minimal binding to plasma proteins, and high lipid solubility.

Some drugs because of these physicochemical requirements are excluded from the CNS when administered systemically. This is true of highly polar and quarternized compounds. These drugs do have effects when administered directly into the CSF. The pharmacological effects, however, may be different and often more toxic than following systemic usage.

The ionization of the drug directly affects the percentage of the drug that penetrates the CNS. The pH of the plasma, therefore, is most important. The altered physiological state of either acidosis or alkalosis directly affects drug penetration and egression from the CNS. For basic drugs, such as the narcotic analgesics, an elevated or slightly alkaline pH allows more of the drug to be present in the un-ionized form and, therefore, increases penetration of the CNS. An acidotic condition will have an opposite effect. A favorable plasma-to-brain ratio may be achieved by creating an acidic plasma pH so that more of the drug (basic narcotic analgesic) is in the ionized form and will not penetrate the CNS. The drug that is present in the CNS may then move out due to the favorable concentration gradient. The subjective effects as well as renal excretion of these drugs are highly dependent upon pH (see below).

The factors described are most important in getting the drug to the site of action. However, concentration alone does not achieve the pharmacological effect that takes into consideration the parameters of affinity and intrinsic activity. Recent data[17] show that a good correlation between the duration of action of morphine and the levels of the drug in the CNS of guinea pigs did not exist.

C. Biotransformation

Metabolic transformation of the narcotic analgesics and antagonists is the primary means of limiting the pharmacological action of these drugs. The rate of elimination of these drugs is also dependent upon the physicochemical properties of the metabolite. The ability to penetrate cellular barriers may involve the rate of biotransformation. Biotransformation affects the probable toxicity of the drug, and the metabolite may have pharmacological activity. It is, therefore, obvious that metabolism is an important facet of drug activity within the organism. The primary pathways whereby the narcotic drugs are metabolized are (1) conjugation, (2) dealkylation, (3) hydrolysis, and (4) oxidation and reduction.

This subject is described in detail elsewhere in this text, therefore it is the purpose here simply to present the basic mechanisms of biotransformation associated with narcotic agonists and antagonists. The interested reader may consult several fine reviews[1-3] and especially the detailed account of detoxication mechanisms by R. T. Williams.[18]

1. Conjugation

Conjugation appears to be the major route for detoxifying narcotic analgesics. The most common conjugate formed is the glucuronide.

Narcotic analgesics that appear to form glucuronide conjugates are morphine,[19-23] codeine (6-OH position),[24] norcodeine,[25] levallorphan,[26] dihydromorphine,[27] levorphanol,[28] meperidine,[29] cyclazocine,[30,31] and norcyclazocine.[31] The evidence for the existence of the glucuronide conjugate is often predicated upon acid or glucuronidase hydrolysis rather than upon chemical isolation of the conjugate.

Phenolic and some hydroxy compounds also may form ethereal sulfates as an ester of sulfuric acid or aryl sulfuric acid. Fujimoto[32] has isolated morphine ethereal sulfate from the urine of chickens and cats and nalorphine ethereal sulfate from cat urine.[33]

2. Dealkylation

Dealkylation represents an important biotransformation route for narcotic analgesics and antagonists. This pathway includes the oxidative removal of N-methyl, N-ethyl, and N-alkyl groups as well as O-methyl, O-ethyl, and O-alkyl groups.[34]

N-dealkylation has been shown to occur with almost all narcotic analgesics studied.[1-3] N-dealkylation has been implicated in analgesia[35] and in the development of tolerance.[36] Let it suffice to say that neither hypothesis remains tenable[1-3] but for a time they provoked intensive experimental activity.[37]

O-dealkylation has taken on significance since the early work of Adler[38,39] in rats and the subsequent identification of morphine following codeine administration in man.[40,41] The important factor here was whether the pharmacological action of codeine is mediated through the release of morphine.[42-47]

The relative rates at which N-demethylation and O-demethylation proceed are different, with N-demethylation proceeding at a more rapid pace than O-demethylation.[48] The primary significance that may be attributed to this microsomal metabolic event is biotransformation of the drug and subsequent termination of narcotic drug action.

3. Hydrolysis

Narcotic analgesics may be metabolized by hydrolysis. The hydrolytic pathway is most important in the biotransformation of such narcotic analgesics as heroin, dihydrocodeine enolacetate, diacetyl-N-alkylnormorphine, 6-monoacetyl morphine, meperidine, alphaprodine, and anileridine.

Heroin is hydrolyzed to 6-MAM and to morphine.[49,50] This biotransformation of heroin occurs quite rapidly in all animals studied. Heroin or 6-MAM appear to function primarily as the agents whereby the drug gains access to the CNS or across the multiple biological membranes. The pharmacological action of heroin appears to be elicited primarily by morphine.

Meperidine (pethidine) is hydrolyzed to meperidinic acid, normeperidinic acid, normeperidine, and the conjugates of meperidinic and normeperidinic acid. It also seems that normeperidinic acid arises primarily from the hydrolysis of normeperidine[51] and, as usual, the primary hydrolytic site for the metabolism of these drugs remains the liver.[52,53]

4. Oxidation and Reduction

Oxidation is a common reaction for the biotransformation of narcotic analgesics containing alcoholic or aldehyde groups that are oxidized to acids, oxidatively dealkylated, or deaminated. [1-3] The oxidation of alkyl side chains, hydroxylation, and a variety of related oxidative processes may also occur. [18]

Reduction, although not as common as oxidation, does occur with the narcotic analgesics. Aldehyde groups may be converted to alcohols, ketones may be reduced to alcohols, and double bonds may be reduced. [18]

The reactions described above as oxidative or reducing are by no means exhaustive and may be represented by the following narcotic analgesics, i.e., morphine, levallorphan, anileridine, meperidine, codeine, cyclazocine, methadone, and propoxyphene. [1-3,18]

D. Excretion

Drug action may be terminated through mechanisms involving metabolism (biotransformation), tissue or bone storage (short and long time periods), and tolerance development. Ultimately, however, the drug must be eliminated from the organism. This may be accomplished by the lungs, saliva, perspiration, biliary system, intestines, and kidney. The narcotic analgesics are eliminated to some extent by each of these routes[1-3]; however, the kidney appears to be the prime excretory pathway for the narcotic analgesics. The renal excretory mechanism (glomerular filtration, renal reabsorption, and tubular secretion) involving the narcotic analgesics has received very little attention. Baker and Woods[54] studied the renal excretion of morphine and concluded that the primary mechanism for the elimination of morphine was glomerular filtration. Recent studies using ^{3}H-dihydromorphine[55] have shown that secretion of the free and conjugated dihydromorphine is an important mechanism for the renal elimination of this drug. Obviously to correct the paucity of information concerning the renal mechanism, whereby the narcotic analgesics are excreted, requires further research.

Biliary excretion of narcotic drugs requires special mention since it appears to represent a relatively important pathway for the elimination of narcotic drugs. Woods[21] quite nicely demonstrated that biliary excretion was a primary pathway in the metabolism of morphine and the major source of fecal morphine. March and Elliott[56] recovered in one rat using ^{14}C-labeled morphine as much as 63% of the administered dose in the bile. Mulé and Gorodetzky[30] found that ^{3}H-cyclazocine was concentrated in the bile; in fact 14% of the dose was accounted for in the 24-hr bile. Recently Yeh and Woods[44] reported that the mean biliary excretion following ^{14}C-codeine administration was 46% of which 43% was conjugated morphine. It is also interesting to note that although the conjugate form of the narcotic analgesic is several fold higher in the bile in comparison to the free drug, very little conjugate is present in the feces. [30,44,57] This suggests that most of the conjugate drug is either absorbed from the intestinal tract or hydrolyzed by intestinal enzymes.

III. DRUG DISPOSITION IN THE CNS

Table I contains the data on drugs for which information is available concerning the disposition within the CNS in various animal species.

A. Morphine

The narcotic analgesic that through the years has received the most attention concerning any aspect of the drug dependence problem is morphine. This fact is especially true in regard to the CNS disposition of drugs. Morphine distribution was studied[58] in both nontolerant and tolerant monkey following a relatively large dose (30 mg/kg as free base) of the drug. There was apparently no difference in the brain level of the drug at 90 min or 4 hr after administration between the tolerant and nontolerant monkey. The drug was estimated by a chemical method[89] that was sensitive only to a level of 5 μg/g of tissue. In all probability the true concentration of morphine in the brain was significantly less than 5 μg/g. Less than 10 μg/g of brain was reported for "bound" morphine. In the dog[21,59] following a subcutaneous injection of 30 mg/kg the maximal drug level in brain was observed at 4 hr (4.6 μg/g); however, this was apparently not much different from the 90-min value[59] of 4.3 μg/g. At 16 hr a concentration of 0.9 μg/g was observed. The availability of ^{14}C-labeled morphine in radiochemically pure form allowed for detailed studies in CNS tissue with sufficient sensitivity to provide reliable data. It was observed that concentrations of ^{14}C-morphine[60] were significantly lower in the CNS of tolerant dogs as compared to nontolerant dogs at each time interval after a single 2 mg/kg as free base injection of the drug. In fact in the gray matter the levels of morphine at 8 hr in the tolerant dogs were 52% lower than those obtained in the nontolerant dogs. The levels of morphine in the white matter were generally lower than gray matter in either nontolerant or tolerant dogs. A comparison of the levels of free morphine in tolerant and nontolerant dogs after consecutive injection (2 mg/kg as free base) indicated higher levels of drug at 35 min, similar at 4 hr and lower at 8 hr for the tolerant dog in comparison to the nontolerant dog. Generally the subcortical areas of the CNS that were primarily gray matter contained higher levels of the drug than those which consisted of white matter. The peak concentration in the subcortical areas was as in the cerebral cortex at 4 hr with little difference observed between 35 min and 4 hr. The levels of drug were similar to those in the cerebral cortex at each time interval and usually lower in the tolerant dogs after a single injection of the drug.

Studies in the dog also were initiated to ascertain whether nalorphine would alter the disposition of labeled morphine in the CNS of the nontolerant[61] or tolerant dog.[62] It appeared that the levels of morphine in the CNS increased by about 40% if nalorphine was administered after morphine in the nontolerant dog. If the drugs (nalorphine and morphine) were administered together there was no change, and pretreatment with nalorphine resulted in an increase or no change in the CNS levels of the drug. In the

TABLE I
Disposition of Narcotic Analgesics and Antagonists
in the Central Nervous System

Drug	Species	Dose (mg/kg)	Route[a]	Peak Brain Concentration (μg/g)	Ref.
Morphine	monkey	30	s.c.	5 at 90 min and 4 hr	(58)
	dog	30	s.c.	5 at 90 min	(21)
		30	s.c.	4.6 (4.1–5.1) at 4 hr	(59)
		2	s.c.	0.230 at 4 hr	(60–62)
	rabbit	20	i.p.	1.8 at 1 hr	(63)
	guinea pig	10	s.c.	0.080	(17, 74)
	rat	150	s.c.	5 at 90 min and 4 hr	(21)
		5	s.c.	0.26 at 1 hr	(43)
		2	s.c./or i.p.p.	0.072 at 1 hr	(64)
		75	i.v.	9.3 at 1 hr	(65)
		500	i.p.	12.9 at 18 min	(66)
		200	s.c.	1.4–4.3 at 15 min	(67)
		500	i.p.	14 at 18 min	(67)
		50	i.v.	3.4 at 15 min	(67)
		600–780	i.p.	20–40 at 20–60 min	(68)
		100	i.v.[b]	26 at 15 min	(69)
		100	i.v.	7.1	(70)
		20	i.p.	0.81 at 30 min	(71)
		500	i.p.	13–14 at 14 min	(72)
		5	s.c.	0.179 at 30 min	(73)
		10	i.p.	0.183 at 30 min	(73)
Dihydro-morphine	dog	2	s.c.	0.155 at 4 hr	(76)
	cat	2	s.c.	0.368 at 2 hr	(75)
		30	s.c.	7.65 at 2 hr	(75)
	rat	1	s.c.	0.020 at 1 hr	(78)
		2	s.c.	0.055 at 1 hr	(78)
		4	s.c.	0.105 at 1 hr	(78)
		100	s.c.	3–5 at 1 hr	(92)
		5	s.c.	0.109 at 1 hr	(76)
		27	i.m.	1 at 1 hr	(77)
	mouse	20	s.c.	0.400 at 1 hr	(76)
Normorphine	rat	200	s.c.	from trace to 2 at 15 min	(67)
		500	i.p.	9.8 at 18 min	(67)
		50	i.v.	3.6 at 15 min	(67)
		28	i.p.	0.72 at 30 min	(71)
		51	i.p.	1.3 at 30 min	(71)
Heroin	rat	75	i.p.p.	0.9 at 15 min	(50)
	mouse	37.5	i.v.	8.2 at 2.5 min	(49)
Codeine	rat	25	s.c.	8.5 at 1 hr	(43)
		30	i.p.	7.9 at 30 min	(73)
		60	s.c.	9.0 at 30 min	(73)
		2	s.c.	0.68 at 40 min	(79)

TABLE I (Continued)

Drug	Species	Dose (mg/kg)	Route[a]	Peak Brain Concentration (μg/g)	Ref.
Meperidine	dog	20	i.v.	11.3 at 40 min	(51)
	rat	100	s.c.	17 at 2 hr	(86)
Levorphanol	monkey	2	s.c.	1.9 at 1 hr	(82)
	dog	2	s.c.	2.5 at 30 min	(80)
	rat	2	s.c.	0.142 at 90 to 120 min	(85)
Methadone	rat	3	s.c.	0.68 at 30 min	(43)
		20	i.p.	4 at 1 hr	(87)
		15	s.c.	17 at 3 hr	(88)
		10	s.c.	22 at 1 hr	(90)
Nalorphine	dog	30	s.c.	14 at 90 min	(63)
		2	s.c.	1.12–1.05 at 30 to 60 min	(81)
	rat	150	s.c.	7 at 90 min	(25)
Cyclazocine	dog	1.25	s.c.	1.21 at 1 hr	(30)
Pentazocine	cat	1	i.m.	1.52 at 30 min	(83, 84, 103)

[a] s.c. = subcutaneous; i.p. = intraperitoneal; i.p.p. = intrapopliteal; i.v. = intravenous; and i.m. = intramuscular.
[b] Eviscerated.

tolerant dogs a different picture emerged. A statistically significant decrease in the CNS concentration of morphine occurred after nalorphine administration (42–56%) at 65 min. Nalorphine also caused a reduction of 5–73% in morphine levels of the heart, lung, liver, and kidney in these dogs. It is obvious that a different effect was obtained with nalorphine in the tolerant and nontolerant dogs. The question is whether the mechanism of antagonism differs with different physiological states or simply that the biochemical and physiological changes induced by morphine are of sufficient magnitude to be reflected in alteral disposition of the drug and yet not reflect receptor site activity.

Siminoff and Saunders[63] studied the disposition of morphine in the rabbit brain of both nontolerant and tolerant animals. Maximal levels were achieved at 1 hr after giving the drug in both nontolerant and tolerant rabbits (1.8–1.9 μg/g) and then slowly declined to a level of 0.35 μg/g for the tolerant rabbit and 0.20 μg/g for the nontolerant animal at 8 hr.

The rat, of course, has been used extensively in studies designed to ascertain the disposition of morphine (Table I). It is obvious, however, that relatively large discrepancies exist with respect to concentrations of the drug at a given time interval. Part of the variation is due to dose, route of administration, and time of sacrifice. Some of the difficulty also may be related to specificity, sensitivity, and reliability of the method or technique utilized to assay for the drug in biological material. Although all the reference data available to the author is recorded in Table I, the discussion concerning the rat will be confined to the data that appears to satisfy the requirements of sensitivity, specificity, chemical or radioactive reliability, and the usage of pharmacological dose levels of the drug. In this regard the data of Miller and Elliott[43] show a peak concentration of ^{14}C-morphine at 1 hr in the cerebral hemisphere of 0.26 μg/g, which then slowly fell to a level of 0.18 μg/g at 150 min. Other areas of the rat brain showed higher levels of drug at 1 hr that is, spinal cord (0.63 μg/g), hypothalamic area (0.58 μg/g), and cerebellum (0.40 μg/g). The concentration of morphine declined slowly in these areas and was still higher than the observed values in cerebral hemisphere at 150 min. It appeared from these studies that there was a relatively good correlation between analgesia and peak concentration of morphine as well as a correlation of the duration of analgesia with brain levels of drug.

A similar study was initiated by Johannesson and Woods[73] using ^{14}C-morphine. These investigators found a mean concentration in the brain of rats of 0.179 μg/g after 5 mg/kg subcutaneously and 0.183 μg/g after 10 mg/kg interperitoneally at 30 min following administration of the drug. It appears from this data that doubling the initial dose has very little effect on the final brain drug levels or that the route of administration played a significant role with regard to disposition during a short time interval. Tolerance development to morphine did not seriously alter the level of ^{14}C-morphine in the brain of the tolerant rat (0.170 μg/g at 30 min).[73] Analgesia was studied in these animals in both the tolerant and nontolerant states as well as in comparison to codeine. As expected morphine was significantly more potent than codeine, but a difference in total potency was apparently somewhat dependent upon the route of administration.

In the guinea pig the maximal concentration of ^{3}H-morphine was observed at 1 hr following a 10 mg/kg subcutaueons injection.[17,74] There was no difference between the levels of drug in the nontolerant and tolerant guinea pigs in the whole brain nor in the subcellular fractions of the brain. About 68% of the administered dose was observed in the supernatant fraction of both the tolerant and nontolerant guinea pigs. A similar although somewhat lower level of drug was observed in the liver supernatant fraction (48–52%) of these animals. Nalorphine when administered to these guinea pigs did not alter the subcellular localization in either the nontolerant or tolerant guinea pig. Utilizing the guinea pig an attempt was made to correlate analgesia and drug levels in the brain of nontolerant and tolerant guinea pigs.[17] The data clearly indicated that a good correlation did not exist. In the tolerant animals there was apparently sufficient morphine in the brain to elicit an analgesic response, but upon testing for analgesia a significant reduction in effect occurred. It is quite obvious that an alteration probably at the cellular level of organization must have resulted following chronic drug administration.

The cat is often considered to act in a rather unique manner to narcotic analgesics. Whereas most animals, that is, dogs, rats, and guinea pigs, show overt depression, the cat appears to be stimulated. Chernov and Woods[75] were interested in ascertaining whether the physiological disposition of morphine-^{14}C was in any way related to the pharmacological action of the drug in this species. Following a 2 mg/kg (as free base) subcutaneous injection these investigators sacrificed cats at intervals of 0.5, 1, 2, 4, and 8 hr. Concentrations of 0.105, 0.273, 0.368, 0.348, and 0.130 μg/g were obtained at the stated time intervals. The peak level of drug was at the 2 hr time period after drug administration. Administration of 30 mg/kg of ^{14}C-morphine to the cat and sacrificing at 0.5, 1, 2, 4, and 8 hr provided drug levels in the cerebral cortex gray matter of 3.54, 4.32, 7.65, 5.85, and 2.94 μg/g, respectively. Except for the obviously higher levels in the CNS the pattern of disposition was not essentially different from that observed after 2 mg/kg. The level of drug in the white matter was considerably less than that observed in the gray matter at the early time intervals and the same or higher at the 8 hr time interval. A comparison of the results in the CNS of the cat with those in the dog apparently did not account for the difference in the response following the narcotic drug morphine.

B. Dihydromorphine

The pattern of disposition in the CNS of the dog for dihydromorphine in comparison to morphine was very similar; however, the absolute levels in the various areas were higher for morphine at each time interval in comparison to dihydromorphine.[76] The difference may be related to the inability of dihydromorphine to penetrate the CNS as effectively as morphine. Although the levels in the CNS of dihydromorphine were one third to one-half of those of morphine, the drug is considered equipotent to morphine as an analgesic.[91]

The data obtained in the rat with dihydromorphine indicated that peak concentrations of the drug were achieved at 1 hr after administration[76,78] and that the drug level within the brain increased with increasing dosage, that is, 0.020 μg/g at 1 mg/kg subcutaneously vs 1 μg/g at 27 mg/kg.[77] This increase is somewhat higher than might be proportionally expected in this species. However, at 5 mg/kg the brain level of dihydromorphine was 0.109 μg/g,[76] which is more like the expected level. The discrepancy may be due to differences in the techniques and methods used by the investigators in these studies.

Van Praag and Simon[92] studied the intracellular localization of ^{3}H-dihydromorphine in rat brain and observed that more than 70% of the administered drug was present in the soluble supernatant fraction and between 18 and 20% in the nuclear fraction. The administration of nalorphine to the rat did not alter the tissue levels nor the intracellular distribution of the drug. The data obtained with dihydromorphine was quite similar to that obtained with morphine in the guinea pig.[74]

In the mouse it appeared that the peak concentration of dihydromorphine was achieved at 1 hr after administration of the drug.[76] Following a 20 mg/kg subcutaneous injection of ^{3}H-dihydromorphine, 0.40 μg/g of the free drug and 0.130 μg/g of the conjugate were observed in the brain of the mouse at 1 hr. In comparison to other tissues in the mouse the levels of drug in the brain were 4–63 times lower. Thus, as in other species, a relatively small fraction of the administered dose reached the CNS site of action.

The data obtained with dihydromorphine did not in essence provide any outstanding information over and above that achieved with morphine in regard to the mechanisms of action of these drugs.

C. Normorphine

The N-demethylation of morphine provides the analog normorphine. This drug gained prominence primarily because of the analgesic hypothesis of Beckett, Casy, and Harper[93] and the hypothesis of Axelrod[36] relating N-demethylation to tolerance development. In both cases there is sufficient experimental evidence to seriously doubt the usefulness of either hypothesis.[1–3]

The existence of normorphine as a metabolite of morphine was first demonstrated in the rat by Misra, Mulé, and Woods.[94] Subsequently Milthers[70] reported the *in vivo* transformation of morphine or nalorphine to normorphine in the brain of rats. Misra, Jacoby, and Woods[95] reported the physiological disposition of ^{3}H-normorphine in the dog and monkey. These authors[95] reported following a 2 mg/kg (as free base) subcutaneous injection in either the dog or the monkey a peak plasma level of about 1.3 μg/ml at 15–30 min. Conjugate levels of drug peaked at 1–3 hr in both the dog and monkey. Urinary excretion studies demonstrated that 34–40% was excreted as free normorphine and 56–60% as conjugated drug. Fecal excretion in the unrestrained monkey accounted for 22–39% of the drug. This would indicate that better than 100% of the drug was accounted for in the unrestrained monkey.

Johannesson and Milthers[67] measured normorphine in the brain of rats after subcutaneous, intraperitoneal, and intravenous administration of normorphine. Following 200 mg/kg subcutaneously normorphine was detected only in two out of ten rats at a level of 1–2 μg/g of brain. At 15 min after the injection of normorphine intraperitoneally (500 mg/kg) and sacrificing the rats at about 18 min a concentration of 9.8 μg/g of the brain was observed. Administration of 50 mg/kg of normorphine to rats intravenously provided a mean concentration of 3.6 μg/g of brain at 15 min after drug.

Studies by Johannesson and Schou[71] indicated brain levels of normorphine of 0.72 μg/g after 28 mg/kg intraperitoneally and 1.3 μg/g after 51 mg/kg. The administration of normorphine at 28 and 51 mg/kg produced analgesia in 37.5 and 70% of the rats, respectively. These authors state that the concentration of normorphine in the brain was four to five times higher than the concentration of morphine when the drugs were administered in identically analgesic doses.

D. Heroin

The disposition of heroin unmetabolized within the CNS is more of academic interest than of pragmatic importance since this drug is almost completely biotransformed to morphine and 6-monoacetylmorphine *in vivo*.[1,2] Way, Young, and Kemp[50] quite clearly show that after a 75-mg/kg injection of heroin in the rat a peak level of 0.9 μg/g of heroin occurs in the brain at 15 min and by 60 min the drug is barely detectable. At the same time the level of 6-MAM at 15 min was 6.3 μg/g and fell to a concentration of 1.9 μg/g at 2 hr. Morphine derived from heroin reached a peak of about 1.8 μg/g at 15 min and remained relatively stable through 2 hr where a concentration of about 1.3 μg/g was observed. Thus it would appear that the pharmacological action of heroin may be attributed primarily to morphine and 6-MAM.

In the mouse a similar study was conducted[49] so that following a 37.5 mg/kg injection of heroin intravenously and sacrificing at 2.5 min a level of 8.2 μg/g of heroin was obtained. However, at 7.5 min heroin was barely detectable in the brain. At the 2.5-min sampling period the concentration of 6-MAM was about 23 μg/g and for morphine was less than 2 μg/g. However, the morphine concentration reached a maximum at 15 min (5 μg/g) and was about the same at 30 min. The 6-MAM level dropped rather rapidly so that at 30 min the concentration was the same as for morphine. It appears that heroin penetrates the CNS mostly as 6-MAM and to a lesser degree as morphine. However, it is felt that the persistence of the pharmacological effect of heroin is due to the presence of morphine.

E. Codeine

Miller and Elliott[43] obtained concentrations of between 7 and 9 μg/g for various areas within the CNS of rats after a 25 mg/kg subcutaneous injection of ^{14}C-codeine at 60 min. The drug levels dropped rapidly to a concentration between 1 and 3 μg/g at 150 min. Analgesia was prominent for 30–120 min with recovery at 150 min after codeine administration.

Following the administration of 30 mg/kg intraperitoneally to rats Johannesson and Woods[73] found 7.9 μg/g in the brain at 30 min. At a dose of 60 mg/kg subcutaneously in rats the brain concentration was 9.0 μg/g at 30 min following drug administration. In codeine tolerant rats mean values of 11.5 μg/g were obtained at 30 min after a 30 mg/kg intraperitoneal injection of codeine. Following a 60 mg/kg subcutaneous injection of codeine in codeine tolerant rats, levels of 7.0 μg/g were obtained at 30 min. It is relatively odd that the value at 60 mg/kg should be less than at 30 mg/kg in these tolerant rats. However, the authors claim no statistical difference when the tolerant data was compared to the nontolerant data. Thus tolerance development to codeine did not appear to be related to an alteration of drug levels in the brain. Morphine biotransformed from codeine could not be demonstrated in brain tissue of the rats.[73] It is possible, however, that relatively small quantities of morphine derived from codeine could not be qualitatively estimated with the methods used in this study.

A recent study utilizing N-^{14}C-methyl-codeine[79] showed that following a 2 mg/kg subcutaneous injection of codeine a peak concentration of 0.68

μg/g was achieved 40 min after administration of the drug. The level of free codeine then dropped rapidly to provide a concentration of about 0.08 at 120 min in the rat. Free codeine in plasma peaked at a concentration of about 0.35 μg/ml at 15 min and rapidly fell to a level of 0.050 μg/ml at 2 hr. Morphine (free) derived from codeine in the plasma appeared to remain relatively stable at about 0.020 μg/ml through 2 hr. Conjugate morphine was considerably higher in the plasma (0.090 μg/ml) and appeared to fall to a level of 0.040 μg/ml at 2 hr. These authors,[79] utilizing sensitive radioactive techniques, could not identify free nor conjugate morphine biotransformed from codeine in the rat brain. Conjugated codeine could not be detected in the brain or in the plasma at any time interval.

F. Meperidine

Although meperidine has gained wide usage as an analgesic, there is very little information concerning the disposition of this drug in the CNS. Way and colleagues[86] have obtained data in the rat that indicate that after a 100 mg/kg subcutaneous injection of meperidine about 17 μg/g of the drug was observed in the brain 2 hr after drug administration. Plotnikoff, Elliott, and Way[96] conducted disposition studies in the rat using ^{14}C-labeled meperidine and reported a 0.1 % of the radioactivity in the cerebrum after a 125 mg/kg subcutaneous injection of meperidine. Assuming a body weight of 100 g and a brain weight of about 1 g, a 0.1 % level of the drug would amount to about 12 μg/g of the brain.

Burns et al.[51] investigated the fate of meperidine in man and also the disposition of the drug in the dog. Following a 20 mg/kg intravenous injection of meperidine the animals were sacrificed at 40 min. The highest level of drug appeared in the lung at 13 μg/g and the lowest in the fat (lumbodorsal). A mean meperidine level of 11.3 μg/g was obtained in the brain of the dogs.

The metabolic pathways are fairly well understood for meperidine and consist of normeperidine, meperidinic acid, normeperidinic acid, bound meperidinic acid, and bound normeperidine.[51] It is unfortunate, however, that no data is available on the existence of the metabolites within the CNS, since they may be involved in the total pharmacological effect of this drug within the mammalian organism.

G. Levorphanol

A significant amount of data is available concerning physiological disposition of levorphanol primarily because of the availability of ^{14}C-labeled levorphanol[97] and excellent methods for the estimation of this drug in biological material. Woods et al.[97] found that free levorphanol had a biological half-life in plasma of 75–90 min and conjugated levorphanol was 3 hr in both the dog and the monkey. Approximately 2.5 % of levorphanol administered to the monkey appeared in the urine as free drug, whereas 35 % appeared as the conjugated. In the dog 4.4 and 42 % appeared in the urine as free and conjugated levorphanol. Less than 0.1 % of the dose appeared in the feces of either species. *N*-demethylation was observed by the studies with $^{14}CO_2$, which ranged from 17 to 24% in the monkey, 1 to 1.5% in the dog, and 4.3 to 5.3% in the rat.

Wuepper and Woods[82] investigated the disposition of ^{14}C-levorphanol in the brain of the monkey after a 2 mg/kg (as free base) subcutaneous injection of the drug. At 1 hr values of 1.9 μg/g for gray cerebral cortex and 1.4 μg/g for white cerebral cortex were obtained. Apparently no statistical difference existed between the disposition of levorphanol in the CNS of the nontolerant and tolerant monkey.

In the dog Wuepper, Yeh, and Woods[80] found levels of 2.0 μg/g and 1.3 μg/g in cerebral cortical gray and white matter, respectively, at 30 min after a 2 mg/kg (free base) subcutaneous injection of ^{14}C-levorphanol. The simultaneous administration of 2 mg/kg nalorphine or levallorphan with ^{14}C-levorphanol resulted in the reduction of the brain levels of levorphanol in comparison to controls (1.37–1.34 μg/g for gray matter and 0.68–0.65 μg/g for white matter). Although the narcotic antagonists significantly reduced the brain levels of levorphanol, they had no effect on the plasma levels of levorphanol.

Heng and Woods[85] obtained concentrations of 0.142 μg/g in the brain of rats following a 2 mg/kg (free base) subcutaneous injection of ^{14}C-levorphanol at 90–120 min after drug administration. The administration of 1 mg/kg of nalorphine to these rats apparently had no effect on the CNS levels of levorphanol. This is, of course, in contrast to that observed in the dog after administration of the antagonist.

H. Methadone

Methadone, a good narcotic analgesic, gained usage as a substitute for opiates during gradual withdrawal from narcotic drugs. Currently, however, much interest centers on this drug because of its use in so-called "methadone maintenance programs."[98,99] The utilization of an old drug in a new manner also has stimulated new interest in the disposition and metabolism of this drug. Recently the biotransformation of methadone in man was studied[100] and the metabolite 2-ethyl-1,5-dimethyl-3,3-diphenyl-1-pyrrodine was identified along with the unchanged drug in the urine.

Disposition studies of *l*-methadone in the brain of animals appear only to be available in the rat. In this regard Miller and Elliott[43] utilizing ^{14}C-methadone observed a maximal level of 0.68 μg/g at 30 min after a 3 mg/kg subcutaneous injection of the drug. At this time (30 min) a concentration of 0.85 μg/g was observed in the spinal cord and about the same concentration in the medulla. The levels of methadone slowly decreased to between 0.20 and 0.40 μg/g at 180 min after drug administration. The highest level of methadone in peripheral tissue was observed for the adrenal gland at about 8 μg/g 30 min after drug administration.

Way and colleagues[87] obtained values of 4 μg/g for methadone in brain at 1 hr after a 20 mg/kg interperitonal injection of the drug. It is quite obvious that the data obtained following 10 mg/kg[90] and 15 mg/kg[88] provided levels of methadone in the brain that are at variance with those reported by Miller and Elliott[43] and Way et al.[87] It would appear that the methods and techniques used by Richards et al.[88] and Elliott et al.[90] were simply not specific for methadone and thus provided inordinately high values for the drug in the brain. It is obvious that further studies with this drug in higher animal species are needed, especially with regard to metabolism and disposition within the CNS.

I. Nalorphine

Nalorphine (N-allylnormorphine, nalline) is considered to be an agonist-antagonist drug. The pharmacology of this drug has been adequately reviewed by Woods,[116] Martin,[101] and Archer and Harris.[102]

Woods[59] in an early study using nonradioactive material obtained concentrations of 14 μg/g at 90 min after a 30 mg/kg subcutaneous injection of nalorphine in dogs. The levels dropped rapidly to 2.6 μg/g at 4 hr (72% decrease) and at 8 hr were barely detectable at 0.20 μg/g. Brain-to-plasma ratios of 2.8 and 0.52 were obtained at 90 min and 5 hr, respectively. Studies initiated by Woods and Muehlenbeck[25] in the rat after 150 mg/kg of nalorphine subcutaneously provided values of 7 μg/g at 90 min for the free drug and 5 μg/g for the conjugate. At 4 hr less than 5 μg/g of brain were reported for free drug and 0 concentration for the conjugate.

The preparation of ^{3}H-nalorphine and the development of sensitive and specific techniques for the analysis of nalorphine allowed Hug and Woods[81] to critically study the disposition of nalorphine in dogs. Peak levels of nalorphine in the cerebral cortex gray matter were obtained at 30 and 60 min after 2 mg/kg (free base) subcutaneous injection of nalorphine. These were 1.12 and 1.05 μg/g, respectively. The drug level fell rapidly so that at 2 hr values of 0.062 μg/g were observed. In cerebral cortex white matter lower levels of drug were obtained but peak concentration times corresponded to that observed for gray matter. Thus values of 0.80 μg/g and 0.99 μg/g were reported for the time intervals of 30 and 60 min. At 2 hr a concentration of 0.131 μg/g was obtained. At this time the gray/white ratio was 0.47. The concentration of nalorphine in subcortical areas of the dog brain were similar to those obtained in the cerebral cortex with respect to time-course, and the drug levels, as noted previously, were higher in the more cellular (gray) areas of the CNS than the white (myelinated) areas.

Hug and Woods[81] reported urinary recoveries in the dog of 2–6% free nalorphine 79–84% as conjugated (glucuronide), which accounted for 81–90% of the administered dose. Chromatographic studies indicated that only free and glucuronide conjugated nalorphine was present in the urine of dogs.

J. Cyclazocine

Cyclazocine (2-cyclapropylmethyl-2-hydroxy-5,9-dimethyl-6,7-benzomorphan), a member of the benzomorphan class of agonist-antagonists, was quite thoroughly studied[30,31] because of the potential use of cyclazocine in the treatment of narcotic addicts, its potency in relationship to morphine, and the peculiar abstinence syndrome of this agonist-antagonist drug. The availability of ^{3}H-cyclazocine and the development of specific techniques for its determination provided the basis for a thorough study of this drug.[30]

Maximal concentrations of cyclazocine 1.215 μg/g for cerebral cortical gray matter and 0.895 μg/g in cerebral cortical white matter were obtained 1 hr after a 1.25-mg/kg subcutaneous injection of H^3-cyclazocine in dogs. At 6 hr the levels were 0.036 and 0.056 μg/g for cortex gray and white matter, respectively. Similar studies in the cyclazocine tolerant dog provided values of 1.259 and 1.108 μg/g in gray and white matter, respectively, at 1 hr and 0.117 and 0.100 μg/g in cerebral gray and white matter, respectively, at 6 hr. Subcortical levels of the drug appeared to correlate quite well with cortical concentrations in both the nontolerant and tolerant animals. It did not appear that a single subcortical area contained extraordinary levels of the drug. No drug was found in the CNS of a 24-hr abstinent dog. In peripheral tissues the values for cyclazocine were high for lung, kidney, spleen, liver, and adrenals. The plasma peak concentration for free cyclazocine occurred at 30 min (0.156 μg/ml) and was barely detectable at 6 hr (0.006 μg/ml) in the nontolerant and tolerant dog. The mean percentage recovery of cyclazocine from urine for the nontolerant dog was 43.7%; for the tolerant dog, 58.5%; and for the abstinent dog, 40.7%.

A metabolic study concerning the fate of ^{3}H-cyclazocine[31] provided evidence for the existence of norcyclazocine and norcyclazocine glucuronide in addition to the free drug and the glucuronide conjugate of cyclazocine. In the urine of the nontolerant, tolerant, and abstinent dogs 2.3–2.7% of ^{3}H-cyclazocine was recovered as free norcyclazocine and an equal amount as the conjugate. In the feces of the dogs 1.5–2.4% was obtained as free norcyclazocine and 0.02–0.7% as conjugated. The metabolite norcyclazocine was not identified in the brain of the dog.

It was concluded from the studies on cyclazocine that the disposition of the drug provided some insight into the potency of the drug but did not help elucidate the latency of abstinence. The metabolite norcyclazocine did not appear to provide any pharmacological activity that might be associated with the total narcotic effect of cyclazocine.

K. Pentazocine

Pentazocine [2′-hydroxy-5,9-dimethyl-2-(3,3-dimethylallyl)-6,7-benzo-morphan] is considered a weak analgesic antagonist and now utilized as a nonnarcotic analgesic (Talwin) with an analgesic potency of approximately one-fourth that of morphine.[104]

Ferrari[83,84,103] investigated the disposition of ^{3}H-pentazocine in the cat after a 1-mg/kg (free base) intramuscular injection and reported concentrations of 1.520 μg/g at 30 min in sensory-motor gray matter. At 2 hr this concentration dropped to 0.616 μg/g. In cerebral white matter concentrations of 1.30 and 0.540 μg/g were obtained at 30 and 120 min in cerebral white matter. There did not appear to be a localization of the drug in the various cortical or subcortical areas of the CNS. An analysis of the whole blood levels of ^{3}H-pentazocine provided values of 0.462, 0.428, and 0.315 μg/ml at 30, 60, and 120 min, respectively, after drug administration.

Peripheral disposition of pentazocine[103] indicated that bile contained the highest level of drug of any tissue or fluid analyzed. However, in terms of the percentage of the dose administered, it appeared as though at 2 hr the GI tract and liver represented relatively high percentages of the absorbed pentazocine, amounting to 27 and 23%, respectively. At 2 hr the bile contained 6% of the absorbed dose. Other tissues and fluids such as urine, blood, kidney, adrenal, spleen, pancreas, heart, lung, thyroid, and brain contained less than 5% of the drug. Muscle, however, contained 13% pentazocine at the 2-hr time interval.

Studies[102] on the metabolism and excretion of pentazocine in rats indicated that 54% of the radioactivity was recovered in urine and 14% in the feces over the 24-hr period. At the end of 96 hr 86% of the radioactive material was recovered from the urine and feces.

IV. TOLERANCE AND PHYSICAL DEPENDENCE

Tolerance to the narcotic drugs and the induction of physical dependence following chronic usage of these drugs represent the basic limitations associated with the usage of narcotic analgesics medically as well as the principal intellectually exciting reason for intensive research with these drugs.

Tolerance and physical dependence are mutually related. However, tolerance to nonnarcotic drugs may exist (that is, hallucinogens, sympathomimetic amines, and phenothiazines) without induction of physical dependence. The converse, however, is not true; that is, if a drug induces physical dependence, then tolerance to this drug occurs concomitantly.

Many postulates[1,2,105−114] have been presented over the years to explain these phenomena. Let it suffice to say that as of the present no theory fully explains or withstands experimental verification of the mechanisms underlying tolerance or physical dependence. Furthermore, it is the purpose of the author to fully explore only the relationship of the physiological disposition or biotransformation of narcotic drugs in relationship to tolerance and physical dependence.

Prior to the advent of sensitive methods to detect nanogram levels of the narcotic drugs within the central nervous system (see Table I) investigators had to be content with a relatively insensitive analysis and could not seriously correlate their findings with the mechanism of narcotic analgesic action. Quite simply one must examine events in the CNS hopefully at the CNS receptor site of activity to gain reliable insight to the mechanism of narcotic drug activity. A concentrated effort was initiated at the University of Michigan at Ann Arbor and continued at the Addiction Research Center at Lexington to ascertain whether an alteration in the disposition or metabolism was involved in the development of tolerance. The results of these studies over a period of years may be summarized as follows:[115] (1) levels of ^{14}C-morphine after a single 2 mg/kg (free base) dose were lower in the CNS of tolerant dogs in comparison to nontolerant dogs; (2) following consecutive 2 mg/kg injections of ^{14}C-morphine the levels of drug in the CNS of tolerant dogs were higher at 35 min, similar at 4 hr, and lower at 8 hr in comparison to nontolerant dogs; (3) no single anatomical area within the CNS appeared to concentrate ^{14}C-morphine, but the gray matter at least at early time intervals contained twice as much drug as the white matter; (4) nalorphine antagonism in the nontolerant dog provided higher levels of ^{14}C-morphine in the CNS or had no effect as compared to control data; (5) nalorphine precipitation of abstinence in the tolerant dog resulted in a decrease in the CNS levels of morphine at 65 min and no statistical change at other time intervals after drug; (6) intracellular studies with ^{3}H-morphine indicated a localization of the drug in the soluble supernatant fraction (68 %) of brain and that neither tolerance or nalorphine antagonism altered the intracellular localization of ^{3}H-morphine; (7) extensive paper chromatographic studies on CNS extracts of nontolerant, tolerant, and nalorphine-antagonized dogs provided no evidence for an N-^{14}C-methyl-labeled metabolite of morphine; and (8) it was concluded that neither disposition or biotransformation of morphine provided any clear insight into the mechanism of tolerance development. Studies conducted with cyclazocine[30] and levorphanol[82] in the CNS essentially support the conclusions derived from the investigations with morphine.

It would appear that unless disposition studies can be conducted with the intracellular receptor sites within the CNS further endeavors along these lines will not yield productive results. In all probability the phenomena of tolerance and physical dependence is related to changes in the biochemical machinery of the cell and not to changes in the drug bathing the neuronal cell structure.

V. SUMMARY

The primary purpose of this review is to focus attention on the disposition and metabolism of narcotic analgesics and antagonists within the CNS. In this regard it was necessary to consider the dynamics of drug disposition that included (1) absorption, (2) disposition, (3) biotransformation, (4) locus of action, and (5) excretion.

Essentially the rate of absorption may be determined by the equation

$$\log D = \log D_0 - \frac{k_a t}{2.30}$$

where D is the concentration of the drug remaining at the site of absorption at time t and k_a is the rate constant for absorption and D_0 is the dose of the drug administered.

The disposition of the drug in the CNS is dependent upon certain chemical and physical characteristics such as (1) ionization, (2) degree of protein binding, (3) lipid solubility, and (4) circulation flow rate characteristics of the region in which the drug is distributed. Thus drugs that are mostly un-ionized, free, and highly lipid soluble will readily penetrate the CNS.

The pathways whereby narcotic analgesics and antagonists are primarily metabolized are (1) conjugation (glucuronide, arylsulfate), (2) dealkylation, (3) hydrolysis, and (4) oxidation and reduction. Theoretically morphine may be metabolized by each of these routes, but to date reliable evidence exists only for conjugation and dealkylation.

The narcotic drugs may be eliminated from the organism through the lungs, saliva, perspiration, bile, intestines, and kidney. The principal pathway is renal, with the biliary and intestinal routes playing an important role.

The physiological disposition primarily in the brain was presented for morphine, dihydromorphine, normorphine, heroin, codeine, meperidine, levorphanol, methadone, nalorphine, cyclazocine, and pentazocine.

Within a confined framework certain generalities may be made concerning the disposition of narcotic drugs in the CNS. These are (1) less than 0.1% of the administered dose generally gained access to the CNS, (2) no extensive localization at a given anatomical site occurred, (3) species variability for the CNS disposition of the drug was not remarkable when reliably derived data were compared, (4) plasma-to-brain ratios were significantly greater than one at early time intervals, (5) a good correlation between CNS disposition and pharmacological effect did not necessarily exist, (6) the concentration and time course within the CNS seemed to be related to the chemical and physical characteristics of the narcotic drug, (7) tolerance and physical dependence did not appear to be the result of an altered disposition or biotransformation of narcotic drugs, and (8) disposition studies have not provided a clear picture of the narcotic agonist and antagonist mechanism of action.

In all probability the mechanism of narcotic agonist and antagonist activity as well as the phenomena of tolerance and physical dependence are related to changes in the biochemical machinery of the neuronal cells and not to changes in the drug molecule.

VI. REFERENCES

1. E. L. Way and T. K. Adler, *Pharmacol. Rev. 12*, 383–445 (1960).
2. E. L. Way and T. K. Adler, "The Biological Disposition of Morphine and its Surrogates," World Health Organization, Geneva, pp. 1–114 (1962).
3. E. L. Way, in The Addictive States (A. Wikler, ed.), Vol. 46, pp. 13–31, Williams & Wilkins, Baltimore (1968).
4. E. Nelson, *J. Pharm. Sci., 50*, 181–192 (1961).
5. B. B. Brodie and C. A. M. Hogben, *J. Pharm. Pharmacol. 9*, 345–380 (1957).
6. J. G. Wagner, *J. Pharm. Sci. 56*, 489–494 (1967).
7. J. H. Fincher, *J. Pharm. Sci. 57*, 1825–1835 (1968).
8. A. Goldstein, L. Arnow, and S. M. Kalman, "Principles of Drug Action," Hoeber Medical Division, Harper & Row, New York (1968).

9. C. G. Farmilo, P. M. Oestreicher, and L. Levi, *Bull. Narcotics 6*, 7–19 (1954).
10. L. B. Mellett, unpublished observations.
11. S. J. Kety, *in* "Handbook of Physiology" (J. Field, H. W. Magoon, and L. E. Hall, eds.), Vol. III, "Neurophysiology," Chapter 71, American Physiological Society, Washington, D.C. (1960).
12. B. B. Brodie, H. Kurr, and L. Shanker, *J. Pharmacol. Exp. Ther. 130*, 20–25 (1960).
13. L. J. Shanker, *Ann. Rev. Pharmacol. 1*, 29–44 (1961).
14. D. P. Rall and C. G. Zabrod, *Ann. Rev. Pharmacol. 2*, 109–128 (1962).
15. L. J. Roth and L. F. Barlow, *Science 134*, 22–31 (1961).
16. W. Feldberg, "A Pharmacological Approach to the Brain, from its Inner and Outer Surface," Williams & Wilkins, Baltimore (1963).
17. S. J. Mulé, *Arch Int. Pharmacodyn. 173*, 201–212 (1968).
18. R. T. Williams, "Detoxication Mechanisms," Wiley, New York (1959).
19. J. M. Fujimoto and E. L. Way, *Fed. Proc. 13*, 58 (1954).
20. R. A. Seibert, C. W. Williams, and R. A. Huggins, *Science 120*, 222 (1954).
21. L. A. Woods, *J. Pharmacol. Exp. Ther. 112*, 158–175 (1954).
22. J. L. Strominger, H. M. Kalckar, J. Axelrod, and E. S. Maxwell, *J. Amer. Chem. Soc. 76*, 6411–6412 (1954).
23. A. E. Takemori, *J. Pharmacol. Exp. Ther. 130*, 370–374 (1960).
24. T. K. Adler, *in* "Current Trends in Heterocyclic Chemistry" (A. Albert, G. M. Bodger, and C. W. Shappee, eds.), Butterworths, London, pp. 151–157 (1958).
25. L. A. Woods and H. E. Muehlenbeck, *J. Pharmacol. Exp. Ther. 120*, 52–57 (1957).
26. G. J. Mannering and L. S. Schanker, *J. Pharmacol. Exp. Ther. 124*, 296–304 (1958).
27. C. C. Hug, Jr., and L. B. Mellett, *J. Pharmacol. Exp. Ther. 149*, 446–453 (1965).
28. H. Fisher and J. P. Long, *J. Pharmacol. Exp. Ther. 107*, 241–246 (1953).
29. N. P. Plotnikoff, E. L. Way, and H. W. Elliott, *J. Pharmacol. Exp. Ther. 117*, 414–419 (1956).
30. S. J. Mulé and C. W. Gorodetzky, *J. Pharmacol. Exp. Ther. 154*, 632–645 (1966).
31. S. J. Mulé, T. H. Clements, and C. W. Gorodetzky, *J. Pharmacol. Exp. Ther. 160*, 387–396 (1968).
32. J. M. Fujimoto and V. B. Haarstad, *J. Pharmacol. Exp. Ther. 165*, 45–51 (1969).
33. J. M. Fujimoto, W. W. Watrous, and V. B. Haarstad, *Proc. Soc. Exp. Biol. Med. 130*, 546–549 (1969).
34. J. R. Gillette, *in* "Progress in Drug Research" (E. Tucker, ed.), Vol. 6, pp. 11–51, Birkhauser Verlag, Basel (1963).
35. A. H. Beckett, *J. Pharm. Pharmacol. 8*, 848–859 (1956).
36. J. Axelrod, *Science 124*, 263–264 (1956).
37. D. H. Clouet, "International Review of Neurobiology," Vol. II, pp. 99–128, Academic Press, New York (1968).
38. T. K. Adler, *Proc. Soc. Exp. Biol. Med. 73*, 401–404 (1950).
39. T. K. Adler, *J. Pharmacol. Exp. Ther. 103*, 337 (1951).
40. T. K. Adler, *J. Pharmacol. Exp. Ther. 106*, 371 (1952).
41. G. J. Mannering, A. C. Dixon, E. M. Baker, and T. Asami, *J. Pharmacol. Exp. Ther. 111*, 142–146 (1954).
42. T. K. Adler, *J. Pharmacol. Exp. Ther. 140*, 155–161 (1963).
43. J. W. Miller and H. W. Elliott, *J. Pharmacol. Exp. Ther. 113*, 283–291 (1955).
44. S. Y. Yeh and L. A. Woods, *J. Pharmacol. Exp. Ther. 166*, 86–95 (1969).
45. P. Paerregaard, *Acta Pharmacol. Toxicol. 14*, 394–399 (1958).
46. L. A. Woods, H. E. Muehlenbeck, and L. B. Mellett, *J. Pharmacol. Exp. Ther. 117*, 117–125 (1956).
47. T. K. Adler, J. M. Fujimoto, E. L. Way, and E. M. Baker, *J. Pharmacol. Exp. Ther. 114*, 251 (1955).
48. A. E. Takemori and G. J. Mannering, *J. Pharmacol. Exp. Ther. 123*, 171–179 (1958).

49. E. L. Way, J. W. Kemp, J. M. Young, and D. R. Grassetti, *J. Pharmacol. Exp. Ther.* *129*, 144–151 (1960).
50. E. L. Way, J. M. Young, and J. W. Kemp, *Bull. Narcotics 17*, 25–33 (1965).
51. J. J. Burns, B. L. Berger, P. A. Lief, A. Wollack, E. M. Papper, and B. B. Brodie, *J. Pharmacol. Exp. Ther.* *114*, 289–295 (1955).
52. F. Bernheim and M. Bernheim, *J. Pharmacol. Exp. Ther.* *85*, 74–77 (1945).
53. E. L. Way, R. Swanson, and A. Gimble, *J. Pharmacol. Exp. Ther.* *91*, 178–184 (1947).
54. W. P. Baker and L. A. Woods, *J. Pharmacol. Exp. Ther.* *120*, 371–374 (1957).
55. C. C. Hug, L. B. Mellett, and E. J. Cafruny, *J. Pharmacol. Exp. Ther.* *150*, 259–269 (1965).
56. C. H. March and H. W. Elliott, *Proc. Soc. Exp. Biol. Med.* *86*, 494–497 (1954).
57. J. Cochin, J. Haggart, L. A. Woods, and M. H. Seevers, *J. Pharmacol. Exp. Ther.* *111*, 74–83 (1954).
58. L. B. Mellett and L. A. Woods, *J. Pharmacol. Exp. Ther.* *116*, 77–83 (1956).
59. L. A. Woods, *J. Pharmacol. Exp. Ther.* *120*, 58–62 (1956).
60. S. J. Mulé and L. A. Woods, *J. Pharmacol. Exp. Ther.* *136*, 232–241 (1962).
61. S. J. Mulé, L. A. Woods, and L. B. Mellett, *J. Pharmacol. Exp. Ther.* *136*, 242–249 (1962).
62. S. J. Mulé, *J. Pharmacol. Exp. Ther.* *148*, 393–398 (1965).
63. R. Siminoff and P. R. Saunders, *J. Pharmacol. Exp. Ther.* *124*, 252–254 (1958).
64. T. K. Adler, H. W. Elliott, and R. George, *J. Pharmacol. Exp. Ther.* *120*, 475–487 (1957).
65. J. C. Szerb and D. H. McCurdy, *J. Pharmacol. Exp. Ther.* *118*, 446–450 (1956).
66. T. Johannesson, *Acta Pharmacol. Toxicol.* *19*, 286–292 (1962).
67. T. Johannesson and K. Milthers, *Acta Pharmacol. Toxicol.* *19*, 241–246 (1962).
68. T. Johannesson, *Acta Pharmacol. Toxicol.* *19*, 23–35 (1962).
69. K. Milthers, *Nature 195*, 607 (1962).
70. K. Milthers, *Acta Pharmacol. Toxicol.* *19*, 235–240 (1962).
71. T. Johannesson and J. Schou, *Acta Pharmacol. Toxicol.* *20*, 165–173 (1963).
72. T. Johannesson and K. Milthers, *Acta Pharmacol. Toxicol.* *20*, 80–89 (1963).
73. T. Johannesson and L. A. Woods, *Acta Pharmacol. Toxicol.* *21*, 381–396 (1964).
74. S. J. Mulé, C. M. Redman, and J. W. Flesher, *J. Pharmacol. Exp. Ther.* *157*, 459–471 (1967).
75. H. I. Chernov and L. A. Woods, *J. Pharmacol. Exp. Ther.* *149*, 146–155 (1965).
76. C. C. Hug, doctoral dissertation, University of Michigan, Ann Arbor, pp. 1–205 (1963).
77. C. F. Blane and H. Dobbs, *Brit. J. Pharmacol. Chemother.* *30*, 166–172 (1967).
78. J. H. Samer and L. A. Woods, *J. Pharmacol. Exp. Ther.* *148*, 176–184 (1965).
79. S. Y. Yeh and L. A. Woods, *J. Pharmacol. Exp. Ther.* *166*, 86–95 (1969).
80. K. D. Wuepper, S. Y. Yeh and L. A. Woods, *Proc. Soc. Exp. Biol. Med.* *124*, 1146–1149 (1967).
81. C. C. Hug and L. A. Woods, *J. Pharmacol. Exp. Ther.* *142*, 248–256 (1963).
82. K. D. Wuepper and L. A. Woods, *Fed. Proc. 19*, 272 (1960).
83. R. Ferrari, *Pharmacologist 7*, 148 (1965).
84. R. Ferrari and J. D. Connolly, *Pharmacologist 8*, 224 (1966).
85. J. Heng and L. A. Woods, *Fed. Proc. 17*, 376 (1968).
86. E. L. Way, A. I. Gimble, W. P. McKelway, H. Ross, Chen-Yo Sung, and H. Ellsworth, *J. Pharmacol. Exp. Ther.* *96*, 477–484 (1949).
87. E. L. Way, Chen-Yo Sung, and W. P. McKelway, *J. Pharmacol. Exp. Ther.* *97*, 222–228 (1949).
88. J. C. Richards, G. E. Boxer, and C. C. Smith, *J. Pharmacol. Exp. Ther.* *98*, 380–391 (1950).
89. L. A. Woods, J. Cochin, E. G. Fornfeld, and M. H. Seevers, *J. Pharmacol. Exp. Ther.* *111*, 64–73 (1954).

90. H. W. Elliott, F. N. H. Chang, I. A. Abdou, and H. H. Anderson, *J. Pharmacol. Exp. Ther. 95*, 494–501 (1949).
91. N. B. Eddy, H. Halbach, O. J. Brenden, *Bull. W. H. O. 14*, 353–402 (1956).
92. D. Van Praag and E. J. Simon, *Proc. Soc. Exp. Biol. Med. 122*, 6–11 (1966).
93. A. H. Beckett, A. F. Casy, and N. J. Harper, *J. Pharm. Pharmacol. 8*, 874–883 (1956).
94. A. L. Misra, S. J. Mulé, and L. A. Woods, *J. Pharmacol. Exp. Ther. 132*, 317–322 (1961).
95. A. L. Misra, H. I. Jacoby, and L. A. Woods, *J. Pharmacol. Exp. Ther. 132*, 311–316 (1961).
96. N. P. Plotnikoff, H. W. Elliott, and E. L. Way, *J. Pharmacol. Exp. Ther. 104*, 377–386 (1952).
97. L. A. Woods, L. B. Mellett, and K. J. Anderson, *J. Pharmacol. Exp. Ther. 124*, 1–8 (1958).
98. V. P. Dole and M. E. Nyswander, *J. Amer. Med. Ass. 193*, 647–650 (1965).
99. V. P. Dole and M. E. Nyswander, *Brit. J. Addict. 63*, 55–57 (1968).
100. A. H. Beckett, J. F. Taylor, A. F. Casy, and M. M. A. Hassan, *J. Pharm. Pharmacol. 20*, 754–762 (1968).
101. W. R. Martin, *Pharm. Rev. 19*, 464–521 (1967).
102. S. Archer and L. S. Harris, *in* "Progress in Drug Research" (E. J. Jucker, cd.), pp. 261–320, Birkhauser Verlag, Basel (1965).
103. R. A. Ferrari, *Toxicol. Appl. Pharm. 12*, 404–416 (1968).
104. A. S. Kates and J. Telford, *J. Pharmacol. Exp. Ther. 143*, 157–164 (1964).
105. J. Cochin, *in* "Enzymes in Mental Health" (G. J. Martin and B. Kisch, eds.), pp. 27–42, Lippincott, Philadelphia (1966).
106. M. H. Seevers and L. A. Woods, *Amer. J. Med. 14*, 546–557 (1953).
107. N. B. Eddy, H. Halbach, H. Isbell, and M. H. Seevers, *Bull. W. H. O. 32*, 721–733 (1965).
108. L. B. Mellett and L. A. Woods, *Prog. Drug Res. 5*, 157–267 (1963).
109. M. H. Seevers and G. A. Deneau, *Arch Int. Pharmacodyn. 140*, 514–520 (1962).
110. W. R. Martin, *in* "The Addictive States" (A. Wikler, ed.), Vol. 46, pp. 206–225, Williams & Wilkins, Baltimore (1968).
111. J. H. Jaffe and S. K. Sharpless, *in* "The Addictive States" (A. Wikler, ed.), Vol. 46, 226–246, Williams & Wilkins, Baltimore (1968).
112. J. Axelrod, *in* "The Addictive States" (A. Wikler, ed.), Vol. 46, 247–264, Williams & Wilkins, Baltimore, (1968).
113. A. Goldstein and D. B. Goldstein, *in* "The Addictive States" (A. Wikler, ed.), Vol. 46, 265–267, Williams & Wilkins, Baltimore (1968).
114. N. D. Eddy, *in* "Origin of Resistence to Toxic Agents" (M. G. Sevay, R. D. Reid, and O. E. Reynolds, eds.), pp. 223–243, Academic Press, New York (1955).
115. S. J. Mulé, *in* "Scientific Basis of Drug Dependence" (H. Steinberg, ed.), pp. 97–109, Churchill Ltd., London (1969).
116. L. A. Woods, *Pharm. Rev. 84*, 175–198 (1956).

Chapter 5

TRANSPORT IN THE CENTRAL NERVOUS SYSTEM*

Carl C. Hug, Jr.

Department of Pharmacology
The University of Michigan Medical School
Ann Arbor, Michigan

I. INTRODUCTION

Most of the prominent and important effects of narcotic analgesics on the body are a result of their action in the central nervous system (CNS). The sites and mechanisms of action of these drugs in the nervous system are unknown. One approach to identifiying a site or mechanism of action of a drug is to study its distribution within the responsive tissues. In the case of the narcotic analgesics the significance and specificity of sites of drug localization in the CNS can be evaluated by correlating the rise and fall of drug concentration at various sites with the onset and duration of pharmacological actions (for example, analgesia), by studying the effects of narcotic antagonists (for example, nalorphine or naloxone) on the distribution of the drug, and by determining the effects of tolerance development on the tissue concentrations of narcotic analgesics. This approach to identifying a site of narcotic action in the nervous system has proven to be unrewarding in the intact animal (for example, see Mulé et al.,[1,2,7] Hug et al.,[3,4] Johannesson et al.,[5,6] and Chernov and Woods[8] and see Way and Adler[9] for earlier references). First, unique sites of localization were not detected. Second, changes in localization induced by nalorphine or tolerance development were small, often neither biologically nor statistically significant. Third, in the intact animal there are numerous factors that influence the uptake of drugs by the nervous system, and the reproducibility and interpretation of experiments thereby are made difficult. Finally, the techniques employed would not be likely to detect sites of localization at the cellular or subcellular levels.

* Supported in part by grant MH-08580 from the National Institute of Mental Health.

Another aspect of narcotic analgesic action in the central nervous system is the apparent limitation on access of morphine-type drugs to their sites of action in the CNS, that is, the blood–brain barrier phenomenon. The levels of morphine in the CNS are lower than those in plasma and most other tissues (see references listed above). As newborn animals develop they become less sensitive to the actions of morphine (increased LD_{50}), and less morphine enters the CNS at a given plasma level of the drug.[10] The mechanisms underlying the blood–brain barrier phenomena are unknown for morphine-like drugs.

The purpose of this chapter is to review some of the more recent studies of the uptake and localization of narcotic analgesics in nervous tissue. The emphasis of most of the studies reviewed is on the role of biological transport in determining the entry and fate of narcotic analgesics in the central nervous system.

A. Concepts of Biological Transport

Christensen[11] has defined biological transport as the "mode by which a solute passes [across a membrane] from one phase to another, appearing in the same state in both phases." There are two basic modes of biological transport: simple diffusion and carrier-mediated transport.*

In simple diffusion the solute moves down its concentration gradient in penetrating a membrane. The solute may diffuse along aqueous-filled channels (pores) that traverse the membrane, or it may leave the aqueous phase and dissolve in the lipid matrix of the membrane at one surface, diffuse through this matrix to the opposite surface of the membrane, and enter the aqueous phase at the latter surface. Diffusion through pores is unlikely for molecules with a minimum radius greater than 4 Å, which is the radius estimated for pores in most cellular membranes. All of the narcotic analgesics have molecular radii far in excess of this minimum value (for example, see Beckett et al.[12]). Thus in the following discussion simple diffusion refers primarily to diffusion of narcotic analgesics through the lipid matrix of the membrane.

Carrier-mediated transport refers to a combination of the solute with a component (carrier) of the membrane at one surface and transfer of the solute–carrier complex to the opposite surface with release of the solute from the carrier into the aqueous phase at the latter surface. There are two types of carrier-mediated transport: passive and active. Passive carrier-mediated transport or facilitated diffusion is distinguished from simple diffusion by a greater rate of equilibration, chemical specificity, saturability, competition for transport among chemically similar compounds, and countertransport function (that is, an increased rate of movement of a solute

---

* Pinocytosis is a concept of transport based on electron microscopic observations. The functional aspects of pinocytosis in relation to solutes of small molecular weight (that is, most drugs) have not been clarified experimentally. *Group translocation* and other forms of vectorial intermediary metabolism do not apply to most drugs that are chemicals foreign to the body.

in one direction through the membrane is produced by an increased movement of the same or another solute using the same carrier in the opposite direction). Facilitated diffusion is an equilibrating system that yields no net transfer of solute once concentrations of the solute on both sides of the membrane are equal. Facilitated diffusion depends on a downhill concentration gradient to produce a net transfer of solute.

Active transport possesses the same characteristics as facilitated diffusion except that active transport can work against a concentration gradient or, in other words, can produce an uphill concentration gradient of the solute at the steady state. Active transport depends on continuous availability of energy-yielding compounds (for example, ATP) in order to produce and maintain solute concentration gradients.

Transfer of solutes across tissue membranes (that is, transcellular transport) is based on the cellular membrane transport processes described above.[13] In several types of tissue (for example, intestine, kidney, and choroid plexus) solutes have been shown to be accumulated by tissue cells prior to or in association with transfer of the solute across the tissue membrane, and the kinetics of accumulation and transfer have been similar.

B. Accumulation of Solutes in Cells and Tissues

In addition to active transport, there are other mechanisms by which cells and tissues concentrate solutes. Solutes may become bound to non-diffusible components of the cell. Solutes that are weak acids or bases and penetrate the membrane only in the un-ionized form may be concentrated on the side of the membrane where the pH favors a greater proportion of the ionized form of the solute.[14] Metabolism of the solute intracellularly to a nonpermeant form would lead to an apparent accumulation of the drug in the cell if the method of analysis did not distinguish between the solute and its nonpermeant metabolite. In some cases the total uptake of a solute by a tissue may represent the combined effects of two or more mechanisms. In the interpretation of experiments of tissue accumulation of solutes, one must be careful to evaluate the contribution of the various mechanisms to the total uptake.

A definitive statement of the role of an active transport system in the accumulation of a solute can be based only on measurements of the actual solute concentrations on both sides of the membrane. Because often it is not possible to determine the solute concentration on one side of a membrane (for example, in intracellular water), the contribution of active transport is frequently estimated indirectly. However, each indirect method has inherent disadvantages that preclude more than tentative conclusions. For example, inhibition of metabolic energy production or utilization may be used to estimate the contribution of active transport to solute accumulation by a tissue. But metabolic inhibitors also may interfere with solute binding to tissue components by occupying the binding sites or by decreasing the metabolic energy available for maintenance of the binding sites. Although the combination of several indirect methods may provide strong evidence of the participation of an active transport system, it cannot provide absolute proof of this participation.

II. TRANSPORT IN THE CEREBROSPINAL FLUID SYSTEM

A. Isolated Choroid Plexus

The first autoradiographic localization of radioactive morphine in the central nervous system revealed high concentrations of radioactivity in the choroid plexus of the lateral cerebral and fourth ventricles of the rat.[15] Little attention was paid to this early observation until Takemori and Stenwick[16] and Hug[17] independently observed that choroid plexus tissue accumulated narcotic analgesics *in vitro* and suggested a functional role for the choroid plexus in the disposition of these drugs in the intact animal.

The process of accumulation of narcotic analgesics by the choroid plexus has the characteristics of an active transport system. Morphine, dihydromorphine, codeine, nalorphine, levorphan, dextrorphan, and *l*-methorphan were accumulated by isolated rabbit choroid plexus against an apparent concentration gradient by a metabolically dependent process. Experiments with certain of the analgesics listed above demonstrated that the process was saturable, competitively inhibited by narcotic analgesics and their antagonists, stereospecific, dependent on aerobic metabolism, and functioned optimally at pH 7.4 and 37°C.

Takemori and Stenwick[16] studied the kinetics of release of morphine from choroid plexus that had previously been incubated with morphine *in vitro*. The run-out process had a slow and a fast component. The slow component was increased in rate while the fast component was unaffected in the presence of a metabolic inhibitor (fluoride or iodoacetate) or a narcotic antagonist (levallorphan). The slow component may represent the active uptake process counteracting the release of morphine from the tissue. The fast component probably represents simple diffusion of morphine out of the choroid plexus (loaded with morphine) into the morphine-free medium.

Several quaternary and primary amines have been found to be accumulated in isolated choroid plexus by a process exhibiting many of the characteristics of an active transport system.[18-20] Hug[17] concluded that the same process handled both the narcotic analgesics and the other organic bases. His conclusion was based on demonstrations of reciprocal competitive inhibition of dihydromorphine and hexamethonium uptake and of counter transport of dihydromorphine into choroid plexus that had been loaded with hexamethonium. The demonstration of counter transport is the strongest point of evidence for involvement of the organic base transport system in the uptake of dihydromorphine by the isolated choroid plexus. Dihydromorphine uptake also was inhibited competitively by other organic bases including choline, decamethonium, mepiperphenidol, and *N*-methyl-nicotinamide. Takemori and Stenwick[16] did not observe inhibition of morphine uptake by hexamethonium, possibly because they used a maximum hexamethonium-to-morphine ratio of only four and observed the interaction after 1 hr of incubation. Hug observed hexamethonium inhibition of dihydromorphine uptake using hexamethonium-to-dihydromorphine concentration ratios of 20–200 during incubation periods of less than 1 hr; the degree of inhibition became insignificant when incubation periods lasted 1 hr or more.

Takemori and Stenwick[16] reported that the uptake of morphine was depressed when calcium or magnesium were omitted from the usual Krebs–Ringer incubation media, but it was unaffected by the omission of phosphate. We have found somewhat different results with dihydromorphine (Hug, unpublished data; Table I). In contrast to the narcotic analgesics, hexamethonium uptake was unaffected by the omission of calcium but was depressed in the absence of magnesium and phosphate.[18] However, the omission of calcium in addition to the omission of magnesium and phosphate produced additional inhibition of hexamethonium uptake. The biological significance of these differences in the effects of omission of certain ions on the uptake of various organic bases is not yet apparent.

Another difference in the observations for hexamethonium and dihydromorphine is that p-aminohippuric acid (1 and 10 mM) did not affect the uptake of hexamethonium (0.1 mM) but 1 mM p-aminohippuric acid or probenecid did depress the accumulation of dihydromorphine (0.01 mM). Probenecid also reduced the accumulation of several narcotic analgesics by renal cortical slices.[17] The significance of the interaction of organic

TABLE I

Influence of Incubation Medium on the Accumulation of Organic Bases by Isolated Choroid Plexus from Rabbits

Substances Omitted from Krebs–Ringer Medium	Percent of Control T/M Ratio		
	Morphine[16] (1 mM)	Dihydromorphine^a (0.01 mM)	Hexamethonium[18] (0.1 mM)
Calcium	50	64 ± 7^b	100
Magnesium	68	104 ± 16^c	78
Phosphate	110	46 ± 11^d	17
Ca, Mg, PO$_4$	—	—	1
Potassium	—	61 ± 10^d	—
Sodium	—	38 ± 8^d	—
Glucose	—	58 ± 13^d	—
All (incubated in sucrose)	—	27 ± 5^b	—

^{a} Isolated choroid plexus tissue was incubated for 30 min at 37°C.[17] Sodium chloride was added in osmotic equivalents to replace the ions omitted.

^{b} $P < 0.001$.

^{c} The results were extremely variable with eight tissues showing a decrease of 16–64% and eight tissues showing an increase of 3–94%.

^{d} $P < 0.01$.

TABLE II

Binding of Dihydromorphine to the Particulate Fraction of Homogenates
of Rabbit Choroid Plexus[a]

Experimental Treatment	n	Particulate/Supernatant Concentration Ratio
None	7	2.1 ± 0.2
Incubation at 0°C	6	2.1 ± 0.2
Addition of		
Iodoacetate 1 mM	4	3.3 ± 0.3
Hexamethonium 1 mM	4	3.1 ± 0.4
Decamethonium 1 mM	7	2.2 ± 0.3
Nalorphine 1 mM	7	2.4 ± 0.3

[a] Rabbit choroid plexus was homogenized in Krebs–Ringer medium, and aliquots were incubated with 0.02 mM dihydromorphine for 30 min at 37°C. The particulate fraction was sedimented at 105,000g. Results were expressed as the ratio of the concentration of dihydromorphine in the particulate fraction (nmol/g) to that in the supernatant (nmol/ml).

acids with narcotic analgesics is not apparent. It should be noted that probenecid lacks specificity in its actions and has been reported to inhibit the transport of a variety of compounds in addition to the typical organic acids (see Berndt[21] for references).

Binding of dihydromorphine to homogenates of rabbit choroid plexus was limited and was unaffected in the presence of metabolic inhibitors or organic bases (Hug, unpublished data; Table II). Although binding to tissue components may contribute to the accumulation of narcotic analgesics by choroid plexus tissue, binding alone would not account for the major portion of the uptake. And binding cannot explain the phenomenon of countertransport. Incidentally, quaternary[18] and primary[19] organic amines also were bound to the particulate fraction of rabbit choroid plexus homogenates. In each case the extent of binding appeared to be insufficient to account for the accumulation of these substances in the intact tissue. Also, the binding to the particulate fraction of choroid plexus homogenates was unaffected by the presence of metabolic inhibitors whereas accumulation of the organic bases was markedly depressed in the intact tissue.

It should be noted that chromatographic studies[17] and specific analysis procedures[16,17] ruled out metabolism of narcotic analgesics in isolated choroid plexus tissue from rabbits. Also, the extent of accumulation of the narcotic analgesics was too great to be accounted for by pH-dependent partitioning of the compounds between the tissue and medium.[17]

In one series of experiments there was no difference in the uptake of dihydromorphine by choroid plexus from rabbits of different ages ranging from 1.5 to 8 weeks of age (Hug, unpublished data).

The potential biological significance of active transport of narcotic analgesics by choroid plexus tissue becomes apparent by analogy to the transport of other compounds by this tissue. A variety of compounds including inorganic ions, organic bases, sugars, amino acids, and proteins have been shown to be actively accumulated by choroid plexus tissue *in vitro* (see Lorenzo and Cutler[22] for references). In most instances the concentration of the compound in cerebrospinal fluid of the intact animal was much lower than the concentration of the compound in plasma. Many of these compounds have been shown to be actively transported out of CSF, presumably by way of the choroid plexus. Thus it is postulated that low CSF concentrations of such compounds are maintained by the active transport of the compounds from CSF to plasma by the choroid plexus (see Davson[23] and Cutler and Lorenzo[24] for references). In the case of the narcotic analgesics, their transport by the choroid plexus may be responsible for the apparent blood-CSF barrier to these compounds and for the rapid disappearance of narcotic analgesics from CSF when they are injected directly into this fluid.[25] Maintenance of low CSF concentrations of morphine-like drugs would permit the CSF to act as a "sink" for diffusion of the drugs out of brain tissue,[23,26] and in this manner the choroid plexus also would contribute to the blood-brain barrier phenomenon for morphine-like drugs. Additional discussion of these possibilities is presented in the next section of this chapter.

The possibility that narcotic analgesics may alter the CSF concentrations of endogenous and exogenous compounds by competing with them for transport by the choroid plexus is recognized, but so far there are few data with which to evaluate this possibility.

B. Ventriculo-Cisternal Perfusion

Pappenheimer, Heisey, and Jordan[27] first demonstrated the active transport of organic compounds (acids) from cerebrospinal fluid (CSF) to blood in goats and speculated that the choroid plexus was the site of transport. Schanker *et al.*[28] demonstrated that organic bases were removed from CSF by a carrier-mediated process in intact rabbits and that these same organic bases (for example, hexamethonium) were transported by rabbit choroid plexus *in vitro*.[18] Thus one might anticipate that morphine and other narcotic analgesics, which are accumulated by choroid plexus *in vitro* (see Section IIA above), would be removed from CSF by a carrier-mediated transport process. The data of Mulé *et al.*[1,2,7] on CSF levels of morphine in dogs given the drug subcutaneously show that CSF generally has a lower concentration of morphine than brain or plasma. This finding could be explained by transport of morphine out of CSF into blood.

Asghar and Way[29] reported that ^{14}C-morphine was removed from artificial CSF perfusing the cerebral ventricular system of the rabbit by an active transport process. They observed that the rate of disappearance of radioactivity from the fluid was saturable; was inhibited by nalorphine, codeine, and heroin; and was decreased by ouabain, oxygen deprivation, and death. Unfortunately Hug *et al.*[30] have been unable to confirm these results for either morphine or dihydromorphine in the rabbit even though

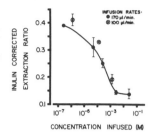

Fig. 1. Artificial CSF containing different concentrations of p-aminohippuric acid (PAH) was infused into the left lateral cerebral ventricles of mongrel dogs anesthetized with pentobarbital sodium. The fluid flowing from the cisterna magna was analyzed for PAH and inulin. Inulin served as a dilution marker. Inulin-corrected extraction ratios were calculated as follows:

$$\frac{\text{Outflow inulin conc.}}{\text{Inflow inulin conc.}} - \frac{\text{Outflow PAH conc.}}{\text{Inflow PAH conc.}}$$

Each point represents the mean $\pm$ range for two experiments.

essentially the same experimental procedure was used. However, with this experimental procedure Hug *et al.*[30] were able to demonstrate a reduction in the absorption of p-aminohippuric acid from CSF upon the addition of another organic acid, probenecid, and thereby confirm the earlier work of Pollay and Davson[31] in the rabbit. The discrepancies in the reports of Asghar and Way[29] and of Hug *et al.*[30] have not yet been explained.

In more extensive studies of the absorption of organic compounds from artificial CSF perfusing the cerebral ventricular system of dogs we have demonstrated a saturable process for the removal from CSF of p-aminohippuric acid (Fig. 1), decamethonium (Fig. 2), and the amino acids cycloleucine and α-aminoisobutyric acid. The absorption of p-aminohippuric acid from CSF was inhibited by probenecid and ouabain.

In contrast to the results with the above-mentioned compounds, the absorption of dihydromorphine from the ventricular fluid was *not* saturable over a 10,000-fold concentration range (Fig. 3), which includes the low concentrations found in CSF after systemic administration of 2 mg/kg doses (Hug, unpublished results) as well as levels sufficiently high to saturate the choroid plexus uptake process *in vitro*.[17] The absorption of dihydromorphine was not inhibited by other narcotic analgesics or narcotic antagonists in a range of molar concentration ratios of 10,000 to 1 (antagonist to dihydromorphine). It was not affected in the presence of ouabain nor by any one of several quaternary organic bases. (It should be noted that dihydromorphine was accumulated by a metabolically dependent process in

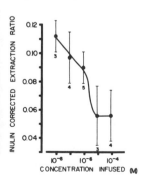

Fig. 2. Artificial CSF containing different concentrations of decamethonium was infused at a rate of 170 μl/min. Each point represents the mean $\pm$ standard error for the number of experiments indicated below the point. See the legend for Fig. 1 for additional details.

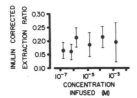

Fig. 3. Artificial CSF containing different concentrations of dihydromorphine was infused at a rate of 170 μl/min. Each point represents the mean $\pm$ range for two or three experiments. See the legend for Fig. 1 for additional details.

choroid plexus isolated from dogs.[17]) In similar experiments with morphine, levorphan, or nalorphine, there was no indication of the function of a carrier-mediated process in the absorption of these narcotic analgesics from the ventriculocisternal perfusion fluid in dogs.

The lack of evidence for a carrier-mediated process in the absorption of narcotic analgesics from CSF in dogs is not readily explained. Perhaps the proportion of drug absorbed from the perfusion fluid by carrier-mediated transport in the choroid plexus *in vivo* was too small (in relation to that absorbed into brain tissue and plasma by other routes) to be detected in these experiments. The major portion of the dihydromorphine infused into the lateral ventricle of two dogs was recovered in the cisternal outflow (54–77%) and another 8–10% was recovered in the whole brain (Hug, unpublished results). The remaining 15–36% presumably was absorbed into the blood. However, we are not yet able to estimate the amounts absorbed by the several possible pathways from cerebrospinal fluid to blood (that is, via the choroid plexus, arachnoid villi, and brain parenchymal capillaries).

III. TRANSPORT IN BRAIN TISSUE

A. Cerebral Cortical Slices

Scrafani and Hug[32] reported that narcotic analgesics including morphine, dihydromorphine, codeine, nalorphine, dextrorphan, levorphan, and *d*- and *l*-methorphan were accumulated against an apparent concentration gradient by rat cerebral cortical slices. In the case of dihydromorphine the accumulation was shown to be time and temperature dependent, optimal at pH 8.2, depressed in a nitrogen atmosphere or in the presence of metabolic inhibitors including ouabain, saturable, and competitively inhibited by codeine and nalorphine. The accumulation was evident in cerebral cortical slices from the mouse, rat, rabbit, and monkey. Other species were not tested.

We[33,34] have now extended our studies of the uptake of narcotic analgesics by brain slices. Metabolically dependent uptake of dihydromorphine occurs in slices of thalamus, midbrain, cerebellum, pons, and medulla. The degree of uptake is less in slices from areas containing fewer neuronal cell bodies such as the brain stem. The uptake process for dihydromorphine was depressed by lowering the concentration of either sodium or potassium in the medium but was unaffected by changes in calcium or magnesium concentrations. The inhibitory effects of nalorphine and metabolic inhibitors were additive. Metabolic inhibitors that depressed the accumulation of dihydromorphine did not alter the intracellular-extracellular pH gradient of cortical slices. These new data lend further support to the hypothesis that narcotic analgesics are accumulated in brain tissue, at least in part, by an active transport system.

Additional experiments have been performed in an attempt to determine the role of active transport in the actions of narcotic analgesics on nervous tissue. The uptake process has an element of stereospecificity since levorphan accumulation was twice that of dextrorphan at concentrations of 1 μM

TABLE III
Effect of Organic Bases and Amino Acids on the Uptake of Dihydromorphine by Cerebral Cortical Slices[a]

Organic Base	Concentration (mM)	n	Percent of Control ±S.E.M.	P-Value
Quaternary Amines				
Carbachol	1	12	98±4	N.S.[b]
Choline	1	12	102±4	N.S.
Decamethonium	1	14	101±5	N.S.
Hexamethonium	1	16	94±4	N.S.
Mepiperphenidol	1	4	111±4	N.S.
Secondary + Tertiary Amines				
Atropine	1	11	74±2	<0.001
Cocaine	1	4	80±2	<0.01
Histamine	1	12	86±2	<0.001
	0.1	8	96±2	N.S.
Norepinephrine	1	12	77±2	<0.001
	0.1	8	98±2	N.S.
Physostigmine	1	11	80±3	<0.001
Reserpine	0.01	4	98±1	N.S.
Serotonin	1	11	81±2	<0.001
	0.1	7	95±1	<0.05
Amino acids				
l-Glutamine	0.1	4	100±2	N.S.
l-Histidine	0.1	4	100±3	N.S.
l-Lysine	1	8	97±3	N.S.
	0.1	4	99±3	N.S.
l-Phenylalanine	0.1	4	95±5	N.S.
l-Tyrosine	0.1	4	101±7	N.S.

[a] Slices were preincubated 15 min and then the organic base and dihydromorphine (1 μM final concentration) were added simultaneously and the incubation was continued for 10 min. The control T/M was 1.8±0.1 (n=26).

[b] N.S. = not significant.

or less in the medium. Above 5 μM the uptake of the isomers was the same. Nalorphine and dihydromorphine were mutually competitive in their accumulation by cerebral cortical slices. In each case the degree of inhibition was small and statistically significant only in concentrations of 1 mM of the inhibitor. Naloxone also depressed the accumulation of dihydromorphine but was no more potent than nalorphine. A variety of nonquaternary organic bases were as potent as nalorphine and naloxone in decreasing dihydromorphine uptake (Table III). The characteristics of dihydromorphine accumulation by cerebral cortical slices did not differ in slices taken from rats physically dependent on morphine sulfate (33 mg/kg every 8 hr for more than 2 months).

Several items in the data summarized above suggest that active transport of narcotic analgesics into nervous tissue is *not* the determining factor in the action of these drugs in the brain of intact animals. First of all, the concentrations of narcotic analgesic antagonists required to depress dihydromorphine uptake by brain slices were approximately 300 times greater than those required to antagonize narcotic analgesic actions *in vivo*.[3,35]* Second, although there appears to be an element of stereospecificity in the uptake process, it does not parallel quantitatively the potency differences among isomers of narcotic analgesics; levorphan is about 50 times more potent than dextrorphan in most of their pharmacologic actions.[36] Third, there was no change in the uptake of dihydromorphine by tissue tolerant to its action, and the degree of inhibition of the uptake by nalorphine was not different in slices from control or tolerant animals. Finally, the uptake process was inhibited by a variety of nonquaternary organic bases, some of which have little or no effect on narcotic analgesic actions.

On the other hand, it is worth noting that morphine and dihydromorphine are equally potent (on a molar basis) in their pharmacological actions (see Hug and Mellett[4] or Hug[39] for references). Morphine in concentrations of 20 μM or less significantly reduced the uptake of acetylcholine,[37] carbachol,[38] and decamethonium[38] by cerebral cortical slices. The accumulation of these quaternary amines by cerebral cortical slices appeared to result from their active transport into the tissue. The accumulation was inhibited competitively by morphine as well as several other organic bases (for example, physostigmine, atropine, *d*-tubocurarine, and strychnine). In light of the evidence for active transport of dihydromorphine into cerebral cortical slices, it is plausible to conclude that the narcotic analgesics and the other organic bases utilize the same transport system to enter brain tissue. Theoretical implications of narcotic analgesic interactions with other organic bases in such a transport system are several (for example, relation to a mechanism of action and basis of potentiation of narcotic analgesic action by nonanalgesic compounds). Morphine does inhibit the release of ace-

---

* Transport of narcotic analgesics and inhibition of this transport by narcotic antagonists may play a more significant role in *in vitro* experiments when drug concentrations are in the millimolar range.

tylcholine from brain and nerves of the guinea pig ileum (see Chapter 11 by Weinstock), but this action has not as yet been related to the interaction of these compounds in a common transport system.

In relation to blood–brain barrier phenomena, it is worth noting that the K_m for dihydromorphine uptake by cerebral cortical slices ($K_m=20$ μM or 6 $\mu g/ml$) approximates the concentrations of dihydromorphine found in the brains of intact rats given small analgesic doses (2 to 5 mg/kg) of the drug.[39,48] Plasma levels ranged from 0.3 to 0.7 $\mu g/ml$ during the first hour (onset of maximal analgesia), while the brain levels ranged from 0.05 to 0.15 $\mu g/g$ during the same period. Perhaps the saturable uptake process is responsible in part for the limited entry of dihydromorphine into brain in the intact animals. However, until careful studies of the kinetics of DHM entry into brain *in vivo* are completed, the role of transport in the blood–brain barrier phenomena for morphine-like drugs remains a matter for speculation.

B. Homogenates of Brain

Binding of a drug to tissue components may lead to accumulation of the drug in the tissue (that is, $T/M>1$). Tissue binding and active transport of drugs frequently have certain characteristics in common (for example, saturability and competition among structural analogs). Our methods of analysis of radioactive narcotic analgesics[41] measure the total drug (chemi-

TABLE IV

Characteristics of Dihydromorphine*a* Uptake by Cerebral Cortical Slices and by the Total Particulate Fraction of Homogenates of Whole Brain

Characteristic	Cerebral Cortical Slices (32,33)	TPF of Whole Brain Homogenates (33,40)
Saturability	K_m 20 μM	K_d 11.6 mM
Nalorphine inhibition	K_i 3.7 mM (competitive)	K_i 8.1 mM (competitive)
Metabolic inhibition	yes (noncompetitive)	no
Nalorphine + metabolic inhibition	additive	same as nalorphine alone
Ouabain inhibition	yes	no
Na/K dependence	yes	no
Optimal pH	8	8
Effect of 0°C	slowed, steady state uptake decreased (vs 22 or 37°C)	slowed, same steady state uptake (vs 22°C)

a Maximal binding of dihydromorphine by the total particulate fraction occurred in medium consisting of 320mM sucrose, 0.01mM $CaCl_2$ and 25mM Tris buffer. The addition of ions or the use of the Krebs–Ringer bicarbonate medium (the medium usually used for slice studies) decreased the degree of binding. Binding was maximal at pH 8, but we routinely performed these studies using incubation media at pH 6.8 that approximates intracellular pH.

cally unchanged) in tissues and do not distinguish between drug bound to tissue and that free in solution in tissue water. Our early studies[32] showed that T/M values for dihydromorphine in brain slices were greater than unity even in the presence of high concentrations of metabolic inhibitors. Also, when large concentrations of dihydromorphine (much greater than those required to saturate the metabolically dependent uptake process) were present in the medium the T/M values remained above unity. These findings led us to suggest that tissue binding contributed to the total accumulation of narcotic analgesics in brain slices. Therefore, we compared the uptake of dihydromorphine by cerebral cortical slices and by the total particulate fraction (sedimented at 105,000 g for 30 min) of homogenates of rat brain.[42] The characteristics of dihydromorphine binding to the total particulate fraction of rat brain homogenates are summarized in Table IV, where they are compared to the characteristics of dihydromorphine uptake by cerebral cortical slices. From these data we have concluded that in addition to a metabolically dependent uptake process, presumably active transport, binding of narcotic analgesics to tissue components contributes to their accumulation in isolated brain tissue.

C. Synaptosomes

When brain tissue is homogenized nerve endings are severed from their axons and the broken membranes of the nerve endings reseal to form osmotically active particles referred to as synaptosomes.[43] Synaptosomes can be isolated by centrifugation of brain homogenates on gradients of sucrose.[44,45] They retain most of the metabolic activities associated with nerve endings including the ability to transport, store, synthesize, and metabolize neurohumors and their precursors.[46-49]

The uptake of dihydromorphine by synaptosomes has been investigated by two groups of investigators. Hug and Oka[50] incubated cerebral cortical slices with dihydromorphine and found about 25% of the radioactivity (in the slice) associated with the synaptosomal fraction; approximately 65% of the radioactivity was in the soluble fraction. Ouabain and nalorphine decreased the uptake of dihydromorphine by all fractions of the tissue slice, but only nalorphine was effective in reducing the uptake in isolated synaptosomes. In contrast to dihydromorphine, norepinephrine uptake both in the tissue slice and in isolated synaptosomes was reduced by ouabain.

Scrafani, Williams, and Clouet[51] incubated homogenates of rat brain with dihydromorphine and found that the drug was accumulated primarily in the synaptosomal fraction. The accumulation was decreased when the homogenate was incubated at 0°C and was increased by the addition of ATP. The authors concluded that the uptake was in part metabolically dependent. They speculated that the norepinephrine transport system of the synaptosome may be involved since dihydromorphine inhibited the uptake of norepinephrine by synaptosomes.

The question of metabolic dependence of synaptosomal uptake of narcotic analgesics requires additional experiments for a satisfactory answer.

Hug and Oka attributed the uptake by isolated synaptosomes primarily to a binding phenomenon since, when the synaptosomes were ruptured by osmotic shock, the synaptic membranes, synaptic vesicles, and myelin fractions of the synaptosomes showed a two- to threefold greater affinity for dihydromorphine than did the other subcellular fractions. Also, the binding of dihydromorphine to these fractions was decreased by nalorphine by about the same amount as was the uptake of dihydromorphine by isolated synaptosomes.

It should be noted, however, that nalorphine reduced dihydromorphine binding to all subcellular fractions by approximately the same amount, between 9 and 24%.[50] The fairly uniform effect of nalorphine on dihydromorphine binding in all fractions suggests that there is nothing unique about the binding of dihydromorphine to any one fraction. The finding that a number of non-narcotic organic bases[52] distribute among subcellular fractions of brain in a pattern similar to that for dihydromorphine raises a question about the specificity of the binding sites for narcotic analgesics.

The biological significance of synaptosomal accumulation of narcotic analgesics remains to be determined. One possibility is that synaptosomal uptake of morphine-type analgesics is related to an action of these drugs on nerve endings and neurotransmitter function. Several authors have presented evidence for an effect of morphine-type analgesics on the disposition of catecholamines,[53] serotonin,[54] and acetylcholine.[55] Whether changes in one or more of these mediators are responsible for the pharmacological actions of narcotic analgesics remains a moot question. Perhaps further studies of narcotic analgesic interactions with synaptosomes will at least clarify the basis of narcotic analgesic effects on neurotransmittor disposition and function. Such information would be useful in designing experiments to answer the more fundamental question of the role of neurotransmitters in the actions of narcotic analgesics.

IV. REFERENCES

1. S. J. Mulé and L. A. Woods, *J. Pharmacol. Exp. Ther. 136*, 232–241 (1962).
2. S. Mulé, L. A. Woods, and L. B. Mellett, *J. Pharmacol. Exp. Ther. 136*, 242–249 (1962).
3. C. C. Hug, Jr., and L. A. Woods, *J. Pharmacol. Exp. Ther. 142*, 248–256 (1963).
4. C. C. Hug, Jr., and L. B. Mellett, *Univ. Mich. Med. Bull. 29*, 165–174 (1963).
5. T. Johannesson and K. Milthers, *Acta Pharmacol. Toxicol. 20*, 80–89 (1963).
6. T. Johannesson and L. A. Woods, *Acta Pharmacol. Toxicol. 21*, 381–396 (1964).
7. S. J. Mulé, *J. Pharmacol. Exp. Ther. 148*, 393–398 (1965).
8. H. I. Chernov and L. A. Woods, *J. Pharmacol. Exp. Ther. 149*, 146–155 (1965).
9. E. Leong Way and T. K. Adler, "The Biological Disposition of Morphine and Its Surrogates," World Health Organization, Geneva (1962).
10. H. J. Kupferberg and E. L. Way, *J. Pharmacol. Exp. Ther. 141*, 105–112 (1963).
11. H. N. Christensen, "Biological Transport," pp. 3, 109, Benjamin, New York (1962).
12. A. H. Beckett, A. F. Casy, N. J. Harper, and P. M. Phillips, *J. Pharm. Pharmacol. 8*, 860–873 (1956).
13. H. N. Christensen, "Biological Transport," pp. 36–40, Benjamin, New York (1962).

14. L. S. Schanker, *Pharmacol. Rev. 14*, 501–530 (1962).
15. J. W. Miller and H. W. Elliott, *J. Pharmacol. Exp. Ther. 113*, 283–291 (1955).
16. A. E. Takemori and M. W. Stenwick, *J. Pharmacol. Exp. Ther. 154*, 586–594 (1966).
17. C. C. Hug, Jr., *Biochem. Pharmacol. 16*, 345–359 (1967).
18. Y. Tochino and L. S. Schanker, *Amer. J. Physiol. 208*, 666–673 (1965).
19. Y. Tochino and L. S. Schanker, *Biochem. Pharmacol. 14*, 1557–1566 (1965).
20. Y. Tochino and L. S. Schanker, *Amer. J. Physiol. 210*, 1229–1233 (1966).
21. W. O. Berndt, *Biochem. Pharmacol. 15*, 1947–1956 (1966).
22. A. V. Lorenzo and R. W. P. Cutler, *J. Neurochem. 16*, 577–585 (1969).
23. H. Davson, "Physiology of the Cerebrospinal Fluid," Little, Brown, Boston (1967).
24. R. W. P. Cutler and A. V. Lorenzo, *Science 161*, 1363–1364, (1968).
25. T. K. Adler, *Fed. Proc. 23*, 283 (1964).
26. W. H. Oldendorf, *Bull. Los Angeles Neurol. Soc. 32*, 169–180 (1967).
27. J. R. Pappenheimer, S. R. Heisey, and E. F. Jordan, *Amer. J. Physiol. 200*, 1–10 (1961).
28. L. S. Schanker, L. D. Prockop, J. Schou, and P. Sisodia, *Life Sci. 1*, 515–521 and 659 (1962).
29. M. K. Asghar and E. L. Way, *Pharmacologist 9*, 218 (1967).
30. C. C. Hug, Jr., J. T. Scrafani, and T. Oka, *Bull. Prob. Drug Dependence*, Appendix 28, pp. 5392–5400 (1968).
31. M. Pollay and H. Davson, *Brain 86*, 137–150 (1963).
32. J. T. Scrafani and C. C. Hug, Jr., *Biochem. Pharmacol. 17*, 1557–1566 (1968).
33. C. C. Hug, Jr., J. T. Scrafani, and T. Oka, submitted for publication in *Biochem. Pharmacol.*
34. J. T. Scrafani, "Uptake of Narcotic Analgesics into Brain," doctoral dissertation, The University of Michigan, Ann Arbor (1969).
35. L. Grumbach and H. I. Chernov, *J. Pharmacol. Exp. Ther. 149*, 385–396 (1965).
36. J. Hellerbach, D. Schnider, H. Besendorf, and B. Pellmont, "Synthetic Analgesics," Part IIA, "Morphinans," pp. 74–77, International Series of Monographs in Organic Chemistry, Pergamon Press, London (1966).
37. J. Schuberth and A. Sundwall, *J. Neurochem. 14*, 807–812 (1967).
38. D. B. Taylor, R. Creese and T. C. Lu, *J. Pharmacol. Exp. Ther. 165*, 310–319 (1969).
39. C. C. Hug, Jr., "Tritium-Labelled Dihydromorphine," doctoral disseration, The University of Michigan, Ann Arbor (1963).
40. J. H. Sanner and L. A. Woods, *J. Pharmacol. Exp. Ther. 148*, 176–184 (1965).
41. C. C. Hug, Jr., L. B. Mellett, and E. J. Cafruny, *J. Pharmacol. Exp. Ther. 150*, 259–269 (1965).
42. T. Oka, J. T. Scrafani, and C. C. Hug, Jr., *Pharmacologist 11*, 269 (1969).
43. V. P. Whittaker, I. A. Michaelson, and R. J. A. Kirkland, *Biochem. J. 90*, 293–303 (1964).
44. L. T. Potter and J. Axelrod, *J. Pharmacol. Exp. Ther. 142*, 291–298 (1963).
45. L. L. Iversen and S. H. Snyder, *Nature 220*, 796–798 (1968).
46. V. P. Whittaker, *Proc. Nat. Acad. Sci. 60*, 1081–1091 (1968).
47. R. W. Colburn, F. K. Goodwin, D. L. Murphy, W. E. Bunney, Jr., and J. M. Davis, *Biochem. Pharmacol. 17*, 957–964 (1968).
48. D. F. Bogdanski, A. Tissari, and B. B. Brodie, *Life Sci. 7*, 419–428 (1968).
49. I. Diamond and E. P. Kennedy, *J. Biol. Chem. 244*, 3258–3263 (1969).
50. C. C. Hug, Jr., and T. Oka, *The Pharmacologist 11*, 293 (1969).
51. J. T. Scrafani, N. Williams, and D. H. Clouet, *Pharmacologist 11*, 256 (1969).
52. J. M. Azcurra and E. DeRobertis, *Int. J. Neuropharmacol. 6*, 15–26 (1967).
53. E. W. Maynert and G. I. Klingman, *J. Pharmacol. Exp. Ther. 135*, 285–295 (1962).
54. F. Shen, H. H. Loh, and E. L. Way, *Pharmacologist 10*, 322 (1968).
55. K. Hano, H. Kaneto, T. Kakunaga, and N. Moribayashi, *Biochem. Pharmacol. 13*, 441–447 (1964).

II. *The Metabolic Disposition of Narcotic Analgesic Drugs*

Chapter 6

BIOTRANSFORMATIONS

Joseph T. Scrafani* and Doris H. Clouet*

New York State Research Institute for Neurochemistry and Drug Addiction Ward's Island, New York

I. INTRODUCTION

Although narcotic analgesic drugs, like other drugs, are foreign to the animal body, they have enough chemical resemblance to normal body constituents to act as substrates for enzyme-catalyzed reactions. The resulting metabolites may be more or less active pharmacologically than the parent compound or may be completely inactive. The importance of such biotransformations is manyfold: (1) the effective exposure of drug to tissue is defined by drug catabolism, be the metabolite more active, less active, or inactive, (2) the competition between drug and a natural substrate for an enzymatic site may change the balance of the chemical equilibrium of a tissue, (3) the identity of a more active metabolite of the drug may be discovered through a study of the metabolism of the parent drug, and (4) the biotransformation usually converts a lipid-soluble drug to a more water-soluble catabolite, thus easing and hastening the elimination of the drug from the body.

The identities of the major metabolites of all but the newest narcotic analgesic drugs have been known for many years. Because the metabolism of morphine and its surrogates has been reviewed in excellent detail by Way and Adler in 1960[1] and, especially, in 1962,[2] the reader is referred to these reviews for a description of the early studies. In this chapter, the pre-1960 literature on the metabolism of opioids is alluded to only if necessary for the continuity of the discussion. Instead there is a general discussion on the nature and sites of the metabolic reactions of the drugs and a section in which newer information on the metabolism of individual drugs is reviewed.

* Present address: New York State Narcotic Addiction Control Commission Testing and Research Laboratory, 80 Hanson Place, Brooklyn, New York.

II. GENERAL PATTERNS OF METABOLISM OF NARCOTIC DRUGS

A. Chemical Reactions

Narcotic analgesic drugs belong as a class in the general category of large organic bases and vary widely in the type and number of substituents capable of reacting chemically. Common to all is a substituted nitrogen, usually in a ring configuration. The N-alkyl (or substituted alkyl) group is removed enzymatically in an oxidative reaction catalyzed by an enzyme system found in the liver microsomal fraction. The N-alkyl groups so removed include methyl (morphine), ethyl (N-ethylnormorphine), allyl (nalorphine), phenethyl (phenazocine), and cyclopropylmethyl (cyclazocine). The liver N-dealkylase system is one of a number of such systems in liver that catalyze a variety of oxidative reactions of naturally occurring as well as foreign compounds, for which one requirement seems to be lipid solubility of the substrate. A direct correlation between lipid solubility and the rate of N-demethylation by liver microsomes in an *in vitro* system has been shown by McMahon for a series of tertiary amines including *dl*-propoxyphene, although other polar substituents on the drug molecules are able to affect this correlation.[3] The nonring nitrogen of some narcotic drugs has two alkyl groups that may be hydrolyzed at different rates. In the case of acetylmethadol, the tertiary amine is N-demethylated more rapidly than the monodemethylated secondary amines.[4] The nature of the alkyl group is also related to the rate of N-demethylation. Nalorphine is N-dealkylated at a faster rate than morphine by liver preparations.[5] When both nalorphine and morphine are present, they seem to compete, although an analysis of the nature of the inhibition is complicated by the difference in rate of dealkylation that alters the relative substrate concentrations during the reaction.[6] Both the pharmacologically active and the inactive isomers are N-dealkylated by the liver system with a slightly faster rate for the active isomer in the demethylation of *dl*-methadone and levo- and dextrorphan.[7]

Other reactions catalyzed by the liver microsomal enzyme systems include O-demethylation (of codeine) and O-deethylation (of ethylmorphine). Esters are hydrolyzed in the liver and in other tissues (meperidine, heroin, and acetylmethadol). The two acetyl groups of heroin are removed hydrolytically at different rates; the 3-acetyl group is hydrolyzed faster than the 6-acetyl group.[8]

Reactions specific for a single drug, such as the formation of a cyclic metabolite of methadone or the hydroxylation of the methyl groups of pentazocine, are described in Section IV of this chapter.

A major pathway in the biotransformation of narcotic drugs is conjugation of the drug with glucuronic, acetic, or sulfuric acids. In the liver, glucuronides of compounds with a free hydroxyl, carboxyl, amino, or sulfhydryl group are formed enzymatically by the transfer of glucuronic acid from uridine diphosphate glucuronic acid.[9] Glucuronides are also formed in the kidney. (See Chapter 17 for a discussion.) Anileridine is conjugated with acetic acid to form acetylanileridine.[10] Morphine and nalorphine have been isolated from urine as ethereal sulfates.[11]

Although these metabolic transformations are usually called detoxications, the metabolites also may be active pharmacologically. The result of such reactions tends toward an increase in water solubility and, therefore, excretability by the kidney rather than any predictable change in biological activity.[12]

B. Effects of Species, Sex, and Dose

The rate and metabolic pathway of biotransformation of narcotic analgesic drugs depend both on the dose and on the animal being observed. To achieve comparable doses in different animal species is difficult when the dose is related to the body weight of the animal, since blood levels and CNS levels are determined by such factors as the rate of drug absorption, metabolism, and excretion as well as by the amount of drug administered. When dogs and monkeys are compared, the same dose of normorphine (2 mg/kg subcutaneously) results in higher blood levels of the free and conjugated drug in a shorter time in the monkeys and in different patterns of urinary excretion of the drug.[13] An additional factor is revealed in these experiments, namely, the level of physical activity of the animal during the period of observation, since monkeys kept in restraining chairs metabolize less of the drug than do animals that are unrestrained. In the same two species, more N-demethylation occurs in monkeys than in dogs *in vivo*, as shown by the rate of expiration of $^{14}CO_2$ from the labeled N-methyl group of morphine.[14] Not only are the rates of drug metabolism species specific, but so also may be the identity of the metabolites of a drug. In Chapter 17, Fujimoto discusses the variation in conjugated metabolites of morphine, or other narcotic drugs, excreted in the chicken, rat, and rabbit.

Within a species there are variations in drug metabolism related to sex, age, and strain of the animal under study.[15] The rate of N-demethylation

TABLE I
N-Demethylation of Ethylmorphine by Liver Microsomes from Four Strains of Rats

	Male		Female	
Strain	30 Days Old	100 Days Old	30 Days Old	100 Days Old
Long–Evans	22.6[a]	26.4	14.6	5.0
Wistar	14.8	23.0	13.9	5.7
Holtzman	6.2	21.6	4.2	8.9
Sprague–Dawley	23.3	24.3	3.3	4.6

[a] Nanomoles of formaldehyde formed from the N-methyl group per milligram microsomal protein per 15 min. (Adapted from Furner *et al.* by courtesy of *Biochemical Pharmacology*.[17])

of morphine by liver preparations from male and female rats differs, although there is no sex difference in activity in the livers from guinea pigs or mice.[16] The rates of N-demethylation of ethylmorphine by liver preparations from male and female rats of four strains at two ages are shown in Table I.[17] It is evident that the liver microsomal preparations from mature female rats have less demethylating activity than from male rats. Otherwise the differences between sexes and strains have no discernible pattern. When the activity in the livers of domestic strains of rats is compared with that in wild strains, the N-demethylase activity in the liver microsomes from wild strains is less than half as active as it is from the domestic strains.[18] The activity of the drug-metabolizing enzymes in rats is sensitive to sex hormone treatment, increased with testosterone, and decreased by estrogen treatment in castrated adult animals, so that lower levels of enzyme activity would be expected in young males or in female rats of all ages.[19]

The nutritional state of the experimental animal also has an effect on the activity of the drug metabolizing enzymes. A lack of substrates for the reduction of TPN in the livers of starved rats can lead to spurious low values for meperidine N-demethylase activity if the assay medium is not supplemented with oxidizable substrates and dehydrogenases for TPNH generation.[20] In addition, it has been found that animals on a protein-free diet have low levels of drug metabolizing activity in the liver.[21,22]

C. Effects of Route of Administration

The number of barriers to be crossed by a narcotic analgesic drug in its passage from its site of entrance into the body to its receptor(s) in the CNS is partially determined by the route of administration. In general, oral administration results in the slowest rate of action and the slowest rate of achieving peak brain levels of the drug followed by the subcutaneous and intraperitoneal routes. Intravenous injection results in high blood levels with subsequent brain levels related to the presence or absence of a blood–brain barrier to the drug. An injection into the CSF or directly into brain tissue bypasses both liver catabolism and the blood–brain barrier, requiring lower doses to achieve higher brain levels than after systemic injection.

There have been a number of studies in the last 10 years in which the route of administration is an essential component of the experimental design. To compare the relative analgesic potency of morphine and normorphine, these drugs have been administered to rats by three routes. After the subcutaneous or intraperitoneal injection of the same dose of the two drugs, less normorphine is found in the brains of rats 15 min later, while after intravenous injection, the brain levels of the two drugs are equal.[23] The brain levels correlate with the pharmacological potency of the drugs that are equal only after intravenous (or intracerebral) administration.

Analgesic potency, as measured by the minimal analgesic dose, for etorphine is 2 μg/kg intravenously, 2.5 μg/kg intramuscularly, and 5–10 μg/kg sublingually.[24] The equivalent doses of dihydromorphine are similarly related for the intravenous and intramuscular routes of administration but are relatively higher when the drug is given sublingually. An examination of the blood levels of the two drugs given sublingually provides an explanation of this difference in relative potency; etorphine reaches a peak in blood 20–80 min after the sublingual application, indicating a rapid passage across the buccal mucosa into the vascular system.

A placental barrier can be demonstrated for certain narcotic drugs. The entrance of dihydromorphine into the blood after intramuscular injection is rapid in the pregnant rat, but the passage of the drug into the brain, or into the fetal blood and brain, is much delayed, compared to etorphine, which reaches a high level not only in blood but also in maternal and fetal brains within an hour after drug treatment.[25] The remarkable potency of etorphine must be, in part, related to the ease with which the drug passes various barriers. The lack of a blood–brain barrier in the fetus itself for dihydromorphine is demonstrated in experiments in which the peak level in the fetal brain is much higher than that in the maternal brain.[26]

There seems to be no blood–brain barrier for meperidine, even though equipotent doses of the drug are lower in young animals. The brain: plasma ratios are the same for 16- and 32-day-old rats, suggesting that an increased metabolism or excretion may account for the higher doses required in the older animals.[27] A lack of barrier to meperidine is also suggested in experiments with pregnant guinea pigs, in which EEG responses to the drug are seen simultaneously in the maternal and fetal brains.[28]

The intracerebral route of drug administration has been utilized to show that the quaternary base, N-methylmorphine, has pharmacological activity once the blood–brain barrier has been bypassed.[29] A comparison of the relative potencies of the parent narcotic analgesic drug and its metabolites can be made by introducing the parent drug both systemically and intraventricularly. For example, the major pharmacological effects of codeine after its injection systemically are attributed to its demethylated metabolites, morphine and norcodeine, as a result of a comparison of the effects of all three compounds after intraventricular administration.[30] In rats the results of a similar study suggest that the pharmacological response to heroin may be mainly assigned to its monodeacetylated metabolite, 6-monoacetylmorphine (MAM), since heroin has only a transient existence, being rapidly transformed to MAM, which reaches higher brain levels than either heroin or the dideacetylated metabolite, morphine.[8]

Microinjections into various areas of the central nervous system have been used to localize specific pharmacological effects of the narcotic analgesic drugs. (This subject is reviewed in Chapter 13.)

Other routes of administration include pellet implantation, a technique utilized to great advantage in the laboratory of Huidobro[31]; intravenous self-administration of opiates in rats[32] and in monkeys[33]; self-administration in drinking water[34,35]; continuous intravenous infusion[36]; and sustained release capsules.[37]

III. SITES OF METABOLISM

A. Liver

In studies of the patterns of urinary excretion after the administration of narcotic drugs, some of the metabolic reactions have been identified as oxidative. These reactions occur in the liver, catalyzed by an enzyme system localized in the microsomal fraction.[38] The liver microsomes (the endoplasmic reticulum) have emerged as the major site of the metabolism of not only opiates but most other categories of drugs. The reactions are oxidative but require TPNH as well as oxygen and the microsomal cytochrome respiratory system.[39] There is evidence that drugs bind to a component of the cytochrome P_{450} system in two types of binding, type I to the lipoprotein of the apoenzyme and type II as a ligand of iron in the heme of cytochrome P_{450}, inducing a change in configuration manifested by a shift in the wavelength of ultraviolet absorption.[40] The N- and O-dealkylations, hydroxylations, and oxidation-reduction reactions of narcotic analgesic drugs are catalyzed by this system of enzymes. There is evidence that the microsomal cytochrome system is common to all of the drug-metabolizing enzymes, while the existence of substrate- or reaction-specific enzymes is still in doubt.[39]

A characteristic of the drug-metabolizing enzymes of the liver microsomes is that the activity is inducible by the administration of single or multiple doses of inducer drugs.[39] Morphine and other narcotic drugs are not inducer drugs.[39] However, the activity of the enzymes that catalyze the biotransformation of morphine and other narcotic drugs is inducible by inducer drugs such as phenobarbital,[4,43] although not by a type II inducer, methylcholanthrene.[44] The type of binding to the microsomal cytochromes exhibited by narcotic drugs is not known, although the inducibility of the demethylation of morphine and meperidine by phenobarbital, and lack of inducibility by methylcholanthrene, suggests that the drugs are type I.[42−44] However, while N-dealkylase activity is induced by phenobarbital administration, the O-demethylation of codeine is not induced.[45]

In the livers of male rats, the chronic administration of morphine produces a substantial decrease in the activity of N-demethylase.[46] This response is probably related to the peculiarity of the activity of the enzyme system in male rat liver already discussed, since it does not occur in other species.[16,47]

There is some evidence that glucuronide formation in the liver is catalyzed by the same drug-metabolizing enzymes; namely, that the glucuronidation of o-aminophenol, sulfadimethoxine, and bilirubin is inducible by either phenobarbital or by polycyclic hydrocarbons.[39,48] However, there is no direct evidence that the formation of glucuronides of narcotic drugs or antagonists is catalyzed by the microsomal P_{450} system, although the reaction certainly occurs in the liver. The synthesis of morphine glucuronide in liver requires UDPG, a liver supernatant dehydrogenase and DPN[49] and another enzyme, a microsomal transferase, to transfer the glucuronic acid moiety from UDPGA to morphine.[50]

There are other biotransformations of narcotic drugs in liver that have not been established as oxidative reactions involving the microsomal system. The hydrolysis of meperidine and normeperidine, the N-methylation of methadone to a quaternary compound, and the deacetylation of heroin all are catalyzed by liver preparations.[51-53]

B. Other Tissues

A number of metabolic reactions have been found to occur in the central nervous system; the N-dealkylation (though not necessarily oxidative) of morphine, codeine, nalorphine, and meperidine;[54,55] the remethylation of normorphine to morphine;[56] and the hydrolysis of heroin.[53] Heroin is hydrolyzed in many tissues including kidney, lung, and blood in addition to liver and brain, possibly by a common lipase or esterase.[53] The remethylation of normorphine to morphine also occurs in lung tissue.[57]

The conjugation of morphine and other agonists and antagonists in the kidney is discussed in Chapter 17 by Fujimoto, who also describes the effect of variations of urinary pH on the patterns of excretion in man and animals.

In the placenta there is drug-metabolizing activity and also an endoplasmic reticulum similar to that in the liver.[58] The rate of catabolism of meperidine by human placenta preparation is two-thirds that of rat liver, although there is no N-demethylation by the placental preparation.[58] In rats and rabbits the activity of the drug-metabolizing enzymes of the placenta is inducible by inducer drugs.[59]

There is a report that there is a low level of drug metabolism in the lung, gastrointestinal tract, and kidney (in addition to liver) and that an undetectable activity may be induced to detectable levels in thyroid, skin, spleen, and thymus. Unfortunately, neither the drugs nor the reactions are specified in this report.[60]

IV. METABOLISM OF INDIVIDUAL DRUGS

A. *d,l-α*-Acetylmethadol

The *l*-isomer of α-acetylmethadol is a narcotic agonist related structurally to methadone, with a long duration of action. The lack of correlation between the duration of pharmacological activity and the presence of the drug in tissue has led to suggestions that a metabolite is an active form of the drug.[61,62] The removal of the N-methyl groups is catalyzed by liver drug-metabolizing enzymes *in vitro* and *in vivo* at two different rates of demethylation: the first methyl is removed rapidly with a half-life of 2 hr and the second much more slowly.[4] The secondary amine, noracetylmethadol, is an active agonist, while methadol is less active.[63] Thus *l-α*-noracetylmethadol is suggested as the active metabolite.[4] After the administration of this compound to rats, little drug is found in the urine and almost half of the injected dose is found in the feces. It seems that the drug is glucuronidated since treatment with β-glucuronidase increases the yield of drug from the feces.[4] When noracetylmethadol labeled in the N-methyl group with ^{14}C is administered to rats, there is an unidentified ^{14}C-labeled compound that remains in the body for a long time with a half-life of 15.5 days that is probably related to the 1-carbon pool which the oxidized methyl group enters by way of formaldehyde.[4]

The acetyl groups are removed from both acetylmethadol and noracetylmethadol at a much slower rate than the N-methyl groups.[4] The resulting carbinols have not been identified in tissues nor found to be excreted in bile.

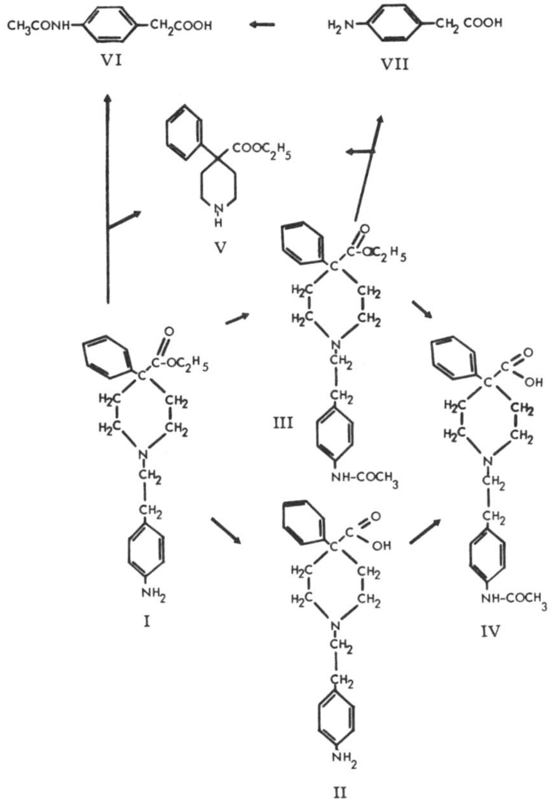

Fig. 1.   The metabolic pathway of anileridine. I = anileridine, II = anileridinic acid, III = acetylanileridine, IV = acetylanileridinic acid, V = normeperidine, VI = p-acetylaminophenylacetic acid, and VII = p-aminophenylacetic acid. (Adapted from Porter[10] and Lin and Way.[64])

B. Anileridine

Anileridine is a potent narcotic analgesic drug related to meperidine both in structure and activity. Its metabolism, as defined by urinary metabolites, is to acetylanileridine, anileridinic acid, and the acetylated acid.[10] Another metabolite is tentatively identified as acetylaminophenylacetic acid.[10] The identity of this compound has been confirmed by Lin and Way, using paper and thin-layer chromatography.[64] In addition, nor-meperidine and p-aminophenylacetic acid have been found in the urine.[64] This rather extensive metabolic pathway is shown in Fig. 1.

C. Codeine

Codeine is a moderately potent narcotic agonist. It is O-demethylated to morphine, N-demethylated to norcodeine, and both metabolites and the parent drug may be conjugated before excretion.[2] Normorphine, too, has been identified as a minor metabolite of codeine.[65] When N-$^{14}CH_3$-labeled codeine is administered to rats, 82% of the injected dose is recoverable in urine, bile, feces, or as ^{14}CO in the expired air.[66] Most of the label is excreted as morphine or its glucuronide, with no evidence of the presence of codeine glucuronide,[66] although this conjugate has been isolated from the urine of many other species.[2] The sex difference in drug-metabolizing enzymes of rat liver is illustrated by the difference in the amount of $^{14}CO_2$ in expired air being less in the female.[41,66]

With several possible candidates for the compound acting pharmacologically, codeine and its metabolites have been examined for their relative metabolism.[30] Both O- and N-demethylation occur rapidly *in vivo* after the subcutaneous injection of codeine to rats, with approximately 30% of methyl groups removed by 24 hr.[30] Since the potency of codeine administered intraventricularly is about 1/100 that of morphine, by the tail-flick test, it is possible that morphine (or its nor derivative) is the active compound.[30] A similar conclusion is reached in studies in which the brain levels of both morphine and codeine are measured after the systemic administration of codeine to rats.[67] Evidence to the contrary is provided in experiments in which nalorphine is used to provoke withdrawal in codeine users.[68] The only correlation of a positive nalline test is with the chronic use of codeine in these subjects.[68] Since nalorphine provokes a pupil diameter change in the codeine-treated dog that does not excrete morphine as a metabolite of codeine, it seems that at least in this pharmacological parameter codeine is the active drug.[68]

In mice, equal amounts of $^{14}CO_2$ are expired after the injection of O- or N-methyl-labeled codeine.[30] *In vitro*, however, the liver N-demethylating activity is about seven times that for O-demethylation.[69] The N-demethylation of codeine, as measured by expired $^{14}CO_2$, increases in the tolerant rat in contrast to that of morphine following morphine injection.[70] This is contrary to effects on liver drug-metabolizing enzymes following drug administration to the animal.[7,20,43,71] The use of respiratory CO_2 as an index of drug metabolism may be inexact because the amount of label in the expired air may represent that in the body one-carbon pool to varying degrees.[4]

Codeine acts similarly to morphine as an inhibitor of drug-metabolizing activity in the liver.[72] The relative demethylation of the O- and N-methyl groups in liver is not constant, leading Elison and Elliott to postulate that the enzyme activities are not identical.[55] A further discussion of liver enzymes may be found in Section IIIA of this chapter.

D. Cyclazocine

Cyclazocine is a narcotic antagonist with agonist activity. Its metabolism has been studied by Mulé and Gorodetzky in the dog.[73] After a dose of 1.25 mg/kg of tritiated cyclazocine, about half of the labeled drug is found in the urine, 4% of the dose as free drug, and 29–36% as a conjugate. The recoveries are similar in the urine from nontolerant and abstinent dogs and from dogs tolerant to cyclazocine.[73] However, the amount of free cyclazocine in the feces of tolerant dogs is higher (16%) than in nontolerant (4%) or abstinent dogs (7%). The recovery of labeled free drug and metabolites is low: only 47–66% in the 5 days after a dose of 1.25 mg/kg. The compounds isolated from the urine include unchanged cyclazocine and norcyclazocine and their glucuronides. A small amount of an unidentified metabolite was isolated by paper chromatography.[74] Table II summarizes the distribution of drugs and metabolites in the urine and feces.

E. Dextromethorphan

Dextromethorphan, the d-isomer of 3-methoxy-N-methylmorphinan, is essentially devoid of analgesic activity and does not produce physical dependence.[75,76] The demethylated derivatives have been established as major metabolites.[2] When tritiated dextromethorphan is administered to rats in a dose of 10 mg/kg, 55% of the label is recovered in the urine and

TABLE II

Excretion of Cyclazocine and its Metabolites in the Urine and Feces of the Dog[a]

Compound	Total Recovery (as Percent of Injected Dose)		
	Nontolerant	Tolerant	Abstinent
Cyclazocine	8.7	20.7	9.9
Cyclazocine glucuronide	34.9	37.8	30.8
Norcyclazocine	4.3	4.7	3.9
Norcyclazocine glucuronide	3.5	2.8	2.4

[a] After a dose of 1.25 mg/kg of ^{3}H-cyclazocine, the recovery in urine and feces over a five-day period is given. (Adapted from Mulé *et al.* by courtesy of the *Journal of Pharmacology and Experimental Therapeutics.*[74])

feces in 4 days, with 65% of this amount in the feces.[77] Only 3% of the label in the urine is due to unchanged drug. The major metabolites are 3-hydroxy-N-methylmorphinan and its glucuronide and sulfate conjugates. A further demethylated metabolite, 3-hydroxymorphinan, is isolated from both urine and feces and, together with its conjugates, comprises only 19% of the labeled excretory compounds.[77]

The same patterns of metabolism are found in the human. Unchanged dextromethorphan, 3-methoxymorphinan, 3-hydroxy-N-methylmorphinan, and 3-hydroxymorphinan have been found in the urine of a one-year-old child who consumed 180 mg of the drug orally.[78] In the adult human all of these metabolites have been found except d-3-methoxymorphinan.[79]

F. Dihydromorphine

Dihydromorphine, morphine reduced at the 7,8 double bond, is chemically and pharmacologically similar to morphine.[80] Its metabolism has been extensively studied in two laboratories.[26,81] After the administration of tritiated dihydromorphine to rats, free and conjugated forms of the drug are found in plasma, kidney, and liver and to a slight degree in brain.[26] The recovery in urine, amounting to 85% of the injected dose, consists of free and conjugated drug, with a minute fraction of dihydromorphine found in the feces.[81] There is a sex-related difference in the amount of conjugated drug excreted in rats, the female rat excreting twice as much conjugate as the male. Not all of the conjugated drug is as glucuronide since hydrolysis by β-glucuronidase results in the liberation of only about half of the bound drug.[81] N-demethylation as a route of biotransformation is confirmed in studies in which N-$^{14}CH_3$-labeled dihydromorphine is administered and the rate of expiration of $^{14}CO_2$ is measured.[82] There is a sex difference in the amount of $^{14}CO_2$ expired, the female rat expiring at a lower rate than the male.[82]

G. Ethylmorphine

Ethyl-3-O-morphine is a derivative of morphine with pharmacological activity similar to that of codeine, of which it is an analog. Its primary importance in this discussion is its use as a substrate for measuring the N-demethylase activity in liver preparations.[17,45,83,84] The kinetics of the demethylation of ethylmorphine have been studied by Rubin, Tephly, and Mannering in naïve and phenobarbital-induced rats.[72,85] They find that the K_m for both groups is the same, 6×10^{-4} M, while the V_{max} is higher in induced animals, indicating an increase in enzyme level in the livers of the induced animals.

H. Heroin

Heroin is a potent agonist with a high addiction liability. Heroin is rapidly 3-deacetylated to 6-monoacetylmorphine in the bodies of all species studied (Fig. 2).[2,8,53,86,87] In brain, a peak level of heroin of 0.9 μg/g is found 15 min after the intrapopliteal injection of 75 mg/kg in the rat, while

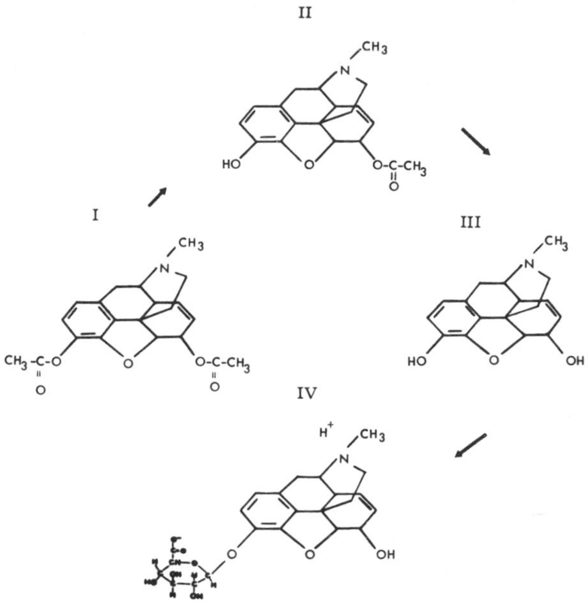

Fig. 2. The metabolic pathway of heroin. I = heroin, II = monoacetylmorphine, III = morphine, and IV = morphine glucuronide. (Adapted from Way et al.[8])

the peak level of 6-monoacetylmorphine, also found 15 min after the injection, is 6.3 μg/g.[8] 6-Monoacetylmorphine is 6-deacetylated to morphine at a slower but still moderately rapid rate, since a peak level of morphine of 1.5 μg/g in brain is sustained in the brains of the heroin-treated rats for a period of 3 hr.[8] These and similar findings have led Way and his colleagues to conclude the following: "Unchanged heroin can exert only minor transient pharmacologic effects and the chief actions of the compound are the consequence of its biotransformation to monoacetylmorphine and morphine. It is postulated further that morphine is probably responsible for most of the pharmacologic effects of heroin and that heroin and monoacetylmorphine function largely as carriers to facilitate access of morphine to its receptor sites in the central nervous system."[8] Morphine glucuronide is found in the urine after heroin administration.[2]

I. Methadone

Methadone is a potent narcotic agonist. The N-demethylation of methadone (Fig. 3, formula I) has been established using liver microsomal preparations and measuring the oxidized methyl group as formaldehyde.[42] However, the structure of the demethylated metabolites remained unknown for many years, although a compound synthesized by Pohland resembled the metabolite isolated from liver incubations with methadone as substrate by Way. The compound is formed when 1,5-dimethyl-3,3-diphenylpyrrolid-2-one is treated with ethyl lithium and is presumed to be a tertiary alcohol or the alkene.[2] The structure has been established as the endocyclic form of the alkene (II) by infrared and nuclear magnetic resonance spectral data.[88] A metabolite isolated from human urine is identical to this synthetic compound in infrared spectral analysis.[88] The only other compound found in the urine in these studies is unchanged methadone, also identified by IR spectral data, which together with the cyclic metabolite accounts for 60% of the administered dose of methadone.[88]

The excretion of metabolites of methadone in the bile has been reported.[89] Unchanged methadone and two unidentified metabolites are found in the bile of rats given methadone. One of these metabolites is water soluble and also is found in *in vitro* incubations with liver preparations.[89] It is possible that this is the quaternary compound, *N*-methylmethadone, isolated by Schaumann from incubations of guinea pig liver and methadone and identified by a paper chromatographic and melting point comparison with the authentic compound.[53]

J. Morphine

In 1960 the metabolic reactions of morphine were described as *N*-demethylation, 3-*O*-methylation, and conjugation as the 3-glucuronide.[2] Further studies on the nature of the 'bound' morphine, on *N*-demethylation, and on new metabolites of morphine are discussed under the appropriate metabolic reaction.

1. Glucuronidation

The site of the glucuronidation of morphine has been postulated as the 3-phenolic hydroxyl position both by the lack of color formation with a phenolic reagent and by the lack of a phenolic bathochromic shift with increasing pH in the UV absorption spectrum by bound morphine before hydrolysis.[2] The presence of a second glucuronide of morphine was suspected for a long time, but only recently has it been isolated and identified chemically.[90] The conjugate is isolated from the urine of rabbits, purified by column chromatography, and identified as the 6-glucuronide by com-

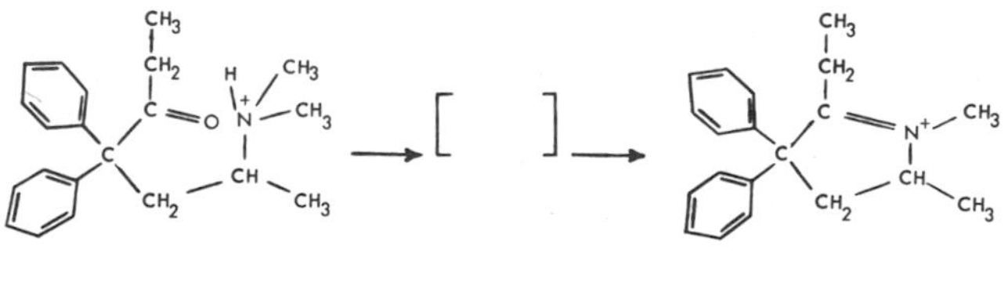

Fig. 3. The metabolic pathway of methadone. I = protonated methadone and II = endocyclic form of metabolite. (Adapted from Beckett *et al.*[88])

parison with synthetic morphine-6-glucuronide.[90-92] This is a minor metabolite, amounting to 0.3% of the administered dose, while morphine-3-glucuronide is excreted in amounts equivalent to 45% of the injected dose in the same animals. Unequivocal evidence of the chemical nature of the 3-glucuronide is also offered by these authors[92] and by Fujimoto and Haarstad.[11]

Glucuronyl transferase of liver microsomes seems to be an inducible enzyme. Its activity is decreased in the livers of male rats during chronic morphine treatment,[93] and its activity is induced by chloroquine.[94] The decreased activity of the enzyme in the livers of tolerant male rats has been described elsewhere in this chapter for other enzymes and attributed to a sex hormone sensitivity of the drug-metabolizing enzymes in male rat liver.[16,47,95] The induction of glucuronyl transferase by chloroquine is reflected by a higher level of morphine glucuronide in the liver and an increase in the LD_{50} of morphine.[94]

2. Formation of Ethereal Sulfates

Morphine ethereal sulfate is isolated from the urine of morphine-treated cats and hydrolyzed to yield authentic morphine and a substance precipitable by barium as barium sulfate. Morphine and sulfate are present as the theoretical amounts for morphine monoethereal sulfate.[96] The same conjugate is isolated from the urine of the chicken and identified by comparison with authentic morphine-3-ethereal sulfate.[11] In the chicken and the cat the ethereal sulfate is a major metabolite of morphine.[11] This compound is most likely to be identical to the metabolite described earlier in studies in the chicken in which the bound drug was formed in the renal tubular cell.[97]

3. N-Demethylation

There have been a large number of papers in the last 10 years describing the N-demethylation of morphine to normorphine and relating it to tolerance[46,98] and pharmacological activity.[99,100] The administration of ^{14}C-labeled (methyl) morphine results in the expiration of $^{14}CO_2$ derived from the methyl group by oxidative demethylation, indicating that demethylation is occurring *in vivo*.[2] The isolation and identification of free and conjugated normorphine in the urine of morphine-treated rats is made by comparison with the authentic compound.[101,102] In rats the total normorphine excretion amounts to 5% of the dose.[101,102] This same excretion pattern is seen when morphine is injected in doses from 10 to 50 mg/kg to rats; the total normorphine amounts to 7% with most in the form of a conjugate and another 2–3% of the injected dose in the feces as free normorphine.[103] Similar results are found in the cat.[104] However, in human urine no normorphine is found after the administration of labeled morphine (quoted in Way and Adler[2]).

It has been reported that N-demethylation also occurs in brain. In eviscerated rats, 5% of the total alkaloid in brain is normorphine while only 1% is in blood, suggesting that the demethylation occurs in the brain.[54,105] In rat brain slices, N-^{14}C-methyl-labeled morphine is converted to [14]CO_2, suggesting that the methyl group is removed oxidatively.[55] However, no microsomal hemes similar to liver P_{450} have been detected in brain, so that the possibility of oxidative demethylation in brain remains conjectural.[106,107]

The N-dealkylation of morphine has been studied *in vitro* in kinetic analyses of substrate competition,[6] lack of identity of N- and O-demethylases,[108] and substrate–enzyme binding dependence on the strength of the methyl C–H bonds.[7,109]

The effects of chronic morphine administration on N-demethylation are discussed in the next section of this chapter.

4. Catechol Formation

The formation of catechols by liver enzymes is observed with many phenolic compounds as substrates.[110] When an incubation mixture includes liver microsomes, NADP, glucose-6-phosphate, catechol-O-methyltransferase, and S-adenosylmethionine, morphine and other opioids are converted to catechol derivatives that are subsequently methylated.[110] This reaction is low with morphine as a substrate (1.2 mμmoles/g liver) but much higher with phenazocine as substrate (28.5 mμmoles/g liver). This metabolite has not been demonstrated *in vivo*.

5. N-Oxide Formation

Morphine-N-oxide has been isolated from the urine of patients receiving mixtures of morphine and amiphenizole or 1,2,3,4,-tetrahydro-9-aminoacridine.[111] Its identification has been made by comparing the isolated compound with the authentic compound. Since the N-oxide is formed when morphine and amiphenizole are mixed, there is a possibility that the N-oxide is formed chemically in the urine. However, it is also possible that the second drug acts by inhibiting the breakdown of morphine-N-oxide formed *in vivo* or by inhibiting a preferred pathway of morphine metabolism. In this regard the N-oxides have been suggested as intermediates in the oxidative N-demethylation of drugs.[112] Extensive kinetic studies by McMahon suggest that the N-oxide is an alternative pathway rather than an intermediate in the oxidative N-demethylation pathway.[113]

6. O-Methylation

Morphine is converted to codeine by 3-methylation.[108] A small amount of codeine is isolated from normal rats given morphine and a larger amount from the urine of Gunn rats that have a deficiency in glucuronide-forming enzymes.[114] S-adenosylmethionine is presumed not to be a methyl donor since labeled S-adenosylmethionine does not provide labeled codeine when morphine is incubated with it and liver homogenate.[114]

K. Nalorphine

Nalorphine is the prototype of the narcotic antagonist, although it does have some agonist activity.[115] Conjugated metabolites appear in the urine of the dog and rat after drug administration, and the drug is de-allylated *in vitro* by liver drug-metabolizing enzymes.[2] The product of de-allylation, normorphine, has been found in brain and liver after the administration of nalorphine to rats or cats.[54,104,105,116,117] Evidence for the extrahepatic deallylation of nalorphine is that normorphine is found in the brains of hepatectomized rats receiving nalorphine.[54,105]

In dogs, the major metabolite in the urine of tritiated nalorphine is nalorphine glucuronide identified by the action of β-glucuronidase in hydrolyzing the conjugated form.[118] Nalorphine is present only to 2–6% of the administered dose.[118] Two conjugates have been identified chemically in the urine of the cat and the rabbit, nalorphine-3-ethereal sulfate and nalorphine-3-glucuronide, respectively.[119]

When nalorphine is given intracisternally, it is much less effective in antagonizing morphine than when given systemically.[120] Also, when nalorphine is given intraventricularly to mice tolerant to morphine after pellet implantation, the withdrawal signs are minimal.[121] These observations suggest that nalorphine itself may not be the active antagonist.[121]

L. Naloxone

Naloxone is an antagonist with little or no agonist activity. The patterns of urinary excretion of naloxone and its metabolites have been studied in the chicken, rat, and rabbit by Fujimoto,[122] who discusses this subject in detail in Chapter 17. In rabbit urine, naloxone-3-glucuronide has been identified; in chicken urine, a 6-alcohol of naloxone-3-glucuronide; and there are other as yet unidentified metabolites, some of which may be ethereal sulfates.[122]

M. Normorphine

Normorphine, a metabolite of morphine, has been implicated in several hypotheses concerning both the mechanism of analgesia and the development of tolerance.[46,98–100] Normorphine is an agonist, weaker than morphine both in pharmacological activity and addiction liability.[118,123,124] After a dose of 2 mg/kg to dogs, about 70% is recovered in the urine, half free drug and half a conjugate.[13] In monkeys, less free drug is found in the urine. The physical activity of the monkeys determines to some extent the amount of conjugate appearing in the urine and also the duration of the pharmacological response.[13,125] In monkeys 30% of the administered dose of normorphine is found in the fecal excretion, mainly as the free drug.[13]

The methylation of normorphine to morphine has been described in liver and in brain *in vitro*[107,126] and *in vivo* in brain.[56,107] A partially purified *N*-methyltransferase has been prepared from brain or liver that requires *S*-adenosylmethionine as a methyl donor.[107,126] A similar but not identical enzyme has been isolated and purified from rabbit lung that catalyzes the transfer of the methyl group of S-adenosylmethionine to a variety of amines including normorphine.[57,127] There is some evidence for the methylation of normorphine *in vivo* in brain. When normorphine and ^{14}C-methyl-labeled methionine are injected intracisternally in rats, a small amount of ^{14}C-labeled morphine is isolated.[56] In another type of experiment, morphine has been isolated from the brains of eviscerated rats after the administration of normorphine.[54]

N. Pentazocine

Pentazocine is mainly a narcotic antagonist, with some agonist activity.[115] The metabolism of tritium-labeled pentazocine *in vitro* by preparations from the livers of monkeys, mice, and rats has been compared.[128] The enzymes of all three species oxidize the methyl group of the dimethylallyl side chain of pentazocine to isomeric alcohol derivatives. There are quantitative species differences in the amounts of each isomeric alcohol produced and, in monkey liver, one of the alcohols is further oxidized to a carboxylic acid.[128]

The metabolism of pentazocine *in vivo* is very similar. In monkeys, 70% of the injected label could be recovered from the urine in the first 48 hr after doses ranging from 0.3 to 34 mg/kg.[128] The unchanged drug, the alcohols, and the carboxylic acid derivatives found *in vitro* also are found in the urine during this time. Some conjugated form of the drug also has been detected but not yet identified. Similar metabolites are found in human urine.[128]

In other studies of the urinary excretion of pentazocine and metabolites in man, only a small amount of the unchanged drug is found in urine samples.[129] This extensive metabolism of pentazocine in humans has been confirmed by Berkowitz and Way, who found only 13% of the excretory products to be free pentazocine.[130] Another small amount is present as the β-glucuronide.

O. *d*-Propoxyphene

d-Propoxyphene is a narcotic agonist related to methadone in structure and with a low addiction liability.[131] The only metabolite found in human urine is the di-*N*-demethylated compound, which has been isolated from urine as a dinitro derivative and identified by elemental analysis and IR spectral analysis.[132] This metabolite is found in urine only to the extent of 25% of a dose administered 48 hr earlier.[133] When ^{14}C-labeled *d*-propoxyphene is administered to rats, again the only metabolite found in the urine is desmethylpropoxyphene, containing 17% of the administered label in the first 24-hr sample.[3] The slow appearance of the demethylated metabolite in the urine is difficult to understand in light of the studies in rats, mice, and guinea pigs, in which the rate of demethylation *in vivo*, measured by the rate of expiration of $^{14}CO_2$, exceeds 50% in the first 5 hr after the intraperitoneal injection of 50 μmoles/kg.[3]

V. METABOLISM IN TOLERANT ANIMALS

A general conclusion reached by many investigators who have compared the metabolism of narcotic analgesic drugs in nontolerant and tolerant animals is expressed by Mulé as follows: "It is concluded that neither distribution nor metabolism of morphine provides any further insight into the mechanism of tolerance development."[134] This conclusion is supported by an examination of the urinary excretion of the metabolites of codeine in codeine-tolerant rats and guinea pigs[65] and of dextromethorphan after the chronic administration of this drug to rats[77] as well as by earlier studies in morphine-tolerant animals (see Way and Adler[2]). Only slight differences in patterns of excretion are seen in tolerant animals other than changes due to the need of tolerant animals to dispose of increased amounts of the drug and metabolites as the dose of drug consumed increases. Also, tolerance does not make any difference in the rate of expiration of $^{14}CO_2$ derived from the N-$^{14}CH_3$ group of morphine in the whole animal,[70] although the percent of injected dose does decrease as the administered dose increases.[82] A comparison of the pathways of metabolism in tolerant and nontolerant animals using tracer radio-labeled drugs has the inherent drawbacks that there is isotope dilution by the unlabeled drug still present in the chronically treated animals and that patterns of metabolism and excretion may depend on saturable processes and thus be drug load dependent. There is one report that a new metabolite of morphine is found in tolerant rats, although there is no suggestion as to its structure as yet.[135]

The effect of chronic narcotic drug treatment on the activity of drug-metabolizing enzymes in male rat liver is a substantial decrease in the

TABLE III
Liver Meperidine N-Demethylase Activity in Morphine-Treated Male Rats also Treated with Inducer Drugs

Additional Treatment	Meperidine N-Demethylase Activity[a]	
	Control	Morphine
None	104 ± 10	32 ± 7
Reserpine	99 ± 10	47 ± 3
Phenobarbital	185 ± 13	98 ± 20
Chlorpromazine	106 ± 9	65 ± 5
19-Nortestosterone	125 ± 3	97 ± 6
19-Nortestosterone + phenobarbital	150 ± 8	121 ± 9

[a] Micromoles per hour per gram microsomal-supernatant protein. The doses are: morphine, 30 mg/kg daily for 10 days; reserpine, 0.25 mg/kg on days 1 and 6; chlorpromazine, 2 mg/kg daily; phenobarbital, 60 mg/kg on days -2, 2, 5, and 8; and 19-nortestosterone, 5 mg/kg on days -1, 2, 5, and 8 of morphine treatment. Enzyme activity is measured by the color reaction of formaldehyde from the oxidized methyl group. (Adapted from Clouet and Ratner by courtesy of the *Journal of Pharmacology and Experimental Therapeutics*.[43])

enzymatic activity.[41] Many narcotic drugs have this effect on the activity of enzymes catalyzing the catabolism of the injected drug as well as that of other narcotic drugs,[42,71,136] and also they decrease the metabolism of nonnarcotic drugs such as hexobarbital.[43,137] The administration of inducer drugs such as phenobarbital or chlorpromazine to rats results in an increase in the activity of enzymes in the liver microsomal system catalyzing the demethylation of morphine, meperidine, acetylmethadol, dihydromorphine, and codeine.[43,71,138] The responses of the enzymatic *N*-demethylase activity in male rat liver is shown in Table III in which the induction by phenobarbital, reserpine, or chlorpromazine; the inhibition by morphine; and the maintainance of control levels by a combination of morphine and inducer drug are illustrated. In these animals there is no relationship between the rate of development of tolerance to morphine and the level of liver drug metabolizing enzymes.[43] Instead, this phenomenon of morphine inhibition of liver drug-metabolizing enzymes seems to be related to the sex difference in drug-metabolizing activity in rat liver, since neither the sex difference nor the inhibition by morphine is found in other species.[16,47]

VI. REFERENCES

1. E. L. Way and T. K. Adler, *Pharmacol. Rev. 12*, 383–445 (1960).
2. E. L. Way and T. K. Adler, *in* "The Biological Disposition of Morphine and Its Surrogates," World Health Organization, Geneva, Switzerland (1962).
3. R. E. McMahon, *J. Med. Pharm. Chem. 4*, 67–70 (1961).
4. R. E. McMahon, H. W. Culp, and F. J. Marshall, *J. Pharmacol. Exp. Ther. 149*, 436–445 (1965).
5. J. Axelrod and J. Cochin, *J. Pharmacol. Exp. Ther. 121*, 107–112 (1957).
6. L. Leadbeater and D. R. Davies, *Biochem. Pharmacol. 13*, 1569–1576 (1964).
7. C. Elison, H. W. Elliott, M. Look, and H. Rapoport, *J. Med. Chem. 6*, 237–246 (1963).
8. E. L. Way, J. M. Young, and J. Kemp, *Bull. Narcotics 17*, 25–33 (1965).
9. R. E. Dohrmann, *Arzneim.-Forsch. 18*, 854–862 (1968).
10. C. C. Porter, *J. Pharmacol. Exp. Ther. 120*, 447–451 (1957).
11. J. M. Fujimoto and V. B. Haarstad, *J. Pharmacol. Exp. Ther. 165*, 45–51 (1969).
12. H. Remmer and H. J. Merker, *Science 142*, 1657–1658 (1963).
13. A. L. Misra, H. I. Jacoby, and L. A. Woods, *J. Pharmacol. Exp. Ther. 132*, 311–316 (1961).
14. L. B. Mellett and L. A. Woods, *Proc. Soc. Exp. Biol. Med. 106*, 221–223 (1961).
15. A. H. Conney and J. J. Burns, *Advan. Pharmacol. 1*, 31–58 (1962).
16. R. Kato and K. Onodo, *Japan. J. Pharmacol. 16*, 217–218 (1966).
17. R. L. Furner, T. E. Gram, and R. E. Stitzel, *Biochem. Pharmacol. 18*, 1635–1641 (1969).
18. J. G. Page and E. S. Vessel, *Pharmacology 2*, 321–336 (1969).
19. R. Kato and J. R. Gillette, *J. Pharmacol. Exp. Ther. 150*, 285–291 (1965).
20. D. H. Clouet, *J. Pharmacol. Exp. Ther. 144*, 354–361 (1964).
21. A. Morello, *Nature 203*, 785–786 (1964).
22. E. Bresnick, *Mol. Pharmacol. 2*, 406–410 (1966).
23. T. Johannesson and K. Milthers, *Acta Pharmacol. Toxicol. 19*, 241–246 (1962).
24. H. E. Dobbs, G. F. Blane, and A. L. A. Boura, *Eur. J. Pharmacol. 7*, 328–332 (1969).
25. G. F. Blane and H. E. Dobbs, *Brit. J. Pharmacol. 30*, 166–172 (1967).
26. J. H. Sanner and L. A. Woods, *J. Pharmacol. Exp. Ther. 148*, 176–184 (1965).
27. J. F. Taylor, *Fed. Proc. 29*, 686 (1970).

28. W. A. Bleyer and M. G. Rosen, *Electroencephalog. Clin. Neurophysiol.* 24, 249–258 (1968).
29. R. S. Foster, D. J. Jenden, and P. Lomax, *J. Pharmacol. Exp. Ther.* 157, 185–195 (1967).
30. T. K. Adler, *J. Pharmacol. Exp. Ther.* 140, 155–161 (1963).
31. C. Maggiolo and F. Huidobro, *Acta Physiol. Lat. Amer.* 11, 70–78 (1961).
32. J. R. Weeks and R. J. Collins, *in* "The Addictive States" (A. Wikler, ed.), Williams & Wilkins, Baltimore (1968).
33. C. R. Shuster, *Fed. Proc.* 29, 2–5 (1970).
34. A. Wikler, W. R. Martin, F. T. Pecor, and C. G. Eades, *Psychopharmacolgia 5*, 55–78 (1963).
35. J. R. Nichols, *in* "The Addictive States" (A. Wikler, ed.), Williams & Wilkins, Baltimore (1968).
36. B. M. Cox, M. Ginsburg, and O. H. Osman, *Brit. J. Pharmacol.* 33, 245–256 (1968).
37. H. O. Andersen and H. Holmen-Christensen, *Dan. Tidsskr. Farm.* 43, 117–126 (1969).
38. J. Axelrod, *J. Pharmacol. Exp. Ther.* 114, 430–438 (1955).
39. A. H. Conney, *Pharmacol. Rev.* 19, 317–366 (1967).
40. F. B. Schenkman and R. Sato, *Mol. Pharmacol.* 4, 613–620 (1968).
41. J. Axelrod, *J. Pharmacol. Exp. Ther.* 117, 322–330 (1956).
42. H. Remmer and B. Absleben, *Klin. Wochschr.* 36, 332–333 (1958).
43. D. H. Clouet and M. Ratner, *J. Pharmacol. Exp. Ther.* 144, 362–372 (1964).
44. N. E. Sladek and G. J. Mannering, *Mol. Pharmacol.* 5, 186–199 (1969).
45. W. J. George and T. R. Tepley, *Mol. Pharmacol.* 4, 502–509 (1968).
46. J. Cochin and J. Axelrod, *J. Pharmacol. Exp. Ther.* 125, 105–110 (1959).
47. J. A. Castro and J. R. Gillette, *Biochem. Biophys. Res. Commun.* 28, 426–430 (1967).
48. H. Remmer and H. J. Merker, *Ann. N.Y. Acad. Sci.* 123, 79–97 (1965).
49. J. L. Strominger, E. S. Maxwell, J. Axelrod, and H. M. Kalckar, *J. Biol. Chem.* 224, 79–90 (1957).
50. A. E. Takemori, *J. Pharmacol. Exp. Ther.* 130, 370–374 (1960).
51. N. P. Plotnikoff, E. L. Way, and H. W. Elliott, *J. Pharmacol. Exp. Ther.* 117, 414–419 (1956).
52. O. Schaumann, *Arch. Exp. Pathol. Pharmakol.* 239, 311–320 (1960).
53. E. L. Way, *Fed. Proc.* 26, 1115–1118 (1967).
54. K. Milthers, *Nature 195*, 607 (1962).
55. C. Elison and H. W. Elliott, *Biochem. Pharmacol.* 12, 1363–1366 (1963).
56. D. H. Clouet, *Biochem. Pharmacol.* 12, 967–972 (1963).
57. J. Axelrod, *J. Pharmacol. Exp. Ther.* 138, 28–33 (1962).
58. G. R. Van Petten, G. H. Hirsch, and A. D. Cherrington, *Can. J. Biochem.* 46, 1057–1061 (1968).
59. C. Finzi and F. Conti, *Atti Accad. Med. Lomb.* 23, 294–298 (1969).
60. Drug Research Board, *Clin. Pharmacol. Therap.* 10, 607–634 (1969).
61. C. Y. Sung and E. L. Way, *J. Pharmacol. Exp. Ther.* 110, 260–270 (1954).
62. R. M. Veatch, T. K. Adler, and E. L. Way, *J. Pharmacol. Exp. Ther.* 145, 11–19 (1964).
63. C. M. Gruber and A. Baptisti, *Clin. Pharmacol. Ther.* 4, 172–181 (1963).
64. S.-C. C. Lin and E. L. Way, *J. Pharmacol. Exp. Ther.* 150, 309–315 (1965).
65. H. F. Kuhn and H. Friebel, *Med. Exp.* 7, 255–261 (1962).
66. S. Y. Yeh and L. A. Woods, *J. Pharmacol. Exp. Ther.* 166, 86–95 (1969).
67. T. Johannesson and J. Schou, *Acta Pharmacol. Toxicol.* 20, 165–173 (1963).
68. H. W. Elliott, N. Nomof, and K. D. Parker, *Clin. Pharmacol. Ther.* 8, 78–85 (1967).
69. A. E. Takemori and G. J. Mannering, *J. Pharmacol. Exp. Ther.* 123, 171–179 (1958).
70. T. K. Adler, *J. Pharmacol. Exp. Ther.* 156, 585–590 (1967).

71. T. Johannesson, L. A. Rogers, J. R. Fouts, and L. A. Woods, *Acta Pharmacol. Toxicol. 22*, 107–111 (1965).
72. A. Rubin, T. R. Tephly, and G. J. Mannering, *Biochem. Pharmacol. 13*, 1053–1057 (1964).
73. S. J. Mulé and C. W. Gorodetzky, *J. Pharmacol. Exp. Ther. 154*, 632–645 (1966).
74. S. J. Mulé, T. H. Clements and C. W. Gorodetzky, *J. Pharmacol. Exp. Ther. 160*, 387–396 (1968).
75. W. M. Benson, P. L. Stefko, and L. O. Randall, *J. Pharmacol. Exp. Ther. 109*, 189–200 (1953).
76. H. Isbell and H. F. Fraser, *J. Pharmacol. Exp. Ther. 107*, 524–530 (1953).
77. J. J. Kamm, A. B. Taddeo, and E. J. VanLoom, *J. Pharmacol. Exp. Therap. 158*, 437–444 (1967).
78. R. Versie, A. Noirfalise, M. Neven, and R. Malchair, *Ann. Med. Leg. Criminol. Police Sci. 42*, 561–565 (1962).
79. K. Willner, *Arzneim.-Forsch. 13*, 26–29 (1963).
80. J. H. Sanner and L. A. Woods, *J. Pharmacol. Exp. Ther. 148*, 176–184 (1965).
81. C. C. Hug, Jr., and L. B. Mellett, *J. Pharmacol. Exp. Ther. 149*, 446–453 (1965).
82. S. Y. Yeh and L. A. Woods, *J. Pharmacol. Exp. Ther. 169*, 168–174 (1969).
83. J. T. Wilson, *Nature 225*, 861–863 (1970).
84. M. W. Anders and G. J. Mannering, *Mol. Pharmacol. 2*, 318–327 (1966).
85. A. Rubin, T. R. Tephly, and G. J. Mannering, *Biochem. Pharmacol. 13*, 1007–1016 (1964).
86. E. L. Way, J. W. Kemp, J. M. Young, and D. R. Grassetti, *J. Pharmacol. Exp. Ther. 129*, 144–154 (1960).
87. W. R. Martin and H. F. Fraser, *J. Pharmacol. Exp. Ther. 133*, 388–399 (1961).
88. A. H. Beckett, J. F. Taylor, A. F. Casy, and M. M. A. Hassan, *J. Pharm. Pharmacol. 20*, 754–762 (1968).
89. H. W. Elliott and C. Elison, *J. Pharmacol. Exp. Ther. 131*, 31–37 (1961).
90. H. Yoshimura, K. Oguri, and H. Tsukamoto, *Biochem. Pharmacol. 18*, 279–286 (1969).
91. H. Yoshimura, K. Oguri, and H. Tsukamoto, *Chem. Pharm. Bull. 16*, 2114–2119 (1968).
92. H. Yoshimura, K. Oguri, and H. Tsukamoto, *Tetrahedron Lett. 4*, 483–486 (1968).
93. A. E. Takemori and G. A. Glowacki, *Biochem. Pharmacol. 11*, 867–870 (1962).
94. E. Sanchez, L. Tampier, and J. Mardones, *Eur. J. Pharmacol. 7*, 106–110 (1969).
95. W. F. Bousquet, B. D. Rupe, and T. S. Miya, *Biochem. Pharmacol. 13*, 123–125 (1964).
96. L. A. Woods and H. I. Chernov, *Pharmacologist 8*, 206 (1966).
97. D. G. May, J. M. Fujimoto, and C. E. Intrurrisi, *J. Pharmacol. Exp. Ther. 157*, 626–635 (1967).
98. J. Axelrod, *Science 124*, 263–264 (1956).
99. A. H. Beckett, A. F. Casy, and N. J. Harper, *J. Pharm. Pharmacol. 8*, 874–884 (1956).
100. A. H. Beckett, A. F. Casy, N. J. Harper, and P. M. Phillips, *J. Pharmacol. 8*, 860–873 (1956).
101. A. L. Misra, S. J. Mulé and L. A. Woods, *Nature 190*, 82–83 (1961).
102. A. L. Misra, S. J. Mulé and L. A. Woods, *J. Pharmacol. Exp. Ther. 132*, 317–322 (1961).
103. K. Milthers, *Acta Pharmacol. Toxico. 19*, 149–155 (1962).
104. L. Tampier and A. Penna-Herreros, *Arch. Biol. Med. Exp. 3*, 146–147 (1966).
105. K. Milthers, *Acta Pharmacol. Toxicol. 19*, 235–240 (1962).
106. A. Inouye and Y. Shinagawa, *J. Neurochem. 12*, 803–813 (1965).
107. D. H. Clouet, *Life Sci. 1*, 31–34 (1962).
108. C. Elison and H. W. Elliott, *J. Pharmacol. Exp. Ther. 144*, 265–275 (1964).
109. C. Elison, H. Rapoport, R. Laursen, and H. W. Elliott, *Science 134*, 1078–1079 (1961).

110. J. Daly, J. K. Inscoe, and J. Axelrod, *J. Med. Chem. 8*, 153–157 (1965).
111. J. T. C. Woo, G. A. Gaff, and M. R. Fennessy, *J. Pharm. Pharmacol. 20*, 763–767 (1968).
112. M. S. Fish, N. M. Johnson, E. P. Lawrence, and E. C. Horning, *Biochim. Biophys. Acta 18*, 564–565 (1955).
113. R. E. McMahon, *J. Pharm. Sci. 55*, 457–466 (1966).
114. B. R. Schmid, J. Axelrod, L. Hammaker, and L. Swarm, *J. Clin. Invest. 37*, 1123–1127 (1958).
115. N. B. Eddy *in* "The Addictive States" (A. Wikler, ed.), Williams & Wilkins, Baltimore (1968).
116. E. M. Fuentes, *An. Fac. Quim. Farm., Univ. Chile 12*, 238–243 (1960).
117. S. Tamaka, *Nippon Yakurigaku Zasshi 57*, 513–519 (1961).
118. C. C. Hug, Jr., and L. A. Woods, *J. Pharmacol. Exp. Ther. 142*, 248–256 (1963).
119. J. M. Fujimoto, W. M. Watrous, and V. B. Haarstad, *Proc. Soc. Exp. Biol. Med. 130*, 541–549 (1969).
120. M. F. Lockett and M. M. Davis, *J. Pharm. Pharmacol. 10*, 80–85 (1958).
121. C. Maggiolo and F. Huidobro, *Nature 211*, 540–541 (1966).
122. J. M. Fujimoto, *J. Pharmacol. Exp. Ther. 168*, 180–186 (1969).
123. L. Lasagna and T. J. deKornfield, *J. Pharmacol. Exp. Ther. 124*, 260–263 (1958).
124. P. D. Orahovats and F. G. Lehman, *J. Pharmacol. Exp. Ther. 122*, 58A–59A (1958).
125. S. G. Holtzmann and J. E. Villarreal, *J. Pharmacol. Exp. Ther. 166*, 125–133 (1969).
126. D. H. Clouet, M. Ratner, and M. Kurzman, *Biochem. Pharmacol. 12*, 957–966 (1963).
127. J. Axelrod, *Life Sci. 1*, 29–30 (1962).
128. K. A. Pittman, D. Rosi, R. Cherniak, A. J. Merola, and W. D. Conway, *Biochem. Pharmacol. 18*, 1673–1678 (1969).
129. A. H. Beckett and J. F. Taylor, *J. Pharm. Pharmacol. 19*, 505–506 (1967).
130. B. Berkowitz and E. L. Way, *Clin. Pharmacol. Ther. 10*, 681–689 (1969).
131. H. F. Fraser and H. Isbell, *Bull. Narcotics 12*, 9–14 (1960).
132. H. M. Lee, E. G. Scott, and A. Pohland, *J. Pharmacol. Exp. Ther. 125*, 14–18 (1959).
133. M. E. Amundsen, M. L. Johnson, and J. A. Manthey, *J. Pharm. Sci. 54*, 684–686 (1965).
134. S. J. Mulé *in* "The Scientific Basis of Drug Dependence" (H. Steinberg, ed.), Churchill, London (1969).
135. I. L. Gutierrez, *Ann. Fac. Quim. Farm. 18*, 187–192 (1966).
136. G. J. Mannering and A. E. Takemori, *J. Pharmacol. Exp. Ther. 127*, 187–190 (1959).
137. H. Remmer, *Arch. Exp. Pathol. Pharmakol. 238*, 35 (1960).
138. L. Shuster and R. V. Hannan, *Can. J. Biochem. 43*, 899–906 (1965).

III. The Effects of Narcotic Analgesic Drugs on General Metabolic Systems

Chapter 7

INTERMEDIARY AND ENERGY METABOLISM

A. E. Takemori

Department of Pharmacology College of Medical Sciences
University of Minnesota,
Minneapolis, Minnesota

I. INTRODUCTION

The effects of morphine and its analogs on the oxidative metabolism of various tissues in the body have been studied by many investigators, and earlier studies on this subject have been summarized by Krueger, Eddy, and Sumwalt[1] and Reynolds and Randall.[2] While many of the studies were extensive, the experiments were performed on tissues other than brain. The interpretation of data from such studies with respect to analgesia, narcosis, or tolerance that occur in the central nervous system becomes difficult. A brief summary of the effects of narcotic analgesics on the metabolism of various preparations of the brain and on the concentrations of cerebral glycolytic intermediates are presented in this chapter.

II. EFFECT OF NARCOTICS *IN VITRO* ON RESPIRATION AND GLYCOLYSIS OF VARIOUS PREPARATIONS OF BRAIN

In 1932, Quastel and Wheatley[3] studied the effects of hypnotics and anesthetics on the respiration of brain minces in the presence of various substrates and advanced the theory that drugs produced narcosis by inhibiting energy-yielding oxidative processes of the neurons. Since this theory was proposed, many investigations have been performed to see whether narcotic analgesics affected tissue respiration of the cerebral cortex and whether inhibition of tissue respiration related to analgesia and/or narcosis.

The effects of various opiates and opioids and their analogs on cerebral metabolism are summarized in Table I. The preponderant finding is that when opiates and opioids exert an inhibitory effect on oxygen uptake of

TABLE I

Effect of Narcotic Analgesics and Their Analogs *In Vitro* on the Cerebral Metabolism of Various Substrates[a]

Drug	Species	Parameter Measured	In Vitro Preparation	Substrate Employed	Concentration of Drug In Vitro (M)	Results	Reference
Alpha-prodine	rat	O_2 uptake	cerebral cortical slices	glucose	1×10^{-2}	42% inhibition	(4)
				lactate	1×10^{-2}	26% inhibition	(4)
				succinate	1×10^{-2}	11% inhibition	(4)
				pyruvate	1×10^{-2}	53% inhibition	(4)
				$\alpha + \beta -$glycero-phosphate	1×10^{-2}	32% inhibition	(4)
				glutamate	1×10^{-2}	16% inhibition	(4)
				oxaloacetate	1×10^{-2}	no effect	(4)
Betaprodine	rat	O_2 uptake	cerebral cortical slices	glucose	1×10^{-2}	41% inhibition	(4)
				lactate	1×10^{-2}	33% inhibition	(4)
				succinate	1×10^{-2}	13% inhibition	(4)
				pyruvate	1×10^{-2}	56% inhibition	(4)
				$\alpha + \beta -$glycero-phosphate	1×10^{-2}	15% inhibition	(4)
				glutamate	1×10^{-2}	10% inhibition	(4)
				oxaloacetate	1×10^{-2}	no effect	(4)
Codeine	beef	O_2 uptake	brain homogenate	succinate	5×10^{-3}	5% inhibition	(49)
					1×10^{-2}	15% inhibition	(49)
					2×10^{-2}	35% inhibition	(49)
					4×10^{-2}	65% inhibition	(49)
Dextro-methorphan	rat	O_2 uptake	cerebral cortical slices	glucose	1×10^{-2}	no effect	(10)
Dex-trorphanol	rat	O_2 uptake	cerebral cortical slices	glucose	1×10^{-2}	no effect	(10)

Drug	Animal	Measure	Preparation	Substrate	Concentration	Effect	Ref.
Levallorphan	rat	O_2 uptake	cerebral cortical slices	glucose	1×10^{-2}	no effect	(10)
Levorphanol	rat	O_2 uptake	cerebral cortical slices	glucose	1×10^{-2}	no effect	(10)
Meperidine	rat	O_2 uptake	cerebral cortical slices	glucose	5×10^{-4}	no effect	(8)
					1×10^{-3}	no effect	(8)
					2×10^{-3}	no effect	(8)
					2.5×10^{-3}	no effect	(5)
					5×10^{-3}	40% inhibition	(5)
					7.5×10^{-3}	90% inhibition	(5)
					1×10^{-2}	95% inhibition	(5)
			K^+-stimulated cerebral cortical slices	glucose	5×10^{-4}	23% inhibition	(8)
					1×10^{-3}	44% inhibition	(8)
					2×10^{-3}	50% inhibition	(8)
	guinea pig	O_2 uptake	cerebral cortical slices	glucose	5×10^{-4}	no effect	(45)
		O_2 uptake	electrically stimulated cerebral cortical slices	glucose	1×10^{-5}	32% inhibition	(45)
					5×10^{-5}	54% inhibition	(45)
					2.5×10^{-4}	64% inhibition	(45)
					5×10^{-4}	84% inhibition	(45)
	rat	anaerobic glycolysis	brain homogenate (CO_2 production)	glucose	2.6×10^{-3}	23–39% stimulation	(11)
	beef	O_2 uptake	brain homogenate	succinate	1×10^{-2}	100% inhibition	(5)
					5×10^{-3}	10% inhibition	(49)
					1×10^{-2}	25% inhibition	(49)
					2×10^{-2}	55% inhibition	(49)
					4×10^{-2}	95% inhibition	(49)
Methadone	rat	O_2 uptake	cerebral cortical slices	glucose	5×10^{-5}	no effect	(8)
					1×10^{-4}	no effect	(7, 8)

TABLE I (cont'd)

Drug	Species	Parameter Measured	In vitro Preparation	Substrate Employed	Concentration of Drug In Vitro (M)	Results	Reference
Methadone	rat	O_2 uptake	cerebral cortical slices	glucose	1.6×10^{-4}	20% stimulation	(5)
					2×10^{-4}	32% stimulation	(8)
					3×10^{-4}	28% stimulation	(7)
					3×10^{-4}	25% stimulation	(5)
					5×10^{-4}	53% stimulation	(8)
					5.3×10^{-4}	33% stimulation	(7)
					1×10^{-3}	80% stimulation	(5)
					1×10^{-3}	no effect	(8)
					2×10^{-3}	95% inhibition	(5)
					2×10^{-3}	42% inhibition	(7)
					2×10^{-3}	57% inhibition	(8)
					5×10^{-3}	95% inhibition	(5)
					1×10^{-2}	75% inhibition	(4)
				lactate	1×10^{-3}	74% inhibition	(5)
					1×10^{-2}	44% inhibition	(4)
				pyruvate	1×10^{-3}	80% inhibition	(5)
					1×10^{-2}	58% inhibition	(4)
	human	O_2 uptake	cerebral cortical slices	glucose	5×10^{-4}	stimulation	(50)
					2×10^{-3}	strong inhibition	(50)
	rat	O_2 uptake	cerebral cortical slices	succinate	1×10^{-2}	20% inhibition	(4)
				$\alpha+\beta-$glycero-phosphate	1×10^{-2}	34% inhibition	(4)
				glutamate	1×10^{-2}	no effect	(4)
				oxaloacetate	1×10^{-2}	no effect	(4)
			K^+-stimulated	glucose	5×10^{-5}	no effect	(8)

Preparation	Substrate	Concentration	Effect	Ref.
cerebral cortical slices		1×10^{-4}	50% inhibition	(8)
		2×10^{-4}	66% inhibition	(8)
		5×10^{-4}	no effect	(8)
		1×10^{-3}	34% inhibition	(8)
		2×10^{-3}	100% inhibition plus 35% inhibition of unstimulated uptake	(8)
brain homogenate	glucose	$5 \times 10^{-3}(dl)$	65% inhibition	(9)
		$5 \times 10^{-3}(d)$	50% inhibition	(9)
		$5 \times 10^{-3}(l)$	56% inhibition	(9)
	succinate	4.4×10^{-4}	13% stimulation	(51)
		8.8×10^{-4}	14% stimulation	(51)
		1.76×10^{-3}	32% inhibition	(51)
		$7 \times 10^{-3}(dl)$	74% inhibition	(9)
		$7 \times 10^{-3}(d)$	66% inhibition	(9)
		$7 \times 10^{-3}(l)$	76% inhibition	(9)
	pyruvate	5.8×10^{-4}	16–27% inhibition	(51)
		1.16×10^{-3}	32–56% inhibition	(51)
		2.3×10^{-3}	89% inhibition	(51)
	ascorbate	$5 \times 10^{-3}(dl)$	53% inhibition	(9)
		$5 \times 10^{-3}(d)$	53% inhibition	(9)
		$5 \times 10^{-3}(l)$	76% inhibition	(9)
beef O_2 uptake brain homogenate	succinate	1×10^{-2}	95% inhibition	(49)
		2×10^{-2}	95% inhibition	(49)
		4×10^{-2}	95% inhibition	(49)
rat anaerobic glycolysis (CO_2 production) brain homogenate	glucose	6×10^{-4}	18% stimulatioon	(11)
		1.2×10^{-3}	21–35% inhibition	(11)
		2×10^{-3}	59% inhibition	(5)
		2.4×10^{-3}	55–75% inhibition	(11)

TABLE I (Continued)

Drug	Species	Parameter Measured	In Vitro Preparation	Substrate Employed	Concentration of Drug In Vitro (M)	Results	Reference
Morphine	rat	anaerobic glycolysis (CO_2 production)	brain homogenate	glucose	$5 \times 10^{-3}(dl)$	175% stimulation	(9)
					$5 \times 10^{-3}(d)$	241% stimulation	(9)
					$5 \times 10^{-3}(l)$	162% stimulation	(9)
			brain "cell-free extract"	glucose	2.4×10^{-3}	25–47% inhibition	(11)
				glycogen	2.4×10^{-3}	no effect	(11)
				hexose diphosphate	2.4×10^{-3}	no effect	(11)
	guinea pig	O_2 uptake	brain mince	glucose	3.2×10^{-3}	32% inhibition	(3)
				lactate	3.2×10^{-3}	30% inhibition	(3)
				pyruvate	3.2×10^{-3}	30% inhibition	(3)
				glutamate	3.2×10^{-3}	no effect	(3)
				succinate	3.2×10^{-3}	24% inhibition	(3)
	rat	O_2 uptake	brain mince	glucose	2.1×10^{-4}	no effect	(52)
					8.4×10^{-4}	no effect	(52)
					3.2×10^{-3}	no effect	(53)
					3.2×10^{-3}	8–40% inhibition	(54)
				lactate	3.2×10^{-3}	40% inhibition	(53)
				pyruvate	3.2×10^{-3}	no effect	(53)
				citrate	3.2×10^{-3}	no effect	(53)
				α-ketoglutarate	3.2×10^{-3}	no effect	(53)
				succinate	3.2×10^{-3}	no effect	(53)
				fumarate	3.2×10^{-3}	no effect	(53)
				malate	3.2×10^{-3}	no effect	(53)

cerebral cortical slices	glucose	1×10^{-4}	no effect	(6)
		5×10^{-4}	no effect	(44)
		1×10^{-3}	no effect	(6, 38)
		2×10^{-3}	no effect	(50)
		3.2×10^{-3}	no effect	(53)
		5×10^{-3}	no effect	(5)
		1×10^{-2}	no effect	(4, 5, 6)
	lactate	3.2×10^{-3}	10% inhibition	(53)
	pyruvate	3.2×10^{-3}	10% inhibition	(53)
	citrate	3.2×10^{-3}	no effect	(53)
	α-ketoglutarate	3.2×10^{-3}	30% inhibition	(53)
	succinate	3.2×10^{-3}	no effect	(53)
	fumarate	3.2×10^{-3}	no effect	(53)
	malate	3.2×10^{-3}	no effect	(53)
electrically stimulated cerebral cortical slices	glucose	1×10^{-5}	no effect	(6)
		1×10^{-3}	50–60% inhibition	(6)
K$^+$-stimulated cerebral cortical slices	glucose	1×10^{-3}	40–60% inhibition	(38, 44, 55, 56)
		2×10^{-3}	50% inhibition in Ca^{++}-free medium	(39)
brain homogenate	glucose	3.2×10^{-3}	20% inhibition	(53)
	lactate	3.2×10^{-3}	20% inhibition	(53)
	pyruvate	3.2×10^{-3}	20% inhibition	(53)
	citrate	3.2×10^{-3}	no effect	(53)
	α-ketoglutarate	3.2×10^{-3}	no effect	(53)
	succinate	3.2×10^{-3}	no effect	(53)
	fumarate	3.2×10^{-3}	no effect	(53)
	malate	3.2×10^{-3}	no effect	(53)

TABLE I (Continued)

Drug	Species	Parameter Measured	In Vitro Preparation	Substrate Employed	Concentration of Drug In Vitro (M)	Results	Reference
	beef	O_2 uptake	brain homogenate	succinate	5×10^{-3}	17% inhibition	(49)
					1×10^{-2}	30% inhibition	(49)
					2×10^{-2}	55% inhibition	(49)
					4×10^{-2}	85% inhibition	(49)
	human	O_2 uptake	cerebral cortical slices	glucose	2×10^{-3}	no effect	(50)
	rat	anaerobic glycolysis (CO_2 production)	brain homogenate	glucose	2.8×10^{-3}	no effect	(11)
					5.6×10^{-3}	no effect	(11)
		anaerobic glycolysis (lactate formation)	cerebral cortical slices	glucose	1×10^{-3}	no effect	(12)
			K^+-depressed cerebral cortical slices	glucose	1×10^{-3}	32% stimulation	(12)
		aerobic glycolysis (lactate formation)	cerebral cortical slices	glucose	1×10^{-4}	no effect	(6)
					1×10^{-3}	no effect	(6, 12)
					1×10^{-2}	no effect	(6)
			K^+-stimulated cerebral cortical slices	glucose	1×10^{-3}	no effect	(12)

aerobic glucose uptake	cerebral cortical slices	glucose	1×10^{-3}	57% stimulation	(12)
	K^+-stimulated cerebral cortical slices	glucose	1×10^{-3}	31% stimulation	(12)
anaerobic glucose uptake	cerebral cortical slices	glucose	1×10^{-3}	no effect	(12)
	K^+-depressed cerebral cortical slices	glucose	1×10^{-3}	171% stimulation	(12)
glucose utilization	cerebral homogenate	glucose	1×10^{-6}	17% stimulation	(12, 15)
			1×10^{-5}	44% stimulation	(12, 15)
			1×10^{-3}	100% stimulation	(12, 15)
$^{14}CO_2$ formation	cerebral homogenate	6-^{14}C-glucose	1×10^{-3}	57% stimulation	(15)
		1-^{14}C-glucose	1×10^{-3}	33% stimulation	(15)
		2-^{14}C-glucose	1×10^{-3}	53% stimulation	(15)
		6-^{14}C-glucose	1×10^{-6}	22% stimulation	(18)

a From Takemori.[62]

cerebral slices or homogenates, extremely high concentrations are required. The concentrations are much, much higher than those one expects to find in the brain *in vivo*. For example, alpha- and betaprodine did not inhibit respiration of cerebral slices until a concentration of $1 \times 10^{-2} M$ was used[4] and morphine did not even show any inhibition at this concentration.[4-6]

Methadone appeared to have a different effect on respiration of cerebral cortical slices. Concentrations of methadone below $1 \times 10^{-4} M$ had no effect,[7,8] concentrations between 1.6×10^{-4} and $5.3 \times 10^{-4} M$ stimulated respiration, and concentrations above $1 \times 10^{-3} M$ inhibited respiration.[5,7,8] The amine portion of the methadone molecule appeared responsible for this diphasic action, but this action is exerted by many agents and is not specific for any one class of drugs.[7] In addition, the pharmacologically less active *d*-form of methadone inhibited the respiration of brain homogenates just as well as the *l*-form did.[9]

Randall and his co-workers[10] found that $1 \times 10^{-2} M$ of neither the analgesically active *l*-form nor the inactive *d*-form of morphinans had any effect on the oxygen uptake of cerebral cortical slices. The *l*- and *d*-forms of the narcotic antagonist, 3-hydroxy-*N*-allylmorphinan, also had no influence on tissue respiration. In view of these findings and the fact that morphine did not alter cerebral respiration and other opioids inhibited respiration only at extremely high concentrations (Table I), the authors concluded that there was no relation between the analgesic effect of a drug and the effect of the drug on the oxygen consumption of the brain.

Anaerobic glycolysis as measured by either CO_2 or lactate formation can be stimulated or inhibited by narcotic analgesics depending on the concentration of the agent *in vitro* (Table I). Meperidine stimulated glycolysis of brain homogenates at a concentration of $1.6 \times 10^{-3} M$[11] and completely inhibited glycolysis at $1 \times 10^{-2} M$.[5] Glycolysis was stimulated by methadone at $6 \times 10^{-4} M$[11] and inhibited by concentrations above $1.2 \times 10^{-3} M$.[5,11] Watts[9] found that concentrations of $5 \times 10^{-3} M$ of *d*- and *l*-isomers of methadone stimulated glycolysis of brain homogenates by 241 and 162%, respectively. However, this stimulation was thought to be attributable to the fact that the control rate of glycolysis fell off after 10 min and methadone allowed the glycolysis to proceed at the original rate. Greig[11] has shown that $2.4 \times 10^{-3} M$ methadone inhibited glycolysis of brain "cell-free extracts" with glucose as the substate but failed to inhibit glycolysis when glycogen or hexose diphosphate was used as the substrate. Also, the addition of ATP to the medium decreased the inhibitory effect of methadone. Therefore she concluded that methadone inhibited hexokinase by competing with ATP for the enzyme.

Although aerobic glycolysis as measured by lactate formation was not altered by morphine,[6,12] aerobic glucose uptake of cerebral slices was stimulated by the drug at $1 \times 10^{-3} M$. Increased glucose utilization by cerebral homogenates also was seen with concentrations as low as $1 \times 10^{-6} M$ morphine.[12] There was a positive correlation between glucose utilization and concentration of morphine in the range of 1×10^{-3} to $1 \times 10^{-6} M$.

Various barbiturates and hypnotics *in vitro* also have been shown to increase aerobic glycolysis of brain homogenates,[13,14] however, the effect of morphine was peculiar in that the increased glucose utilization was observed only after a 30-min lag period.[15] Further investigation of this effect by a systematic analysis of glycolytic intermediates in the incubation medium revealed that morphine decreased the concentration of glyceraldehyde-3-P and increased the level of 3-P glycerate, indicating a facilitation of the glyceraldehyde-3-P dehydrogenase step of glycolysis.[16] It also was found that morphine maintained the concentration of ATP higher and the concentration of AMP lower than those in the control incubation medium. Since AMP was found to inhibit the activity of glyceraldehyde-3-P dehydrogenase appreciably, morphine may have deinhibited this step by keeping the AMP concentration low and thereby stimulated glycolysis. The high ATP concentration in the presence of morphine also provides more substrate for the phosphofructokinase step. Evidence for the stimulation of this step is the observance of an increased concentration of fructose-1,6-diphosphate in the presence of morphine.[16] Since fructose-1,6-diphosphate is the most potent stimulator of phosphofructokinase,[17] its accumulation autocatalytically stimulates the P-fructokinase step and further enhances the glycolytic rate. These results also explain the lag period associated with the morphine effect because the changes in the concentrations of the adenine nucleotides in the incubation medium do not become manifested until 30 min of incubation.[16]

The $^{14}CO_2$ formation from glucose-6-^{14}C, glucose-1-^{14}C, or glucose-2-^{14}C by cerebral homogenates was also increased by morphine.[15] In view of the fact that morphine did not alter oxygen consumption of homogenates appreciably, the inhibition of CO_2 fixation in brain tissue might be considered. In this regard Berl and co-workers[19] have shown that a considerable amount of CO_2 fixation occurs in the brain.

Again, the concentrations of narcotics used *in vitro* to study glycolysis are much above what is found in the brain *in vivo*. However, concentrations of morphine (1×10^{-5} and $1 \times 10^{-6} M$) used to study glucose utilization[12,15] and $^{14}CO_2$ formation[18] by rat cerebral homogenates do approach concentrations attained in the brain after pharmacological doses in this particular species.[20-22] The significance of this is unclear, but the fact that increased glucose utilization in homogenates is seen with concentrations of morphine much lower than that used with slices suggests a permeability barrier against morphine in the cerebral slices. This suggestion would be unlikely according to Scrafani and Hug,[23] who recently showed that narcotic analgesics are actively transported into cerebral cortical slices.

III. EFFECT OF NARCOTICS *IN VITRO* ON STIMULATED CEREBRAL CORTICAL SLICES

The respiratory rates of the brain *in vivo* are considerably higher than those observed in various preparations *in vitro*.[24-27] Elliott[28] estimated that respiratory rates *in vitro* of whole brain are approximately one-half those *in vivo*. It is possible to stimulate the respiration of cerebral cortical slices so that it approximates the respiratory rates found *in vivo* by increasing the K^+ ion concentration of the medium[29,30] or by applying electrical pulses.[31,32] This effect does not occur with homogenates; thus the phenomenon is somehow associated with the integrity of the neuronal membranes. Readers interested in the mechanism of the stimulatory effect can refer to two excellent books by McIlwain[32] and Quastel and Quastel[33] for further discussion. The spontaneous activity of the brain *in vivo* cannot be duplicated *in vitro* even under the best experimental conditions; however, the artificial stimulations do make the cerebral slices more sensitive to various central nervous system depressants.

Since the original observation of the stimulatory effect of K^+ ions on the respiration of cerebral slices by Ashford and Dixon[29] in rabbits and Dickens and Greville[30] in rats, several workers have studied the effect of depressant agents such as hypnotics[31,34] and ethanol[34-37] on the oxidation of K^+-stimulated cerebral slices. Effects of a large variety of drugs also have been studied on electrically stimulated cerebral slices by many investigators and most notably by McIlwain and his colleagues.[32] The general finding was a marked inhibition of the oxidation of stimulated slices with depressant agents at concentrations that were relatively ineffectual on the oxidation of unstimulated slices.

The effect of analgesics on stimulated cerebral cortical slices is summarized in Table I. Morphine, methadone, and meperidine markedly inhibited the K^+-stimulated respiratory rate of cerebral cortical slices at concentrations that had no effect on the oxygen uptake of unstimulated slices.[8,38] The inhibitory effect was restricted to the portion of oxygen uptake that was stimulated by K^+ ions.

Elliott and co-workers[39] were unable to inhibit the K^+-stimulated oxygen uptake of cerebral slices unless the medium was free of Ca^{2+} ions and the concentration of morphine was $2 \times 10^{-3}M$. Quastel and Quastel[33] have pointed out the importance of K^+/Ca^{2+} ratio rather than K^+ ion alone in stimulating the cerebral slices and quoted the fact that removal of Ca^{2+} from the medium containing the usual amount of K^+ ions can also stimulate to the same extent as high concentration of K^+ ion. On the other hand, Elliott and Bilodeau[40] reported that although respiration of cerebral slices is higher in the absence than in the presence of Ca^{2+} ions, the effect of K^+ ions was more marked in the presence of Ca^{2+}. The importance of Ca^{2+} ions may be implicated *in vivo* by the finding by Kakunaga and co-workers[41,42] that Ca^{2+} introduced intracisternally to mice antagonized morphine analgesia.

The concentration of narcotics employed to inhibit K^+-stimulated cerebral slices is still quite high, but there is no doubt the tissue becomes more sensitive to the inhibitory effect of the drugs. McIlwain has claimed that electrical stimulation makes cerebral slices much more sensitive to the effects of drugs than chemical stimulation. For instance, atropine at $1 \times$

$10^{-3}M$ will inhibit electrically stimulated cerebral slices but not K^+-stimulated ones.[43] Morphine appears to inhibit electrically stimulated slices at lower concentrations than those required to inhibit K^+-stimulated slices.[6,44] Meperidine can inhibit electrically stimulated slices at concentrations as low as $1 \times 10^{-5}M$,[45] which easily approach concentrations attained in the brain *in vivo*.[46,47]

Ashford and Dixon[29] showed that K^+ ions not only stimulated respiration of cerebral slices but markedly inhibited anaerobic glycolysis. Interestingly, morphine completely reversed the K^+-depressed glucose uptake and partially reversed the K^+-depressed lactate formation at concentrations $(1 \times 10^{-3}M)$, which had no effect on normal anaerobic glucose uptake or lactate formation.[12] Lower concentrations were not employed, but here again is a situation where an effect of a narcotic is manifested only under conditions of high K^+ ions.

The concentrations of narcotics employed to inhibit stimulated cerebral slices are again quite high, with the possible exception of the effect of meperidine on electrically stimulated slices. The possibility exists that the narcotics may not have reached proper compartmental concentrations within the neuron. Our data indicate that uptake of N-$^{14}CH_3$-morphine by normal cerebral slices does not significantly differ from that by K^+-stimulated slices.[48] Thus the possibility that stimulation of the cerebral slices might increase the permeability of the slices to narcotics is unlikely. The knowledge that the stimulated slices do become more sensitive to the inhibitory effects of narcotics and that the inhibition is restricted to the stimulated portion of the oxygen consumption may help to yield valuable data that may relate to the mechanism of the action of narcotics.

IV. EFFECT OF MORPHINE ADMINISTRATION ON CEREBRAL METABOLISM

In many of the earlier studies, the effect of narcotics were tested only *in vitro*. The effect of narcotic administration *in vivo* on the cerebral metabolism *in vitro* is shown in Table II. Studies in which attempts were made to relate tolerance or physical dependence to cerebral metabolism also are summarized in Table II.

Gross and Pierce[54] showed that after very high acute doses of morphine were given to rats the oxygen uptake of their brain minces in the absence of substrate was slightly higher than that of normal brain minces. The extra oxygen consumption due to added glucose of minces from morphinized rats was 10–50% less than that of minces from normal rats. The brain minces of rats made tolerant to 500 mg/kg respired the same as those of normal rats. They attributed the effect of morphinization on the no substrate oxidation to the level of blood glucose. The increase in blood glucose level seen after an acute dose of morphine was not seen after the rats had become tolerant. Since the initial respiratory rates of the minces from normal and acutely morphinized rats varied greatly, the authors

TABLE II

Effect of Acute and Chronic Morphine Administration on Cerebral Metabolism

Species	Injection Schedule[a]	Parameter Measured	In Vitro Preparation	Substrate Employed	Results	Reference
Rat	500 mg/kg morphine sulfate s.c. 1 hr before sacrifice	O_2 uptake	brain mince	glucose	no substrate uptake 20–30% higher than uptake of mince from normal rats; extra O_2 uptake due to added glucose 10–50% less than that of mince from normal rats	(54)
	100 mg/kg morphine sulfate s.c. progressively increased for 5 weeks in 100 mg/kg increments to 500 mg/kg and maintained at this dose at least 3 weeks	O_2 uptake	brain mince	glucose	no substrate uptake as well as extra uptake due to added glucose same as those of mince from normal rats	(54)
Rabbit	20 mg/kg morphine (calculated as free base) i.v. and sacrificed when maximum respiratory depression ensued	aerobic glucose uptake	cerebral slices	glucose	same as that of slices from control rabbits; addition of $7 \times 10^{-6} M$ morphine into the medium had no effect	(60)
			K^+-stimulated cerebral slices	glucose	same as that of K^+-stimulated slices from control rabbits	(60)
		$^{14}CO_2$ formation from labeled glucose	cerebral slices	^{14}C-glucose	same as that of slices from control rabbits	(60)

Rat	15 mg/kg morphine sulfate s.c. 1 hr before sacrifice	aerobic glucose uptake	K+-stimulated cerebral slices	^{14}C-glucose	same as that of K+-stimulated slices from control rabbits	(60)
			cerebral cortical slices	glucose	52% increase over uptake of slices from control rats	(12)
			K+-stimulated cerebral cortical slices	glucose	41% increase over K+-stimulated uptake of slices from control rats	(12)
		aerobic lactate formation	cerebral cortical slices	glucose	24% increase over formation by slices from control rats	(12)
			K+-stimulated cerebral cortical slices	glucose	not significantly different from formation of K+-stimulated slices of control rats	(12)
		anaerobic glucose uptake	cerebral cortical slices	glucose	86% over uptake of slices from control rats	(12)
			K+-depressed cerebral cortical slices	glucose	264% increase over K+-depressed uptake of slices from control rats	(12)
		anaerobic lactate formation	cerebral cortical slices	glucose	22% increase over formation by slices from control rats	(12)
			K+-depressed cerebral cortical slices	glucose	91% increase over formation by K+-depressed slices from control rats	(12)
		glucose utilization	cerebral homogenate	glucose	44% increase over utilization of homogenate from control rats	(61)

TABLE II (continued)

Species	Injection Schedule[a]	Parameter Measured	In Vitro Preparation	Substrate Employed	Results	Reference
	15 mg/kg morphine sulfate i.p., b.i.d. progressively increased for 3 weeks in 15 mg/kg increments to 45 mg/kg b.i.d.	O_2 uptake	K^+-stimulated cerebral cortical	glucose	$1 \times 10^{-3}M$ morphine *in vitro*, which had previously inhibited K^+-stimulated uptake in slices from control rats, had no effect	(12, 44)
		aerobic glucose uptake	cerebral cortical slices	glucose	93% increase over uptake of slices from control rats	(12)
			K^+-stimulated cerebral cortical slices	glucose	not significantly different from uptake of K^+-stimulated slices of control rats	(12)
	15 mg/kg morphine sulfate i.p., b.i.d. for 7–10 days	O_2 uptake	K^+-stimulated cerebral cortical slices	glucose	$1 \times 10^{-3}M$ morphine or $1 \times 10^{-4}M$ methadone *in vitro* had no effect; $1 \times 10^{-3}M$ meperidine *in vitro* caused 28% inhibition of K^+-stimulated uptake	(8, 38)

same as above and 10 mg/kg nalorphine i.p. 30 min before sacrifice	O₂ uptake	morphine-adapted K⁺-stimulated cerebral cortical slices (see above)	glucose	reversal, i.e., $1 \times 10^{-8}M$ morphine caused 47% inhibition and $1 \times 10^{-4}M$ methadone and $1 \times 10^{-3}M$ meperidine caused 58% inhibition of K⁺-stimulated uptake	(8)
chronically morphinized for 1 month using morphine sulfate i.p., and rats were tolerant to 100 mg/kg/day	O₂ uptake	K⁺-stimulated cerebral cortical slices	glucose	in Ca²⁺-free media, $2 \times 10^{-3}M$ morphine *in vitro*, which had previously inhibited K⁺-stimulated uptake of slices from control rats, had very little effect	(39)
20 mg/kg morphine s.c. progressively increased for 6 weeks in 20 mg/kg increments to 120 mg/kg	O₂ uptake	K⁺-stimulated cerebral cortical slices	glucose	$1 \times 10^{-3}M$ morphine *in vitro*, which had previously inhibited K⁺-stimulated uptake of slices from control rats, had no effect	(55, 56)

[a] s.c. = subcutaneous; i.v. = =intravenous; and i.p. = intraperitoneal.

felt that the inhibition of the extra oxygen uptake due to added glucose in acutely morphinized rats was only apparent and not a true inhibition.

Morphine at a concentration of $1 \times 10^{-3} M$ *in vitro* that had previously inhibited the K^+-stimulated oxygen uptake of cerebral cortical slices from control rats failed to alter the stimulated respiratory rate of slices of rats that were made tolerant to the analgesic and depressant effects of morphine sulfate.[38,44] It appeared that the cells of the cerebral cortex had become adapted during the chronic morphinization so that they no longer responded to the inhibitory effect of morphine *in vitro*. Hano and co-workers[55,56] also showed this cellular adaptation to morphine in rats made tolerant to 120 mg/kg, and Elliott and his colleagues[39] working with Ca^{2+}-free medium reported the cellular adaptation in rats made tolerant to 100 mg/kg.

When the rate of development of this cellular adaptation was studied, the adaptation began to appear after only two or three days of morphine administration to rats, and complete adaptation was seen by the sixth or seventh day[38] (Fig. 1). Morphine sulfate (15 mg/kg) was administered i.p. to rats twice daily and saline was administered to the control group of rats at the same schedule. The effect of $1 \times 10^{-3} M$ morphine *in vitro* on the KCl-stimulated oxygen uptake was measured after various days of injections. Rats were abruptly withdrawn from morphine injections after the seventh day, and the course of recovery was followed. Each point in Fig. 1 represents the mean $\pm$ standard error of five to six rats. The slices became fully sensitive to the depressant effect of morphine after a week of withdrawal. The morphine-adapted cerebral slices also were less sensitive to other analgesics such as methadone and meperidine in concentrations that normally produced marked inhibition of the K^+-stimulated respiration of the cerebral slices. The morphine-adapted cerebral slices

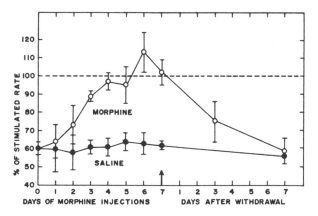

Fig. 1. Effect of morphine administration and withdrawal on the course of cellular adaptation to morphine in cerebral cortical slices of rats. Morphine, open circles, control, closed circles. From Takemori.[38]

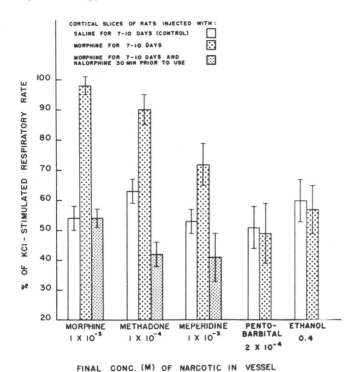

CORTICAL SLICES OF RATS INJECTED WITH:
SALINE FOR 7–10 DAYS (CONTROL) ☐
MORPHINE FOR 7–10 DAYS ▦
MORPHINE FOR 7–10 DAYS AND
NALORPHINE 30 MIN PRIOR TO USE ▦

% OF KCl - STIMULATED RESPIRATORY RATE

FINAL CONC. (M) OF NARCOTIC IN VESSEL

Fig. 2. Effect of several drugs on the KCl-stimulated cerebral cortical slices of morphine-tolerant rats.

were completely cross adapted to the depressant effect of methadone and partially cross adapted to the depressant effect of meperidine[8] (Fig. 2). Morphine sulfate (15 mg/kg) was administered i.p. to rats twice daily for 7–10 days, and saline was injected to the control group of rats at the same schedule. The dose of nalorphine hydrochloride employed was 10 mg/kg i.p. The effect of various drugs *in vitro* on the KCl-stimulated oxygen uptake was measured. The results are expressed in relation to the fully stimulated respiratory rate, which was held at unity for calculation of percentages. The bars represent mean $\pm$ standard error of 10–15 rats. Administration of 10 mg/kg of nalorphine to chronically morphinized rats 30 min before they were sacrificed abolished any established adaptation to morphine, methadone, or meperidine.[8] In the case of methadone, the slices exhibited a significant increase in responsiveness to the depressant effect of the narcotic. Some indication of this hypersensitivity also was seen with meperidine. This observation may relate to the post-tolerance sensitivity referred to by Seevers and Woods.[57]

The cellular adaptation to narcotic analgesics was fairly specific. Other classes of central nervous system depressants such as pentobarbital and ethanol when employed in appropriate concentrations *in vitro* inhibited the K^+-stimulated respiratory rate of cerebral slices from normal rats to about the same extent as morphine. However, the morphine-adapted cerebral slices were not cross adapted to the depressant effects of either pentobarbital or ethanol (Fig. 2). In this regard Wallgren and Lindbohm[58] could not adapt the cerebral slices to the depressant effect of ethanol *in vitro* by the chronic administration of ethanol to rats. Additionally, Turnbull and Stevenson[59] recently reported that cerebral slices from barbital-dependent rats were not adapted to the depressant effect of barbital *in vitro*.

The anaerobic and aerobic glucose uptake and lactate formation of cerebral slices taken from rats administered 15 mg/kg morphine sulfate were significantly increased over those of slices taken from control rats,[12] but this effect was not observed with cerebral slices from morphinized rabbits.[60] Dodge[61] also showed that the glucose utilization of cerebral homogenates from morphinized rats was significantly higher than that of cerebral homogenates from control rats. The glucose uptake of K^+-stimulated cerebral slices of morphinized rats was higher than that of control rats, but the lactate formation did not differ. Both the anaerobic glucose uptake and lactate formation of K^+-depressed cerebral slices from morphinized rats were higher than those of control rats, but the increase in glucose uptake was much greater. Thus cerebral tissue of rats appears to metabolize glucose by a pathway(s) other than that leading to lactate under the influence of morphine.

The aerobic glucose uptake of cerebral slices taken from chronically morphinized rats was approximately double that of cerebral slices taken from control rats. However, the aerobic glucose uptake of K^+-stimulated cerebral slices from chronically morphinized rats was not significantly different from that of K^+-stimulated cerebral slices of control rats.[12]

V. EFFECT OF NARCOTICS ON CERTAIN CEREBRAL ENZYMES *IN VITRO*

The effect of narcotics on many enzymes of various tissues have been studied, but data concerned only with cerebral enzymes are summarized in Tables III and IV. Examination of the effects *in vitro* as shown in Table III reveals that there is very little direct effect of narcotics on most cerebral enzymes. There were no significant effects of morphine on the activities of ATPase, hexokinase, glucose-6-*P* dehydrogenase, phosphofructokinase, aldolase, lactic dehydrogenase, or glyceraldehyde-3-*P* dehydrogenase. Although Wang and Bain[63] found that NAD-cytochrome *c* reductase of brain homogenates of rats was the most sensitive of the cytochrome enzymes to the inhibitory effect of morphine, the minimal effective concentration of morphine was very large $(1 \times 10^{-3}M)$. Other analgesics, including the antagonist nalorphine, which had a free phenolic group on

TABLE III

Effect of Narcotics on Certain Cerebral Enzymes *In Vitro*

Enzyme	Species	Narcotic Added *In Vitro*	Concentration (M)	Results	Reference
ATPase	rat	morphine	1×10^{-3}	no effect	(67)
	rat	morphine	?	no effect	(68)
Hexokinase	rat	morphine	?	no effect	(68)
Glucose-6-*P* dehydrogenase	rat	morphine	1×10^{-6} and 1×10^{-3}	no effect	(64)
Phosphofructokinase	rat	morphine	1×10^{-3}	no effect	(16)
Aldolase	rat	morphine	?	no effect	(68)
Glyceraldehyde-3-*P* dehydrogenase	rat	morphine	1×10^{-3}	no effect	(16)
Lactic dehydrogenase	rat	morphine	?	no effect	(68)
Malic dehydrogenase + NAD-cytochrome *c* reductase	rat	morphine	1×10^{-3}	20% inhibition	(63)
		meperidine	1×10^{-3}	59% inhibition	(63)
		l-methadone	1×10^{-3}	8% inhibition	(63)
NAD-cytochrome *c* reductase	dog	morphine	1×10^{-3}	47% inhibition	(63)
		meperidine	1×10^{-3}	17% inhibition	(63)
		l-methadone	1×10^{-3}	17% inhibition	(63)
	rat	morphine	1×10^{-3}	58% inhibition	(63)
		meperidine	1×10^{-3}	26% inhibition	(63)
		l-methadone	1×10^{-3}	20% inhibition	(63)
		nalorphine	1×10^{-3}	67% inhibition	(63)
		6-monoacetylmorphine	1×10^{-3}	40% inhibition	(63)
		desomorphine	1×10^{-3}	39% inhibition	(63)
		dihydromorphinone	1×10^{-3}	35% inhibition	(63)
		heroin	1×10^{-3}	18% inhibition	(63)
		ethylmorphine	1×10^{-3}	9% inhibition	(63)
		codeine	1×10^{-3}	10% inhibition	(63)
		l-3-OH-*N*-methylmorphinan	1×10^{-3}	10% inhibition	(63)
		d-3-OH-*N*-methylmorphinan	1×10^{-3}	10% inhibition	(63)

TABLE IV

Effect of Acute and Chronic Morphine Administration on Certain Cerebral Enzymes

Enzyme	Species	Treatment of Animals[a]	Results	Reference
ATPase	rat	30 mg/kg morphine sulfate i.p. 1 hr before sacrifice	no effect on activity in the present of no ions, Mg^{2+}, or Mg^{2+} + Na^+ + K^+	(65)
		2.5 mg/kg morphine sulfate i.p. progressively increased to 30 mg/kg for 55–60 days	43% decrease in activity in the absence of ions but no effect in the presence of Mg^{2+} or Mg^{2+} + Na^+ + K^+	(65)
Hexokinase	rat	45 mg/kg morphine hydrochloride s.c. 1 hr before sacrifice	no effect	(69)
	mouse	45 mg/kg morphine hydrochloride s.c. 1 hr before sacrifice	about 16% increase in activity	(69)
Glucose-6-P dehydrogenase	rat	15 mg/kg morphine sulfate s.c. 1 hr before sacrifice	about 26% increase in activity	(64)
Phosphorylase a	rat	45 mg/kg morphine hydrochloride s.c. at various times before sacrifice	no effect	(70)

	mouse	45 mg/kg morphine hydrochloride s.c. at various times before sacrifice	about 28% decrease in activity	(70)
Phosphorylase (total)	rat	45 mg/kg morphine hydrochloride s.c. at various times before sacrifice	no effect	(70)
	mouse	45 mg/kg morphine hydrochloride s.c. at various times before sacrifice	no effect	(70)
Aldolase	rat	Morphine s.c. dose progressively increased to 250 mg/kg in 10 weeks	slight increase	(68)
Malic dehydrogenase	rat	20 mg/kg morphine sulfate s.c. progressively increased to 240 mg/kg in 9 weeks	no effect	(66)
NAD-cytochrome c reductase	rat	20 mg/kg morphine sulfate s.c. progressively increased to 240 mg/kg in 9 weeks	no effect	(66)
Cytochrome oxidase	rat	20 mg/kg morphine sulfate s.c. gressively increased to 240 mg/kg in 9 weeks	no effect; 25% decrease in activity upon withdrawal	(66)

[a] i.p. = intraperitoneal and s.c. = subcutaneous.

the 3-position of the phenanthrene ring, had the same inhibitory effect on the activity of NAD-cytochrome c reductase.[63]

VI. EFFECT OF ACUTE AND CHRONIC MORPHINE ADMINISTRATION ON CERTAIN CEREBRAL ENZYMES

Again morphine at pharmacologic doses has very little influence on most cerebral enzymes studied (Table IV). Even when significant changes in the activities of the enzymes do occur, the changes are relatively small and are not specifically related to the narcotic analgesics. For example, the activity of cerebral glucose-6-P dehydrogenase increased by about 26% after rats had been administered 15 mg/kg morphine. Although the effect on this enzyme was not observed in tissues other than the brain, treatment with other central nervous system depressants such as hypnotics and anesthetics also increased the activity of cerebral glucose-6-P dehydrogenase to the same extent as morphine.[64]

Upon chronic administration of morphine, only the activity of ATPase was decreased,[65] and this effect could be observed only when the assay was performed in the absence of ions. In the presence of Mg^{2+} or Mg^{2+}, Na^+, and K^+, morphine had no effect on ATPase. The activity of NAD-cytochrome c reductase, which was inhibited by morphine *in vitro*, was not altered in chronically morphinized rats.[66]

The effect of morphine administration on cerebral enzymes that are involved in the metabolism of endogenous susbstances such as acetylcholine, 5-hydroxytryptamine, and catecholamines are discussed in other chapters of this book.

VII. EFFECT OF ACUTE AND CHRONIC MORPHINE ADMINISTRATION ON CEREBRAL GLYCOLYTIC INTERMEDIATES

The effect of morphine administration on the levels of glycolytic intermediates and energy-rich compounds in the brain have been studied by several investigators.[61,68−72] Abood et al.[68] reported that brains of rats who were administered 50 mg/kg of morphine sulfate 5 hr before sacrifice contained more glycogen, fructose diphosphate, pyruvic acid, lactic acid, ATP, and ADP and less creatine-P and glucose-6-P than those of control rats. In contrast, Estler and Ammon[69] showed that after 45 mg/kg of morphine hydrochloride in rats there was a small but significant increase in the ATP concentration of brain while no significant change from control in the concentrations of glycogen, lactate, and creatine-P was observed. In the mouse, however, the authors observed a significant decrease in glycogen, lactate, and creatine-P concentrations and an increase in ATP concentration of brain after administration of 45 mg/kg morphine hydrochloride. In an earlier study, Estler and Heim[72] reported that the administration of 5, 15, or 45 mg/kg of morphine hydrochloride in mice resulted in a decrease of ADP concentration in brain in addition to the changes seen above. Only the increase in ATP and the decrease in creatine-P caused by morphine administration were antagonized by 0.5, 1.0, or 5.0 mg/kg of the narcotic antagonist levallorphan. Levallorphan alone did not alter the levels of these compounds in the brain. The authors noted the fact that morphine was a depressant in the rat and a stimulant in the mouse and concluded that the changes in the metabolites of brain are due to changes in functional activity and not to a specific action of morphine on the metabolism of the brain.[69]

Abood et al.[71] also investigated the effects of 10 weeks of chronic morphine administration on cofactors and glycolytic substrates of brain. Chronic morphinization resulted in a fall in glucose-6-P, fructose diphosphate, triose phosphates, and pyruvic acid and a large rise in lactate. ATP levels were increased to about twice that of control levels, but the creatine-P concentration did not change. The authors concluded that the energy reservoirs of the brains of the chronically morphinized rats were similar to those of the controls and allowed the addicted animal to adjust to activity and stress better than the acutely morphinized rats.

The question arises of whether the changes in glycolytic intermediates seen by the above investigators are related to a possible effect of morphine on the control points of glycolysis. Lowry et al.[73,74] examined the possible control points of glycolysis during a study of the effect of phenobarbital on cerebral glycolysis of mice. They found during ischemia, that is, the time between decapitation and freezing, glucose-6-P and fructose-6-P concentrations fell precipitously while the concentration of fructose diphosphate increased greatly. This indicated a facilitation of the phosphofructokinase step, and the authors proposed the following theory of glycolytic control. Whenever ATP formation does not keep up with ATP usage, ADP and P_i concentrations must increase, which in turn produces a greater percentage increase in AMP concentration through the activity of adenylate kinase. The increased levels of P_i, ADP, and particularly AMP enhances phosphofructokinase activity. There were no obvious indications that any step beyond the phosphofructokinase step was affected. When mice were administered 125–150 mg/kg of phenobarbital one hour before sacrifice, the concentration of fructose diphosphate decreased and that of the hexose monophosphates increased. This apparent inhibition of the phosphofructokinase step was accompanied by a rise in brain glucose concentration and a fall in lactate concentration. These results indicated a slowing of the glycolytic flux by the barbiturate.

According to the data of Abood et al.,[71] morphine appears to stimulate the phosphofructokinase step since they observed a fall in glucose-6-P and a rise in fructose diphosphate and lactate concentrations in brains of morphinized rats. It should be noted, however, that these authors reported the average values for only two animals and employed a rather large dose of morphine (50 mg/kg). Additionally, the tissues were not frozen quickly by immersion in liquid nitrogen but were frozen between two pieces of dry ice. Lowry *et al.*[73] have since shown that huge changes in glycolytic substrate and cofactor concentrations occur during the brief period of ischemia prior to freezing.

We have compared the effects of pentobarbital, morphine, and nalorphine administration on cerebral concentrations of glycolytic substrates and cofactors using more pharmacologic doses for rats than previous investigators. The quick freezing method consisting of immersion of the decapitated head immediately into isopentane cooled in liquid nitrogen[73] was used.

Pentobarbital at a dose of 35 mg/kg increased the cerebral concentration of glucose, glucose-6-*P*, and fructose-6-*P* and decreased the pyruvate and lactate concentrations (Fig. 3). This apparent inhibition of cerebral glycolytic flux was accompanied by an increase in ATP and creatine-*P* levels (Fig. 4). These results are similar to those of Lowry *et al.*[73,74] who concluded that phenobarbital decreased cerebral glucose utilization in mice by inhibiting the phosphofructokinase step of glycolysis.

Morphine at a dose of 15 mg/kg also increased the cerebral concen-

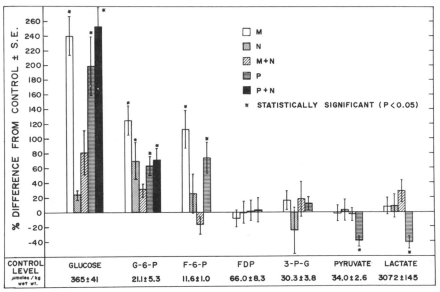

Fig. 3. Effect of morphine, nalorphine, and pentobarbital administration on cerebral glycolytic intermediates of rats. Treatments of the animals in the various groups were as follows. M = 15 mg/kg morphine sulfate injected s.c. 1 hr before sacrifice, N = 10 mg/kg nalorphine hydrochloride i.p. 30 min before sacrifice, M + N = 15 mg/kg morphine sulfate s.c. 1 hr before sacrifice and 10 mg/kg nalorphine hydrochloride i.p. 30 min before sacrifice, P = 35 mg/kg sodium pentobarbital i.p. 30 min before sacrifice, and P + N = 35 mg/kg sodium pentobarbital and 10 mg/kg nalorphine hydrochloride i.p. 30 min before sacrifice. The control group consisted of animals injected with saline solution in the same volume and at the same time before sacrifice. The bars represent mean ± standard error of 5–10 rats. From Dodge.[61]

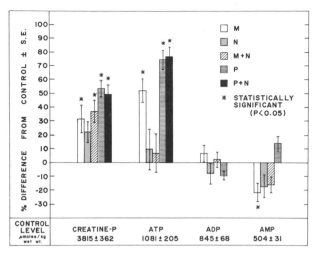

Fig. 4. Effect of morphine, nalorphine, and pentobarbital administration on creatine-P and adenine nucleotide concentrations in brain of rats. Treatment of the animals in the various groups were exactly the same as those described in the legend of Fig. 3. The determinations were made on the same animals presented in Fig. 3, and the bars represent mean ± standard error of 5–10 rats. From Dodge.[61]

trations of glucose, glucose-6-P, and fructose-6-P but had no effect on either pyruvate or lactate concentrations (Fig. 3). Concentrations of ATP and creatine-P were increased, and the AMP concentration was significantly lowered (Fig. 4). That the morphinized animals did not show a fall in lactate and pyruvate levels would suggest that morphine does not slow glucose metabolism. It was discussed earlier that morphine influenced glucose metabolism *in vitro*, at least in part, by stimulating the glyceraldehyde-3-P dehydrogenase step. The low AMP concentration in the presence of morphine was thought to increase the glyceraldehyde-3-P dehydrogenase step indirectly since AMP inhibited the action of this enzyme. Cerebral AMP concentration was significantly decreased after morphine administration (Fig. 4), but 3-P-glycerate concentration was not altered (Fig. 3). These results indicate that although similar alterations in adenine nucleotides were observed *in vitro* and *in vivo*, compensatory mechanisms are such that morphine does not influence the glyceraldehyde-3-P dehydrogenase step *in vivo*.

Further evidence that the effect of morphine on glycolysis differs from that of pentobarbital is the fact that nalorphine reversed all the changes on glycolytic intermediates produced by morphine except the increase in creatine-P. Nalorphine had no influence on the changes produced by pentobarbital and also had no effect of its own on the levels of glycolytic intermediates (Figs. 3 and 4). It is difficult to explain why the change in creatine-P was not reversed by nalorphine; however, Estler and Heim[72] were able to show in mice that levallorphan can reverse the increase in creatine-P as well as the increase in ATP produced by morphine.

The effects of chronic morphinization on cerebral glycolytic intermediates also were studied. Rats were injected 10 mg/kg morphine sulfate twice daily, which was increased in 5 mg/kg increments daily to 30 mg/kg on the fifth day. The dose was then increased in 10 mg/kg increments daily for 7 days to a dose of 100 mg/kg twice daily. When the brains of these rats were analyzed for glycolytic intermediates, there was no significant effect on the concentrations of any of the substances previously studied in Figs. 3 and 4.

VIII. CONCLUDING REMARKS

Among the numerous explanations offered for the phenomenon of tolerance to narcotic analgesics, increased biotransformation of the drug and cellular adaptation have the greatest likelihood for clarifying this problem.[75] Cellular adaptation becomes an attractive explanation for the formation of tolerance since the biotransformation of narcotics appears unrelated to the development of tolerance.[57,76-78] In fact, enzymic studies show less N-demethylating capacity[79,80] and less glucuronyl transferase activity[81,82] in livers of tolerant rats than in those of nontolerant rats.

The use of stimulated cerebral cortical slices has provided a useful technique for showing a qualitative difference between cerebral tissues of tolerant animals and those of nontolerant animals. The cerebral tissue can be adapted to the depressant effect of morphine, and this adaptation can be reversed by nalorphine. Although these observations parallel the general pharmacologic effects of morphine and nalorphine, the concentration of morphine employed ($1 \times 10^{-3}M$) to depress the K^+-stimulated respiration of cerebral cortical slices is still quite high compared to the concentration theoretically attainable in the central nervous system following the administration of ordinary doses. The fact remains, however, that until stimulated cerebral slices were utilized, one could not even demonstrate the inhibitory effect of narcotics on cerebral oxidation except at extremely high concentrations. Perhaps the effect of narcotics becomes evident only after proper compartmentation of the drug in the neuron. Some indication of this is that morphine affects glucose utilization of homogenates at much lower concentrations *in vitro* than that of cerebral slices. If the oxidative mechanisms of the neurons are affected by narcotics, presumably further compartmentation of the drugs into the mitochondrial structures may be necessary. Therefore the concentrations of narcotics added *in vitro* may not necessarily reflect the concentration at the inhibitory site.

Electrically stimulated cerebral cortical slices have been reported to be more sensitive to the inhibitory action of analgesics than chemically stimulated slices.[45] If this is indeed true, further work on cellular adaptation with electrically stimulated slices may be fruitful.

The study of the actual mechanism of the cellular adaptation of cerebral slices to narcotics becomes difficult because the depressant effect of narcotics occurs only on stimulated oxidation and slices were reported to be the only cerebral preparation capable of being stimulated. However, Sugawara and Utida[83] showed that excess K^+ ions in the medium also stimulated the respiratory rate and aerobic glycolysis of brain mitochondria from guinea pigs. Krall and co-workers[84] reported that K^+ ions stimulated oxidative phosphorylation of brain mitochondria. Abood and co-workers[85,86] earlier showed that electrical pulses stimulated respiration and decreased oxidative phosphorylation of brain mitochondria. Perhaps an answer to the mechanism of cellular adaptation and of the inhibitory effect of narcotics on oxidative metabolism of the cerebral cortex can be more closely approached by studying the effects of narcotics on stimulated brain mitochondria. In this regard Nukada and Andoh[87] have shown an inhibition of dinitrophenol-stimulated brain mitochondria by ethanol at concentrations that had no effect on unstimulated mitochondria.

It can be stated that morphine has a definite effect on cerebral glucose metabolism. The effects on the glycolytic intermediates can be reversed by nalorphine, and the formation of tolerance to these effects can be demonstrated. These effects also are probably different from those produced by barbiturates. Barbiturates appear to increase glucose content of brain by slowing glycolytic flux at the phosphofructokinase step. The possibility that morphine may increase glucose transport into the brain must be entertained since morphine does not appear to slow the glycolytic flux. The increase in cerebral glucose concentration may not be attributable to an increased blood glucose concentration because neither morphine nor pentobarbital affected blood glucose levels at the dosages employed. Whether this effect of morphine on cerebral glucose metabolism is related to its pharmacologic effect awaits elucidation.

IX. REFERENCES

1. H. Krueger, N. B. Eddy, and M. Sumwalt, "The Pharmacology of the Opium Alkaloids," Public Health Reports, Suppl. No. 165, U.S. Government Printing Office, Washington, D.C. (1943).
2. A. K. Reynolds and L. O. Randall, "Morphine and Allied Drugs," University of Toronto Press, Toronto (1957).
3. J. H. Quastel and A. H. M. Wheatley, *Proc. Roy. Soc. London, Ser. B* **112**, 60–79 (1932).
4. F. W. Schueler and E. G. Gross, *Proc. Soc. Exp. Biol. Med.* **69**, 566–569 (1948).
5. H. W. Elliott, A. E. Warrens, and H. P. James, *J. Pharmacol. Exp. Ther.* **91**, 98–102 (1947).
6. J. L. Bell, *J. Neurochem.* **2**, 265–282 (1958).
7. J. W. Miller and H. W. Elliott, *J. Pharmacol. Exp. Ther.* **110**, 106–114 (1954).
8. A. E. Takemori, *J. Pharmacol. Exp. Ther.* **135**, 252–255 (1962).
9. D. T. Watts, *Arch. Biochem.* **25**, 201–207 (1950).
10. L. O. Randall, J. Kruger, C. Conroy, B. Kappell, and W. M. Benson, *Arch. Exp. Pathol. Pharmakol.* **220**, 26–39 (1953).
11. M. E. Greig, *Arch. Biochem.* **17**, 129–137 (1948).

12. A. E. Takemori, *J. Pharmacol. Exp. Ther.* *145*, 20–26 (1964).
13. M. E. Greig, *J. Pharmacol. Exp. Ther.* *91*, 317–323 (1947).
14. J. L. Webb and K. A. C. Elliott, *J. Pharmacol. Exp. Ther.* *103*, 24–34 (1951).
15. A. E. Takemori, *Biochem. Pharmacol.* *16*, 87–97 (1966).
16. P. W. Dodge and A. E. Takemori, *Biochem. Pharmacol.* *18*, 1873–1882 (1969).
17. J. V. Passoneau and O. H. Lowry, *Biochem. Biophys. Res. Commun.* *7*, 10–15 (1962).
18. A. E. Takemori, unpublished observations.
19. S. Berl, G. Takagaki, D. D. Clarke, and H. Waelsch, *J. Biol. Chem.* *237*, 2570–2573 (1962).
20. J. C. Szerb and D. H. McCurdy, *J. Pharmacol. Exp. Ther.* *118*, 446–450 (1956).
21. J. W. Miller and H. W. Elliott, *J. Pharmacol. Exp. Ther.* *113*, 283–291 (1955).
22. T. K. Adler, H. W. Elliott, and R. George, *J. Pharmacol. Exp. Ther.* *120*, 475–487 (1957).
23. J. T. Scrafani and C. C. Hug, Jr., *Biochem. Pharmacol.* *17*, 1557–1566 (1968).
24. A. Geiger and J. Magnes, *Amer. J. Physiol.* *149*, 517–537 (1947).
25. S. S. Kety, B. D. Polis, C. S. Nadler, and C. F. Schmidt, *J. Clin. Invest.* *27*, 500–510 (1948).
26. S. S. Kety and C. F. Schmidt, *J. Clin. Invest.* *27*, 476–483 (1948).
27. C. F. Schmidt, S. S. Kety, and H. H. Pennes, *Amer. J. Physiol.* *143*, 33–52 (1945).
28. K. A. C. Elliott, "The Biology of Mental Health and Disease Symposium," pp. 70–73, Hoeber, New York (1952).
29. C. A. Ashford and K. C. Dixon, *Biochem. J.* *29*, 157–168 (1935).
30. F. Dickens and G. D. Greville, *Biochem. J.* *29*, 1468–1483 (1935).
31. H. McIlwain, *Biochem. J.* *53*, 403–412 (1953).
32. H. McIlwain, "Biochemistry and the Central Nervous System," Churchill Ltd., London (1966).
33. J. H. Quastel and D. M. J. Quastel, "The Chemistry of Brain Metabolism in Health and Disease," Charles C. Thomas, Springfield, Illinois (1961).
34. J. J. Ghosh and J. H. Quastel, *Nature (London)* *174*, 28–31 (1954).
35. V. C. Sutherland, C. H. Hine, and T. N. Burbridge, *J. Pharmacol. Exp. Ther.* *116*, 469–479 (1956).
36. D. W. Clarke and R. L. Evans, *Can. J. Biochem. Physiol.* *37*, 1525–1526 (1959).
37. H. Wallgren and E. Kulonen, *Biochem. J.* *75*, 150–158 (1960).
38. A. E. Takemori, *J. Pharmacol. Exp. Ther.* *135*, 89–93 (1962).
39. H. W. Elliott, N. Kokka, and E. L. Way, *Proc. Soc. Exp. Biol. Med.* *113*, 1049–1052 (1963).
40. K. A. C. Elliott and F. Bioldeau, *Biochem. J.* *84*, 421–428 (1962).
41. K. Hano, H. Kaneto, and T. Kakunaga, *Jap. J. Pharmacol.* *14*, 227–229 (1964).
42. T. Kakunaga, H. Kaneto, and K. Hano, *J. Pharmacol. Exp. Ther.* *153*, 134–141 (1966).
43. H. McIlwain, *Brit. J. Pharmacol.* *6*, 531–539 (1951).
44. A. E. Takemori, *Science* *133*, 1018–1019 (1961).
45. H. McIlwain, *Biochem. Pharmacol.* *13*, 523–529 (1964).
46. E. L. Way, A. I. Gimble, W. P. McKelway, H. Ross, C. Y. Sung, and H. Ellsworth, *J. Pharmacol. Exp. Ther.* *96*, 477–484 (1949).
47. J. J. Burns, B. L. Berger, P. A. Lief, A. Wollack, E. M. Papper, and B. B. Brodie, *J. Pharmacol. Exp. Ther.* *114*, 289–298 (1955).
48. R. O'Dea and A. E. Takemori, unpublished observations.
49. D. T. Watts, *J. Pharmacol. Exp. Ther.* *95*, 117–121 (1949).
50. H. W. Elliott, V. C. Sutherland, and E. B. Boldrey, *Fed. Proc.* *8*, 288 (1949).
51. M. E. Greig and R. S. Howell, *Arch. Biochem.* *19*, 441–448 (1948).
52. S. B. Wortis, *Arch. Neurol. Psych.* *33*, 1022–1029 (1935).
53. M. H. Seevers and F. E. Shideman, *J. Pharmacol. Exp. Ther.* *71*, 373–382 (1941).
54. E. G. Gross and I. H. Pierce, *J. Pharmacol. Exp. Ther.* *53*, 156–168 (1935).
55. K. Hano, H. Kaneto, and T. Kakunaga, *Folia Pharmacol. Jap.* *58*, 114–115§ (1962).

56. K. Hano, H. Kaneto, and T. Kakunaga, *Folia Pharmacol. Jap. 59*, 71§ (1963).
57. M. H. Seevers and L. A. Woods, *Amer. J. Med. 14*, 546–557 (1953).
58. H. Wallgren and R. Lindbohm, *Biochem. Pharmacol. 8*, 423–424 (1961).
59. M. J. Turnbull and I. H. Stevenson, *J. Pharm. Pharmacol. 20*, 884–885 (1968).
60. R. Siminoff and P. R. Saunders, *J. Neurochem. 5*, 354–358 (1960).
61. P. W. Dodge, doctoral dissertation, University of Minnesota, Minneapolis (1967).
62. A. E. Takemori, *in* "The Addictive States" (A. Wikler, ed.), pp. 53–73, Williams & Wilkins, Baltimore (1968).
63. R. I. H. Wang and J. A. Bain, *J. Pharmacol. Exp. Ther. 108*, 354–361 (1953).
64. A. E. Takemori, *J. Neurochem. 12*, 407–415 (1965).
65. S. K. Ghosh and J. J. Ghosh, *J. Neurochem. 15*, 1375–1376 (1968).
66. R. I. H. Wang and J. A. Bain, *J. Pharmacol. Exp. Ther. 108*, 349–353 (1953).
67. A. E. Takemori, *Fed. Proc. 24*, 548 (1965).
68. L. G. Abood, *Fed. Proc. 9*, 252 (1950).
69. C.-J. Estler and H. P. T. Ammon, *J. Neurochem. 11*, 511–515 (1964).
70. C.-J. Estler, *Int. J. Neuropharmacol. 6*, 241–243 (1967).
71. L. G. Abood, E. Kun, and E. M. K. Geiling, *J. Pharmacol. Exp. Ther. 98*, 373–379 (1950).
72. C.-J. Estler and F. Heim, *J. Neurochem. 9*, 219–225 (1962).
73. O. H. Lowry, J. V. Passonneau, F. X. Hasselberger, and D. W. Schulz, *J. Biol. Chem. 239*, 18–30 (1964).
74. P. D. Gatfield, O. H. Lowry, D. W. Schulz, and J. V. Passonneau, *J. Neurochem. 13*, 185–195 (1966).
75. N. B. Eddy, *in* "Origins of Resistance to Toxic Agents" (M. G. Sevag, R. D. Reid, and O. E. Reynolds, eds.), pp. 223–243, Academic Press, New York (1955).
76. E. L. Way and T. K. Adler, *Bull. W. H. O. 26*, 261–284 (1962).
77. E. L. Way and T. K. Adler, *Bull. W. H. O. 27*, 359–394 (1962).
78. L. B. Mellett and L. A. Woods, *Fortschr. Arzneim.-Forsch. 5*, 156–267 (1963).
79. J. Axelrod, *Science 124*, 263–264 (1956).
80. G. J. Mannering and A. E. Takemori, *J. Pharmacol. Exp. Ther. 127*, 187–190 (1959).
81. A. E. Takemori, *J. Pharmacol. Exp. Ther. 130*, 370–374 (1960).
82. A. E. Takemori and G. A. Glowacki, *Biochem. Pharmacol. 11*, 867–870 (1962).
83. H. Sugawara and S. Utida, *Sci. Pap. Coll. Gen. Educ., Univ. Tokyo 11*, 139–151 (1961).
84. A. R. Krall, M. C. Wagner, and D. M. Gozansky, *Biochem. Biophys. Res. Commun. 16*, 77–81 (1964).
85. L. G. Abood, R. W. Gerard, and S. Ochs, *Amer. J. Physiol. 171*, 134–139 (1952).
86. L. G. Abood, *Amer. J. Physiol. 176*, 247–252 (1954).
87. T. Nukada and N. Andoh, *Jap. J. Pharmacol. 17*, 325–326 (1967).

Chapter 8

PHOSPHOLIPID METABOLISM

Salvatore J. Mulé

New York State Narcotic Addiction Control Commission
New York, New York

I. INTRODUCTION

Phospholipids are predominant and essential molecules of animal cell membranes[1,2] and as such exert an important structural role within the membrane.[3,4] The amphipathic nature of these molecules allows the polar (hydrophilic) sector and the nonpolar (hydrophobic) sector to position itself between an aqueous-nonaqueous interface. This type of bimodal molecule is therefore surface active since the strongly polar or charged groups orient the molecule toward water or other polar molecules and at the same time the nonpolar groups are oriented away from the polar environment. In an aqueous environment these molecules aggregate to form micelles; thus the reactive unit is not the individual phospholipid molecule but the phospholipid micelle. Actually it has been suggested[1] that the lipids of the cell membranes may be small micelles having a central lipophilic core about 40 Å in diameter. These micelles might be hexagonally arranged in sheets in the plane of the membrane, and protein could be present on both sides of the flexible micellar sheet.

The functional role of the phospholipids is still obscure.[2] These bimodal molecules, however, have been studied in relationship to protein synthesis,[5,6] ion transport,[7] protein secretion,[8] mitochondrial electron transfer,[9] phagocytosis,[10,11] biological membrane function,[12] and the sarcoplasmic reticulum of muscle.[13] It is also quite obvious that any chemical molecule that would alter the micellar to lamellar structure of the cell membrane would alter cell properties, for example, transport of molecules and ions, cell mobility, cell permeability, cell reactivity, and cell integrity.

Some studies have been initiated to ascertain the effect of certain drugs on phospholipid metabolism.[14] It is certain that increased awareness of the importance of phospholipids in membranes will stimulate further investigations of this nature. Our interest is, of course, the effect of narcotic analgesics and antagonists on phospholipids primarily within the central nervous system. Some experiments[15-19] have been initiated in this direction with the hope of elucidating the mechanism whereby these drugs act as analgesics as well as of gaining insight into the phenomenon of tolerance and physical dependence. It is the purpose of this report to review critically the role of narcotic drugs as they relate to phospholipid metabolism and function.

II. CHEMISTRY OF GLYCERIDES

A knowledge of glyceride chemistry is important to understand the phospholipids in relationship to their metabolism and interaction with narcotic drugs. The interested reader is referred to the monographs by Lovern,[20] Hanahan,[21] and Van Deenan and De Haas[22] for a more detailed study of glyceride chemistry.

A. Neutral Glycerides

The neutral glycerides are compounds in which the alcoholic groups of glycerol are esterified with fatty acids. If (Fig. 1A) esterification occurs only at the α'-carbon a monoglyceride is formed, subsequent esterification at the β and α-carbons produces a diglyceride and a triglyceride, respectively. The triglycerides are most abundant in mammalian tissues, with only trace amounts of the other glycerides present.

B. Phosphoglycerides

The phosphoglycerides are the most abundant naturally occurring complex lipids and comprise more than 70% of a given tissues phospholipids. The general structure of the difattyacylglycerylphosphoryl group with substituents appears in Fig. 1B. The primary alcoholic group is in ester linkage with the phosphate at the α position.

The β-carbon atom, adjacent to the α-carbon, is usually esterified with a fatty acid. The remaining carbon is designated α' and is generally in ester linkage with a fatty acid; however, in the case of plasmalogens this becomes a vinyl ether linkage and with ether analogs of the phosphatides, an ether linkage. If the X-substituent at the α position is choline, serine, or ethanolamine, the compounds are called, respectively, phosphatidylcholine, phosphatidylserine, and phosphatidylethanolamine.

Phosphatidylcholine often represents more than half of the phospholipids isolated from mammalian tissues. This phosphatide is usually present in the tissue as a zwitterion at physiological pH. The strongly basic character of choline makes phosphatidylcholine quite soluble in alcohol.

Phosphatidylethanolamine has an amine group that is less basic than choline and thus is less soluble in alcohol. This phosphatide also exists as a zwitterion at physiologic pH and is quite abundant in animal tissues.

The amino acid serine when substituted on the phosphatidyl group produces phosphatidylserine, which normally in nature is found as the salt

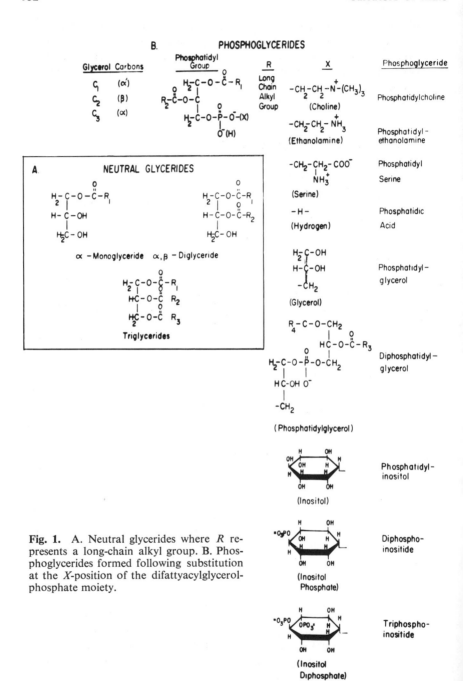

Fig. 1. A. Neutral glycerides where *R* represents a long-chain alkyl group. B. Phosphoglycerides formed following substitution at the *X*-position of the difattyacylglycerol-phosphate moiety.

of either Na^+, K^+, Ca^{2+}, or Mg^{2+}. Phosphatidylserine is practically insoluble in alcohol, primarily because both the carboxylic and amine groups of serine are weakly dissociated in alcohol.

Phosphatidic acid appears in mammalian tissues in trace quantities but is quite important because of its probable role in ion transport[7,23] and as an intermediate in the biosynthesis of phosphoglycerides. Phosphatidic acid is often found in tissues as either the Ca^{2+} or Na^+ salt. It is a somewhat unstable and sparingly soluble in alcohol.

Substitution of glycerol at the phosphoryl group of phosphatidic acid produces the simplest member of the polyglycerophosphatides, known as "phosphatidylglycerol." This phosphatide is present in plants and does not achieve appreciable importance in animal tissues. The most important member of this class is diphosphatidylglycerol (cardiolipin), which exists primarily in the mitochondria of cells. Linoleic appears to be the abundantly represented fatty acid in this compound, which is rather unique for the phospholipids.

The inositol containing glycerophosphatides are acidic and usually represent a small percentage (less than 10%) of the total phospholipids in tissues. These phosphoinositides are generally found in nature as salts of ions (Na^+, K^+, Mg^{2+}, and Ca^{2+}). The di- and tiphosphatides are present in trace quantities except for brain tissue. The phosphoinositides appear to be sparingly soluble in alcohol.

Phosphorous-containing lipids that are not represented in Fig. 1, but nevertheless deserve discussion are sphingomyelins, plasmalogens, lysophosphatides and alkyl ether phosphatides.

The sphingomyelins are phosphatides in which sphingosine is bound by an amide link to a fatty acid and by ether linkage to phosphorylcholine. Sphingomyelin occurs primarily in brain and is soluble in warm ethanol.

The lysophosphoglycerides are simply phosphatides with only one fatty acid in ester linkage with the alcoholic group of glycerol, the most common being lysophosphatidylcholine (lysolecithin) with the free alcoholic group at the β position in the molecule.

The plasmalogens are a class of phosphoglycerides in which the α'-carbon is in vinyl ether linkage with an aldehyde. The most commonly found member of this group is ethanolamine plasmalogen. Serine and choline plasmalogen are also present in mammalian tissue but in trace amounts.

Glycerol-ether phospholipids contain an alkyl ether linkage at the α' position in the molecule. Ether analogs of ethanolamine and choline are present in tissues but generally in trace quantities.

III. BIOSYNTHESIS OF GLYCERIDES

Most tissues are capable of biosynthesizing glycerophosphatides, triglycerides, and sphingolipids. During the past few years the metabolic pathways concerning the biosynthesis of complex lipids have been elucidated. Excellent reviews on this subject for the interested reader are those

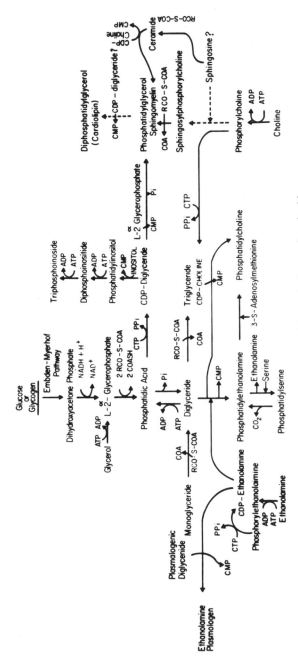

Fig. 2. Biosynthesis of glycerides and glycerophosphatides.

of Kennedy,[24] Ansell and Hawthorne,[2] and Rossiter.[25] The present discussion is confined to a cursory review of the lipid biosynthetic pathways that allows for an intelligent understanding of the interaction of narcotic drugs with lipids.

The known pathways for glycerides synthesis are shown in Fig. 2.

A. Glyceride–Glycerol Intermediates

Primary intermediates in the biosynthesis of phosphoglycerides and triglycerides are (1) L-α-glycerophosphate (2) diglyceride (3) cytidine diphosphate diglyceride and (4) cytidine diphosphate choline (ethanolamine). Of course ATP, enzymes, and fatty acyl-CoA derivatives are also necessary for the formation of these compounds.

The two pathways for the formation of L-α-glycerophosphate involve either ATP, glycerokinase, and Mg^{2+} or, alternatively, glucose or glycogen may be broken down (Embden–Meyerhof) to dihydroxyacetone phosphate and glycerophosphate via $NADH+H^+$ and glycerophosphate dehydrogenase.

Phosphatidic acid may be formed by three different pathways: (1) through fatty acyl-CoA esterification of the alcoholic groups of L-α-glycerophosphate (2) through phosphorylation of α-β-diglyceride with ATP and diglyceride kinase and (3) through the formation of lysophosphatidic acid via ATP and monoglyceride with subsequent fatty acid acylation of lysophosphatidic acid to form phosphatidic acid. Phosphatidic acid phosphatase hydrolyzes phosphatidic acid to diglyceride, which serves as an important intermediate in the formation of triglycerides and phosphoglycerides.

Phosphatidic acid may react with cytidine triphosphate (CTP) and phosphatidic acid transferase to form cytidine diphosphate diglyceride (CDP-diglyceride).

Ethanolamine and/or choline may react with ATP to yield their respective phosphoryl derivatives. The phosphorylated bases react with CTP to yield the cytidine diphosphate choline or ethanolamine derivatives.

Diglyceride is an important intermediate that may be formed through an enzymatic reaction of monoglyceride with fatty acyl-CoA and monoglyceride acyltransferse. The free hydroxyl group of diglyceride may then be esterified enzymatically with fatty acyl-CoA and diglyceride acyltransferase to form triglyceride.

B. Glycerophosphates

Phosphatidylinositol is primarily formed through the reaction of CDP-diglyceride with myoinositol. The di- and tri-substituted polyphosphoinositides are formed through sequential transfer of phosphoryl groups from ATP to phosphatidylinositol.

Phosphatidylglycerol is biosynthesized by a transfer of a phosphatidyl group from CDP-diglyceride to L-α-glycerophosphate, which is dephosphorylated to yield phosphatidylglycerol.

There is no direct evidence for the formation of diphosphatidyl-glycerol (cardiolipin), but it is suggested that a further transfer of phosphatidyl group from CDP-diglyceride to the glycerol moiety of phosphatidylglycerol would yield diphosphatidylglycerol.

Phosphatidylethanolamine is formed through a series of reactions that begin with the formation of phosphorylethanolamine by an enzyme-catalyzed reaction of ATP and ethanolamine kinase with ethanolamine. The enzyme ethanolamine-phosphate-cytidyl transferase catalyzes the reaction of phosphorylcholine with CTP to produce CDP-ethanolamine. Diglyceride reacts with CDP-ethanolamine, and ethanolamine transferase catalyzes the formation of the final product phosphatidylethanolamine. Phosphatidylethanolamine may also be produced by the decarboxylation of phosphatidylserine.

Phosphatidylserine is apparently formed by an enzymatically catalyzed exchange reaction between ethanolamine of phosphatidylethanolamine and the amino acid serine.

Phosphatidylcholine is formed by two primary pathways. The first pathway is analogous to that described for phosphatidylethanolamine, that is, the formation of phosphorylcholine by the reaction of ATP-choline kinase with choline. Phosphorylcholine is converted to CDP-choline via CTP and choline-phosphate-cytidyl transferase. Diglyceride reacts with choline phosphotransferase and CDP-choline to yield phosphatidylcholine. The alternate pathway utilizes S-adenosyl methionine and phosphatidyl-ethanolamine. The methyl group of S-adenosyl methionine methylates the amino group of phosphatidylethanolamine to yield first the N-monomethyl derivative, then the N,N-dimethyl derivative, and lastly phosphatidyl-choline.

Sphingomyelin may be formed through ceramide (N-acylsphingosine) by an enzyme catalyzed reaction with CDP-choline. Spingomyelin may also be produced by the acylation of sphingosylphosphorylcholine. The biosynthesis of the latter compound has not yet been described.

The biosynthesis of plasmalogen compounds is not completely understood, so information concerning their formation is limited. It is believed that choline plasmalogen may be formed[26] by the transfer of phosphoryl choline from CDP-choline to plasmalogenic diglyceride. Ethanolamine plasmalogen may be synthesized by an analogous reaction utilizing ethanolamine (Fig. 2).

Little is known concerning the biosynthesis of the glycerol ether phosphatides. However, it is possible that they are formed via the reduction of plasmalogens.

C. Catabolism

The enzymes involved in the hydrolysis of glycerophosphatides are found in snake venous, plants, bacteria, and animal tissues. The enzymes that degrade neutral glycerides are referred to as "lipases," those that break down glycerophosphatides are called "phospholipases." Detailed reviews are given by Kates,[27] Ansell and Hawthorne,[2] and Rossiter.[25]

Four major groups of phospholipases are known and designated A, B, C, and D. Although there is still some confusion concerning the exact site of activity of these enzymes, the following statements appear to be generally accepted: (1) phospholipase A promotes the hydrolysis of fatty acid ester bonds at the β position and does not affect the ester links at the α' position (2) phospholipase B (lysophospholipase) catalyzes the hydrolysis of the remaining fatty acid ester bond (α position) or hydrolyzes both fatty acid ester bonds to yield the glycerylphosphoryl base and free fatty acid (3) phospholipase C catalyzes the hydrolysis of the ester bond between the glyceride–glycerol group and the phosphoryl group to yield diglyceride and phosphoryl base and (4) phospholipase D attacks the link between the phosphoryl group and the base to yield phosphatidic acid and the free base.

There are apparently enzymes that are also capable of hydrolyzing the vinyl ether groups of the plasmalogens to yield either lysophosphatides or upon further hydrolysis the glycerylphosphoryl base.

IV. PHYSIOLOGICAL FUNCTION OF GLYCERIDES

A. Biological Membranes

Membranes compartmentalize the biological organism into functional units. The plasma membrane separates the cell from the extracellular environment and regulates the passage of substances between the compartments. The cells are further subdivided into specialized compartments by intracellular membranes. The membranes have been characterized by the Danielli–Davson model[28] and more recently by the unit membrane model of Robertson.[29] These membranes are characterized by a bimolecular phospholipid leaflet sandwiched between two layers of protein. The following basic assumptions concerning membrane structure exist: (1) the phospholipids are hydrophobically bound to each other (2) the polar heads of the phospholipids oppose the protein layers and the alkyl section of the molecules are grouped together and (3) the protein layers are extended polypeptide chains about 20 Å wide.

B. Active Transport

It appears that the cell membranes function to protect the cellular environment, which of necessity requires the location of special transport systems to regulate the translocation of molecules across the cell membranes. In this capacity the phospholipids of the membrane have been directly implicated in the transport of proteins following the stimulation of the pancreas by secretogens[8] and the extrusion of Na^{2+} from the avian salt gland following stimulation with acetylcholine.[7,23]

It appears that when protein secretion is stimulated in the pancreas with acetylcholine or pancreozymin an increase in the metabolism of phosphatidic acid, phosphatidylinositol and phosphatidylethanolamine occurred. Studies indicated that the increased turnover of phosphatidylinositol occurred in both the smooth and rough endoplasmic reticulum and that the effect was not primarily associated with the extrusion of zymogen granules from the cell. Rather the data indicated that the phospholipids were involved in the process of transmembrane transport of these proteins across intracellular membranes during the segregation of these proteins in the preparation of zymogen granules. Thus the phospholipids appear to be involved in a phase of the secretory process other than final extrusion through the outer cell wall.

Although the exact role played by the phospholipids in transmembrane transport is unknown, much knowledge has been obtained by investigating the transport of Na^+ in the avian salt gland. In these experiments acetylcholine stimulated the turnover of phosphatidic acid and sodium chloride secretion. The mechanism as visualized by Hokin and Hokin[7,23] and referred to as the phosphatidic acid cycle consisted of the synthesis of phosphatidic acid from diglyceride and ATP at the inner surface of the membrane. The sodium phosphatidate then diffused through the membrane and was hydrolyzed at the outer surface with release of sodium and phosphate ion. The phosphate is transported back across the membranes, and the cycle is renewed. The system in effect acts as an ATPase. There are, of course, many unknown factors associated with this cycle as well as much criticism.[30,31] A major drawback of this cycle is the observation that more than two Na ions must be transported per cycle turn in order to meet the requirements of a Na/O_2 ratio well above 12.

C. Protein Synthesis

Recently some evidence suggests that phospholipids play a role in protein biosynthesis. Hunter and Goodsall[32] incubated protoplasts from bacillus megaterium with ^{14}C-amino acids, and radioactive lipids were isolated with the amino acids. Kinetic studied indicated that the lipo-amino acids were precursors of protein. The isolated lipid could donate 20–25% of its amino acid to protein. Hunter and Godson[33] suggested that the protein was formed in the protoplasts at a phospholipid membrane.

Isolated membrane fractions of *Escherichia coli* were shown to incorporate $^{32}P_i$ into phospholipids, which following hydrolysis revealed amino acids in the phospholipid fraction.[34]

Hendler[35] incubated homogenates of hen oviduct with labeled amino acids and showed that during incubation the lipid fraction quickly gained and lost radioactivity. Inhibition of protein synthesis inhibited the formation of the lipo–amino acid complexes. The author[35] suggests that the complexes might be intermediates in protein biosynthesis.

Although the data seem to indicate a relationship between protein biosynthesis and phospholipids, the actual relationship still remains to be vigorously documented.

D. Mitochondrial Function

The studies of Fleisher and colleagues (Green and Fleisher,[9] Fleisher et al.,[36] and Green and Tzagoloff[12]) have established a relatively large body of evidence that indicates the need of phospholipids for some of the basic mitochondrial functions.

Fleisher established the fact that it was not only necessary to show that removal of phospholipid from the mitochondria caused a loss of activity but that activity could be restored by reintroduction of the phospholipids.

After establishing the need for lipid in electron transfer in each of the four complexes of the electron transfer chain it was shown that electron transfer in lipid extracted particles could be achieved by reconstituting with micellized mixtures of phospholipids. It was not necessary to replace the native set of phospholipids.

It appears that the general role of the phospholipids is to control mitochondrial membrane formation. In the form of membranes the enzymic complexes within the inner mitochondrial membrane are active. In the absence of lipid the membranes cannot be formed, and the lipid-depleted enzymic complexes exist only in bulk form.

The data obtained from the mitochondrial studies indicate that lipid is essential for the formation of complexes from any membrane that may polymerize to water soluble aggregates when lipid free, so that whenever the molecularization of a complex is lipid dependent, lipid will be essential for activity.

E. Phagocytosis

Phagocytosis is a process whereby certain cells engulf foreign particles. During engulfment such cells as the polymorphonuclear leukocytes (PMN) exhibit an increase in O_2 consumption glucose uptake and glycogen breakdown (Sabarra and Karnovsky[37]). Karnovsky[38] implicated the lipids in phagocytosis and expressed the following reasons for investigating the role of lipids in a phagocytizing cell: (1) Lipids are important components of biological membranes, and there are gross changes in cellular membranes during ingestion of particles. The plasma membrane invaginates to form a phagocytic vacuole (2) The cell may be made to perform a defined physiological function under controlled conditions (3) The secreting acinar cells of the pancreas demonstrate an increased metabolic activity in some phospholipids (reverse pinocytosis).

Studies were conducted[37,39] and evidence obtained that lipids participated in phagocytosis. Apparently acetate-^{14}C, glucose-^{14}C, and orthophosphate-^{32}P were incorporated into lipids of phagocytizing cells more rapidly than in resting cells. Separation of individual ^{32}P-labeled phosphatides indicated that the greatest increase in the incorporation of ^{32}P occurred in the acidic phosphatides that is, phosphatidic acid, phosphatidylinositol, and phosphatidylserine. No effect was observed with phosphatidylcholine.

Studies on the role of phospholipids in phagocytosis by Sastry and Hokin[11] provided data similar to Karnovsky and Wallach[39] except for the effect on phosphatidylserine and the percentage values obtained with the minor phosphatides. These results indicated that the increased synthesis of phosphatidic acid on induction of phagocytosis was brought about by activation of diglyceride kinase plus lysophosphatidic acid acylase. The increased incorporation of ^{32}P into phosphatidylinositol may not be solely a result of the higher specific activity of phosphatidic acid, a precursor of phosphatidylinositol, since labeled inositol incorporation into phosphatidylinositol also was increased in cells undergoing phagocytosis.

V. EFFECT OF MORPHINE AND NALORPHINE *IN VITRO* AND *IN VIVO*

A. *In Vitro* Studies

A series of experiments were initiated (Mulé[16]) whereby $^{32}P_i$ or a ^{14}C precursor of the phospholipids were incorporated into cerebral cortex slices from the guinea pig. The phospholipids were isolated, and the effect of morphine or nalorphine on the incorporation of the labeled precursor into the phosphatide was ascertained. Tables I and II summarize the effect of these drugs on certain phosphoglycerides. Morphine ($10^{-2}M$) was quite effective in stimulating the incorporation of $^{32}P_i$ into all the glycerides except phosphatidylcholine, where an inhibition of 56% was obtained. Intermediate concentrations of morphine (5×10^{-3} to $10^{-5}M$) had a variable effect on the incorporation of $^{32}P_i$ into phosphatides. At $10^{-6}M$ morphine a statistically significant effect was obtained only with phosphatidic acid. Nalorphine, although primarily an antagonist, had an effect on phospholipid metabolism quite similar to that observed for morphine. A maximal effect was obtained at $10^{-2}M$ with some of the phospholipids being affected at intermediate levels of the drug (10^{-3} to $10^{-5}M$). At $10^{-6}M$ nalorphine stimulated the turnover of phosphatidylinositol, phosphatidylcholine, and phosphatidic acid.

Nalorphine at various concentrations from 10^{-2} to $10^{-6}M$ had no truly physiological antagonistic effect on phospholipid turnover in the presence of $10^{-2}M$ morphine, although an apparent inhibitory effect occurred at the high levels of nalorphine (10^{-2} to $10^{-4}M$). This apparent inhibitory effect might well represent a toxic effect of nalorphine and morphine rather than a true antagonism.

Table III presents the results obtained with various ^{14}C-labeled precursors of phosphatides in the presence of $10^{-2}M$ morphine. It is obvious that morphine does not simply stimulate the turnover of phosphate in the glycerophosphatides but affects the turnover of the entire phospholipid molecule. A stimulation was observed with glycerol-^{14}C incorporation into

TABLE I

The Incorporation of $^{32}P_i$ into Phospholipids of Guinea Pig Cerebral Cortex Slices in the Presence of Morphine[a]

Phospholipid	Radioactivity[b] (counts/min/100 mg)					
	Control	Morphine 10^{-2} M	Percentage Change	Morphine 10^{-6} M	Percentage Change
Phosphatidylinositol	5785 ± 432 (30)	15852 ± 1542 (9)	174[c]	6861 ± 1228 (10)	19
Phosphatidylcholine	11708 ± 1008 (34)	5200 ± 1032 (8)	−56[c]	11011 ± 1965 (10)	−6
Phosphatidylserine	3160 ± 315 (28)	5817 ± 953 (9)	84[c]	3209 ± 583 (7)	1
Phosphatidylethanolamine	2752 ± 268 (31)	5663 ± 997 (8)	106[c]	2929 ± 663 (10)	6
Phosphatidic acid	3647 ± 156 (32)	12259 ± 1775 (9)	236[c]	4512 ± 374 (9)	24[c]

[a] From Mulé.[(16)]
[b] All radioactivity was corrected to a specific activity of 10^5 counts/min/μg of phosphate for the orthophosphate of the medium. The values are means ±SEM with the number of observations in parentheses.
[c] $P < 0.05$ as compared to control.

TABLE II

Effect of Nalorphine on the Incorporation of $^{32}P_i$ into Phospholipids of Guinea Pig Cerebral Cortex Slices[a]

Phospholipid	Radioactivity[b] (counts/min/100 mg)				
	Control	Nalorphine $10^{-2}M$	Percentage Change	Nalorphine $10^{-6}M$	Percentage Change
Phosphatidylinositol	5709 ± 1491 (9)	10403 ± 1163 (7)	82[c]	8909 ± 595 (4)	56[c]
Phosphatidylcholine	8084 ± 802 (9)	3350 ± 560 (7)	−59[c]	14823 ± 1165 (4)	83[c]
Phosphatidylserine + phosphatidylethanolamine	2715 ± 414 (9)	2790 ± 483 (7)	3	3705 ± 343 (4)	36
Phosphatidic acid	3854 ± 297 (9)	15368 ± 1779 (7)	299[c]	5137 ± 212 (4)	33[c]

[a] From Mulé.[(16)]
[b] All radioactivity was corrected to a specific activity of 10^5 counts/min/μg of phosphate for the orthophosphate of the medium. The values are means ±SEM with the number of observations of parentheses.
[c] $P < 0.05$ as compared to control.

TABLE III

The Incorporation of ^{14}C Precursors into Phospholipids of Guinea Pig Cerebral Cortex in the Presence of $10^{-2}M$ Morphine[a]

Phospholipid	Radioactivity (counts/min/100mg)[b]								
	^{14}C-Glycerol		Percentage Change	^{14}C-Myo-Inositol		Percentage Change	^{14}C-Methyl Choline		Percentage Change
	Control	Morphine		Control	Morphine		Control	Morphine	
Phosphatidylinositol	598 ± 79 (7)	1287 ± 160 (7)	115[c]	8929 ± 816 (9)	15407 ± 1253 (8)	72[c]			
Phosphatidylcholine	1014 ± 105 (7)	704 ± 91 (7)	−30[c]				18305 ± 852 (6)	12022 ± 1045 (5)	−34[c]
Phosphatidylserine + phosphatidylethanolamine	419 ± 44 (7)	592 ± 72 (7)	41[c]						
Phosphatidic Acid	473 ± 72 (7)	1691 ± 314 (7)	257[c]						
Diphosphoinositide				837 ± 95 (8)	2146 ± 243 (8)	156[c]			
Triphosphoinositide				764 ± 246 (8)	1189 ± 292 (7)	56[c]			

[a] From Mulé.[16]
[b] All radioactivity was corrected to a specific activity of 10^6 counts/min/ mole of either glycerol, myo-inositol or choline in the medium. The values are means ±SEM with the number of observations in parenthesis.
[c] $P < 0.05$ as compared to control.

the phosphatides, except for phosphatidylcholine where an inhibition of 30% was obtained. A similar inhibition was observed for this phospholipid when choline-^{14}C was the incorporating labeled precursor. The effect of morphine on the incorporation of myo-inositol-^{14}C into inositol containing phospholipids also was quite similar to that obtained with orthophosphate-^{32}P incorporation.

The cortical slices were analyzed for phospholipid-phosphorus, morphine-^{14}C, and nalorphine-^{3}H following incubation.[16] The phospholipid-P content in the presence of 10^{-2} or $10^{-3}M$ morphine was 6–9% higher than control ($P<0.05$), suggesting morphine stimulated *de novo* synthesis of phospholipids.

Since the effect of morphine and nalorphine were observed predominently at high drug levels, it was important to determine how effectively these drugs penetrated the slice. A concentration ratio of 1:1 (slice:medium) was obtained with $10^{-2}M$ morphine-^{14}C. Ratios of 2.75:1 and 3.0:1 were obtained with 10^{-4} and $10^{-6}M$ morphine, respectively, in the medium. Nalorphine-^{3}H concentrations in the medium of 10^{-2}, 1.66×10^{-3}, and $6.6 \times 10^{-5}M$ provided slice-to-medium ratios of 0.16, 1.72, and 2.5 to 1, respectively. The cerebral cortex slices apparently concentrated the drugs, and this effect increased with decreasing levels of drug in the medium.

B. *In Vitro—In Vivo* Studies

In Table IV appear the data obtained followig the administration of 40 mg/kg (free base) of morphine to guinea pigs, removing the cerebral cortex, and incubating the slices with ortho-^{32}P in the Krebs–Henseleit medium.[17] The results indicate that *in vivo* morphine effectively stimulates, the incorporation of ^{32}P$_i$ into the phosphatides. Although the addition of $10^{-2}M$ morphine to the slices was quite effective in stimulating the incorporation of ^{32}P$_i$ into the phospholipids, additions of 10^{-4} or $10^{-6}M$ morphine did not exert an effect greater than that observed with the same levels of morphine *in vivo*. If the animals were sacrificed 4 hr after morphine there was (Mulé[17]) a variable effect with a stimulatory incorporation of ^{32}P$_i$ noted for phosphatidylinositol, phosphatidylcholine, phosphatidylserine, and phosphatidylethanolamine. Essentially no effect was obtained on phospholipid metabolism with *in vivo* levels of the drug if the animals were sacrificed 10 hr after morphine administration.

If the guinea pigs were made tolerant to morphine the effect on phospholipid metabolism was abolished with *in vivo* drug levels (Table V). It appears that complete adaptation to the phospholipid effect of morphine occurred following tolerance development. However, the phospholipids did respond to the addition of $10^{-2}M$ morphine, but to a lesser extent than that observed for nontolerant cerebral cortex phosphatides. No effect was obtained following the addition of either 10^{-4} or $10^{-6}M$ morphine to the cerebral cortex slices from the tolerant guinea pigs.

Essentially no effect of morphine was obtained on phospholipid metabolism with cerebral cortex slices from guinea pigs 24 hr or 3 months

TABLE IV

The Incorporation of $^{32}P_i$ into Phospholipids of Cerebral Cortex Slices from Nontolerant Guinea Pigs[a]

Phospholipid	Control	Radioactivity with Additions of Morphine *In Vitro* (counts/min/μmole P/100 mg)[b]					
		None	Percentage Change	10^{-2} M	Percentage Change	10^{-6} M	Percentage Change
Phosphatidylinositol	542 ± 68 (8)	867 ± 106 (6)	60[c]	1858 ± 39 (5)	243[c]	782 ± 128 (6)	44[c]
Phosphatidylcholine	1718 ± 279 (8)	2808 ± 216 (6)	63[c]	945 ± 130 (6)	−45[c]	2586 ± 178 (6)	50[c]
Phosphatidylserine + phosphatidylethanolamine	395 ± 64 (8)	692 ± 118 (6)	75[c]	592 ± 44 (4)	50[c]	629 ± 85 (6)	59[c]
Phosphatidic acid	770 ± 99 (8)	1150 ± 162 (6)	49[c]	2095 ± 231 (5)	172[c]	1076 ± 144 (6)	40[c]

[a] From Mulé.[17] Nontolerant guinea pigs were sacrificed 1 hr after a 40 mg/kg (free base) s.c. injection of morphine.
[b] All radioactivity was corrected to a specific activity of 10⁵ counts/min/μg of phosphate for the orthophosphate of the medium. The values are means ±SEM with the number of observations in parentheses.
[c] $P < 0.05$ as compared to control.

TABLE V
The Incorporation of $^{32}P_i$ into Phospholipids of Cerebral Cortex Slices from Morphine-Tolerant Guinea Pigs[a]

Phospholipid	Control	Radioactivity with Additions of Morphine *In Vitro* (counts/min/μmole P/100 mg[b])					
		None	Percentage Change	$10^{-2}M$	Percentage Change	$10^{-6}M$	Percentage Change
Phosphatidylinositol	1009 ± 245 (12)	723 ± 140 (4)	−28	1623 ± 320 (9)	−61	902 ± 114 (9)	−11
Phosphatidylcholine	1771 ± 339 (12)	1424 + 198 (9)	−19	943 ± 122 (9)	−47	2089 ± 202 (9)	18
Phosphatidylserine	367 ± 55 (12)	357 ± 54 (9)	−3	446 ± 45 (9)	21	311 ± 29 (9)	−15
Phosphatidylethanolamine	264 ± 51 (12)	318 ± 54 (9)	−13	477 ± 38 (9)	23	341 ± 32 (9)	−6
Phosphatidic acid	979 ± 90 (12)	918 ± 142 (9)	−6	2350 ± 71 (9)	140	1200 ± 100 (9)	22

[a] From Mulé.[17] Morphine tolerant guinea pigs were sacrificed 1 hr after the last 40 mg/kg (free base) s.c. injection of morphine.
[b] All radioactivity was corrected to a specific activity of 10^5 counts/min/μg of phosphate for the orthophosphate of the medium. The values are means ±SEM with the number of observations in parentheses.

abstinent from drugs. The phosphatides were responsive, however, to the addition of $10^{-2}M$ morphine and provided percentage changes that were similar to those obtained for the nontolerant animals, except for the appreciable quantities of $^{32}P_i$ incorporated into the polyphosphoinositides. The addition of either 10^{-4} or $10^{-6}M$ morphine to these slices provided little if any effect on phospholipid metabolism.

VI. SUBCELLULAR EFFECT OF MORPHINE *IN VIVO*

A. Brain

Morphine (40 mg/kg body weight as free base) markedly increased (87–119%) the incorporation of $^{32}P_i$ into total phospholipids (Fig. 3) at 16 and 24 hr after intraperitoneal (i.p.) $^{32}P_i$ administration.[19] No statistically significant effect was obtained at either 4 or 48 hr. Following subfractionation of the cerebral cortex (Table VI) a statistically significant turnover of each phospholipid in the homogenate and soluble supernatant fraction at 48 hr was obtained. Significant changes were obtained with certain phospholipids in the various fractions at time intervals of 16, 24, and 48 hr after i.p. $^{32}P_i$. Phosphatidylcholine appeared to be most affected by morphine in each subcellular fraction. This is an interesting and important fact since phosphatidylcholine represents from 25 to 40% of the total lipid phosphorus in the brain.

The effect of morphine on the phospholipids from subfractions of the mitochondria following discontinuous sucrose gradient centrifugation was equivocal, that is, both stimulatory and inhibitory at the various time intervals. A similar effect was obtained with the mitochondrial fraction phospholipids subjected to hypo-osmotic shock.

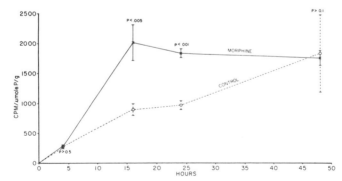

Fig. 3. Total radioactivities in the phospholipids extracted from the cerebral cortex after a 150 μc/100 g i.p. injection of $^{32}P_i$. Morphine (40 mg/kg as free base) administered 1 hr prior to sacrifice. The control guinea pigs received a comparable volume of 0.9% saline. Each value is the mean $\pm$ SEM of four to eight observations.

TABLE
Effect of Morphine on the Incorporation of $^{32}P_i$ into Phospholids from

Phospholipid	Time (hr)	Homogenate			Crude Nuclear		
		C^b	M^c	Percentage Change	C	M	Percentage Change
Phosphatidylinositol	16	10 ± 3	16 ± 5	60^d	17 ± 2	17 ± 4	0
	24	11 ± 4	11 ± 3	0	14 ± 3	17 ± 5	14
	48	10 ± 4	22 ± 3	120^d	17 ± 1	20 ± 4	18
Phosphatidylcholine	16	32 ± 4	49 ± 3	53^d	76 ± 8	94 ± 7	24^d
	24	31 ± 6	53 ± 5	71^d	45 ± 4	38 ± 6	-16
	48	59 ± 4	157 ± 5	166^d	47 ± 3	121 ± 10	157^d
Phosphatidic acid	16	9 ± 6	11 ± 2	22^d	11 ± 1	11 ± 4	0
	24	7 ± 4	8 ± 2	14	5 ± 3	5 ± 2	0
	48	9 ± 1	281 ± 7	211^d	5 ± 2	11 ± 2	120^d

[a] $^{32}P_i$ (150 μc/100 g) was injected i.p., and the animals were sacrificed at the given time.
[b] C, control. Each guinea pig received 0.9% saline s.c. hr prior to sacrifice. Each value is the mean
$\pm$ SEM for 6–18 observations.

B. Liver

Morphine did not alter statistically the incorporation of intraperitoneally administered $^{32}P_i$ into phospholipids as determined by total radioactivity in total lipid phosphorous (Fig. 4). However, with respect to individual phospholipids the incorporation of $^{32}P_i$ into phosphatidic acid was significantly stimulated at 24 and 48 hr (34–78%). A significant stimulation was observed with phosphatidylinositol (59%) at 24 hr. No significant effect was observed at any other time interval with the other phospholipids from guinea pig liver.

Table VII summarizes the data obtained with certain phospholipids from guinea pig liver subcellular fractions. The greatest stimulation

TABLE
Effect of Morphine on the Incorporation of $^{32}P_i$ into Phospholipids from

Phospholipid	Time (hr)	Homogenate			Crude Nuclear		
		C^b	M^c	Percentage Change	C	M	Percentage Change
Phosphatidylinositol	16	118 ± 14	150 ± 8	27	99 ± 11	127 ± 20	28
	24	118 ± 10	107 ± 16	-9	153 ± 8	108 ± 12	-29^d
	48	75 ± 8	111 ± 11	48^d	69 ± 7	101 ± 11	46^d
Phosphatidylcholine	16	1688 ± 20	1789 ± 21	6	1295 ± 30	1294 ± 82	0
	24	1466 ± 62	1398 ± 40	-5	1253 ± 41	1291 ± 62	3
	48	1343 ± 41	2212 ± 54	65^d	1046 ± 26	1861 ± 38	78^d
Phosphatidic acid	16	46 ± 6	29 ± 8	-37	88 ± 21	41 ± 16	-53^d
	24	40 ± 8	24 ± 6	-40^d	56 ± 12	38 ± 8	-32
	48	41 ± 12	74 ± 4	89^d	56 ± 11	131 ± 14	134^d

[a] $^{32}P_i$ (150 μc/100g) was injected i.p., and the animals were sacrificed at the given time.
[b] C, control. Each guinea pig received 0.9% saline s.c. hr prior to sacrifice. Each value is the mean
$\pm$ SE of six observations.

VI

Cerebral Cortex Subcellular Fractions[a] (counts/min/μmole P/g)

Crude Mitochondrial			Microsomal			Soluble Supernatant		
C	M	Percentage Change	C	M	Percentage Change	C	M	Percentage Change
13 ± 4	9 ± 2	31	7 ± 2	15 ± 3	114[d]	84 ± 3	119 ± 11	42[d]
8 ± 2	9 ± 3	12	12 ± 2	15 ± 4	25	70 ± 4	25 ± 2	−64
9 ± 1	9 ± 3	0	8 ± 3	14 ± 3	75[d]	42 ± 4	147 ± 6	250[d]
42 ± 4	41 ± 5	−2	19 ± 5	58 ± 10	205[d]	133 ± 10	158 ± 12	19[d]
41 ± 5	56 ± 3	−12	57 ± 4	79 ± 8	38[d]	105 ± 8	36 ± 6	−66[d]
32 ± 6	106 ± 10	231[d]	65 ± 5	147 ± 6	126[d]	110 ± 7	495 ± 8	350[d]
6 ± 1	8 ± 4	33	6 ± 2	10 ± 5	67	70 ± 5	10 ± 4	−86[d]
5 ± 1	4 ± 2	−20	5 ± 2	5 ± 3	0	44 ± 3	112 ± 11	154[d]
5 ± 2	4 ± 2	−20	6 ± 1	9 ± 3	50	31 ± 2	156 ± 8	403[d]

[c] M, morphine. Each guinea pig received 40 mg/kg (free base) s.c./hr prior to sacrifice. The values are the means ±SE for 6–18 observations.
[d] $P > 0.05$ as compared to control.

apparently occurred at 48 hr after $^{32}P_i$. This effect was most pronounced with phospholipids from the soluble supernatant fraction and for phosphatidylinositol, which varied from 51 to 160% at the various time intervals.

An analysis of the phospholipids from the smooth and rough microsomal fractions of liver revealed a stimulatory effect for phosphatidic acid ranging from 33 to 468% at intervals through 48 hr after $^{32}P_i$. There did appear to be a greater stimulation in the incorporation of $^{32}P_i$ into phosphatides from the rough microsomal liver fractions in comparison to the smooth microsomal fraction. A variable stimulatory or inhibitory effect was obtained with the phospholipids from the microsomal supernatant fraction at the various time intervals.

VII

Liver Subcellular Fractions[a] (counts/min/μmole P/g)

Crude Mitochondrial			Microsomal			Soluble Supernatant		
C	M	Percentage Change	C	M	Percentage Change	C	M	Percentage Change
93 ± 11	145 ± 6	56[d]	153 ± 16	200 ± 20	31[d]	109 ± 15	165 ± 18	51[d]
100 ± 18	78 ± 8	−22	161 ± 14	120 ± 12	−25	127 ± 9	203 ± 12	60[d]
56 ± 7	94 ± 10	68[d]	88 ± 7	151 ± 10	72[d]	73 ± 11	190 ± 16	160[d]
1197 ± 60	1546 ± 72	29	1952 ± 102	1949 ± 91	0	1455 ± 91	2459 ± 15	68
1068 ± 34	1004 ± 62	−6	1591 ± 40	1401 ± 62	−12	1648 ± 102	1536 ± 48	−7
1014 ± 42	1858 ± 81	83[d]	1335 ± 68	2503 ± 10	87[d]	1187 ± 60	2388 ± 84	101[d]
138 ± 20	69 ± 8	−50[d]	13 ± 10	21 ± 12	62	29 ± 4	22 ± 8	−24
76 ± 16	59 ± 4	−22	20 ± 6	27 ± 10	35	25 ± 11	27 ± 6	8
82 ± 12	228 ± 20	178[d]	27 ± 4	58 ± 9	115[d]	35 ± 8	54 ± 6	54[d]

[c] M, morphine. Each guinea pig received 40 mg/kg (free base) s.c. hr prior to sacrifice. The values are the means ±SEM of six observations.
[d] $P < 0.05$ as compared to control.

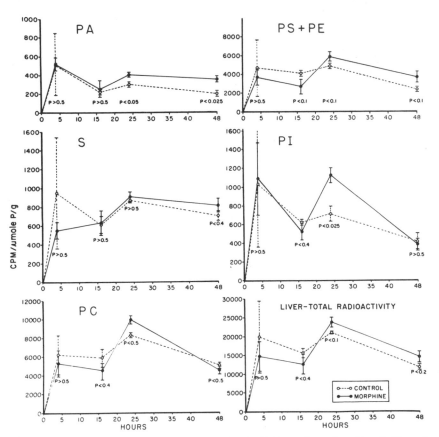

Fig. 4. $^{32}P_i$ incorporation into individual phospholipids and total radioactivities obtained from liver after 150 μc/100 g i. p. injection of $^{32}P_i$. Morphine (40 mg/kg as free base) administered 1 hr prior to sacrifice. The control guinea pigs received a comparable volume of 0.9 % saline. Each value represents the mean obtained from six observations ± SEM. PA, phosphatidic acid; S, sphingomyelin; PC, phosphatidylcholine; PS + PE, phosphatidylserine plus phosphatidylethanolamine; PI, phosphatidylinositol.

VII. THEORETICAL CONSIDERATIONS OF NARCOTIC DRUG ACTION

A. Receptors—Phospholipids—Ions

It seems quite obvious that narcotic analgesics must exert their pharmacological effect through an interaction with a receptor in the central nervous system. It is further obvious that the location of the receptor be in or on nerve cells. However, the extract structure and essential chemical composition of the receptor is not so obvious. Thus its assumed identity is often described in such terms as biochemical substance (macromolecule or fraction thereof), enzyme complex, protein, sequence of amino acids, lipoprotein, lipid, glycolipid, phospholipid, etc.

Beckett and Casy,[40,41] on the basis of structure–activity relationships of a series of narcotic analgesic compounds, postulated a narcotic analgesic receptor that consisted of a charged anionic site (65×8.5 Å) with the focus of charge at the anionic site being separated by a cavity and a flat surface. Thus a cationic analgesic drug carrying a charge at physiological pH could form an ionic bond with the anionic receptor that then might be further reinforced by van der Waals forces. Steric requirements in the vicinity of the anionic site would reduce bonding forces upon increased width of the cation. Drug-receptor bonding may be completely prevented if the width of the cation exceeds 8 Å. The aromatic ring of the narcotic analgesic may be associated with the flat surface of the receptor and stabilized there by van der Waals forces. The cavity in the receptor allows for close association of the cationic nitrogen and the anionic site as well as for a three dimensional structure of the receptor surface. Although Beckett's narcotic analgesic receptor is hypothetical and has obvious limitations (Mellett and Woods[42]) it does provide a model whereby a realistic attack upon receptor identity within the central nervous system may proceed.

On the basis of the probable receptor and the molecular structure of phospholipids it seemed quite possible that these compounds could act as central receptors for narcotic drugs. Let us examine this possibility in some detail. The phosphatides are amphipathic in character, which allows for orientation of the polar end of the molecule toward water and the hydrocarbon portion toward organic solvents. Such orientation occurs in the bimolecular leaflet model for cell membranes along with additional protein at the intracellular and extracellular surfaces (Danielli and Davson[28] and Robertson[29]). The phosphatides contain both an acid and basic group within the molecule. It is this highly acidic phosphoryl group (PO_4) that makes these compounds quite attractive as CNS receptors for narcotic drugs. Although it would seem that these molecules exist to some extent in the form of an ampholyte, this does not appear to be the case either in the crystalline form or in certain solvents (Dervichian[43]).

Let us focus attention on the acidic portion of the molecule that has a pK_a of about 4.0. At physiological pH the percentage of the phosphatide ionized would be at least 99%. At the same pH (7.4) morphine would be 80% ionized. It is, therefore, obvious that the phosphate groups of the phospholipids could well be Beckett's[40] postulated anionic site on the analgesic receptor. It is, of course, difficult to ascertain whether the size of the group at the physiological site in the membrane would conform with the size hypothesized by Beckett. However, the dimensions present in crystal lattice from are not in conflict with Beckett.[43]

Although, the phosphatides may act as sites for binding narcotic drugs, it is the events that must be initiated following ionic binding that elicits pharmacological activity. Therefore, the physiological function of the phospholipids appear of prime importance at this point. Tobias[44] studied changes of opacity, rigidity, diameter, length, axoplasmic flow, and surface contour of axons accompanying electrical activity. Following treatment of the axons with phospholipases all the normal membrane electrical characteristics disappeared. The phospholipids were considered essential for normal membrane electrical activity. An essential role was denied proteins in this regard because proteases that hydrolyze peptide bonds did not affect threshold, conduction velocity, action potentials, membrane resistance, or capacitance of the axons. It was further shown that the phospholipids were hydrophilic in contact with KCl and hydrophobic in contact with $CaCl_2$ and that the light-scattering, dimensional structure, rigidity, and surface contour changes were attributed to hydration changes associated with K^+ and Ca^{2+} changes in the membrane. The fact that phospholipids may act as carriers for ions and be directly involved in ion transport was postulated by Hokin and Hokin[7] and supported by a series of elegant experiments culminating in the phosphatidic acid cycle. Goldman[45] further postulated that the phosphate groups present in bimolecular leaflets might act as ion-exchange gates for the control of ion flow through cell membranes during neuronal excitation. It has been shown that phospholipids bind monovalent as well as polyvalent cations[45-48] with an apparent affinity for Ca^{2+} that greatly exceeded that for either Na^+ or K^+ (5000–10,000-fold greater).[49]

Recently it was shown[58] that morphine as well as nalorphine could effectively inhibit the binding of Ca^{2+} to phospholipids. The inhibition of Ca^{2+} transported by the phospholipids was related to ionization of the drug. Furthermore it was demonstrated that morphine-^{14}C could bind (69%) to phosphatidic acid and that this binding could be inhibited by divalent ions. A postulated series of reactions for the transport of Ca^{2+} and its inhibition by narcotic analgesics was presented as follows:[50]

(1) $PA \rightarrow HPA^- + H^+$
(2) $HPA^- \rightarrow PA^{2-} + H^+$
(3) $PA^{2-} + Ca^{2+} \rightarrow CaPA$ [in the presence of a narcotic analgesic (NA)]
(4) $PA^{2-} + Ca^{2+} + NA \rightarrow CaPA + NAPA$

This series of reactions could occur with other divalent ions or, for that matter, with monovalent ions. It is difficult to interject stoichiometry into the reaction since it is not known whether the events would occur with a single polar group or a cluster of polar groups within the membrane. It is visualized that during the neuronal resting state the polar groups would be occupied by ions. The total number occupied depends upon the concentration of the ions, the binding constant of the ions, and the configurational accessibility of the phospholipid polar groups. Thus pharmacological action of morphine may be interpreted as a displacement of ions or a successful competition with ions for binding sites on the phosphatides. Subsequent to this event a distinct alteration in ion conductance, membrane permeability, electrical activity, and neuronal excitability occurs. Such changes may lead to a reduction in pain perception centrally (analgesia).

It is possible to present in this regard a rather large body of evidence that supports a narcotic–phospholipid–polyvalent cation interrelationship:

1. Narcotic analgesics inhibit the binding and thus the transport of Ca^{2+}.[50]
2. Morphine alters phospholipid metabolism in the cerebral cortex *in vitro* and *in vivo*.[15–18]
3. Ca^{2+} rather selectively inhibits the analgesia produced by morphine.[51]
4. Ca^{2+} antagonizes the morphine inhibition of gut contractions produced by coaxial stimulation.[52]
5. Morphine increases Mg^{2+} levels in the plasma and inhibits Ca^{2+} transport across the rat intestine.[53]
6. Morphine causes a marked urinary excretion in Ca^{2+}.[54]
7. O_2 uptake inhibition by morphine of KCl-stimulated cortex slices may be shown only in a Ca^{2+}-free medium.[55]

Although the ion exchanger hypothesis for narcotic drug action is most attractive, it is difficult to explain the lack of specificity, at least for the inhibition of Ca^{2+} transport, since local anesthetics,[56] as well as narcotic depressants and stimulants, may produce this effect.[50,57] In order to achieve greater emphasis for the phospholipid effect of narcotic analgesics, further studies involving *in vivo* levels of phospholipids, ions, and electrical activity within the neuron must be initiated in the acute, chronic, and abstinent physiological state.

VIII. SUMMARY

The phospholipids are unquestionably essential molecules of animal cell membranes, and their chemistry, as well as biosynthesis, is fairly well understood. The physiological function of phosphatides, however, remains to be clarified.

Recent studies indicate that the phospholipids may be involved in the following actions: (1) regulation of the passage of substances between the intracellular and extracellular compartment (2) transmembrane transport of ions or proteins (3) protein synthesis (4) mitochondrial function and (5) phagocytosis.

The chemical nature of the phosphatides as well as the narcotic analgesics made the possibility of an interaction between these molecules quite likely. It was clearly shown that morphine could alter phospholipid metabolism *in vitro*, *in vivo*, and within intracellular organelles. The most interesting functional aspect of the morphine phospholipid effect is whether the phospholipids might act as the CNS receptors for narcotic drugs. In this capacity it was hypothesized that morphine would interfere with phospholipid ionic binding sites associated with neuronal membranes and thus alter neuronal excitability culminating in a reduction in pain perception centrally (analgesia). There are, of course, many aspects of this mechanism that require detailed analysis at a neuronal level where physiological events occur after acute or chronic administration of narcotic drugs.

IX. REFERENCES

1. J. A. Lucy and J. T. Dingle, *in* "Metabolism and Physiological Significance of Lipids" (R. M. C. Dawson and D. N. Rhodes, eds.), pp. 384–397, John Wiley, New York (1964).
2. G. B. Ansell and J. N. Hawthorne, *in* "Phospholipids," Chapter 10, pp. 244–277, Elsevier, Amsterdam, (1964).
3. H. J. Deuel, *in* "The Lipids," Vol. 2, Chapter 7, pp. 707–816, Interscience, New York (1955).
4. J. C. Dittmer, *in* "Comparative Biochemistry" (M. Florkin and H. S. Mason, eds.), Vol. 3, pp. 231–264, Academic Press, New York (1962).
5. G. D. Hunter and G. N. Godson, *Nature 189*, 140–141 (1961).
6. R. W. Hendler, *Biochim. Biophys. Acta 60*, 90–97 (1962).
7. L. E. Hokin and M. R. Hokin, *Fed. Proc. 22*, 8–18 (1963).
8. L. E. Hokin and M. R. Hokin, *in* "Functionelle und morphologische Organisation der Zelle," Sekretion und Exkretion, pp. 49–67, Springer-Verlag, Berlin (1965).
9. D. E. Green and S. Fleischer, *in* "Metabolism and Physiological Significance of Lipids" (R. M. C. Dawson and D. N. Rhodes, eds.), pp. 581–617, John Wiley New York (1964).
10. M. L. Karnovsky and D. F. H. Wallach, *J. Biol. Chem. 236*, 1895–1901 (1961).
11. P. S. Sastry and L. E. Hokin, *J. Biol. Chem. 241*, 3354–3361 (1966).
12. D. E. Green and A. Tzagoloff, *J. Lipid Res. 7*, 587–601 (1966).
13. A. Martonosi, J. Donley, and R. A. Halpin, *J. Biol. Chem. 243*, 61–70 (1968).
14. G. B. Ansell, *in* "Advances in Lipid Research" (R. Paoletti and D. Kritchensky, eds.), Vol. 3, pp. 139–170, Academic Press, New York (1965).
15. M. Brossard and J. H. Quastel, *Biochem. Pharmacol. 12*, 766–768 (1963).
16. S. J. Mulé, *J. Pharmacol. Exp. Ther. 154*, 370–384 (1966).
17. S. J. Mulé, *J. Pharmacol. Exp. Ther. 156*, 92–100 (1967).
18. S. J. Mulé, *in* "The Addictive States" (A. Wikler, ed.), Vol. 46, Chapter 3, pp. 32–52, Williams & Wilkins Baltimore (1968).
19. S. J. Mulé, *Biochem. Pharmacol.* (in press).
20. J. A. Lovern, *in* "The Chemistry of Lipids of Biochemical Significance," Methuen, London (1955).
21. D. J. Hanahan, *in* "Lipid Chemistry," John Wiley, New York (1960).
22. L. L. M. Van Deenan and G. H. De Haas, *Ann. Rev. Biochem. 35*, 157–194 (1966).
23. L. E. Hokin and M. R. Hokin, *in* "Drugs and Membranes" (C. A. M. Hogben and P. Lindgreen, eds.), Vol. 4, pp. 23–40, Pergamon, Oxford, England (1963).
24. E. P. Kennedy, *Fed. Proc. 20*, 934–940 (1961).
25. R. J. Rossiter, *in* "Metabolic Pathways" (D. M. Greenberg, ed.), Vol. 2, pp. 69–115, Academic Press, New York (1968).
26. J. Y. Kiyasu and E. P. Kennedy, *J. Biol. Chem. 235*, 2590–2594 (1960).
27. M. Kates, *in* "Lipide Metabolism" (K. Bloch, ed.), pp. 165–184, John Wiley, New York (1960).
28. J. F. Danielli and H. Davson, *J. Cell. Comp. Physiol. 5*, 495–508 (1934).
29. J. D. Robertson, *in* "Intracellular Membraneous Structures" (S. Seno and E. V. Cowdry, eds.), pp. 379–401, Chugoku Press, Ltd., Okayama, Japan (1964).
30. J. Jarnefelt, *Biochim. Biophys. Acta 59*, 655–662 (1962).
31. H. Yoshida, T. Nukoda, and H. Fujisowa, *Biochim. Biophys. Acta 48*, 614–615 (1961).
32. G. D. Hunter and R. A. Goodsall, *Biochem. J. 78*, 564–570 (1961).
33. G. D. Hunter and G. N. Godson, *Nature 189*, 140–141 (1961).
34. P. B. Hill, *Biochim. Biophys. Acta 57*, 386–398 (1962).

35. R. W. Hendler, *Biochim. Biophys. Acta 60*, 90–92 (1962).
36. S. Fleisher, G. Brierly, H. Klovwen, and D. B. Slauterbach, *J. Biol. Chem. 237*, 3264–3272 (1962).
37. A. J. Sbarra and M. L. Karnovsky, *J. Biol. Chem. 235*, 2224–2229 (1960).
38. M. L. Karnovsky in "Metabolism and Physiological Significance of Lipids" (R. M. C. Dawson and D. N. Rhodes, eds.), pp. 501–508, John Wiley, London (1964).
39. M. L. Karnovsky and D. F. H. Wallach, *J. Biol. Chem. 236*, 1895–1901 (1961).
40. A. H. Beckett, A. F. Casy, N. J. Harper, and P. M. Phillips, *J. Pharm. Pharmacol. 8*, 860–873 (1956).
41. A. H. Beckett and A. F. Casy, *Bull. Narcotics IX*, 37–54 (1957).
42. L. B. Mellett and L. A. Woods, *in* "Progress in Drug Research" (E. Jucker, ed.), Birkhäuser Verlag, Basel and Stuttgart (1963).
43. D. G. Dervichian, *in* "Biochemical Problems of Lipids" (G. Popjak and E. LeBreton, eds.), pp. 263–342, Interscience, New York (1956).
44. J. M. Tobias, D. P. Agin, and R. Pawlowski, *Circulation 26*, 1145–1150 (1962).
45. D. E. Goldman, *Biophys. J. 4*, 167–188 (1964).
46. M. B. Abramson, R. Katzman, C. E. Wilson, and H. P. Gregor, *J. Biol. Chem. 239*, 4066–4072 (1964).
47. R. E. Rojas and J. M. Tobias, *Biochim. Biophys. Acta 94*, 394–404 (1965).
48. J. E. Garvin and M. C. Karnovsky, *J. Biol. Chem. 221*, 211–222 (1956).
49. H. Hanson and R. M. C. Dawson, *Eur. J. Biochem. 1*, 61–68 (1967).
50. S. J. Mulé, *Biochem. Pharmacol. 18*, 339–346 (1969).
51. T. Kakunaga, H. Kaneto, and K. Hano, *J. Pharmacol. Exp. Ther. 153*, 134–141 (1966).
52. J. G. Nutt, *Fed. Proc. 27*, 753 (1968).
53. C. Marchand and M. Vachon (personal communication).
54. C. Marchand and G. Denis, *J. Pharmacol. Exp. Ther. 162*, 331–337 (1968).
55. H. W. Elliott, N. Kokka, and E. L. Way, *Proc. Soc. Exp. Biol. Med. 113*, 1049–1052 (1963).
56. M. Blaustein, *Biochim. Biophys. Acta 135*, 653–668 (1967).
57. H. Hauser and R. M. C. Dawson, *Biochem. J. 109*, 909–916 (1968).

III. *The Effects of Narcotic Analgesic Drugs on General*
Metabolic Systems

Chapter 9

PROTEIN AND NUCLEIC ACID METABOLISM

Doris H. Clouet*

New York State Research Institute for
Neurochemistry and Drug Addiction
Ward's Island, New York†

I. INTRODUCTION

The mechanisms by which genetic information is transmitted during reproduction and cell division and by which nucleic acids are replicated and
nucleic acids and proteins are synthesized and degraded and by which these
processes are controlled are so complex chemically that there are, consequently, many sites at which foreign compounds including drugs may have
a direct effect on the metabolism of these macromolecules. Among the
metabolic reactions that have been suggested as possible sites of action of
narcotic analgesic drugs or related compounds are the following: (1) The
transcription from DNA may be disturbed by the binding or intercalation of
the drug to the double-stranded DNA, (2) the coding properties of DNA
may be faulty through drug-induced mutation of coding DNA, (3) the synthesis of messenger RNA may be impaired by prior drug effects on DNA,
(4) the synthesis of RNA may be affected by a drug-induced alteration of
precursor pools, (5) the synthesis of protein may be inhibited by prior drug
effects on RNA or on polysomal structure or function, or (6) the enzymatic
or binding properties of protein may be impaired by a direct drug: protein
interaction. In addition to these direct effects, alterations in the metabolism
of proteins or nucleic acids may be consequent to the effect of the drug on
other metabolic processes such as transport or energy supply. In this chapter
such effects of narcotic drugs on protein or nucleic acid metabolism are

---

* The author wishes to thank her colleagues Milton Ratner and Norman Williams for
  their contribution to the work and NIMH for support of some of the work performed
  in the author's laboratory (MH 14013).
† Present address: New York State Narcotic Addiction Control Commission Testing
  and Research Laboratory, 80 Hanson Place, Brooklyn, New York.

described, with particular emphasis on mammalian systems, since similar effects in single cells are described in Chapter 15.

II. PROTEIN METABOLISM

A. Effects on Protein Metabolism in the Liver

1. Microsomal Drug-Metabolizing Enzymes

The administration of certain inducer drugs such as phenobarbital or zoxazolamine to animals produces an increase in the activity of the drug-metabolizing enzymes (DME) localized in the liver microsomal fraction.[1,2] Narcotic analgesic drugs are not inducer drugs and instead produce a decrease in the activity of some DME in liver, either after a single injection[3] or after chronic drug administration.[4] The administration of morphine to male rats partially inhibited the increase in liver DME activity induced by phenobarbital, chlorpromazine, or reserpine as well as similar induction by the anabolic steroids 19-nortestosterone and chloromethyltestosterone[3] (Table I).

Since a concomitant increase in the protein synthetic activity of liver microsomes has been found after the administration of phenobarbital or 3-methylcholanthrene,[5,6] it is possible that morphine acts to inhibit the induction of DME by inhibiting protein synthesis in liver microsomes. This possibility has not been investigated directly. However, the changes in the level of protein in the combined microsomal-supernatant fraction of liver paralleled the changes in liver DME activity.[3]

TABLE I
Morphine and Liver Drug Metabolizing Enzyme Activity

| | Enzyme Activity[a] | | | |
| | Control | | Morphine | |
Treatment	Meperidine	Hexobarbital	Meperidine	Hexobarbital
None	103.2	53.3	37.2	18.7
Phenobarbital	185.0	84.4	97.5	95.4
Nortestosterone	107.6	62.7	61.1	34.4
Phenobarbital + nortestosterone	149.5	105.1	121.0	86.8

[a] Enzyme activity equals micromoles substrate metabolized per hour per gram McS protein (assayed as described in Clouet and Ratner[3]). There were three rats in each group, assayed individually. The doses were morphine, 30 mg base/kg body weight daily for 10 days; 19-nortestosterone, 5 mg/kg every third day beginning on the day before morphine was started; phenobarbital, 60 mg/kg given 2 days prior and on the second, fifth, and eighth day of morphine. (Adapted from Clouet and Ratner by courtesy of the *Journal of Pharmacology and Experimental Therapeutics*.[3])

During the induction of the DME there is a vast proliferation of the endoplasmic reticulum of the liver.[7] The rate of synthesis of both nuclear and microsomal RNA is increased at this time, as measured by the incorporation of labeled uridine into RNA fractions.[8] Messenger RNA, in particular, seems to be synthesized, since the addition of RNA isolated from the livers of induced animals produced an increase in the incorporation of amino acids into protein in an *in vitro* system.[9] An opposite effect has been reported after the narcotic drug levorphanol, in that it produced an inhibition of RNA synthesis in rat liver cells.[10] This inhibition resembled the effect described earlier in *Escherichia coli*[11] in being selective for ribosomal RNA synthesis. The same investigator found that the activity of DNA-dependent RNA polymerase was not affected by the addition of levorphanol *in vitro*.[10]

2. Isolated Hepatic Microsomes

The administration of a single dose of morphine (30 mg/kg body weight of morphine base) to rats has a biphasic effect on the ability of the liver to incorporate ^{14}C-leucine into protein when measured in isolated microsomal–supernatant fractions prepared from rats killed at various times after the injection of the drug. The amino acid incorporation was lowest 2 hr after the injection (57% of the control level) and highest 6 hr after the injection (150%).[12] When the microsomes and soluble fractions from the livers of morphine-treated rats were assayed with the appropriate preparation from untreated animals, the microsomes showed the same biphasic response as the combined microsomal–supernatant fraction.[12] The effect of chronic morphine administration was explored only in a group of rats consuming morphine in drinking water (Table II). Liver microsomes isolated from these rats had an increased capacity to incorporate leucine into protein after 11 days of drug and returned to control level by 30 days.[12] However, the effect of morphine administered in drinking water may be quite different from the effect produced by chronic injection of the drug with the cyclic periods of narcosis and withdrawal.

TABLE II
Liver Microsomal Incorporation of ^{14}C-Leucine into Protein

Days of Treatment	Morphine	Control
11 (11.5)	$26,600 \pm 2100^a$	$13,200 \pm 800$
15 (12.4)	$25,400 \pm 1700$	$15,800 \pm 1600$
30 (23.5)	$14,600 \pm 600$	$13,800 \pm 1100$

^{a} Incorporation of ^{14}C-leucine by isolated liver microsomal fractions as counts per minute per 20-min incubation/10^5 counts/min in soluble pool $\pm$ S.D. (assayed as described in Ratner and Clouet[12]). The average amount of morphine base consumed in milligrams per day is given in parentheses. (Adapted from Ratner and Clouet with the permission of *Biochemical Pharmacology*.[12])

TABLE III
Inhibition of Amino Acid Incorporation by Narcotic Drugs Added
In Vitro

Drug	Percent Inhibition[a]			
	10^{-4} M	10^{-3} M	5×10^{-3} M	10^{-2} M
Morphine	3	3	12	12
Meperidine	—	5	—	13
Levorphanol	—	1	—	6
Nalorphine	—	(+2)	12	44
Levallorphan	5	21	55	96

[a] Percent inhibition as compared to the amount of ^{14}C-leucine incorporated into protein by isolated brain ribosomal fractions (assay described in Clouet[13]), without drug as 100%. (Adapted from Clouet.[13])

B. Effects on Protein Metabolism in Brain

1. Cell-tree Amino Acid Incorporating Systems

The addition of narcotic drugs to an incorporating system that included brain ribosomes and brain supernatant fraction had no effect on the incorporation of ^{14}C-leucine into protein until concentrations of drug higher than 10^{-3} M were used (Table III). In the special case of levallorphan, it was found that sRNA in the supernatant fraction was precipitated by the drug and that this inhibition was reversed by the addition of excess supernatant fraction.[13]

Ribosomes isolated from the brains of morphine-treated rats had a decreased ability to synthesize protein that was both dose-dependent and time-dependent. After a single injection of morphine at doses from 10–60 mg/kg, there was a significant inhibition of the capacity to synthesize protein 2 hr after the injection of drug.[14] At a dose of 60 mg/kg the inhibition lasted from 1 to 3 hr after the injection. After five daily doses of 30 mg/kg morphine there was an increased ribosomal synthetic activity as compared to the first injection.[14]

The various components of the incorporating system were examined in order to identify the nature of the inhibition. The functional activity of the brain ribosomes isolated from morphine-treated rats was normal in regard to the response to the synthetic messenger RNA, poly U. When the levels of endogenous brain messenger RNA, brain ribonuclease, and RNA ase inhibitor were measured, animal to animal variation was greater than differences in treatment groups so that no specific effect of morphine on these components was discernible.[15] However, brain polysomal preparations from drug-treated rats were less stable than control ribosomes.[13] When the polysomes were first incubated in an amino acid-incorporating system and then centrifuged through a 20–50% sucrose gradient, some of the newly labeled peptides on the polysomes and the polysomes themselves sedimented to the bottom of the tube (Fig. 1). The antagonist levallorphan had no effect on polysomal stability. This lability of brain ribosomes isolated from narcotic-treated rats resembled the polysomal lability in HeLa cells incubated in the presence of levorphanol.[16]

2. Protein Synthesis in Brain In Vivo

The assay of the ability of isolated ribosomes to promote protein synthesis is performed under optimal conditions that are not necessarily the conditions prevalent in the tissue from which the ribosomes are isolated. To measure protein synthesis in brain *in vivo*, labeled amino acids were injected into the CSF by way of the *cisterna magna* and the animal was killed a short time later (usually 30 min). The brain proteins were isolated and examined for radioactivity. After a dose of 60 mg/kg morphine, protein synthesis, as evidenced by the specific radioactivity of the brain proteins, declined for a period of at least 6 hr, with the greatest decreases in the labeling of microsomal and supernatant protein.[17] A calculation of brain protein turnover showed a significant ($P<0.001$) decrease in turnover from 0.27 to 0.23%/hr.[17] Protein synthesis also was depressed in the brains of rats assayed 1.5 hr after a fifth daily injection of morphine.[17]

One of the pharmacological effects of morphine in rats is hypothermia, which may lower metabolic rates in general. The administration of chlorpromazine produces an inhibition of protein synthesis in the brains of mice, which is ascribed to the hypothermia produced by the drug.[18] The inhibition of protein synthesis produced by morphine, however, is also found in mosphine-treated animals that are kept normothermic in a 30° incubator.[17]

An inhibition of protein synthesis in the CNS after morphine administration is at first glance contradictory to the studies in which protein synthesis seems to be required in order for tolerance to chronic drug administration to develop. (See Chapter 20 for a detailed discussion of this phenomenon.) However, 24 hr after the administration of a single dose of morphine, the incorporation of ^{14}C-lysine into brain protein is increased above the level in untreated rats (Fig. 2). Since the amount of protein in brain microsomal-supernatant fraction increases during chronic morphine treatment,[19] it is possible that the biphasic effect of the drug on the rate of protein synthesis in brain does not result in a net deficit of protein in the brain and may actually result in higher levels of some proteins. An exploration of the nature of the protein synthesized a day after an initial injection of morphine would be interesting in view of the effects of inhibitors of protein synthesis on the development of tolerance to morphine.[20,21] The tolerance that develops rapidly during an infusion of morphine in rats can be inhibited by treatment of the animals with inhibitors of protein synthesis such as puromycin or cycloheximide.[22] However, in the experimental animals used for Fig. 2, in which protein synthesis in brain was inhibited

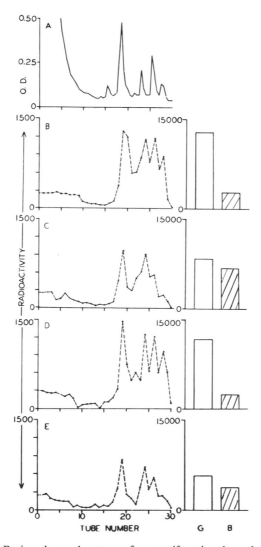

Fig. 1. Brain polysomal patterns after centrifugation through a 20–50% sucrose gradient for 16 hr. The brain ribosomal preparations were incubated in a ^{14}C-leucine-incorporating system before centrifugation. A is a typical pattern of optical density at 254 mμ with top of tube to the left. B–E are records of radioactivity from similar gradients from rats receiving saline (B); morphine, 60 mg/kg intraperitoneally (C); levallorphan, 100 mg/kg (D); and morphine, 1 mg/kg intracisternally (E). On the right is shown the distribution of radioactivity: G is the amount on gradient and B is the amount sedimenting to the bottom of the tube. (Adapted from Clouet.[13])

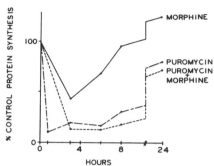

Fig. 2. Incorporation of ^{14}C-lysine into brain proteins *in vivo*. Each rat received either 0.8 mg/kg puromycin in saline or an equal volume of saline at time 0, and half of the rats received 60 mg/kg morphine at the same time. The rats (eight for each point) were given ^{14}C-lysine intracisternally 30 min before sacrifice at intervals after morphine or saline, and the brain proteins were isolated and examined for radioactivity that was compared to the values for animals receiving only saline as 100%. (Adapted from Clouet.[13])

almost completely for 8 hr, there was no effect on analgesic or hypothermic tolerance to a second dose of morphine either 5 or 24 hr after the first dose. Smith and his colleagues stress the difference between "short-term" and "long-term" tolerance: The first presumably is due to the depletion of neurotransmitters and the second to true tolerance possibly involving the synthesis of protein.[23] It will be very important in new studies in this area to define tolerance by the methods both of its production and of its measurement.

C. Amino Acids in Brain

After the injection of labeled amino acids into the CSF as precursors of brain protein in assays of the rate of protein synthesis, the levels of radioactivity of the free amino acid pool of brain were found to be higher in morphine-treated rats than in saline-injected control animals.[17] Since ^{14}C-leucine was the usual precursor in the studies, this amino acid was studied in detail after its injection into the CSF. In the first 30 min after the introduction of uniformly labeled ^{14}C-leucine into the CSF, ^{14}C-from the carbon skeleton was found in glutamic acid, glutamine, aspartic acid, GABA, and many other amino acids as well as still in leucine.[24] All of the ^{14}C-labeled compounds were present in higher concentrations in the brains of rats killed 2 hr after a single injection of morphine than in untreated controls, indicating that there was a decreased rate of transport of the injected aminoacid from the CSF in the drug-treated rats. The relative radioactivity in each compound was similar for the two treatment groups, indicating that the oxidation of leucine through the citric acid cycle in brain was unchanged 2 hr after the morphine treatment.[24] Additional information obtained from the examination of the brain amino acid pool by amino acid analyzer in these experiments was that the levels of endogenous amino acids in brain were not affected by the drug treatment.[24]

D. Inhibition of Protein Synthesis in Single Cells

This subject is discussed in detail in Chapter 15. It should be noted here only that in mammalian cells grown in tissue culture in the presence of levorphanol or levallorphan there is an inhibition of protein synthesis a few minutes after the addition of the drug.[25,26]

III. NUCLEIC ACID METABOLISM

A. Synthesis of RNA

In *E. coli* the addition of 10^{-3} *M* levorphanol to the culture medium inhibits the incorporation of ^{32}P into RNA.[11] In mammalian cells grown in tissue culture (HeLa cells) the same concentration of this drug inhibits the incorporation of ^{14}C-guanine into ribosomal RNA, *s*RNA, and a rapidly labeled RNA that may be *m*RNA.[16] At a slightly higher concentration of drug (3×10^{-3} *M*), not only was protein and RNA synthesis inhibited but there was a loss of ATP and GTP from the cells.[25] Greene and Magasanik suggest that the changes in the rate of synthesis of the macromolecules may be related to the loss of nucleotides from the cell, since the leaking of nucleotides preceded the inhibition of the synthetic rates.[25]

In rats the rate of incorporation of ^{14}C-orotic acid into brain RNA was measured after the injection of this precursor into the CSF.[27] In these experiments the total brain RNA was separated by the centrifugation of brain homogenate into nuclear-mitochondrial (NMt) and microsomal-supernatant (McS) fractions before the isolation of RNA. In a time curve of ^{14}C-orotic acid into NMt RNA, the labeled precursor was injected at times from 10 to 120 min before the animals were sacrificed. In the brains from morphine-treated rats, there was an inhibition of synthesis of both 18S RNA (tubes 10–15) and 28S RNA (tubes 18–22) in samples from ^{14}C-orotic acid pulses of 30 and 60 min that represent the time periods 90–120

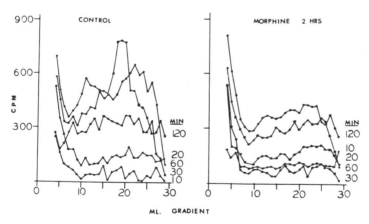

Fig. 3. Incorporation of ^{14}C-orotic acid into brain NMt RNA *in vivo*. ^{14}C-orotic acid (6 μc/rat) was injected intracisternally at times varying from 10 to 120 min before sacrifice of the animal. The rats were killed 2 hr after an injection of morphine or saline. Brain homogenates were prepared and separated into nuclear-mitochondrial (NMt) or microsomal-supernatant (McS) fractions before centrifugation through a 5–20% gradient for 16 hr. The lightest RNA is on the left.

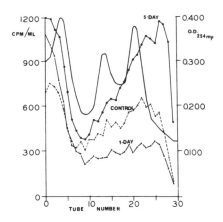

Fig. 4. RNA synthesis after one or five injections of morphine. The rats were killed 2 hr after the first or fifth daily injection of 60 mg/kg morphine and 1 hr after the administration of ^{14}C-orotic acid intracisternally. The experimental details are the same as in Fig. 3. The solid line records the O.D. 254 mμ for the typical gradient, and the lines connecting points describe the radioactivity in the NMt RNA.

min and 60–120 min after the injection of morphine (Fig. 3). The rate of incorporation of orotic acid into McS RNA was unchanged by drug treatment. Two hours after the fifth daily injection of morphine, there was a large increase in the incorporation of precursor ^{14}C-orotic acid into RNA of the NMt fraction, particularly in the >28S forms (Fig. 4). Since these heavy forms of RNA are usually considered pre-ribosomal, these results suggest that there may be a synthesis of new rRNA in the nuclei after the chronic administration of morphine.

A number of investigators have shown that inhibitors of RNA synthesis abolish or decrease the tolerance to chronic treatment with narcotic drugs.[19,23,28–30] It seems, from the results of experiments using HeLa cells in culture, that a primary effect of narcotic drugs is on some step in the synthesis or utilization of RNA in the process of protein synthesis and that the inhibition of protein synthesis is consequent to this.[25]

In rats the effects of actinomycin D, 5-fluoruracil, and 6-mercaptopurine in inhibiting the development of tolerance to the intravenous infusion of morphine suggest that the synthesis of RNA as well as protein is required for the development of tolerance.[30]

B. Synthesis of DNA

There have been no studies on the effect of narcotic drugs on DNA in animals, although it may be inferred from the effect of actinomycin D on the development of tolerance that DNA-dependent RNA polymerase function is necessary for tolerance to develop. In HeLa cells in culture there is a slight inhibition of the rate at which ^{14}C-thymidine is incorporated into DNA in the presence of levorphanol, similar to the effect in bacterial cells.[16,25]

C. Interactions of Narcotic Drugs with Nucleic Acids

One would expect that the ionized forms of the basic narcotic analgesic drugs and the acidic nucleic acids would at least form salt bonds. However,

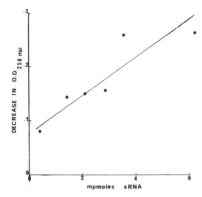

Fig. 5. The spectral absorbance of levallorphan and sRNA. Levallorphan at a concentration of 2.7 μmoles/ml and stripped sRNA at levels of 0.4–6.2 mμmoles of sRNA/ml were scanned in a Beckman DK2A recording spectrophotometer from 250 to 320 mμ. The difference in the absorbance of the drug and sRNA in the same cuvette and the sum of the individual absorbance is shown at 258 mμ.

any such salts must be very soluble since it has proved difficult to demonstrate an interaction of drugs and nucleic acids. Only with levallorphan is a precipitate formed that is visible in the test tube.[14] A reaction between stripped sRNA and levallorphan can be demonstrated spectrophotometrically by comparing the absorbance in the UV of the two compounds individually with the absorbance when both compounds are in the same cuvette. At optical density (O.D.) 258 mμ, the decrease in the combined

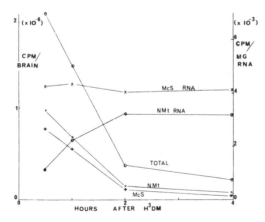

Fig. 6. The binding of tritiated dihydromorphine (^{3}HDM) to brain RNA. Rats were killed at various times after the administration of ^{3}HDM (0.5 mg/kg) intracisternally, and their brains were homogenized and separated into NMt and McS fractions before the isolation of RNA by the phenol extraction method. The left ordinate is the scale for the radioactivity in homogenate and in the NMt and McS fractions, while the right ordinate is the scale for the specific radioactivity of the isolated RNA.

absorbance below the theoretical value depends on the amount of sRNA if the drug concentration is kept constant (Fig. 5). The decrease in absorbance suggests that the ionized complex precipitates from solution to a certain extent.[31] Attempts to show binding of tritium-labeled dihydromorphine (D³HM) to various forms of RNA by equilibrium dialysis and Sephadex sieving were unsuccessful.[31]

The administration of D³HM to rats by way of the CSF in a pharmacologically active dose (50 μg/100 g body weight) results in the presence of the drug in fractions of RNA isolated from the brain (Fig. 6). In these experiments the rats were killed at times from 0.5 to 4 hr after the injection of the drug, and the brains were removed and homogenized. The brain homogenates were separated into NMt and McS fractions by centrifugation before RNA was extracted by the phenol procedure. The amount of drug found in the McS–RNA fraction did not change during the 4 hr, although the total amount of drug in brain decreased markedly during this period. In the NMt–RNA fraction, the level of labeled drug increased to a maximum 2 hr after the injection. Unchanged D³HM could be isolated from the fraction. The binding, however, was not tight since the drug could be removed from the fraction by either ethanol extraction or exhaustive dialysis.[27] The results of these experiments do not eliminate the possibility that the association of the drug and nucleic acids occurred during the homogenization, although the time-dependent binding in the NMt–RNA fraction suggests that the transport of the drug into the nuclei *in vivo* is a limiting factor.

It has been reported that the administration of morphine, methadone, and heroin to pregnant hamsters in very high doses produces fetal abnormalities and decreases the litter size.[32] In these experiments an interaction of the narcotic drug with coding DNA in the fetus is a possibility, although there are many other possibilities related to the physiological changes induced in both pregnant animal and fetus by the administration of the drug.

Thus there is no evidence that the narcotic analgesic drugs and nucleic acids interact directly *in vivo*. Other groups of alkaloids, the *vinca* alkaloids and the *ipecac* alkaloids, have been shown to inhibit RNA synthesis in mammalian cells in tissue culture,[33,34] with the inhibition localized at the aminoacyl sRNA transfer reaction on the ribosome.[33] Some of the evidence described in this section suggests that this step may be a site for the action of narcotic drugs on tissue components.

IV. DISCUSSION

The effects of narcotic drugs on the metabolism of protein and nucleic acids in mammalian tissue may be summarized in the same order as the list of possible metabolic sites of action described in the introduction of this chapter. There is no evidence that coding or transcriptional functions of DNA in liver or brain are altered by the administration, or the addition *in vitro* of narcotic drugs to these tissues. There is a slight inhibition of DNA synthesis in mammalian cells in culture (HeLa cells) in the presence of the same concentration of levorphanol that inhibits RNA synthesis almost completely. A transient inhibition of RNA synthesis, especially of rRNA, is found in brain shortly after the administration of a single dose of morphine to rats. After the fifth daily injection there is an increased rate of synthesis of nuclear RNA. There is some evidence that the polysomal structure is unstable, both in HeLa cells and in the brains of morphine-treated rats. This polysomal lability in rat brain may be causally related both to the transient inhibition of protein synthesis after an initial drug injection and to the increased synthesis of RNA after five daily injections. The effect of morphine on polysomal structure may be mediated by an interaction similar to the specific effect of other groups of alkaloids on the attachment of aminoacyl sRNA to the ribosome. The biphasic responses in the rate of synthesis of both macromolecular species in mammalian nervous tissue cannot be compared directly to responses in single cells, as the single cells are continuously exposed to the narcotic drug without an opportunity to adapt to the periodic exposure to drug that the animal experiences. In the central nervous system the initial response to a single injection of morphine does resemble in some respects the response of single cells upon transfer to a medium containing a narcotic drug.

The importance of changes in the rate of synthesis of RNA and protein to the study of an initial biochemical effect of a narcotic drug is emphasized both by the abolition of the development of tolerance by inhibitors of protein or RNA synthesis and by the prevalence of theories of narcotic drug action that involve the synthesis of new species of the macromolecules or a change in the rate of synthesis of those already present in the tissue.[22,35-37] In view of the biochemical complexity of the central nervous system and its functional organization, both in discrete anatomical areas and in common structural elements distributed throughout the nervous system, it is unlikely that a narcotic drug can produce gross changes in the metabolism of protein or RNA that are measurable in whole brain unless the dose of drug is large enough to influence not only limiting reactions in specific areas but also similar reactions elsewhere in the nervous system. One hopes that this latter effect is pertinent in the experiments in animals described in this chapter. The experimental animal, the young adult male Wistar rat, is unresponsive to a hot plate at 55°C for about 3 hr after an intraperitoneal injection of 60 mg morphine base/kg body weight. Thus it is possible that the transient inhibitions of RNA and protein synthesis in brain following a single injection of morphine are consequent to other responses to drug administration such as decreased respiration or blood supply rather than causal mechanisms for the initiation of pharmacological responses. This drawback to studying the relationship of narcotic drugs to macromolecular synthesis may be overcome by (1) studying the interactions in nerve cells in tissue culture or in model cell systems (such as HeLa cells), (2) examining the turnover of RNA and protein in small regional units of the nervous system, or (3) measuring drug effects on the rate of turnover of single proteins, such as individual enzymes, in the nervous system.

V. REFERENCES

1. H. Remmer, *Naturwissenschaften 45*, 189–197 (1958).
2. A. H. Conney and J. J. Burns, *Nature 184*, 363–370 (1959).
3. D. H. Clouet and M. Ratner, *J. Pharmacol. Exp. Ther. 144*, 362–371 (1964).
4. J. Cochin and J. Axelrod, *J. Pharmacol. Exp. Ther. 125*, 105–112 (1959).
5. H. V. Gelboin and L. Sokoloff, *Science 134*, 611–613 (1961).
6. A. Von der Decken and T. Hultin, *Arch. Biochem. Biophys. 90*, 201–206 (1960).
7. J. R. Fouts, *Biochem. Biophys. Res. Commun. 6*, 373–381 (1961).
8. J. W. Zemp, *Pharmacologist 10*, 179 (1968).
9. L. A. Loeb and H. V. Gelboin, *Nature 199*, 809–811 (1963).
10. S. Sakiyama, *Seikagaku 40*, 214–219 (1968).
11. E. J. Simon, *Nature 198*, 794–795 (1963).
12. M. Ratner and D. H. Clouet, *Biochem. Pharmacol. 13*, 1655–1661 (1964).
13. D. H. Clouet, *in* "Protein Metabolism of the Nervous System" (A. Lajtha, ed.), Plenum Press, New York (1970).
14. D. H. Clouet and M. Ratner, *J. Neurochem. 15*, 17–23 (1968).
15. D. H. Clouet *in* "Drug Abuse: Social and Psychopharmacological Aspects" (J. O. Cole and J. R. Wittenborn, eds.), Charles C Thomas, Springfield, Illinois (1969).
16. W. D. Noteboom and G. C. Mueller, *Mol. Pharmacol. 2*, 534–542 (1966).
17. D. H. Clouet and M. Ratner, *Brain Res. 4*, 33–43 (1967).
18. L. Shuster and R. V. Hannan, *J. Biol. Chem. 239*, 3401–3406 (1964).
19. M. T. Spoerlein and J. Scrafani, *Life Sci. 6*, 1549–1564 (1967).
20. B. M. Cox and O. H. Osman, *Brit. J. Pharmacol. 35*, 373p (1969).
21. A. A. Smith, M. Karmin, and J. Gavitt, *Biochem. Pharmacol. 15*, 1877–1879 (1966).
22. B. M. Cox and M. Ginsburg, *in* "Scientific Basis of Drug Dependence" (H. Steinberg, ed.), Churchill Ltd., London (1969).
23. A. A. Smith, M. Karmin, and J. Gavitt, *J. Pharmacol. Exp. Ther. 156*, 85–91 (1967).
24. D. H. Clouet and A. Neidle, "The Effect of Morphine Administration on the Metabolism of Leucine in Rat Brain," *J. Neurochem. 17*, 1069–1074 (1970).
25. R. Greene and B. Magasanik, *Mol. Pharmacol. 3*, 453–472 (1967).
26. W. D. Noteboom and G. C. Mueller, *Mol. Pharmacol. 5*, 38–48 (1969).
27. D. H. Clouet and N. Williams, *Fed. Proc. 27*, 753 (1968).
28. M. Cohen, A. S. Keats, W. Krivoy, and G. Ungar, *Proc. Soc. Exp. Biol. Med. 119*, 381–384 (1965).
29. I. Yamamoto, R. Inoki, Y. Tamari, and K. Iwatsubo, *Jap. J. Pharmacol. 17*, 140–142 (1967).
30. B. M. Cox, M. Ginsburg, and O. H. Osman, *Brit. J. Pharmacol. 33*, 245–256 (1968).
31. D. H. Clouet and N. Williams, unpublished observations.
32. W. F. Geber and L. C. Schramm, *Pharmacologist 11*, 248 (1969).
33. A. Grollman, *Proc. Nat. Acad. Sci. 56*, 1867–1874.
34. W. A. Creasey and M. E. Markiw, *Biochim. Biophys. Acta 87*, 601–608 (1964).
35. L. Shuster, *Nature 189*, 314–315 (1961).
36. D. B. Goldstein and A. Goldstein *Biochem. Pharmacol. 8*, 48 (1961).
37. H. O. J. Collier, *Nature 205*, 181–182 (1965).

Chapter 10

CATECHOLAMINES AND 5-HYDROXYTRYPTAMINE*

E. Leong Way and Fu-Hsiung Shen

Department of Pharmacology
University of California at San Francisco
San Francisco, California

I. GENERAL INTRODUCTION

Numerous attempts have been made to implicate the pharmacologic effects of morphine or its surrogates with the biogenic amines in the brain. Such premises are the natural consequences of the facts that the amines have a ubiquitous, although sometimes specialized, distribution in the central nervous system that often can be affected by drugs and that certain of these amines have been established to play key roles in the peripheral nervous system as neurotransmitters. While their functions in the CNS are not nearly as clearly defined, it appears reasonable to make inferences by analogy and, despite the gaps, attempt to utilize the existing knowledge of the biogenic amines as a framework for explaining morphine effects.

Among the biogenic agents that have been considered to be involved in the effects of morphine, the most prominent have included acetylcholine, epinephrine, norepinephrine, dopamine, and 5-hydroxytryptamine (5-HT). Since acetylcholine has been considered in another section, the emphasis in this chapter will be on the four amines. The separation of their discussion, however, should not lead the reader to forget that the roles of all these substances may be intimately enmeshed. Consequently, any treatment that affects one biogenic amine in the CNS could have significant consequences on the others that are also present. All of these amines have been studied with respect to their possible relationships with both the acute and

* Portions of the work described in this review were supported by USPHS service grants MH-17017 and GM-01839. We are grateful to H. H. Loh and Gladys Friedler for their comments and suggestions.

chronic actions of morphine. The acute studies in the main have been concerned with analgetic mechanisms and the chronic aspects with tolerance and physical dependence mechanisms. This review considers the possible roles of epinephrine, norepinephrine, dopamine, and 5-HT in each of these mechanisms. There is, suprisingly, no extensive review of this area although various aspects have been covered in some detail and analyzed.[1-3] The most comprehensive is the recent review of the biochemistry of addiction by Dole.[4]

II. EPINEPHRINE

A. Introduction

Between 1930 and 1950 most of the studies attempting to relate morphine effects with the catecholamines were concerned mostly with epinephrine since, at the time, considerably more was known about it than norepinephrine and dopamine. However, recent findings with the latter two substances have resulted in a lessened interest in epinephrine, particularly since, of the three substances, the content of epinephrine in the CNS is seemingly least affected by functional changes or drugs. Owing to the lack of suitable procedures to measure catecholamines in the brain, it is not surprising that earlier workers sought to find clues concerning morphine effects in the adrenal gland where high concentrations of epinephrine were known to exist. However, such experiments cannot be extrapolated to CNS changes as evidenced by subsequent findings indicating that changes in the levels of adrenal epinephrine effected by morphine are not always paralleled by changes in brain epinephrine or norepinephrine levels[19].

B. Acute Effects of Morphine

It should be pointed out beforehand that the early procedures for catecholamines failed to distinguish between epinephrine, and norepinephrine, and the former substance constitutes at least 80% of the catecholamine content in the adrenals.

Single injections of adequate doses of morphine generally effect the release of epinephrine from the adrenal medulla. This is reflected by a fall in epinephrine in the organ and a rise in blood level and urinary excretion. The epinephrine-depleting response of morphine on the adrenals, first reported by Elliott in 1912,[5] has been confirmed by many investigators.[6-20] The extent of the response to morphine varies with species. The cat is more sensitive than the rat or the dog.[6,9] In the rat, four injections of morphine HCl (20 mg/kg) at 1-hr intervals evoked a definite rise in "sympathin" excretion in the 24-hr urine sample, but no change was noted after a single injection.[20]

That there is a central component to the adrenal epinephrine-depleting action by morphine is shown by the fact that the depletion effect can be prevented by sectioning the splanchnic nerve[6,21] or pretreating the animal with phenobarbital.[19,22] Moreover, the hyperglycemic effect of morphine, which has been shown to be due to epinephrine release from the adrenal gland,[10,11,13,24] can be prevented by adrenalectomy,[23] pentobarbital,[24] or by ganglionic blockade with tetraethylammonium chloride.[25]

The profound effect of morphine on adrenal epinephrine levels prompted several investigators to propose that morphine analgesia is mediated wholly or in part through adrenal medullary release of epinephrine.[26-28] The hypothesis was based on findings in the dog and rat that the rise in pain reaction threshold after morphine or its surrogates, meperidine or methadone, was reduced after adrenalectomy. Moreover, epinephrine was found to elevate the pain threshold to heat in humans[27] and to elevate the reaction threshold to electrical stimulation in the dog.[19] Also, various sympathomimetic agents had been reported to increase the threshold to nociceptive stimulation in dogs and cats[27,30] and to potentiate morphine and meperidine analgesia in humans.[26,31] Further, Watts reported that the degree of hyperglycemia (epinephrine release) evoked by the *d*- and *l*-forms of methadone in dogs paralleled their analgetic potencies.[24] While these findings appear consistent with an adrenal adrenergic mechanism for morphine analgesia, substantial evidence in the literature fails to support the hypothesis.

Gross *et al.* were unable to detect rises in epinephrine blood levels after morphine in the dog.[27] Also various sympatholytic agents did not reverse the rise in pain threshold after morphine in the rat[32] and after methadone in the dog.[27] Miller and his associates provided convincing evidence that the release of epinephrine from the adrenal medulla plays little if any role in mediating the increase in pain response threshold after morphine administration.[33] They found that adrenalectomy or adrenal gland demedullation did not decrease the response of rats to morphine; indeed, adrenalectomized rats, maintained by cortisone, DOCA, or saline, were more sensitive to morphine. The threshold dose in rats for the hyperglycemic effect of morphine, a sensitive index for release of epinephrine from the adrenal medulla, was 10–30 times that required for elevating the pain reaction threshold. Also, the ganglionic blocker, tetraethylammonium chloride (TEA), which was shown to prevent morphine hyperglycemia,[25] did not reverse the effect of morphine on the tail-flick reaction time to thermal stimulus. Further, although morphine produced hyperglycemia and elevated plasma "sympathin" levels in dogs, the absolute changes in "sympathin" levels were much less than could be expected to raise the pain reaction threshold. Thus the bulk of the evidence by Miller and his associates indicates that the small discharge of epinephrine from the adrenal medulla, which may occur after morphine administration, can hardly account for the effects of morphine on pain reaction threshold. While their findings do not rule out the possibility that morphine could evoke a local discharge of epinephrine in the central nervous system, Gunne reported that brain epinephrine level in rats was unaffected by a single injection of morphine at doses between 20 to 90 mg/kg.[34]

C. Effects of Chronic Morphinization

In 1929, Abe noted that the adrenal epinephrine-depleting effect of morphine in rabbits disappeared after long-term morphine treatment.[35] A decreased release was suggested by the fact that the amount of epinephrine in the adrenal gland was increased. This was later confirmed by Wada, who also found that the increase in the adrenal vein epinephrine resulting from morphine administration was diminished as the morphine doses were repeated.[11] Maynert treated rats chronically with 100 mg/kg morphine intraperitoneally (i.p.) twice a day and found that the mean level of epinephrine in the adrenal was decreased by 17–51% for the initial 5 days but by the tenth day it was elevated 16%.[19] In another series of experiments, he noted that the increase in adrenal epinephrine content was proportional to the dose of morphine and the duration of treatment. However, Gunne was unable to detect any change in adrenal epinephrine (or norepinephrine) in the dog or rat chronically morphinized for 3 weeks, although during the latter course of his experiments he used doses of morphine that exceeded those used by Maynert.[34,36] It is difficult to reconcile these divergent findings.

With repeated injections of morphine, the urinary excretion of epinephrine and its metabolites increased in rough proportion to the dose of morphine. These findings were noted in rats,[36,37] dogs,[36] and human volunteers.[38] In all three groups it was also found that after the initial rise in urinary epinephrine excretion with increased morphine dosage, a partial tolerance gradually developed as drug administration was continued.

D. Effects of Morphine Withdrawal

In 1932 Tachigawa observed an elevation in levels of epinephrine in the blood of chronically morphinized dogs during withdrawal.[39] Gunne found upon abrupt withdrawal that the adrenal epinephrine level was decreased in the dog after 72 hr and in the rat after 48 hr, but after precipitated withdrawal with nalorphine, only the dog showed a depletion at 4 hr.[34] In agreement with these findings Maynert found that after precipitation of the abstinence syndrome with nalorphine in three species of chronically morphinized animals, adrenal epinephrine was reduced 59% in the dog and 30% in the rabbit and was unchanged in the rat.[17] Maynert pointed out that this difference in response was related to the severity of the abstinence syndrome in each species since dogs exhibited excitement with frequent convulsions while rats appeared sedated. Gunne also found the degree of adrenal epinephrine depletion in the dog proportional to the severity of the withdrawal signs.[36] Central involvement in the effect was suggested by the fact that the epinephrine depletion elicited by precipitated withdrawal with nalorphine could be prevented by section of the splanchnic nerve[36] or pentobarbital pretreatment.[19]

The urinary excretion of epinephrine was increased in chronically morphinized dogs and rats after abrupt withdrawal, with maximum excretion occuring on the second or third day.[36,37] An increased excretion also occurred after precipitated withdrawal.[36] However, Weil-Malherbe and associates found an abrupt fall in catecholamine excretion during the withdrawal phase in human subjects.[38] This discrepancy might be due to the fact that in human volunteers the degree of dependence was less and the withdrawal from morphine was gradual.

E. Conclusions

In summary, the fluctuations of epinephrine levels in the adrenal medulla and urine that occur after chronic morphinization and upon withdrawal do not appear to be causally related to the basic mechanism concerned with tolerance and physical dependence development. The depletion of adrenal epinephrine and the rise in urinary levels that may occur after high doses of morphine appear to be consequential to the pharmacologic events initiated by the drug. Likewise, the changes in urinary epinephrine excretion resulting from morphine withdrawal seem to reflect a general stress response. Even if brain epinephrine is involved in tolerance and dependence mechanisms, it seems highly unlikely that adrenal or urinary epinephrine can provide precise indices of such a relationship.

III. NOREPINEPHRINE

A. Introduction

Among the various catecholamines in the CNS, norepinephrine has been most extensively explored for its possible interactions with morphine. These studies gained impetus not only from its established role as a neurotransmitter in the peripheral nervous system but also because its content in the brain is affected by drugs and stress. Furthermore, norepinephrine appears to be localized at certain sites in the CNS where morphine appears to act preferentially, for example, in the hypothalamus, midbrain, and medulla. Norepinephrine has been studied fairly intensively for its possible role in pain mechanisms using morphine as a tool. These studies will be discussed after considering the acute effects of morphine on brain norepinephrine content. Extensive attempts also have been made to implicate norepinephrine in morphine tolerance and physical dependence mechanisms, and these facets are discussed in the sections on the chronic effects of morphine and on morphine withdrawal.

B. Acute Effects of Morphine

The effects of a single dose of morphine on brain norepinephrine are quite complex and depend on the dose of morphine employed as well as the species. In general, as Martin pointed out in his review the level of brain norepinephrine increases when excitation or convulsions are induced.[40] In mice 20 mg/kg of morphine effected a slight decrease in brain norepinephrine (13%) while 10 or 15 mg/kg were ineffective.[41] In rats a single injection of morphine at the dose of 15–30 mg/kg usually had no effect,[36,42] while 60 mg/kg of morphine may either decrease[17,19,36] or increase[34,42] brain norepinephrine. Freedman et al. observed a biphasic change; the norepinephrine level in rat brain was decreased 4 hr after 60 mg/kg of morphine but increased after 24–48 hr.[43] The norepinephrine content in the brain of the rat, rabbit, and dog were relatively unaffected by 60 mg/kg of morphine.[19] In monkeys the brain norepinephrine was reduced by 3 mg/kg but increased by 30 mg/kg of morphine.[44]

In the cat which generally exhibits profound excitement to morphine, a reduction in norepinphrine levels in the brain occurs. Vogt reported there was a reduction of the midbrain and hypothalamic stores of "sympathin" in cats given 30–60 mg/kg of morphine.[45] Quinn et al. confirmed this observation by noting that 30 mg/kg of morphine caused a 60% depletion of brain stem norepinephrine within 6 hr, with a return to normal by 18 hr.[46] Gunne found that 30 mg/kg of morphine reduced cat brain norepinephrine from 0.46 μg/g to 0.18 μg/g within 2 hr.[36] Maynert also reported a 37% depletion in cat brain norepinephrine by 60 mg/kg of morphine.[19] In unanesthetized cats intravenous and intraventricular but not intracisternal injections of morphine evoked marked hyperglycemia and partial depletion of hypothalamic stores of norepinephrine.[47] Either nalorphine[48] or chlorpromazine[49] will block the reduction in brain norepinephrine induced by morphine. Reis et al. noted a fall in norepinephrine levels between 45 to 80% within 3 hr in the hypothalamus, midbrain, cerebellum, and cervical cords of cats given 15 mg/kg i.p., whereas other regions of the brain were unaffected and an increase of approximately 165–170% was observed in the mammillary bodies and globus pallidus.[50] The decrease in hypothalamic norepinephrine occurred in the medial but not the lateral subdivisions. The most susceptible regions to the influence of morphine appear to be those areas in which norepinephrine levels exhibit daily rhythmic fluctuations. One possible explanation for the morphine-induced decline in brain norepinephrine would be that the release rate of the compound in the norepinephrine-containing neurons exceed its rate of neuronal resynthesis.

There have been numerous attempts to implicate norepinephrine with pain mechanisms and the analgetic effects of morphine. It should be pointed out that the evidence for involvement of norepinephrine (or other biogenic amines) in pain mechanisms is highly circumstantial and, hence, any explanation of morphine effects in such terms represents building on the superstructure rather than the foundation.

The arguments attempting to involve norepinephrine in pain mechanism and with morphine analgesia rest primarily on the following evidence:

1. Norepinephrine per se and certain sympathomimetics exhibit analgetic properties.
2. They also enhance the analgetic effects of morphine.
3. In general various manipulations which elevate brain norepinephrine, produce analgesia, or enhance morphine analgesia.

4. Conversely, lowering brain norepinephrine may produce hyperalgesia or antagonism of morphine analgesia.

The experimental procedures for evaluating the "analgetic" effects of norepinephrine are based mostly on altering the threshold response to various forms of nociceptive stimulus. However, the methods usually lack sensitivity, and in order to magnify the response to norepinephrine, antagonism or enhancement of morphine analgesia has often been used as the end point. Moreover, since obtaining data with norepinephrine per se is restricted by its limited ability to gain access into the CNS after systemic administration, the assessment is generally made only after using various manipulations to elevate or lower the norepinephrine level in the brain. Thus brain norepinephrine may be increased either by stimulating its synthesis with one of its precursors such as DOPA or slowing its degradation with an monoamine oxidase (MAO) inhibitor. Conversely, brain norepinephrine can be lowered by blocking its synthesis with an agent such as α-methyltyrosine (α-MT) or by facilitating release and preventing its re-uptake in the neuron with an agent such as reserpine. These pharmacologic tools provide circumstantial evidence but never definitive answers since the compounds have pharmacologic actions of their own and may also affect other metabolic pathways. Nonetheless, the use of these substances to modify morphine effects often present insight as to the mechanisms that may be involved with analgesia and morphine actions. As a consequence of such applications information has been compiled to indicate that elevation of brain norepinephrine results usually in an elevation of the nociceptive threshold to noxious stimulus and enhancement of morphine analgesia, whereas the opposite responses are seen after lowering brain norepinephrine.

Radouco-Thomas et al. found that the administration of DOPA to elevate brain norepinephrine increased the nociceptive threshold and enhanced the analgetic effects of the morphine surrogates, meperidine and methadone.[51] Also, preventing norepinephrine degradation by catechol-O-methyl-transferase inhibition with pyrogallol raised the threshold to electrical stimulation in the rabbit tooth.[52] Likewise, inhibiting norepinephrine breakdown with MAO inhibitors, such as pargyline, nialamide, or tranylcypromine, effected a rise in pain threshold and enhancement of narcotic analgesia.[51–56] Both the MAO inhibitors and DOPA will often reverse the antianalgetic effects of agents which, presumably, could act by lowering brain norepinephrine.

Many agents have been used to lower brain norepinephrine and modify morphine analgesia, but most extensive studies by far have been carried out with reserpine. Unfortunately, reserpine also lowers brain dopamine and 5-HT as well and, consequently, it has been difficult to isolate and assign the actions of reserpine to changes in the level of one particular brain amine. Furthermore, the experimental reports on reserpine effects on morphine analgesia have been contradictory although the bulk of the evidence favors reserpine antagonism of morphine analgesia.

Schneider first reported in 1954 that reserpine alone, at doses up to 10 mg/kg subcutaneously (s.c.), which failed to alter the tail-flick reaction time in mice, markedly antagonized the prolongation in tail-reaction time elicited by 10 mg/kg morphine s.c.[57] Since then a number of laboratories[53,54,56,58-69] have confirmed these findings, but in contrast to these results there are some reports that morphine analgesia is unaffected[3,61] or potentiated by reserpine.[64,70-73]

The discrepancies among the various workers with respect to reserpine effects on morphine analgesia are difficult to explain. Different experimental conditions relevant to species, method of analgesia assessment, dose, and time of reserpine pretreatment are important factors. However, beyond these considerations there are divergent findings by two groups from the same laboratory using the same procedure, which cannot be explained. Leme *et al.*, using the hot-plate procedure to assess morphine analgesia, claimed that reserpine potentiated morphine effects,[72] whereas 6 years later the opposite results were reported by Verri *et al.*[67].

Tsou and Tu attempted to reconcile the differences in reserpine effects on morphine analgesia by comparing several assay methods for measuring analgesia.[61] They noted reserpine antagonism of morphine analgesia with the tail clip-biting, hot-wire tail-flick, and tail electric shock-motor response procedures but no effect on the morphine response when either the hot-plate paw withdrawal or tail electric shock-vocalization method was used. Ross and Ashford also reported antagonistic effects with reserpine by the tail-clip procedure, but they also found potentiation rather than no effect on morphine analgesia by the hot-plate procedure.[64] In general, the contradictions with respect to reserpine effects on morphine analgesia appear to reside chiefly in studies with the hot-plate procedure. Nott pointed out that the paw elevation response in the hot-plate method requires a high degree of motor coordination, and reserpine may act to prolong the time to the end point by a depressant action on motor activity rather than through an analgetic action or a true interaction with morphine.[66] The simple tail-flicking response avoids these complications. Admittedly, the test is based primarily on a nociceptive reflex, but it yields data that appear to correlate reasonably well with the analgetic actions of narcotic compounds in the clinic, and there is a supraspinal component involved when analgesia is produced by morphine.[74] By using the increase in ED_{50} of morphine required to prolong the tail-flick reaction time to thermal stimulus in mice, we have repeatedly found that 24-hr pretreatment with 2 mg/kg of reserpine results in antagonism of morphine analgesia.[68]

The recent studies by Dewey *et al.* indicate what the length of time of pretreatment with reserpine largely dictates the response that is observed with respect to morphine analgesia.[69] If reserpine, 4 mg/kg, is injected simultaneously with morphine, the dose of morphine to prolong the tail-flick response is reduced. If, however, reserpine is given 16 hr prior to morphine, instead of enhancement of morphine analgesia, antagonism occurs and the antagonism is evidenced by a sixfold increase in the ED_{50} of morphine.

The exact mechanism by which reserpine antagonizes morphine analgesia is unknown. A central rather than peripheral action appears to be involved, since Schaumann pointed out that those rauwolfia alkaloids that do not release biogenic amines from the brain do not affect morphine analgesia.[53] This postulation is further supported by the fact that tetrabenazine, which possesses selective reserpine-like central actions, also reduces morphine analgesia in the tail-clip test.[62]

Conceding that the antagonism of the analgesic effect of morphine by reserpine is likely to be a central action, there is still no general agreement with respect to the mode of the antagonism. Interpretations have been offered implicating the catecholamines, 5-HT, and both or neither of the two substances. We already have pointed out difficulties in dissociating *the* effect of reserpine from its amine-depleting activity, which may be crucial if indeed one amine is involved. Solution of this dilemma would be necessary if a particular amine were to be implicated in pain mechanisms.

As an approach toward resolving the reserpine enigma, the interaction of morphine with reserpine has been examined after attempting to block selectively the synthesis of norepinephrine more than that of 5-HT with various compounds. As outlined by Spector in his review,[75] norepinephrine synthesis can be blocked by any of three steps occurring in the following sequence: (1) hydroxylation of tyrosine to DOPA, (2) decarboxylation of DOPA to dopamine, and (3) hydroxylation of dopamine to norepinephrine. However, owing to the fact that the initial reaction is likely the rate-limiting step, marked reduction of norepinephrine synthesis *in vivo* can be achieved only with compounds that inhibit the enzyme involved in hydroxylating tyrosine, that is tyrosine hydroxylase. The substances that inhibit decarboxylation of DOPA or hydroxylation of dopamine *in vitro* generally have not reduced tissue norepinephrine to any appreciable extent. Of those that do, such as methyldopa and α-methyl-*m*-tyrosine (α-MmT), the lowering of norepinephrine level is now known to result primarily from the displacement of the naturally occurring amine by the amine metabolites of the α-methylated amino acids. Several analogs of tyrosine can act as competitive inhibitors of tyrosine hydroxylase; the most widely used have been α-methyltyrosine (α-MT) or its methylester (α-MMT).

Since information concerning the effects of α-MT on norepinephrine synthesis was not available until recently, very little has been published concerning its effects on pain and morphine analgesia. Verri *et al.* found, uring the hot-plate procedure in mice, that the prolongation in reaction time effected by 10 mg/kg of morphine s.c. was antagonized by two doses (100 mg/kg each i.p.) of α-MT administered 4 and 8 hr before the morphine.[67]

Of the three types of inhibitors of norepinephrine synthesis, the ones inhibiting decarboxylation of DOPA have been studied to the greatest degree simply because they were introduced earlier. Medakovic and Banic reported that α-MmT, 400 mg/kg, injected 90 min before morphine, had no effect on the prolongation in tail-reaction time to radiant heat produced by morphine nor did it alter reserpine antagonism of morphine analgesia when injected 21 hr before administering reserpine.[54] In the mouse, however, α-MmT antagonized morphine anlagesia measured by the hot-plate method and reversed reserpine antagonism of morphine analgesia. Tsou and Tu reported that methyldopa did not reduce morphine analgesia (tail flick) in mice at doses as high as 200 or 400 mg/kg but could eliminate the antimorphine effect of reserpine if given simultaneously or prior to but not after reserpine.[61] Methyldopa was reported also by Ross and Ashford to have no effect in mice on morphine analgesia by the tail-clip procedure or in preventing the inhibitory effect of reserpine on morphine.[64] Although Rudzik and Mennear were able to demonstrate reserpine antagonism of morphine analgesis by the writhing technique in mice, neither α-MmT nor methyldopa altered the ED_{50} value of morphine.[63] They concluded that the effect was due to some intrinsic property of reserpine other than its effect on brain amines. If the presence of catecholamine is a requirement for morphine action, one would expect catecholamine depletion by the decarboxylase inhibitors to enhance instead of abolish or not alter reserpine antagonism of morphine analgesia, and this was not the case. Admittedly if reserpine and methyldopa effect brain lowering of epinephrine by different mechanisms, then their respective actions on morphine would not necessarily have to be additive. Since sympathomimetic agents can enhance morphine analgesia,[26-31] there is always the question of whether the respective amine metabolites of methyldopa or α-MmT might act to oppose rather than enhance the antimorphine effects of reserpine. Such arguments are futile, and the situation will remain obscure until more decisive experiments can be performed to exclude or implicate norepinephrine in pain mechanisms.

With respect to lowering brain norepinephrine by inhibition of dopamine-β-hydroxylase, Watanabe et al. studied the effects of the inhibitior, diethylthiocarbamate on morphine analgesia.[77] They found that in contrast to reserpine, diethylthiocarbamate enhanced the analgetic action of morphine in rats even though it lowered the level of norepinephrine in the brain. Watanabe et al. suggested that since the rate of synthesis of norepinephrine is known to be decreased by diethylthiocarbamate and increased by reserpine, the analgetic effect of morphine may be inversely correlated with the turnover of norepinephrine.[77] These studies are of considerable interest and should be pursued.

C. Chronic Morphinization

Chronic administration of morphine causes no changes in brain norepinephrine levels in the dog.[1,17,21] Martin related this to the fact that no excitatory changes are manifested during the stabilization period.[2] In contrast, rats exhibit stimulatory effects with repeated morphine injections and, with the exception of one study by Neal,[78] the brain norepinephrine level was found to be increased.[1,19,37,43,79,80] Akera and Brody further reported that the increase in whole brain norepinephrine during chronic morphinization was not dose-dependent, nor was brain norepinephrine elevated after chronic treatment with methadone or levorphanol.[79]

Brain norepinephrine level was found to be higher in chronically morphinized rats than in controls following monoamine oxidase (MAO) inhibition.[1,17,19] Brain norepinephrine levels also were reported to be higher in the chronically morphinized rat than in controls following pretreatment with norepinephrine-releasing agents. After 5 mg/kg of reserpine 20 hr before sacrifice, the reduction in brain norepinephrine was less in morphine-tolerant animals.[1] Likewise, methyldopa effected a decrease in brain norepinephrine levels in tolerant animals to a lesser extent than in controls. Both Gunne[1] and Maynert[17,19] suggested that the increase could be the result of increased synthesis of brain norepinephrine, and Maynert further postulated that the increased synthesis was accompanied also by an increased rate of release of norepinephrine.[19] He added that the elevated concentration of free norepinephrine within the brain could play a role in antagonizing the depressant effect of morphine and also might account for the hyperexcitability and viciousness of addicted rats. Compatible with the notion that reduced storage or increased discharge of norepinephrine is a consequence of chronic morphinization is the finding that a significant reduction in the number of small granular vesicles in the terminal adrenergic nerves also occurs.[82] However, Neal using tritium-labeled norepinephrine found that the clearance of the compound after intracisternal administration was the same in control and morphinized rats.[78] Moreover, Akera and Brody reported that although brain norepinephrine was higher in chronically morphinized rats than in controls after the MAO inhibitor, tranylcypromine, the relative increase in both groups was not significant.[29] Also, chronic morphine treatment did not influence the level of catecholamines in the brain after blocking catecholamine resynthesis with the potent inhibitor of tyrosine hydroxylase, α-MMT.[81]

The urinary excretion of norepinephrine or its metabolites was increased in rats,[1,37,79,88] dogs,[1] and humans.[38] Some degree of tolerance development to this effect was noted in the animal experiments.[1,37,80]

D. Withdrawal of Morphine

Abrupt withdrawal caused a depletion in brain norepinephrine level in the dog, and the severity of the abstinence symptoms was proportional to the reduction in norepinephrine.[1] This was true for all parts of the brain, although the brain stem appeared to be more resistant to the depletion than the cerebellum and telecephalon. Induced withdrawal with nalorphine also lowers brain norepinephrine. Nalorphine, which has no marked effect on brain norepinephrine levels by itself, markedly decreases brain norepinephrine in the morphine-dependent dog and rabbit.[1,17–19,34,36] In contrast to the dog, rats did not show any change in brain norepinephrine following either abrupt withdrawal or nalorphine-precipitated withdrawal.[1,17–19,34,37,79,80] The reason for this species variation was not established, but Gunne attempted to relate this to the different withdrawal signs manifested by the two species.[1] He pointed out that dogs appeared excited, and this in turn would reduce brain norepinephrine, while rats showed a mixed state of drowsiness and irritability and, as a consequence, a change in brain norepinephrine might not be expected.

The urinary excretion of norepinephrine after abrupt withdrawal was found to be increased in rats,[1,37,79,80] and dogs[1,18] but unchanged or decreased in man.[38] In the case of rats or dogs, the first peak excretion occurred on the second to thrid day of abrupt withdrawal, followed by a second peak around the fifth or eighth day and a return to normal value by the second week.[1,37,80]

Attempts have been made to modify the abstinence syndrome with agents that might alter brain catecholamine levels, but with the exception of reserpine the results on other agents are contradictory. Huidobro and his associates reported that in mice, enhancement of withdrawal precipitated by nalorphine generally occurs with agents that lower brain norepinephrine.[83,84] Thus precipitated abstinence was rendered more severe by reserpine and certain reserpine-like substances. However, they also noted that methyldopa and α-MmT decreased the severity of precipitated abstinence and made the conclusion that the effects of reserpine were mediated other than by its ability to deplete catecholamine. On the other hand, enhancement of precipitated withdrawal was noted by Gunne in the morphine-dependent dog after pretreatment with 25 mg/kg of methyldopa or α-MmT 16 hr prior to nalorphine; pretreatment with 1 mg/kg reserpine 40 and 16 hr earlier also exacerbated the withdrawal signs.[85] However, agents that tend to elevate brain catecholamines (l-DOPA and the MAO inhibitor, nialamid) also made abstinence more severe, although to a lesser degree than those types known to reduce norepinephrine stores. Fraser and Isbell pointed out in their evaluation of the effects of reserpine and chlorpromazine on morphine abstinence that neither drug appeared to be of value for treatment and, as a matter of fact, reserpine produced a marked aggravation of the signs and symptoms.[86]

Huidobro and his collaborators also reported that attenuation of the withdrawal signs in the mouse result from treatments that tend to enrich the norepinephrine stores.[83,84] They found that precipitated abstinence with nalorphine became less severe after giving certain MAO inhibitors or catecholamine precursors. However, we have already pointed out Gunne's contrary results.[85] Our own unpublished data (Shen, Loh, and Way) also are not consistent with their findings. We have found that the *abrupt* withdrawal signs in the morphine-dependent mouse can be intensified by the injection of the MAO inhibitor, pargyline. Clarification of these apparent contradictory findings must await further studies.

Martin and Eades examined the possible relationship of adrenergic and cholinergic neurons in the CNS to physical dependence on morphine by studying the effects of various pharmacologic agents on the chronic spinal dog.[87] They found that the adrenergic agonists, amphetamine and methoxamine, produced spinal reflex facilitatory changes that were qualitatively similar to those seen in precipitated and withdrawal abstinence. These changes could be completely antagonized by phenoxybenzamine, an alpha receptor antagonist. However, in the morphine-dependent abstinent animal, phenoxybenzamine did not suppress spinal cord signs of abstinence. They concluded, therefore, that although increased utilization of norepinephrine in the CNS may be responsible for, or associated with, certain signs of abstinence, it was not a necessary condition for signs of abstinence to become manifest. On the other hand, their findings with cholinergic antagonists suggested to them that cholinergic neurons do play a role in the genesis of spinal cord signs of the morphine abstinence symptoms.

In morphine-dependent rats treated with α-MMT to inhibit catecholamine synthesis, precipitated withdrawal with nalorphine effected a lowering of brain norepinephrine levels but not those of dopamine. Histofluorescent studies of the brain revealed a general reduction in the norepinephrine stores at norepinephrinergic terminals throughout the CNS, while those at dopaminergic terminals appeared unaffected. Since nonspecific stress induced by immobilization also depleted norepinephrine but not dopamine in inhibitor-treated animals, it was concluded that the changes effected by nalorphine induced withdrawal represented a stress response.[81]

E. Conclusions

The evidence to date suggests that norepinephrine does not play a primary role in mediating morphine effects. The lowering of brain norepinephrine or increased urinary excretion that may occur after acute morphine administration appears to be the consequence of drug-induced effects. Acute administration of morphine effects sympathetic activation centrally and peripherally, leading eventually to a depletion of brain and adrenal stores of catecholamines. This may effect an accelerated synthesis of the compounds in these organs with repeated morphine injection.

To the reviewers it also appears that norepinephrine is not primarily concerned with initiating tolerance and physical dependence development to morphine. The inconsistent changes in brain norepinephrine and in urinary excretion that occur during chronic morphinization appear to be consequential to pharmacologic events induced by morphine. Likewise, those changes occurring after withdrawal of morphine seem to reflect stress resulting from the abstinence syndrome, since the degree of brain depletion and urinary increase of norepinephrine are proportional to the intensity of the abstinence.

IV. DOPAMINE

A. Introduction

The compound has long been known as a precursor of norepinephrine. However, its possible role in control of motor function was not suggested until 1959, when Carlsson found extraordinarily high concentrations of dopamine in the corpus striatum, which forms an important part of the extrapyramidal system.[88] Consistent with the postulate are the findings that the dopamine content of the caudate nucleus and putamen of Parkinsonian patients are reduced often as low as 10% of normal[89] and that the loss of dopamine is due to degeneration of dopaminergic fibers in the substantia nigra.[90] The development of procedures for estimation of norepinephrine and dopamine turnover separately and for differentiating the functionally different biogenic amine neurons in the CNS have led to a reinvestigation of the effect of morphine on catecholamine metabolism in the brain.[91]

B. Acute Effects of Morphine

Little work has been done on the effect of morphine on the brain dopamine content. Takagi and Nakama observed a 32% decrease in brain dopamine concentration in mice with a single injection of 20 mg/kg of morphine.[41] The maximal depleting effect was obtained 15–30 min after administration and a return to control value occurred by 60 min. The time course of dopamine depletion appeared to correspond roughly to that of morphine analgesia as measured by the tail-clip procedure. Nalorphine at a dose that had no effect on brain dopamine level (1 mg/kg) prevented the depletion effect of morphine. The authors suggested that the reduction in brain norepinephrine by morphine is mediated indirectly through a reduction of brain dopamine and that the dopamine released by morphine may act as an inhibitory transmitter of spinal motor neurons with consequantial suppression of the pain reflex. However, Segal and Deneau reported an increase in dopamine in monkey's caudate nucleus after 3 or 30 mg/kg of morphine.[44] Likewise, Sharman reported no change in dopamine in striatum of mice given morphine 45 mg/kg in two doses at a 2-hr interval.[92] Moreover, he found that 0.1 mg/kg of M-99, a potent morphine surrogate, increased the dopamine level 2–4 hr after administering the drug and that this increase was partially prevented by 1 mg/kg of M-285, a strong antimorphine compound.

Recently, Gunne and Jonsson reported that morphine elicited preferential effects on the synthesis of dopamine over norepinephrine in the rat.[81] After inhibiting catecholamine synthesis with α-MMT, the concentrations of both norepinephrine and dopamine were reduced in the brain to approximately one-half the control values over a 4-hr period. When a single injection of morphine (20 mg/kg, i.p.) was given in addition, brain dopamine was further reduced substantially while norepinephrine was unaffected. The marked depletion of dopamine occurred at the dopaminergic terminals of the nucleus caudatus, putamen, nucleus accumbens, and tuberculum olfactorium. These changes were ascribed to increased neuronal activity.

C. Chronic Effects of Morphine

Gunne noted no change in dopamine content in the telencephalon of dogs treated daily for 70–90 days with increasing doses of morphine up to 120 mg/kg.[1] Dopamine also was found unchanged in various brain regions of the morphine-dependent monkey.[44] The urinary excretion of dopamine was increased during the addiction period of two male volunteers chronically treated with morphine for 20–50 days.[38] Sloan and Eisenman also observed an increase in urinary excretion of dopamine in rats during an addiction cycle.[37,80] The depletion of dopamine at the dopaminergic terminal elicited by morphine disappeared with continued administration of morphine.[81] Clouet and associates found a marked enhancement of brain dopamine turnover in rats given repeated injections of morphine.[93]

Takagi and Kuriki reported that tetrabenazine markedly suppressed the development of tolerance to morphine in mice. Daily injections of a dose of morphine that effected complete analgesia by the tail-pinch method (10 mg/kg s.c.) resulted in tolerance development as evidenced by the fact that only 6% of the animals exhibited analgesia to morphine after 9 days.[94] However, when tetrabenazine (40 mg/kg s.c.) was injected daily 2 hr before the morphine, only 64% exhibited analgesia. The suppressive effect of tetrabenazine on morphine tolerance development was reversed by daily administration of DOPA (100 mg/kg); 17% exhibited analgesia when treated with all three drugs. 5-HTP 100 mg/kg had no effect on tetrabenazine-tolerance suppression.

D. Effects of Morphine Withdrawal

A decrease in brain dopamine was found 72 hr after morphine withdrawal in dogs exhibiting moderate to severe abstinence,[1] whereas Segal and Deneau reported an increase in dopamine in caudate nucleus of monkey either 24–48 hr after abrupt withdrawal or after nalorphine-precipitated withdrawal.[44] The urinary excretion of dopamine in rats was increased after withdrawal with peaks occurring on the third and eighth day.[37,80] However, Weil-Malherbe et al. found the urinary excretion of dopamine was decreased in the human volunteer during the withdrawal phase.[38] In the same study mentioned earlier, wherein precipitation of withdrawal with nalorphine effected a lowering of brain norepinephrine in morphine-dependent animals pretreated with α-MmT, brain dopamine was not affected.[81]

E. Conclusions

The early evidence suggests that dopamine may be more intimately involved with morphine effects than norepinephrine. While the available information is still insufficient to allow a precise assignment of the role of dopamine, future studies that attempt to relate regional dopamine turnover in the CNS with morphine effects may be quite fruitful.

V. 5-HYDROXYTRYPTAMINE (5-HT, SEROTONIN)

A. Introduction

Interest in 5-HT has been greatly stimulated by recent studies implicating the compound possibly in sleep mechanisms,[95] temperature regulation,[96] social and sexual behavior,[97,98] and pain mechanisms.[99,100] The work has been greatly facilitated by the discovery that p-chlorophenyl-alanine (PCPA) inhibits 5-HT synthesis with little or no effect on catecholamine synthesis.[101] As one of the biogenic amines in the CNS with a distribution resembling somewhat that of norepinephrine, 5-HT also has been examined for its possible relationship with morphine with respect to analgetic, tolerance, and physical dependence mechanisms.

B. Acute Effects of Morphine

Brodie et al. noted in 1956 that a single dose of 20 mg/kg dose of morphine produced no change in the 5-HT content in rabbit brain.[102] This finding was later confirmed in the dog by Maynert et al., who found that the concentration of 5-HT in the brain stem was not changed by 5, 60, 125, or 200 mg/kg of morphine[103]; the 5-HT content in the rat and rabbit brain were also unaffected. Sloan et al. compared the effects of morphine (15–60 mg/kg) and thebaine (20 mg/kg) in rats and noted that both compounds had no effect on the brain content of 5-HT.[80] However, there is one report that morphine increases[104] and one[105] that it decreased brain serotonin. In studies that have not been published, we found that neither the brain level of 5-HT nor its rate of 5-HT synthesis was altered by a 100 mg/kg dose of morphine.[106]

Although convincing evidence existed indicating that total brain 5-HT was not affected by the acute administration of morphine, there have been, nonetheless, numerous attempts to relate 5-HT changes to morphine effect. The evidence that has been compiled is less voluminous than that for norepinephrine, but what is available is equally equivocal. Several groups of investigators have hypothesized that 5-HT is involved in pain mechanisms and in mediating morphine analgesia. However, the agreement is at the theoretical rather than the experimental level since the postulates have been based on contrary evidence. Both 5-HT enhancement of morphine analgesia and 5-HT antagonism of morphine analgesia have been submitted to support the same contention.

The experimental approaches toward implicating 5-HT in pain mechanisms are similar in design to those used to study norepinephrine and utilize the same criteria—antagonism or enhancement of morphine effects on nociceptive stimulus. As in the case with norepinephrine, the access of 5-HT into the CNS is restricted, and in order to study the effect of 5-HT on morphine analgesia various pharmacologic agents have been employed to increase or decrease brain 5-HT. Brain 5-HT is elevated usually by stimulating synthesis with a precursor, generally 5-hydroxytryptophan (5-HTP), or by inhibiting its metabolism to 5-hydroxyindoleacetic acid (5-HIAA) with a MAO inhibitor. Brain 5-HT is lowered usually either by inhibition of synthesis with PCPA or preventing neuronal re-uptake with reserpine. Consequently, the criticisms that apply for such pharmacologic manipulations in assessing norepinephrine apply equally for 5-HT.

The experiments designed to influence pain or morphine analgesia by attempting to elevate brain 5-HT have yielded conflicting results. Sigg *et al.* reported in 1957 that 5-HTP, at a dose (100 mg/kg) that did not alter the tail-flick reaction time in mice by itself, enhanced and prolonged the analgetic effect of morphine; 5-HT at 5–10 mg/kg was likewise effective.[59] However, since several other amines including the sympathomimetics, amphetamine, and mescaline also enhanced morphine effects, they concluded that the 5-HT effects were nonspecific. Contreras and Tamayo reported that 75 mg/kg of 5-HTP elevated the threshold to electrical stimulus in the rat to a degree comparable to 10 mg/kg of morphine and reversed the antagonism of morphine analgesia produced by reserpine, guanethidine, or tolazoline.[107] Radouco-Thomas *et al.* also reported that 5-HTP increased the nociceptive threshold in the guinea pig, and this was accompanied by a parallel increase in cerebral 5-HT but not in brain catecholamines.[51] Moreover, 5-HTP counteracted reserpine antagonism of morphine analgesia. Nicak reported that the analgetic action of morphine (response to ultrasonic sound) in mice and rats was potentiated by 1, 10, and 20 mg/kg of 5-HT (i.v).[108] Reserpine, 2.5 mg/kg s.c., given 1, 4, and 24 hr prior to 5-HT, abolished the analgetic potentiating effect of 5-HT. While the above experiments sometimes may provide valuable clues, they do not provide decisive answers. Since 5-HT does not pass the blood–brain barrier readily, any results obtained with 5-HT after peripheral administration are always open to question. Increasing levels of brain 5-HT by exogenous supplementation with 5-HTP does not necessarily mean that it will be handled in the same manner as endogenous 5-HTP and, hence, a different pharmacologic response may be observed. Brodie *et al.* have pointed out that the pharmacologic effects of 5-HTP at low and high doses are often opposite and that brain norepinephrine can be depleted by 5-HTP.[109]

In contrast to the above-mentioned findings, Gaddum and Vogt described an antagonistic action of morphine on the effects produced by 5-HT.[110] Administration of the latter substance intraventricularly in the cat produced sedation, but following morphine an excitatory state resulted. Peripherally the effects of morphine and 5-HT appear to be antagonistic, as evidenced by the fact that the stimulatory effects of 5-HT on the inferior mesenteric ganglia are blocked by morphine with a high degree of selectivity.[111] Moreover, the inhibitory action of morphine on the peristaltic reflex is known to be inhibited by 5-HT.[112–115] This may have relevance to central effects since it has been shown that all drugs with a morphine-like action depress the peristaltic reflex proper and the graded effect of the longitudinal muscle—their efficacy being closely correlated with their analgesic potency.[114,115] We have found (Shen, Loh, and Way, unpublished) that 5-HT intracerebrally antagonizes the effect of morphine as evidenced by the increased ED_{50} of morphine on the tail-flick response.

Elevation of brain 5-HT by means of MAO inhibition appears to enhance the effects of morphine. The interaction between MAO inhibitors and narcotic analgesics in mice was studied by Rogers and Thornton.[116] The acute toxicity, respectively, of meperidine, morphine, and phenazocine as well as the antagonist pentazocine, was potentiated by either iproniazid or tranylcypromine. Studies with pentazocine revealed that alteration in the rate of biotransformation was not responsible since the decay of the compound in blood was the same with and without MAO inhibition. On the other hand, a correlation existed between the elevation in brain 5-HT and increased toxicity of meperidine resulting from MAO inhibition, whereas there was no significant correlation between increased meperidine toxicity and brain norepinephrine or dopamine. Rogers and Thornton also cited three papers[117-119] reporting that MAO inhibitors enhanced the toxicity of meperidine. Since morphine and its surrogates have been shown ot release 5-HT from peripheral tissues,[120,121] they suggested that narcotic analgetics also may cause a release of 5-HT in the CNS.

The attempts to relate 5-HT with morphine effects by lowering brain 5-HT have provided data with varying degrees of validity. Medakovic and Banic compared the effects of reserpine and α-MmT on morphine analgesia in rats using the radiant heat tail-flick procedure.[54] Reserpine, 1 mg/kg s.c., 3 hr prior to mrophine nullified the effects of 4 mg/kg of morphine s.c., whereas α-MmT, 400 mg/kg, 90 min before morphine failed to do so. Further, reserpine was equally effective in antagonizing morphine in α-MmT treated animals. Since norepinephrine stores should have been depleted by α-MmT, they argued that by exclusion 5-HT rather than norepinephrine was involved with the actions of morphine. They also noted that reserpine failed to antagonize morphine in animals pretreated with the MAO inhibitor iproniazid; in fact, under these conditions reserpine appeared to enhance morphine effects. In order to reconcile the apparent contradictory data obtained with reserpine, Medakovic and Banic hypothesized that 5-HT exerted dualistic effects and that the response elicited was dependent on the concentration of 5-HT. We have already pointed out the shortcomings of drawing conclusions from such an approach and should like to point out also that their experiments claiming reserpine enhancement of morphine in iproniazid-treated animals did not provide data for morphine in iproniazid controls. Moreover, they more or less ignored their additional data on α-MmT in mice that were not consistent with their findings in the rat.

The experiments designed to assess pain mechanisms and morphine effects by lowering brain 5-HT with PCPA or by making lesions at serotoninergic sites in the brain appear to have greater validity. Tenen reported that PCPA antagonized the effects of morphine on the flinch-jump response to electrical stimulus in the rat and suggested that the antagonism of morphine analgesia might be attributable to a deficiency of 5-HT induced by PCPA.[122] In support of his postulate, he cited previous findings wherein PCPA was found to increase the animal's sensitivity to pain, suggesting that painful stimuli might be attenuated through 5-HT mechanisms.[100] Consistent with these results, Harvey and associates noted that decrease in brain 5-HT induced by lesions in the medial forebrain bundle or by PCPA in the rat resulted in a decreased threshold to electrical foot shock.[99] The increased pain sensitivity in the lesioned animal was associated with a 76% decrease in 5-HT in the telencephalon. The injection of 75 mg/kg 5-HTP i.p. returned both telencephalic 5-HT and pain threshold to normal values, whereas dopa had no effect. He also hypothesized that 5-HT may function to inhibit the effects of a painful stimulus.

In support of the conclusions implicating 5-HT in pain mechanisms, our own unpublished findings in mice (Shen, Loh, and Way, unpublished) indicate that 5-HT intracerebrally exerts a mild elevation of the nociceptive threshold to radiant heat. While this would appear to indicate a synergistic rather than antagonistic action between morphine and 5-HT, we also have found that when 5-HT is given intracerebrally in combination with morphine it *increases* the dose of morphine required to elicit analgesia. Hence, it may be that 5-HT acts as a partial agonist-antagonist of morphine.

C. Chronic Effects of Morphine

Cochin and Axelrod mentioned, although they presented no data, that the chronic daily administration of morphine sulfate in progressively increasing amounts up to 70 mg/kg for 45 days in rats effected no change in brain 5-HT.[123] This was confirmed in mice by Bartlet, who treated the mice chronically with morphine (highest dose, 100 mg/kg) and found a normal concentration of brain serotonin.[124] Maynert *et al.* made a thorough investigation in chronically morphinized dogs, rats, and rabbits and found that the treatment did not alter brain 5-HT levels in any of the three species.[103] Gunne also found no change in brain 5-HT in dogs and rats chronically treated with morphine; the urinary excretion of 5-HT was found to be decreased in chronically morphinized dogs, but the rats excreted more 5-HIAA during the late stage of morphinization.[1,18] Way, Loh, and Shen also found that total brain levels of 5-HT in mice rendered tolerant to and dependent on morphine were not significantly different from those of nontolerant animals.[125] They also pointed out that these measurements reflected the steady state level resulting from equal rates of synthesis and efflux and that it might be more meaningful to measure the rate of brain 5-HT turnover as described by Tozer *et al.*[126]

The procedure involves blocking the conversion of 5-HT to 5-HIAA with the MAO inhibitor, pargyline, and on the assumption that brain 5-HT is converted solely to 5-HIAA, the rate of 5-HT synthesis may be calculated from the initial increase in 5-HT. On comparing the turnover of brain 5-HT in tolerant and nontolerant mice, the mean increase in brain 5-HT in tolerant animals was found to be more than double that in nontolerant controls.[125] A single injection of morphine did not elevate the 5-HT turnover. These findings were confirmed by Haubrich and Blake[127] in the rat using a different procedure to measure 5-HT turnover.[128] With probenecid to block the transport of 5-HIAA from the CNS to blood, they noted that the rate of accumulation of 5-HIAA in animals rendered tolerant to morphine by pellet implantation was considerably greater than that in nontolerant controls. Loh, Shen, and Way reported in a follow-up study that the increased rate of 5-HT synthesis resulting from chronic morphine administration could be blocked by cycloheximide, an inhibitor of protein synthesis, and this was accompanied by an inhibition of tolerance and physical dependence development to morphine.[129] When tolerance and dependence development to daily injections of morphine were prevented by concomitant administration of the specific antagonist, naloxone, the increase in brain serotonin turnover was also prevented (Shen, Loh, and Way, unpublished). Furthermore, they found that when brain 5-HT synthesis was inhibited by PCPA, tolerance and physical dependence development to morphine was prevented in part.[125,130] Inasmuch as PCPA does not completely reduce brain 5-HT to zero, it is not surprising that complete blockade was not achieved.

Reserpine also has been studied with respect to its ability to modify morphine tolerance in mice.[68] The antagonism of morphine analgesia by reserpine appears to be greatly enhanced after tolerance to morphine has developed. Thus in nontolerant mice 2 mg/kg of reserpine prior to morphine increased the ED_{50} morphine on the tail-flick response from 10 to 15 mg/kg. However, in mice rendered tolerant by morphine pellet implantation (ED_{50} 50 mg/kg) the ED_{50} increased even higher to 1540 mg/kg. Thus reserpine elevated the ED_{50} of morphine 1.5 times in nontolerant mice and 30 times in tolerant mice. The time course of the antagonism also was more prolonged in implanted mice. The antagonism by reserpine was abolished in both tolerant and control mice by pretreatment with pargyline. There was no significant difference in brain 5-HT levels in both groups after reserpine, but in the tolerant group recovery from reserpine sedation appeared to correlate with a more rapid return to normal 5-HT levels. While the meaning of this phenomenon is not currently clear, hopefully this magnification of response between tolerant and nontolerant animals by reserpine may provide a useful tool for discerning some of the basic mechanisms involved.

D. Effects of Morphine Withdrawal

Maynert *et al.* found no change in the concentration of 5-HT in the brains of dogs, rats, and rabbits after nalorphine-precipitated withdrawal.[103] This was confirmed by Sloan in rats.[37,80] Gunne also reported no change in brain 5-HT in dogs and rats after abrupt withdrawal or precipitated withdrawal.[1] Way, Loh, and Shen also found that brain 5-HT levels 2 weeks after withdrawal in tolerant mice were not significantly different from nontolerant animals and, further, the elevated turnover rate that existed during the tolerant state had reverted to normal.[125]

The output of 5-HIAA in the urine in both the dog and rat fluctuated considerably and appeared to be secondary to changes in diuresis.[37] The excretion of 5-HIAA in morphine-tolerant dogs was increased during the first day of aburpt withdrawal and in precipitated withdrawal with nalorphine. The excretion of 5-HIAA in rats was subnormal for the first 2 days following abrupt withdrawal and then returned to normal.[37]

Attempts have been made to modify the morphine abstinence syndrome with agents that might alter brain 5-HT, but no clear picture has emerged.

Collier, in the development of his receptor theory of tolerance and dependence, pointed out that the abstinence signs in the dog and rat are very similar to those produced by the administration of 5-HTP.[131] Among the numerous pharmacologic agents studied by Huidobro and his associates, 5-HT alone exerted no effect on nalorphine-precipitated withdrawal in the mouse, but DL-tryptophan and 5-HT following iproniazid attenuated the withdrawal signs.[83,84] Other agents that reduced abstinence intensity included tyrosine, phenylalanine, and MAO inhibitors, whereas reserpine-like substances, LSD, and DOPA enhanced the syndrome. PCPA has been reported to modify naloxone-precipitated withdrawal.[125]

E. Conclusions

The evidence suggesting that 5-HT may have a role in pain mechanisms furnishes an incentive to examine in greater depth the possible relationship of morphine with this putative neurohormone. While it has been shown that morphine does not significantly affect total brain 5-HT, this is insufficient grounds to exclude a relationship. Further work should be directed toward seeking correlations of the two substances in discrete areas in the CNS and with dynamic mechanisms related to 5-HT synthesis, release, or uptake. Justification emanates from increasing evidence that the latter parameters may have greater relevance to drug effects than the steady state levels. Three recent studies may be cited as examples. The central effects of amphetamine has been shown to depend on an intact catecholamine synthesis rather than on high catecholamine concentrations.[132] An intact dopamine synthesis appears to be necessary for amphetamine potentiation of leptazol convulsions.[133] Finally, the intensity and duration of reserpine sedation have been correlated with the initial rate of 5-HT synthesis.[109] A study of the reserpine:morphine interaction along such lines, particularly with the morphine-tolerant animal, might be highly rewarding.

A possible role for 5-HT in both morphine tolerance and physical dependence is suggested by the elevated 5-HT turnover in the brain that ensues with repeated morphine administration. That a common underlying process may operate in both tolerance and dependence development is indicated by a study linking in a quantitative fashion the two syndromes with respect to onset, intensity, and duration.[134] Furthermore, when different types of pharmacologic agents are used to block the increase in brain 5-HT synthesis resulting from chronic morphinization, there is an accompanying loss of tolerance and dependence development to morphine. The inhibition of increased 5-HT synthesis can be achieved with an inhibitor of protein synthesis (cycloheximide),[129] a specific morphine antagonist (naloxone), and an inhibitor of the enzyme involved in the rate-limiting step of 5-HT synthesis (PCPA). Of the three compounds, the results with PCPA have been the least convincing in that only partial blockade of tolerance and physical dependence were achieved—and even these findings have been questioned. It may be that since PCPA does not effect complete blockade of 5-HT synthesis, the residual amount is still biologically active. In any event, there is need to confirm and extend these findings in other animal species (other than the mouse) as well as to design experiments that might reveal a causal relationship between the elevated 5-HT synthesis and the pharmacologic events.

VI. REFERENCES

1. L.-M. Gunne, *Acta Physiol. Scand. 58* (Suppl. 204), 1–91 (1963).
2. W. R. Martin, *Pharmacol. Rev. 19*, 495–496 (1967).
3. T. Johannesson, *Acta Pharmacol. Toxicol. 25*, (Suppl. 3), 57–60 (1967).
4. V. P. Dole, Biochemistry of Addiction, *Annu. Rev. Biochem.* (in press).
5. T. R. Elliott, *J. Physiol. (London) 44*, 374–409 (1912).
6. G. N. Stewart and J. M. Rogoff, *J. Exp. Med. 24*, 709–738 (1916).
7. G. N. Stewart and J. M. Rogoff, *J. Pharmacol. Exp. Ther. 19*, 59–85 (1922).
8. R. Deanesley, *Amer. J. Anat. 47*, 475–509 (1931).
9. T. Hayama, *Jap. J. Med. Sci. 4*, 41 (1932).
10. H. Sato and F. Ohmi, *Tohoku J. Exp. Med. 21*, 411–432 (1933).
11. M. Wada, H. Tanaka, T. Hirano, and Y. Taneiti, *Tohoku J. Exp. Med. 34*, 52–71 (1938).
12. N. Emmelin and R. Stromblad, *Acta. Physiol. Scand. 24*, 260–266 (1951).
13. A. S. Outschoorn, *Brit. J. Pharmacol. 1*, 605–615 (1952).
14. Y. Satake, *Tohoku J. Exp. Med. 60* (Suppl. 2), 1–158 (1954).
15. J. A. Richardson, E. F. Woods, and A. K. Richardson, *J. Pharmacol. Exp. Ther. 122*, 64A (1958).
16. A. F. DeSchaepdryver, "On the Secretion, Distribution, and Excretion of Adrenaline and Noradrenaline," St. Catherine Press, Bruges (1959).
17. E. W. Maynert and G. I. Klingman, *J. Pharmacol. Exp. Ther. 135*, 285–295 (1962).
18. L-M. Gunne, *Nature (London) 195*, 815–816 (1962).
19. E. W. Maynert, in "The Addictive State" (A. Wikler, ed.), Chapter VI, pp. 89–95, Williams & Wilkins, Baltimore (1968).
20. T. B. B. Crawford and W. Law, *Brit. J. Pharmacol. 13*, 35–43 (1958).
21. H. Sibuta, K. Endo, and G. Nagakura, *Tohoku J. Exp. Med. 50*, 1–6 (1949).
22. E. W. Maynert and R. Lavi, *J. Pharmacol. Exp. Ther. 143*, 40–95 (1964).
23. R. C. Bodo, F. W. Cotui, and A. E. Benaglia, *J. Pharmacol. Exp. Ther. 61*, 48–57 (1937).
24. D. T. Watts, *J. Pharmacol. Exp. Ther. 102*, 269–271 (1951).
25. J. L. Morrison, *Fed. Proc. 6*, 359–360 (1947).
26. A. C. Ivy, F. R. Goetzl, and D. Y. Burrill, *War Med. 6*, 67–71 (1944).
27. E. G. Gross, H. Holland, H. R. Carter, and E. M. Christensen, *Anesthesiology. 9*, 459–471 (1948).
28. F. J. Friend and S. C. Harris, *J. Pharmacol. Exp. Ther. 93*, 161–167 (1948).
29. A. C. Ivy, E. R. Goetzl, S. C. Harris, and D. Y. Burrill, *Quart. Bull. Northwestern Univ. Med. School 18*, 298–306 (1944).
30. H. J. Kiessig and G. Orzechowski, *Arch. F. Exper. Pathol. Pharmakol. 197*, 391–404 (1941).
31. M. Nickerson, *Fed. Proc. 6*, 360–361 (1947).
32. N. A. David and H. J. Semler, *Fed. Proc. 11*, 335–336 (1952).
33. J. W. Miller, R. George, H. W. Elliott, C. Y. Sung, and E. Leong Way, *J. Pharmacol. Exp. Ther. 114*, 43–50 (1955).
34. L-M. Gunne, *Nature (London) 184*, 1950–1951 (1959).
35. T. Abe, *Jap. J. Med. Sci. 4*, 100–104 (1929).
36. L-M. Gunne, *Acta Physiol. Scand. 58*, (Suppl. 204), 1–91 (1963).
37. J. W. Sloan and A. J. Eisenman, in "The Addictive States" (A. Wikler, ed.), Chapter VII, pp. 96–105, Williams & Wilkins, Baltimore (1968).

38. H. Weil-Malherbe, E. R. B. Smith, A. J. Eisenman, and H. F. Fraser, *Biochem. Pharmacol. 14*, 1621–1633 (1965).
39. Y. Tachigawa, *J. Orient. Med. 17*, 521–528 (1932).
40. W. R. Martin, *Pharmacol. Rev. 19*, 495–496 (1967).
41. H. Takagi and M. Nakama, *Jap. J. Pharmacol. 16*, 483–484 (1966).
42. J. W. Sloan, J. W. Brooks, A. J. Eisenman, and W. R. Martin, *Psychopharmacologia 3*, 291–301 (1962).
43. D. X. Freedman, D. H. Fram, and N. J. Giarman, *Fed. Proc. 20*, 321 (1961).
44. M. Segal and G. A. Deneau, *Fed. Proc. 21*, 327 (1962).
45. M. Vogt, *J. Physiol. (London) 123*, 451–481 (1954).
46. G. P. Quinn, B. B. Brodie, and P.A. Shore, *J. Pharmacol. Exp. Ther. 122*, 63A (1958).
47. K. E. Moore, L. E. McCarthy, and H. L. Borison, *J. Pharmacol. Exp. Ther. 148*, 169–175 (1965).
48. M. Vogt, *Brit. Med. Bull. 13*, 166–171 (1957).
49. G. P. Quinn and B. B. Brodie, *Med. Exp. 4*, 349–355 (1961).
50. D. J. Reis, M. Rifkin, and A. Corvelli, *Eur. J. Pharmacol. 9*, 149–152 (1969).
51. S. Radouco-Thomas, P. Singh, F. Garcin, and C. Radouco-Thomas, *Arch. Biol. Med. Exp. 4*, 42–62 (1967).
52. J. L. Gardella, I. Izquierdo, and J. A. Izquierdo, "Analgesic Properties of Pyrogallol and Catecholamines. III." International Pharmacological Congress Abstract, p. 83 (1966).
53. W. Schaumann, *Arch. Exp. Pathol. Pharmakol. 235*, 1–9 (1958).
54. M. Medakovic and B. Banic, *J. Pharm. Pharmacol. 16*, 198–206 (1964).
55. R. J. Defalque, *Anesth. Analg. Current Res. 44*, 190 (1965).
56. C. Paeile and C. Munoz, Reserpine antagonism of morphine analgesia, effects of nialamide, dl-dopa and pyrogallol on this antagonism studied by the algesiometric test of the dental pulp in rabbits. "Abstracts of the third International Pharmacological Congress," p. 65–66 (1966).
57. J. A. Schneider, *Proc. Soc. Exp. Biol. Med. 87*, 614–615 (1954).
58. S. Radouco-Thomas, C. Radouco-Thomas, and E. L. Breton, *Arch. Exp. Pathol. Pharmakol. 232*, 279–281 (1957).
59. E. B. Sigg, G. Caprio, and J. A. Schneider, *Proc. Soc. Exp. Biol. Med. 97*, 97–100 (1958).
60. L. B. Witkin, M. Maggio, and W. F. Barrett, *Proc. Soc. Exp. Biol. Med. 101*. 377–379 (1959).
61. K. Tsou and Z-H. Tu, *Acta. Physiol. Sinica 26*, 360–366 (1963).
62. H. Takagi, T. Takashima, and K. Kimura, *Arch. Int. Pharmacodyn. 149*, 484–492 (1964).
63. A. D. Rudzik and J. H. Mennear, *J. Pharm. Pharmacol. 17*, 326–327 (1965).
64. T. W. Ross and A. Ashford, *J. Pharm. Pharmacol. 19*, 709–713 (1967).
65. H. Takagi and M. Nakama, *Jap. J. Pharmacol. 18*, 54–58 (1968).
66. M. W. Nott, *Eur. J. Pharmacol. 5*, 93–99 (1968).
67. R. A. Verri, F. G. Graeff, and A. P. Corrado, *Int. J. Neuropharmacol. 7*, 293–292 (1968).
68. F. H. Shen, H. Loh, and E. L. Way, *Fed. Proc. 28*, 793 (1969).
69. W. L. Dewey, L. S. Harris, J. F. Howes, J. W. Snyder, and O. T. Kirk, "The Mouse Tail-Flick Test," reported to the Committee on Problems of Drug Dependence, February 25 (1969).
70. J. Tripod and E. Gross, *Helv. Physiol. Pharmacol. Acta 15*, 105–115 (1957).
71. L. Tardos and Z. Jobbágyi, *Acta Physiol. Hung. 13*, 171–178 (1958).
72. J. G. Leme and M. R. E. Silva, *J. Pharm. Pharmacol. 13*, 734–742 (1961).
73. P. C. Dandiya and H. K. Menon, *Arch. Int. Pharmacodyn. 141*, 223–232 (1963).
74. S. Irwin, R. W. Houde, D. R. Bennett, L. C. Hendershott, and M. H. Seevers, *J. Pharmacol. Exp. Ther. 101*, 132–143 (1951).
75. S. Spector, *Pharmacol. Rev. 18*, Part I, 599–609 (1966).

76. S. Spector, A. Sjoerdsma, and S. Udenfriend, *J. Pharmacol. Exp. Ther. 147*, 96–95 (1965).
77. K. Watanabe, Y. Matsui, and H. Iwata, *Experentia 25*, 950–951 (1969).
78. M. J. Neal, *J. Pharm. Pharmacol. 20*, 950–953 (1968).
79. T. Akera and T. Brody, *Biochem. Pharmacol. 17*, 675–688 (1968).
80. J. W. Sloan, J. W. Brooks, A. J. Eisenman, and W. R. Martin, *Psychopharmacologia 4*, 261–270 (1963).
81. L.-M. Gunne, J. Jonsson, and K. Fuxe, *Eur. J. Pharmacol. 5*, 338–342 (1969).
82. J. D. P. Graham, J. D. Lever, and T. L. B. Spriggs, *Brit. J. Pharmacol. 37*, 19–23 (1969).
83. C. Maggiolo and F. Huidobro, *Arch. Int. Pharmacol. 138*, 157–168 (1962).
84. F. Huidobro, E. Contreras, and R. Croxatto, *Arch. Int. Pharmacol. 146*, 444–454 (1963).
85. L.-M. Gunne, *Arch. Int. Pharmacol. 157*, 293–298 (1965).
86. H. F. Fraser and H. Isbell, *Arch. Neurol. Psychiat. 76*, 257–267 (1956).
87. W. R. Martin and C. G. Eades, *Psychopharmacologia 11*, 195–223 (1967).
88. A. Carlsson, *Pharmacol. Rev. 11*, 490–493 (1959).
89. H. Ehringer and O. Hornykiewicz, *Klin. Wochenschr. 38*, 1236–1239 (1960).
90. N.-E. Anden, A. Dahlstrom, K. Fuxe, and K. Larsson, *Amer. J. Anat. 116*, 329–333 (1965).
91. N.-E. Anden, A. Carlsson, A. Dahlstrom, K. Fuxe, W.-A. Hillarp, and K. Larsson, *Life Sci. 3*, 523–530 (1964).
92. D. F. Sharman, *Brit J. Pharmacol. 28*, 153–163 (1966).
93. D. H. Clouet and M. Ratner, *Science 168*, 854–856 (1970).
94. H. Takagi and H. Kuriki, *Int. J. Neuropharmacol. 8*, 195–196 (1969).
95. M. Jouvet, in "Advances in Pharmacology," Vol. 6B, pp. 265–279 (S. Garattini and P. S. Shore, eds.), Academic Press, New York (1968).
96. W. Feldberg and R. D. Meyers, *J. Physiol. (London), 177*, 239–245 (1965).
97. B. J. Meyerson, *Acta. Physiol. Scand. 63*, (Suppl. 241), (1964).
98. A. Tagliamonte, P. Tagliamonte, G. L. Gessa, and B. B. Brodie, *Science 166*, 1433–1435 (1969).
99. J. A. Harvey, C. E. Lints, and E. Garbarits, *Pharmacologist. 10*, 211 (1968).
100. S. S. Tenen, *Psychopharmacologia (Berlin) 10*, 204–219 (1967).
101. B. K. Koe and A. Weissman, *J. Pharmacol. Exp. Ther. 154*, 499–516 (1966).
102. B. B. Brodie, P. A. Shore, and A. Pletscher, *Science 123*, 992–993 (1956).
103. E. W. Maynert, G. I. Klingman, and H. K. Kaji, *J. Pharmacol. Exp. Ther. 135*, 296–299 (1962).
104. D. Bonnycastle, M. F. Bonnycastle, and E. G. Anderson, *J. Pharmacol. Exp. Ther. 135*, 17–20 (1962).
105. K. Turker and A. Akcasu, *New Istanbul. Contrib. Clin. Sci. 5*, 89–97 (1962).
106. F. Shen, E. Leong Way, and H. Loh, unpublished.
107. E. Contreras and L. Tamayo, *Arch. Biol. Med. Exp. 4*, 69–71 (1967).
108. A. Nicak, *Med. Pharmacol. Exp. 13*, 43–48 (1965).
109. B. B. Brodie, M. S. Comer, E. Costa, and A. Dlabac, *J. Pharmacol. Exp. Ther. 152*, 340–349 (1966).
110. J. H. Gaddum and M. Vogt, *Brit. J. Pharmacol. 11*, 175–179 (1956).
111. L. Gymerk and E. Bindler, *J. Pharmacol. Exp. Ther. 135*, 344–348 (1962).
112. M. Medakovic, *Arch. Int. Pharmacodyn. 114*, 201–209 (1958).
113. M. Medakovic, *J. Pharm. (London) 11*, 43–48 (1959).
114. E. A. Gyand, H. W. Kosterlitz, and G. M. Lees, *Arch. Exp. Pathol. Phamakol. 248*, 231–246 (1964).
115. H. W. Kosterlitz and G. J. Lees, *Pharmacol. Rev. 16*, 301–339 (1964).
116. K. J. Rogers and J. A. Thornton, *Brit. J. Pharmacol. 36*, 470–480 (1969).
117. M. Nymark and J. Nielsen, *Lancet 2*, 524–525 (1963).
118. G. Brownlee and G. W. Williams, *Lancet 1*, 669 (1963).
119. A. H. Loveless and D. R. Maxwell, *Brit. J. Pharmacol. 25*, 158–170 (1965).

120. B. K. Bhattacharya and G. P. Lewis, *Brit. J. Pharmacol. 11*, 202–208 (1956).
121. T. F. Burks and J. P. Long, *J. Pharmacol. Exp. Ther. 150*, 267–276 (1967).
122. S. S. Tenen, *Psychopharmacologia 12*, 278–285 (1968).
123. J. Cochin and J. Axelrod, *J. Pharmacol. Exp. Ther. 125*, 105–110 (1959).
124. A. L. Bartlet, *Brit. J. Pharmacol. 15*, 140 –146 (1960).
125. E. L. Way, H. Loh, and F. H. Shen, *Science 162*, 1290–1292 (1968).
126. T. N. Tozer, N. H. Neff, and B. B. Brodie, *J. Pharmacol. Exp. Ther. 153*, 177–182 (1966).
127. D. R. Haubrich and D. E. Blake, *Fed. Proc. 28*, 793 (1969).
128. N. H. Neff, T. N. Tozer, and B. B. Brodie, *J. Pharmacol. Exp. Ther. 158*, 214–218 (1967).
129. H. H. Loh, F.-H. Shen, and E. Leong Way, *Biochem. Pharmacol. 18*, 2711–2721 (1969).
130. F.-H. Shen, H. H. Loh, and E. Leong Way, *Pharmacologist 10*, 211 (1968).
131. H. O. J. Collier, *Advan. Drug Res. 3*, 171–188, Academic Press, New York (1966).
132. J. V. Dingell, M. L. Owens, M. R. Norvich, and F. Sulser, *Life Sci. 6*, 1155–1162 (1967).
133. P. S. J. Spencer and T. A. R. Turner, *Brit. J. Pharmacol. 37*, 94–103 (1969).
134. E. Leong Way, H. H. Loh, and F.-H. Shen, *J. Pharmacol. Exp. Ther. 167*, 1–8 (1969).

Chapter 11

ACETYLCHOLINE AND CHOLINESTERASE

Marta Weinstock*

Department of Pharmacology
St. Mary's Hospital School
London, W2, England

I. THE EFFECT OF MORPHINE ON BRAIN ACETYLCHOLINE LEVELS

Since the original observations of Richter and Crossland[1] that the level of total brain acetylcholine is increased by sleep and by general depressants, a number of studies have been made on the effect of morphine on the concentration of this amine in the brain. Giarman and Pepeu[2] found that morphine (50 mg/kg), which produced severe depression in rats, raised the acetylcholine content of the brain by 47%. This effect was shared by the general depressants methylpentynol and pentobarbitone. A similar rise in brain acetylcholine also was obtained with levorphan but not with nor-levorphan or levallorphan.[3] The latter pair of drugs produced neither analgesia nor depression in rats in the doses used.

The increase in acetylcholine levels by morphine has been confirmed by other workers in rats[4] and mice,[5] but all have used doses greatly in excess of those required to produce significant analgesia. However, their collective findings suggested that the rise in acetylcholine may either result from or be the cause of the general depressant effect of narcotic analgesic drugs.

It is well known that the depressant effects of morphine, including the analgesia, can be antagonized by the simultaneous administration of nalorphine [6-8] or levallorphan.[9,10] They can also be abolished by the repeated administration of morphine for several days or weeks.[11,12] Thus one might expect that either of these procedures, simultaneous injection of an antagonist or chronic dosing with morphine, should reduce or abolish the elevation in brain acetylcholine produced by morphine. However, no consistent an-

---

* Present address: Department of Pharmacology, Tel Aviv University, Ramat Aviv, Tel Aviv, Israel.

tagonism has been obtained by the use of either nalorphine or levallorphan. Maynert[4] found that 50 mg/kg nalorphine produced no change in the response of brain acetylcholine levels to morphine (100 mg/kg), although the depressant effects of morphine were abolished. Herken et al.[3] were able to reduce the effect of levorphan but not that of morphine on brain acetylcholine levels by giving the antagonist levallorphan.

Studies of the effect of chronic administration of morphine have yielded conflicting results. Maynert[4] treated rats with morphine twice daily for 9 days, but was unable to demonstrate any significant reduction in the action of the drug on brain acetylcholine levels, even though the depressant effects were no longer present. On the other hand, after treatment of mice with morphine in increasing doses for 6 weeks, Hano and associates[5] showed that the effect on brain ACh was gradually reduced.

Since changes in total brain acetylcholine must reflect the overall effect of the drug on a variety of different nervous pathways, it is doubtful whether such measurements could ever yield specific information concerning any particular action of opiates on the central nervous system. A more fruitful study of the relationship between morphine and acetylcholine metabolism in the brain has been that of Crossland and Slater.[13] By separating brain ACh into "free" and "bound" forms, they were able to differentiate the action of morphine from that of general depressants such as ether and pentobarbitone. Thus the anesthetics caused a rise in both "free" and "bound" acetylcholine, whereas morphine raised the "bound" form but lowered "free" acetylcholine. These findings suggested that the effect of morphine was to prevent acetylcholine release from the "bound" to the "free" form. This differed from the effect of general anesthetics, which could be explained as being a simple consequence of reduced neuronal activity.

Unlike other workers,[4] Crossland and Slater were able to antagonize the effect of morphine on both "free" and "bound" acetylcholine levels by pretreatment with nalorphine. It may be significant that they used a smaller dose of nalorphine (10 mg/kg) than Maynert (50 mg/kg).

It would be interesting to extend the studies of Slater and Crossland to include the effect of chronic morphine administration on the levels of "free" and "bound" amine to see whether the action of morphine is lost. It might also be better to use the antagonist naloxone, which has virtually no agonist properties instead of nalorphine. In addition, other opiates such as methadone and meperidine should be tested to see whether they affect "free" and "bound" acetylcholine levels in a similar way to morphine. These studies would give some further indication of the specificity of the effect for morphine-like analgesic drugs.

The increase in bound ACh and the decrease in free amine theoretically could be brought about by (1) prevention of hydrolysis of ACh by cholinesterase, (2) stimulation of ACh synthesis, or (3) inhibition of ACh release.

The effect of morphine on these various parameters of ACh metabolism are now discussed.

II. EFFECT OF MORPHINE ON ACETYLCHOLINESTERASES

A. *In Vitro* Studies

Bernheim and Bernheim[14] originally reported that morphine can inhibit brain cholinesterase in concentrations of $1 \times 10^{-4}M$, which they considered were sufficiently low for them to speculate that this action of morphine might be responsible for some of the characteristic effects of the drug. Since then several workers have shown that other analgesic drugs also inhibit various preparations of cholinesterase *in vitro;* these include dilaudid,[15] methadone,[16] and morphinan derivatives.[3,17]

Young and his colleagues studied the effect of morphine and several morphinan derivatives on preparations of cholinesterase derived from bovine erythrocytes, rat brain, dog serum, and intestine. They could not demonstrate any correlation between ability of the drug to inhibit cholinesterase from any source and analgesic potency. Similarly, Hein and Powell[18] failed to show any clear relationship between anticholinesterase activity and analgesia or psychotomimetic properties among a series of benzomorphan derivatives, using a purified preparation of bovine acetylcholinesterase.

The analgesic antagonists, nalorphine[19] and levallorphan,[3,17] also were found to be powerful inhibitors of cholinesterase. However, unlike morphine they were not found to increase brain acetylcholine levels.

B. *In Vivo* Studies

In a careful study of the inhibition of brain cholinesterase by morphine, Johanesson showed that the concentration of substrate as well as that of tissue (and therefore of enzyme) were critical when determining the action of reversible inhibitors.[20] Thus, ten times as much morphine was required to inhibit cholinesterase when the substrate (acetylcholine) concentration was 1×10^{-3} g/ml than when it was 2×10^{-5} g/ml. He also found that the concentration of morphine in the brain was 20–40 μg/g 35–60 min after the injection of 600–700 mg/kg. When studying the anticholinesterase activity of brain preparations, he added morphine *in vitro* so that the final amount achieved was identical to that reached in the brain after parenteral administration, that is 20–40 μg/g. With such concentrations, he found an inhibition of the enzyme of 30–60%.

Lavikainen and Mattila[21] and Hano and his colleagues failed to demonstrate an effect of morphine on brain cholinesterase *in vivo*, but Johanesson explained this failure by the fact that both groups of workers used insufficient morphine in the reaction mixtures.

Johanesson compared the anticholinesterase activity of brain preparations of rats treated chronically with morphine and of those given only a single injection of the drug. He could detect no significant difference in the ability to hydrolyze acetylcholine in the two groups of rats.[20] He suggested that the excitation and convulsions caused by large doses of morphine in rats may be due to central cholinesterase inhibition. There are several other findings that could substantiate this suggestion. It is known that nalorphine does not antagonize the convulsant action of morphine.[8] Thus if the stimulant effects were due to cholinesterase inhibition, one would not expect them to be reduced by nalorphine since this drug is itself a powerful inhibitor of cholinesterase.[19] It also would explain the lack of a difference in cholinesterase activity in rats chronically treated with morphine since these animals do not become tolerant to the convulsant action of the drug.

It can be concluded from these studies that the increase in brain acetylcholine produced by injections of morphine is not due to inhibition of cholinesterase for several reasons. First, doses required to inhibit the enzyme are about ten times greater than those that elevate brain acetylcholine.[20] Second, nalorphine and levallorphan both inhibit cholinesterase but do not raise brain acetylcholine levels.[3,4] Also, powerful anticholinesterases such as physostigmine were found to raise both the "free" and "bound" acetylcholine in the brain whereas morphine did not.[13] It also can be concluded that neither the depressant nor analgesic effects of morphine are due to inhibition of cholinesterase. However, the foregoing results suggest that the stimulant effects of the drug may be related to anticholinesterase activity. In this respect it may be significant that nalorphine also produces convulsions in rats and mice and that the lethal effects of morphine are potentiated by neostigmine.[22]

II. THE EFFECT OF MORPHINE ON ACETYLCHOLINE SYNTHESIS

It was shown that levorphan $10^{-3}M$, levallorphan $10^{-3}M$, and dextrorphan $10^{-5}M$ can stimulate acetylcholine synthesis in rat brain *in vitro*. However, morphine did not share this stimulant action but actually depressed the acetylation of choline and a concentration of $10^{-3}M$.[23]

De la Lande and Bentley also found that morphine inhibited acetylcholine synthesis in whole cell preparations of brain in concentrations of $10^{-3}M$.[24] It therefore is not possible to explain the increased levels of brain acetylcholine after morphine in terms of increased synthesis.

IV. THE EFFECT OF MORPHINE ON ACETYLCHOLINE RELEASE

There have been very few reports on the effects of morphine on the release of acetylcholine from nerve endings in the brain. Techniques by which the amounts of acetylcholine released from particular neurons or functional groups of neurons can be measured are not yet available.

Beleslin and colleagues[25,26] have shown that morphine, at a concentration of 1 μg/ml, reduced the release of acetylcholine into the fluid perfusing the cerebral subarachnoid space of anesthetized cats. At a slightly higher concentration (2.5 μg/ml), morphine also reduced the amount of the amine released into the perfusate from the lateral ventricle. The same effect was obtained when the drug (2 mg/kg) was injected intravenously into the cats. It may be significant that these changes were produced by doses of morphine much lower than those required to alter total brain acetylcholine levels. These doses were well within the range required for the analgesic effects of morphine.

One criticism that may be leveled against these studies is that the animals were, of necessity, anesthetized throughout the experiments. It has been shown that anesthetic agents themselves profoundly effect acetylcholine metabolism in the brain,[27] and hence the full action of morphine may have been masked or altered.

All other studies of the effects of analgesic drugs on acetylcholine release have been made on peripheral nerve endings, mainly in isolated organ preparations. In 1957 Schaumann showed that morphine in concentrations of 1×10^{-4} g/ml reduced the amount of acetylcholine released spontaneously from segments of guinea pig ileum incubated in Tyrode's solution.[28] In a preparation of guinea pig ileum stimulated transmurally, Paton showed that the induced contractions were due to the release of acetylcholine from postganglionic nerve endings.[29] Morphine and several other opiates were shown to inhibit these contractions. The mechanism of action of morphine on this preparation was studied by de la Lande and Porter.[30] They found that morphine does not affect the resting release of acetylcholine but that it acts on the neurons by a process akin to the depression of nerve conduction by local anesthetics. They also suggested that morphine-sensitive neurons had relatively long pathways, since subdivision of the gut abolished its action.

Using a large number of narcotic analgesic drugs, Cox and Weinstock showed that there was a highly significant correlation ($r = 0.99$) between analgesic potency and the ability of the drugs to inhibit acetylcholine release.[31] It also was shown that only those isomers of optical enantiomorphs that had analgesic activity inhibited acetylcholine release. The isomers with L configuration, dextrorphan and dextromethorphan, were inactive in both tests.

Nalorphine was found to be approximately equiactive with morphine in inhibiting contractions after transmural stimulation.[31] A number of other narcotic antagonists, such as pentazocine, cyclozocine, and levallorphan, also have been shown to reduce the response to transmural stimulation. There also appears to be quite a good correlation ($r = 0.774$) between the analgesic potency of the drugs in man and their ability to depress contractions of the ileum.[32]

The effect of morphine on the release of acetylcholine also has been studied in a few other preparations. Pelikan showed that morphine impaired transmission in the superior cervical ganglion of the cat and concluded from his observations that this action was due to an inhibition of acetylcholine release at preganglionic nerve endings.[33] Using the rabbit sino-atrial node right atrial preparation, Kennedy and West[34] found that morphine in concentrations of 1×10^{-6} g/ml reduced the release of acetylcholine from the postganglionic nerve endings.

There is thus little doubt that morphine can inhibit acetylcholine release in a variety of preparations. Probably the most significant factor arising from all these studies is that this action of morphine occurs at relatively low concentrations: $1-2.5 \times 10^{-6}$ g/ml in the cerebrospinal fluid of cats,[25,26] 1×10^{-8} g/ml on the stimulated ileum, and 1×10^{-6} g/ml in the rabbit heart. Much less morphine is required to inhibit acetylcholine release than to inhibit either its synthesis or its hydrolysis by cholinesterase. If the brain contains cholinergic nerve endings that are as sensitive to the depressant actions of morphine-like drugs as the nerve endings in guinea-pig ileum, then sufficient drug would be present in brain tissue after administration of analgesic doses[35-37] to depress the release of acetylcholine from these nerve terminals.

Since all opiate analgesics also are powerful respiratory depressants, the question arises as to whether inhibition of acetylcholine release is associated with analgesia or the effect on respiration or both. Schaumann has obtained some evidence that changes in acetylcholine metabolism may also affect the activity of the respiratory center.[38] He found that respiratory depression was effectively antagonized by intravenous physostigmine or intracisternal prostigmine. He suggested that morphine inhibited respiration by reducing the output of acetylcholine at central cholinergic synapses. The cholinesterase inhibitors antagonized the effects of morphine by stabilizing these reduced amounts of the amine.

The results of studies of the effect of anticholinesterase drugs on the analgesia produced by opiates have been somewhat conflicting. If the analgesia in fact results from the reduced amounts of "free" acetylcholine due to insufficient liberation, one might expect that, like respiratory depression, analgesia would be diminished by the administration of cholinesterase inhibitors. In order for this hypothesis to be tested one must be sure that the anticholinesterase drug used is able to inhibit the enzyme in the central nervous system, which implies that it must penetrate the blood–brain barrier. The majority of workers who studied this problem have used prostigmine, which they administered intravenously, but this does not readily enter the brain. Using this drug, no antagonism of morphine analgesia has been found but more often potentiation.[38,40] De Jongh could not potentiate the analgesic action of morphine with prostigmine in guinea pigs, but he did so with atropine.[41] In contrast to these observations Saxena potentiated the effect of morphine on pain threshold with prostigmine and with pilocarpine, while D.F.P. proved to be ineffective. However, he was unable to antagonize the effect of prostigmine and pilocarpine by giving atropine and therefore concluded that the anticholinesterase action of prostigmine is not involved in the potentiation.[42]

Because of the poor penetration of this drug into the central nervous system it is much more likely that prostigmine influences morphine analgesia by altering (in some way) the sensitivity of the area to which the painful stimulus is applied. Alternatively the powerful anticholinesterases may fail to antagonize morphine analgesia because, by preferentially competing with morphine for the enzyme, they enable more morphine to become available to act on other receptors, including those concerned with inhibiting acetylcholine release. Whatever the explanation for these conflicting results, it should be emphasized that no conclusions can be drawn unless one is sure that the anticholinesterase given really does stabilize acetylcholine released in the central nervous system.

V. CONCLUSIONS

The studies on the guinea pig ileum show a good correlation between the ability of narcotic and antagonist analgesic drugs to inhibit acetylcholine release and the ability to diminish pain sensation. Furthermore, the effective concentrations in the ileum were low enough to correlate with amounts occurring in the brain after analgesic doses. No such relationship was found between analgesic activity and cholinesterase inhibition for a variety of compounds. Moreover, while inhibition of acetylcholine release by morphine was antagonized by nalorphine, naloxone, and levallorphan, inhibition of cholinesterase was not. Repeated administration of morphine also resulted in a loss of its effect on acetylcholine release,[29] but its anticholinesterase activity remained undiminished.[20] If such a close relationship could be found between analgesic potency and the inhibition of acetylcholine release from central neurons, coupled with the demonstration that nalorphine and chronic administration of morphine also attenuate the effect, we would be well on the way to understanding how morphine acts in the central nervous system.

VI. REFERENCES

1. D. Richter and G. Crossland, *Amer. J. Physiol. 159*, 247–255 (1949).
2. N. J. Giarman and G. Pepeu, *Brit. J. Pharmacol. 19*, 226–234 (1962).
3. H. Herken, D. Maibauer, and S. Muller, *Arch. Exp. Pathol. Pharmakol. 230*, 313–324 (1957).
4. E. W. Maynert, *Arch. Biol. Med. Exp. 4*, 36–41 (1967).
5. K. Hano, H. Kaneto, T. Kakunaga, and N. Moribayashi, *Biochem. Pharmacol. 10*, 441–447 (1964).
6. E. R. Hart and E. L. McCawley, *J. Pharmacol. Exp. Ther. 82*, 339–348 (1964).
7. K. Unna, *J. Pharmacol. Exp. Ther. 79*, 27–31 (1943).
8. L. A. Woods, *Pharmacol. Rev. 8*, 175–198 (1956).
9. P. J. Costa and D. D. Bonnycastle, *J. Pharmacol. Exp. Ther. 113*, 310–318 (1955).
10. C. A. Winter, P. D. Orahovats, and E. G. Lehman, *Arch. Int. Pharmacodyn. 110*, 186–202 (1957).
11. P. A. J. Janssen and A. H. Jageneau. *J. Pharm. Pharmacol. 10*, 14–21 (1958).
12. J. Cochin and C. Kornetsky, *J. Pharmacol. Exp. Ther. 145*, 1–10 (1964).
13. J. Crossland and P. Slater, *Brit. J. Pharmacol. 33*, 42–47 (1968).
14. F. Bernheim and M. L. C. Bernheim, *J. Pharmacol. Exp. Ther. 57*, 427–436 (1936).
15. C. I. Wright, *J. Pharmacol. Exp. Ther. 72*, 45–46 (1941).

16. G. S. Eadie, F. Bernheim, and D. B. Fitzgerald, *J. Pharmacol. Exp. Ther. 94*, 18–21 (1948).
17. D. C. Young, R. A. Vauder Ploeg, R. M. Featherstone, and E. G. Gross, *J. Pharmacol. Exp. Ther. 114*, 33–37 (1955).
18. G. E. Hein and K. Powell, *Biochem. Pharmacol. 16*, 567–573 (1967).
19. T. R. Blohm and W. G. Willmore, *Proc. Soc. Exp. Biol. Med. 77*, 718–721.
20. T. Johanesson, *Acta. Pharmacol. Toxicol. 19*, 23–35 (1962).
21. P. Lavikainen and M. Mattila, *Ann. Med. Exp. Fenn. 37*, 133–140 (1959).
22. J. C. Szerb, *Arch. Int. Pharmacodyn. 111*, 314–321 (1957).
23. R. W. Morris, *Arch. Int. Pharmacodyn 133*, 236–243 (1961).
24. I. S. de la Lande and G. A. Bentley, *Aust. J. Exp. Biol. 33*, 555–566 (1955).
25. D. Beleslin, R. L. Polak, and D. H. Sproull, *J. Physiol. 177*, 420–428 (1965).
26. D. Beleslin and R. L. Polak, *J. Physiol. 177*, 411–419 (1965).
27. K. A. C. Elliott, R. L. Swank, and N. Henderson, *Amer. J. Physiol. 162*, 469–474 (1950).
28. W. Schaumann, *Brit. J. Pharmacol. 10*, 456–461 (1955).
29. W. D. N. Paton, *Brit. J. Pharmacol. 12*, 119–127 (1957).
30. I. S. de la Lande and R. B. Porter, *Brit. J. Pharmacol. 29*, 158–167 (1967).
31. B. M. Cox and M. Weinstock, *Brit. J. Pharmacol. 27*, 81–92 (1966).
32. E. A. Gyang and H. W. Kosterlitz, *Brit. J. Pharmacol. 27*, 514–527 (1966).
33. E. W. Peliken, *Ann. N.Y. Acad. Sci. 40*, 52–69 (1960).
34. B. L. Kennedy and T. C. West, *J. Pharmacol. Exp. Ther. 157*, 149–158 (1967).
35. J. C. Szurb and D. H. McCurdy, *J. Pharmacol. Exp. Ther. 118*, 446–450 (1956).
36. L. B. Mellett and L. A. Woods, *J. Pharmacol. Exp. Ther. 125*, 47–104 (1959).
37. K. Milthers, *Acta Pharmacol. Toxicol. 19*, 235–240 (1962).
38. W. Schaumann, *Arch. Exp. Pathol. Pharmakol. 233*, 98–11 (1958).
39. D. Slaughter and D. W. Munsell, *J. Pharmacol. Exp. Ther. 68*, 104–112 (1940).
40. S. Flodmark and T. Wramner, *Acta. Physiol. Scand. 9*, 88–96 (1945).
41. D. K. de Jongh. *Acta Physiol. Pharmacol. Neer. 3*, 164–172 (1954).
42. P. N. Saxena, *Ind. J. Med. Res. 46*, 653–658 (1958).

Chapter 12

CORTICOSTEROID HORMONES

Jewell W. Sloan

Addiction Research Center
Lexington, Kentucky

I. INTRODUCTION

The effects of morphine and other narcotic analgesics on adrenal cortical secretion have been studied from three frames of reference: (1) The ability of chronic morphine administration to cause hypertrophy of the adrenal gland was first reported by McKay and McKay[1] and subsequently confirmed by other investigators.[2-5] Morphine was used by Selye[6] to first evoke an alarm reaction and its attendant increase in the adrenal cortical response. In addition to its ability to stimulate the adrenal cortex, Briggs and Munson[7] demonstrated that morphine in the anesthetized rat also could suppress the response to stressful stimuli. (2) Clinical observations and impressions indicated that morphine diminishes both sexual interest and excitability and that during the morphine abstinence syndrome spontaneous orgasms have been observed and reported.[8] (3) A relationship between the analgesic action of morphine and adrenal cortical function has been suggested by the observations that adrenalectomy reduces the analgesic action of morphine[9,10] and that steroids can both increase and decrease the analgesic action of morphine.[11,12] In this regard it is of interest that Craig[13] has reported on a steroid analgesic, 17-α-acetoxy-6-dimethylaminomethyl-21-fluoro-3-ethoxypregna-3,5-dien-20-one hydrochloride (SC-17599), that shares many of the pharmacological actions of morphine.

The purpose of this monograph is to review the effects of morphine and other narcotic analgesics on the adrenal cortical response and to discuss the pharmacological and physiological implications of these studies.

II. STEROID BIOCHEMISTRY AND PHYSIOLOGY

A. Definitions and Terminology

The term "steroid" was first proposed by Callow (1936)[14] for com-

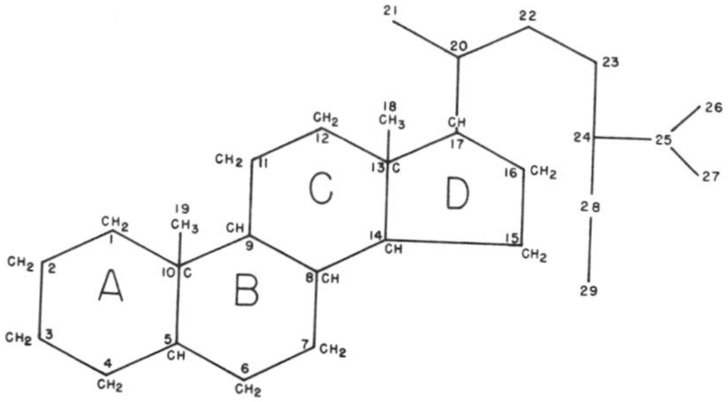

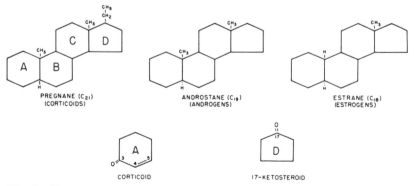

Fig. 1. The representation of the parent nucleus ot steroids (perhydrocyclopentano-phenanthrene) and the numbering of carbon atoms.

pounds that contain the phenanthrene structure (Fig. 1; rings A, C, and B) to which is attached the cyclopentane structure (ring D). There are three general classes of naturally occurring steroids: pregnanes (C-21), androstanes (C-19), and estranes (C-18) (Fig. 2). Corticosteroids are steroids possessing 21 carbon atoms and three or more oxygen atoms, one of which must have a Δ^4-3-ketone in ring A (Fig. 2). The 17-keto-steroids (17-KS) are characterized by a ketonic oxygen at position 17. They are derived from C-18, C-19, and C-21 steroids and thus are secreted not only by the adrenal cortex but by other steroid-producing tissue.[15-17]

The steroids with which this review is concerned are the glucocorticoids, the mineralocorticoids, and the 17-KS. The glucocorticoids have a -O or -OH substituent at C-11 (corticosterone) and an -OH group at C-17 (cortisone and cortisol). The 17-hydroxy corticosteroids are frequently referred to as 17-OHCS. Cortisol (hydrocortisone) is the major glucocorticoid in man, the dog, and the guinea pig while corticosterone is the major glucocorticosteroid in the rat and rabbit.

PREGNANE (C₂₁)
(CORTICOIDS)

ANDROSTANE (C₁₉)
(ANDROGENS)

ESTRANE (C₁₈)
(ESTROGENS)

CORTICOID

17-KETOSTEROID

Fig. 2. The parent compounds of the three general classes of naturally occurring steroids (pregnanes, androstanes, and estranes) are shown as well as the essential configuration of ring A for corticoids and ring D for 17-ketosteroids.

Glucocorticoids are primarily concerned with the regulation of protein, carbohydrate, lipid, and nucleic acid metabolism[18] but also have some effect on water metabolism.[19]

Aldosterone, the principal mineralocorticoid, is characterized by an aldehyde group at C-18 and is concerned with salt and water metabolism and its major biological action is to increase sodium transport by the kidney.[20]

As much as two-thirds of the total urinary 17-KS may be of adrenal origin,[21] and since some are derived from the metabolism of the glucocorticoids they are of interest in assessing the effects of narcotic analgesics on corticosteroid excretion. In women, urinary 17-KS are products of the adrenal cortex only, whereas in men they are products of the adrenal cortex and testis. Esterone is also a 17-KS, but since it is acidic in nature it is removed from urinary extracts by the alkali wash. The gonadal hormones have an important influence on adrenal cortical function in addition to their effects on primary and secondary sex characteristics. Estrogens have been shown to increase ACTH secretion,[22] plasma corticosteroid levels,[23] and adrenal size.[24] The androgens methylandrostenediol and testosterone propionate have been found to prevent adrenal atrophy in rats.[25]

B. Control of Production

1. Glucocorticoids

A variety of afferent impulses to the hypothalamus excite hypothalamic neurons and are believed to cause the release of a corticotropin-releasing factor(s) (CRF) from nerve endings[26-32] in the median eminence. The CRF thought to be released at this site is transported to the poorly innervated adenohypophis through a portal vascular system, where it stimulates the release of the adrenocorticotrophic hormone (ACTH).[33,34] ACTH is released in a regular diurnal rhythm and is carried by the general circulation to the adrenal cortex, where it stimulates the production of glucocorticoids. In man, the minimum plasma levels are seen shortly after midnight and maximum values are attained between 8 and 10 A.M.[35] The extremely vascular adrenal cortical cells secrete glucocorticoid hormones into the blood as soon as they are synthetized, and no appreciable storage occurs.[21] ACTH secretion, in addition to being determined in part by hypothalamic activity, is also regulated by circulating blood levels of glucocorticoids. This hypothesis, first elucidated by Ingle and Kendall,[36] has been confirmed by several lines of experimental evidence. The injection of adrenal cortical extracts[37] or corticosteroids[38-42] inhibits ACTH secretion. The site of glucocorticoid inhibition and CRF release has been studied by steroid implants in various areas of the hypothalamus,[43-47] by surgical isolation of different areas of the basal hypothalamus, median eminence, and pituitary from the rest of the brain,[48,49] and by electrical stimulation of various areas of the brain.[50-54] The results of such studies implicate a poorly defined area of the median eminence as the locus for

CRF release. The site of inhibition is located in unknown higher centers, probably between the amygdala and median eminence.

2. Aldosterone

The regulation of aldosterone secretion differs from that of the gluco-corticoids. While ACTH has some stimulating effect on aldosterone production,[55,56] some of the important physiologic stimuli that affect its release are salt depletion,[57-60] high concentration of potassium ion,[61] angiotensin II,[62] renin,[63] and acute hemorrhage.[64,65] The reader is directed elsewhere for a review of this subject.[20,21,66-68]

III. INDEXES OF ADRENAL CORTICAL ACTIVITY

A. Adrenal Weights

Continued hyperactivity of the adrenal cortex was believed to be associated with an increase in the size of the gland; consequently wet or dry weights were used for measuring the effects of severe and protracted stress.[69]

B. Circulating Eosinophil Counts

Circulating eosinophil counts have long been used to assess adrenal cortical activity. The "alarm" reaction produces an eosinopenia,[70] an effect that is also produced by injecting ACTH[71,72] or corticosteroids.[71,73] In spite of the widespread use of circulating eosinophil counts as an indirect measure of adrenal cortical function, it is not known with certainty how adrenal cortical function and circulating eosinophil counts are related. It is known that cortisone inhibits the production[74-76] and decreases blood levels of histamine.[77-79] In addition, histamine has a chemotactic effect upon eosinophils. In vitro histamine causes the granules of the eosinophil cell to accumulate in the end of the cell in closest proximity to the highest concentration of histamine. It is postulated from these findings that corticosteroids exert their effects on circulating eosinophil cells by reducing circulating histamine, which is the stimulus for the release of eosinophils.[80]

C. Adrenal Ascorbic Acid Content

The adrenal content of ascorbic acid was found to be decreased in the rat by ACTH,[81] and the decrease was related to the dose of administered ACTH.[82] These findings resulted in the widespread use of adrenal ascorbic acid content as a measure of the adrenal cortical response. Although some correlation appears to exist between plasma corticosterone levels and adrenal ascorbic acid content, dissociations are known. Several lines of evidence suggest that the vitamin may be involved in hormone synthesis by exerting an inhibitory influence on steroid hydroxylase systems, and therefore its removal would allow increased steroid biosynthesis.[83-86] In contrast, Sweat and Bryson[87] demonstrated that ascorbic acid prevented the inhibition of 11-β-hydroxylation by adrenochrome, which suggests that its presence might increase steroid biosynthesis.

D. Urinary and Plasma Steroids

Methods have been perfected within recent years with the aid of fluorometry, spectrophotometry, polarography, and column, paper, thin-layer, and gas chromatography, which make it possible to identify many of the individual steroids. Several techniques have been devised for determining steroids in blood and urine. Following preliminary extraction procedures, plasma and urinary 17-OHCS[88] may be determined as Porter–Silber chromogens[89] or as 17-ketogenic steroids. The Porter–Silber reaction requires the presence of a 17,21-dihydroxy-20-ketone, although all corticosteroids do not react with the phenylhydrazine sulfuric acid color reagent. This is a very satisfactory method for measuring cortisol. The blue tetrazolium salts (BTZ) react with a side chain on adrenal steroids but do not require the presence of a 17-hydroxyl group. The difference between the Porter–Silber reaction and the BTZ reaction therefore measures essentially corticosterone. Aldosterone has been measured by polarographic[90] and chromatographic methods. The urinary 17-ketosteroids may be determined following purification[91] by modifications of the Zimmerman reaction,[92] whereby 17-KS produce a violaceous color with m-dinitrobenzene in alkaline solution. Since the 17-KS are derived from both the testis and the adrenal, their determination is of no value in differentiating the secretory activity of the adrenal gland and testis.

IV. DRUG EFFECTS

A. Acute

Extensive studies of the effects of single doses of morphine on the adrenal cortical response in the unanesthetized rat have been conducted. However, data concerning the effects of morphine in other species and other analgesics are scanty. These data are summarized in Table I. In the rat[93-99] and dog[98] single doses of morphine ranging from 3.75 to 30 mg/kg have been shown to increase urinary excretion and plasma levels of corticosteroids. Sobel[99] found that 25 mg/kg of morphine did not effect urinary corticosteroid excretion in the guinea pig. In man, much smaller doses of 0.6 to 1.1 mg/kg had no effect on urinary 17-OHCS,[100-102] although the normally occurring midday depression in plasma 17-OHCS levels was further depressed following doses of 0.2 mg/kg.[103] Meperidine in doses of 10 mg/kg/day was without effect on urinary 17-OHCS excretion in man.[102] Single doses of morphine decreased ascorbic acid content in the rat.[93,104-109] Experimental evidence regarding the dose of morphine necessary to produce a decrease in adrenal ascorbic acid in the rat is not in agreement; thus Nikodijevic and Maickel[93] found a decrease after 3.74

mg/kg while George and Way[107] found a decrease with 22 mg/kg but not with 8 mg/kg. Morphine also has been shown to depress the level of circulating eosinophil counts in the mouse and in man.[110] The adrenal cortical effects produced by morphine are produced by other narcotic analgesics. Thus, $(+)$-methadone in doses of 22–60 mg/kg and $(-)$-methadone in doses of 4.0 mg/kg as well as heroin and meperidine decreased adrenal ascorbic acid levels in the rat.[107,109] With regard to urinary 17-KS excretion, on the other hand, morphine has been shown to have a depressant effect in both the rat[96] and man.[100] Codeine and meperidine depressed circulating eosinophils in the mouse and man.[110]

Nalorphine, the narcotic antagonist, had no effect on urinary 17-OHCS in man[102] but prevented morphine's depleting effect on adrenal ascorbic acid in the rat[105] and had a depressant effect on circulating eosinophil counts in man.[127]

In summary, the data show that in the unanesthetized rat, dog, and mouse single doses of morphine and other narcotic analgesics increase the adrenal cortical response and that this effect is probably antagonizable by nalorphine. In man and the guinea pig single doses of morphine are either without effect or produce a liminal depression of 17-OHCS excretion.

B. Chronic

Effects of chronic morphine administration on steroid excretion are summarized in Table II. Paroli and Melchiorri[111] found that when rats were treated chronically with morphine, urinary steroid excretion appeared to progress through several stages. The first stage observed in rats receiving morphine in doses of 15–75 mg/kg/day lasted about 10 days and was characterized by an increase in corticosterone and a decrease in 17-KS and possibly aldosterone excretion. Rats receiving lower doses of 5 or 10 mg/kg/day had urinary steroid levels equal to or slightly less than controls. During the second stage, which emerged after the second week of treatment and lasted for about 30 days, steroid excretion was decreased. Rats receiving 10 mg/kg/day for an extended period of time (40 days) entered the stage that resembles the abstinence phase. During this stage the urinary excretion of corticosterone and aldosterone was elevated while 17-KS excretion returned to preaddiction levels. In these studies, excepting one group of rats treated for 140 days with morphine, all differences are between the treatment condition and the pretreatment condition and aging is thus not controlled for. In the guinea pig, morphine in doses of 25 mg/kg/day, administered for 12 days, decreased the urinary excretion of 17-OHCS.[99]

Rats treated for 6 days with morphine in doses of 20 mg/kg/day were found to have higher levels of adrenal ascorbic acid than control animals,[106] which contrasts with the acute effects of morphine on this measure.

Chronic treatment with nalorphine produced a slight but nonsignificant decrease in urinary steroid excretion in the rat. Nalorphine antagonized the effects of morphine on steroid excretion when the two drugs were given chronically in equal doses.[111]

TABLE

Single Doses of Narcotic Analgesics and Nalorphine

Species and Strain	Drug	Dose (mg/kg)	Route of Adminis-tration	Collection Period
Rat, Sprague–Dawley	morphine	20–30	i.p.	0.5–5 hr
Rat, Sprague–Dawley	morphine	3.75–30	i.p.	1.25 hr
Rat, Sprague–Dawley	morphine	0.75–1.5	i.p.	1.25 hr
Rat, Holtzman	morphine	100	i.p.	20–80 min
Rat (white)	morphine	20	i.p.	35 min
Rat, Sprague–Dawley	morphine	20–40	i.p.	2 hr 10 min
Rat, Sprague–Dawley	morphine	15	i.p.	
Rat, Sprague–Dawley	morphine	8	s.c.	1.5 hr
Rat, Sprague–Dawley	morphine	22	s.c.	1.5 hr
Rat, hooded or Wistar	morphine	20	s.c.	1–3 hr
Rat	morphine	10		
Rat, Sprague–Dawley	morphine	10	ven. cath.	1 hr
Rat, Sprague–Dawley	morphine	10	ven. cath.	8 hr
Mouse, Swiss albino	morphine	10	i.p.	1–4 hr
Guinea pig	morphine	25	i.p.	6 hr
Dog	morphine	8	i.v.	5–120 min
Man, normal	morphine	0.2	s.c.	4–5 hr (midday)
Man, postaddict	morphine	0.6–1.1	s.c.	24 hr
Man, postaddict	morphine	0.6–0.9	s.c.	24 hr
Man, postaddict	morphine	0.6–0.9	s.c.	2, 4, 6, 8, 10, 12 hr
Man, postaddict	morphine	1.1 (avg)	s.c.	24 hr
Man, normal	morphine	0.1	i.m.	2–4.5 hr
Rat	heroin		i.m.	
Rat	meperidine		i.m.	
Mouse, Swiss albino	meperidine	20	i.p.	2–4 hr
Man, normal	meperidine	0.7	i.m.	2 hr
Man, postaddict	meperidine	10 (avg)	s.c.	24 hr
Rat, Sprague–Dawley	(−)-methadone	1.4	s.c.	1.5 hr
Rat, Sprague–Dawley	(−)-methadone	4.0	s.c.	1.5 hr
Rat, Sprague–Dawley	(+)-methadone	8.0	s.c.	1.5 hr
Rat, Sprague–Dawley	(+)-methadone	22–60	s.c.	1.5 hr
Mouse, Swiss albino	codeine	30	i.p.	2–4 hr
Man, normal	codeine	30	i.m.	2 hr
Rat, Sprague–Dawley	nalorphine	5	s.c.	1.5 hr
Rat, Sprague–Dawley	nalorphine	5	i.p.	10 hr
Rat, Sprague–Dawley	{ nalorphine	5	i.p.	1 hr 5 min }
	{ morphine	30	i.p.	1 hr }
Man, postaddict	nalorphine	1 (avg)	s.c.	24 hr

^{a} ↑ = increase, ↓ = decrease, 0 = no effect, and + = blocked.

I
on the Adrenal Cortical Response[a]

Urine			Plasma		Adrenal Ascorbic Acid	Circulating Eosinophils	Reference
17-KS	17-OHCS	Corticosterone	17-OHCS	Corticosterone			
					↓		(93, 104, 105)
				↑	↓		(93)
				0	0		(93)
				↑			(94)
					↑		(106)
				↑			(95)
↓		↑					(96)
					0		(107)
					↓		(107)
					↓		(108)
					↓		(109)
			↑				(97)
			↓				(97)
						↓	(110)
	0						(99)
			↑				(98)
			↓				(103)
↓							(100)
	0						(101)
			0				(101)
	0						(102)
						↓	(110)
					↓		(109)
					↓		(109)
						↓	(110)
						↓	(110)
	0						(102)
					0		(107)
					↓		(107)
					0		(107)
					↓		(107)
						↓	(110)
						↓	(110)
					0		(104)
					0		(105)
					+		(105)
	0						(102)

TABLE
Chronic Treatment with Narcotic Analgesics

Species and Strain	Drug	Dose (mg/kg/day)	Route of Administration	Treatment Period
Rat, Sprague–Dawley	morphine	5	i.p.	40 days
Rat, Sprague–Dawley	morphine	10	i.p.	3–10 days
Rat, Sprague–Dawley	morphine	10	i.p.	8–10 days
Rat, Sprague–Dawley	morphine	10	i.p.	18–30 days
Rat, Sprague–Dawley	morphine	10	i.p.	63–70 days
Rat, Sprague–Dawley	morphine	10	i.p.	68–70 days
Rat, Sprague–Dawley	morphine	10	i.p.	98–140 days
Rat, Sprague–Dawley	morphine	10	i.p.	133–140 days
Rat, Sprague–Dawley	morphine	15	i.p.	3–5 days
Rat, Sprague–Dawley	morphine	15	i.p.	3–10 days
Rat, Sprague–Dawley	morphine	15	i.p.	18–40 days
Rat, Sprague–Dawley	morphine	15	i.p.	33–40 days
Rat, Sprague–Dawley	morphine	20	i.p.	3–5 days
Rat, Sprague–Dawley	morphine	20	i.p.	18–40 days
Rat, Sprague–Dawley	morphine	25	i.p.	3–10 days
Rat, Sprague–Dawley	morphine	25	i.p.	8–10 days
Rat, Sprague–Dawley	morphine	25	i.p.	18–20 days
Rat, Sprague–Dawley	morphine	25	i.p.	33–40 days
Rat, Sprague–Dawley	morphine	25	i.p.	38–40 days
Rat, Sprague–Dawley	morphine	75	i.p.	3–5 days
Rat, Sprague–Dawley	morphine	75	i.p.	3–10 days
Rat, Sprague–Dawley	morphine	75	i.p.	18–40 days
Rat, white	morphine	20	i.p.	6 days
Guinea pig	morphine	25	s.c.	12 days
Man, 13 postaddicts	morphine	3.4	s.c.	35–144 days
Man, 2 postaddicts	morphine	3.5	s.c.	50–60 days
Man, 7 postaddicts	morphine	0.3–1.4*	s.c.	5 weeks
Man, 7 postaddicts	morphine	3.4**	s.c.	26–28 weeks
Man, 2 postaddicts	morphine	0.6 or 3.7	s.c.	10 or 65 days
Man, 4 postaddicts	normorphine	0.6–1.4	s.c.	14–95 days
Rat, Sprague–Dawley	nalorphine	50	i.p.	40 days
Rat, Sprague–Dawley	{ morphine	20	i.p.	3–40 days }
	{ nalorphine	20	i.p.	3–40 days }

[a] ↑ = increase, ↓ = decrease, 0 = no change, ← or → = change not statistically significant, * = ascending dose, ** = stabilization dose, † = early morning, and †† = midday.

II
and the Adrenal Cortical Response[a]

Collection Period	Urine				Plasma	Adrenal Ascorbic Acid	Circu-lating Eosino-phils	Refer-ence
	Corti-coste-rone	17-OHCS	Aldos-terone	17-KS	17-OHCS			
72 hr	↓		↓	↓				(111)
72 hr			↓					(111)
72 hr	0			↓				(111)
72 hr	↓			↓				(111)
72 hr			0					(111)
72 hr	0			↓				(111)
72 hr	↑			↓				(111)
72 hr			↑	↓				(111)
72 hr	↑			↓				(111)
72 hr			↓					(111)
72 hr	↓			↓				(111)
72 hr			↓					(111)
72 hr	↑			↓				(111)
72 hr	↓			↓				(111)
72 hr			↓					(111)
72 hr	↑			↓				(111)
72 hr	↓			↓				(111)
72 hr			↓	↓				(111)
72 hr	↓			↓				(111)
72 hr	↑			0				(111)
72 hr			0					(111)
72 hr	↓			↓				(111)
6th day						↑		(106)
6 hr		↓						(99)
24, 48, or 72 hr				↓			0	(100)
24, 48, or 72 hr		↓			↓ ↑ ↑ ↑↑		0	(101)
24 hr		↓						(112)
24 hr		↓						(112)
48–96 hr		↓						(113)
48–96 hr		↓						(113)
72 hr	↓		↓	↓				(111)
72 hr								(111)
72 hr		0		0				(111)

In man, chronic morphine treatment decreased urinary 17-KS and 17-OHCS during both a 3-week period of rapid dose escalation and a 1- to 5-month period of stabilization where the dose was maintained at about 3.4 mg/kg/day.[100,101] There was a tendency for the development of tolerance to this effect with the prolonged administration of morphine. In addition to urinary 17-OHCS, early morning plasma levels of 17-OHCS were decreased, which contrasts with the diurnal pattern seen during the preaddiction period when the early morning plasma 17-OHCS were increased.[101] In addition, there was a midday rise during the addiction period, whereas single doses of morphine produced a further decrease in the midday levels of plasma 17-OHCS.[103] In another study,[112] human subjects received ascending doses of morphine over a 5-week period and were stabilized at about 3.4 mg/kg for 29 weeks. An analysis of the acquired data by a paired group comparison showed that the urinary excretion of 17-OHCS during the first two weeks of the ascending dose phase (0.3–1.3 mg/kg/day) was slightly but not significantly less than the urinary 17-OHCS excretion seen during the preaddiction control period. The levels seen during the twenty-sixth and twenty-eighth weeks were intermediate to those seen during the preaddiction and ascending dose phases, which indicates that partial tolerance may have developed. There was individual variability, however, and some subjects had highly significant decreases in urinary 17-OHCS; partial tolerance appeared to develop to this effect of morphine as the treatment period continued. In addition to the effects of urinary 17-OHCS, urine volumes were found to increase during the stabilization period. If aldosterone excretion is affected in man as it is in the rat by the chronic administration of morphine, then the decrease in aldosterone production would be associated with a relative diuresis, which could account for the increased urine excretion.

Normorphine, which is morphine-like in most respects, also produced a depression of urinary 17-OHCS in man during a treatment period of 14–95 days.[113]

The effects of morphine early in the course of chronic administration are similar to those of single doses. With continuing administration the effect is predominently one of suppression of corticosteroid excretion to which partial tolerance develops.

C. Abstinence

The effects of withdrawal on the adrenal cortical response are summarized in Table III. In the rat the withdrawal of morphine produces an early, or primary, abstinence syndrome that emerges very rapidly.[4,5,114–118] The maximum physiological signs occur between the sixteenth and twenty-fourth hours and subside between the third and fourth days. Urinary 17-KS, 17-OHCS, and aldosterone levels were shown to be maximally increased in the rat by the third day of abstinence.[111] Changes in steroid excretion occurring earlier than 24 hr were not detected since urines were pooled in daily samples. In man the primary abstinence syndrome emerges later than in the rat. The peak effects are seen somewhere around 48 hr following abrupt withdrawal.[119-121] The physiological signs of the abstinence syndrome are less intense, however, when the usual method of gradual withdrawal is used.[122] Urinary 17-KS and urinary and plasma 17-OHCS are significantly increased, and circulating eosinophil counts are decreased when morphine is either abruptly or gradually withdrawn.[100,101] When normorphine was abruptly withdrawn the abstinence syndrome was slow in onset and mild in intensity compared to morphine. Urinary 17-OHCS excretion, on the other hand, was elevated on the first day of abstinence. The peak urinary levels of corticosteroids were found on the third day when the abstinence scores were highest.[113] Plasma 17-OHCS levels and circulating eosinophil counts returned to normal when the abstinence syndrome was terminated with morphine.[101] In man, treatment with either cortisone or ACTH has no ameliorative effect on primary abstinence and in fact appears to intensify the physiological signs. Similarly the eosinopenia seen during abstinence is either unaffected or enhanced.[123] The urinary levels of 17-OHCS are not elevated during protracted abstinence.[112]

In summary, both the rat and man show an enhanced excretion of corticosteroids during early or primary abstinence from morphine and normal levels during protracted abstinence.

Nalorphine, as well as other narcotic antagonists, precipitates a syndrome in a variety of species that closely resembles the abstinence or withdrawal syndrome although the physiological symptoms appear much more rapidly. The peak intensity is reached in a matter of minutes, and the effects subside within an hour or two.[124-128] Theories of the modes of action of nalorphine and other opioid antagonists have been recently reviewed.[129] Data concerning the effects of nalorphine-precipitated abstinence on the adrenal cortical response in the morphine-dependent animal are summarized in Table III. In the morphine-dependent rat[111] it has been demonstrated that nalorphine produces an increased excretion of urinary corticosterone, aldosterone, and 17-KS.

D. Adrenal and Gonadal Responsivity

Neither acute nor chronic treatment with morphine appears to impair the ability of the human adrenal gland to respond to ACTH. However, because both single doses and chronically administered morphine affect both blood and urine levels of corticosteroids, there is a question of how responsivity to ACTH should be expressed. Thus plasma 17-OHCS levels during the early morning hours in two morphine-dependent subjects were significantly less than the levels seen at the same hours during the control period; however, the peak response to ACTH at the second or third hour was as high during addiction as during the control period. The increase in plasma 17-OHCS levels following ACTH was greater during chronic morphine treatment than during the control period. An enhanced response to ACTH also was seen with regard to urinary levels of 17-OHCS.[101] The drop in circulating eosinophil counts was the same as during preaddiction

TABLE
Acute and Protracted Abstinence

Species and Strain	Drug	Dose (mg/kg/day)	Route of Administration	Treatment Period	Type of Withdrawal	Drugs Received Drug
Rat, Sprague–Dawley	morphine	30	i.p.	40 days	abrupt	
Rat, Sprague–Dawley	morphine	30	i.p.	40 days	abrupt	
Rat, Sprague–Dawley	morphine	75	i.p.	40 days	abrupt	
Rat, Sprague–Dawley	morphine	25	i.p.		abrupt	
Rat, Sprague–Dawley	morphine	25	i.p.		abrupt	
Rat, Sprague–Dawley	morphine	75	i.p.		abrupt	
Man, postaddict	morphine	3.4	s.c.	35–144 days	abrupt	morphine
Man, postaddict	morphine	3.4	s.c.	35–144 days	protracted	
Man, postaddict	morphine	4.2	s.c.	137 days	abrupt	morphine
Man, postaddict	morphine	4.2	s.c.	137 days	abrupt	{ morphine morphine
Man, postaddict	morphine	3.6	s.c.	137 days	abrupt	morphine
Man, postaddict	morphine	3.6	s.c.	137 days	abrupt	{ morphine morphine
Man, postaddict	morphine	3.1–4.2	s.c.	125–160 days	abrupt	
Man, postaddict	morphine	3.5	s.c.	50–60 days	abrupt	{ morphine morphine
Man, postaddict	morphine	0.6–3.7	s.c.	10–65 days	abrupt	
Man, postaddict	normorphine	0.6–1.4	s.c.	14–95 days	abrupt	
Man, postaddict	morphine	3.5	s.c.	34 weeks	gradual	{ morphine pentobarbital
Man, postaddict	morphine	3.5	s.c.	34 weeks		
Man, postaddict	morphine	3.5	s.c.	34 weeks	total	
Man, postaddict	morphine	3.5	s.c.	34 weeks	total	
	morphine	1.7–3.4	s.c.	2 weeks	abrupt	
Man, postaddict	{ morphine	1.7–3.4	s.c.	3 weeks	abrupt	
Man, postaddict	cortisone	4.3	i.m.	2 days	abrupt	
Man, postaddict	morphine	3.4–4.3	s.c.	2 weeks	abrupt	
Man, postaddict	{ morphine	3.4–4.3	s.c.	3 weeks	abrupt	
	ACTH	1.4	i.m.	2 days	abrupt	

a ↑ =increase, ↓ =decrease, O=no Effect, and ← or →=change not statistically significant.

in one subject but delayed by 1 hr in the other following the administration of ACTH. In another study in man the actual levels of urinary 17-KS after an infusion of ACTH were not as high during morphine addiction as during the preaddiction period, although when the values were expressed as percentage change from the preinfusion level, the increases during morphine treatment and preaddiction were of the same order of magnitude. The maximal decrease in circulating eosinophil counts in two of three subjects following ACTH administration was delayed about 2 hr when the patients were dependent on morphine. When chorionic gonadotropin was administered to the morphine-dependent patients, urinary 17-KS excretion was increased. The actual level was not as high as the level reached in the preaddiction period. Expressed as a percentage change from the preinfusion level, however, the increase during morphine dependence was greater than during the preaddiction period.[100]

Rats treated chronically with morphine in doses of 15 or 25 mg/kg/day for 28 days had increased urinary corticosterone and 17-KS levels after ACTH treatment. The actual levels were lower than those attained after

III

and the Adrenal Cortical Response[a]

During Abstinence			Urine			Plasma			
Dose (mg/kg/day)	Route of Administration	Abstinence Period	Corticosterone	17-OHCS	Aldosterone	17-KS	17-OHCS	Circulating Eosinophils	Reference
		3–5 days	↑		↑				(111)
		8–13 days	↑		0				(111)
		3–13 days	↑	↑	↑				(111)
		3 days	↑	0	↑				(111)
		5 days	0	0	↑				(111)
		1–5 days	↑		↑				(111)
0.04–0.2	s.c.	3 days			↑			↓	(100)
		5–35 days			0			0	(100)
0.04–0.06	i.v.	24–42 hr				↑		↓	(101)
0.4 / 1.0	i.v. / s.c.	48 hr				↓		→	(101)
0.19	i.v.	24–42 hr				↑		↓	(101)
0.4 / 1.0	i.v. / i.v.	48 hr				↓		→	(101)
		20–48 hr				↑		↓	(101)
0.05–0.1 / 0.05–0.06	s.c. / i.v.	1–15 days		↑					(101)
		24–48 hr		↑					(113)
		24–72 hr		↑					(113)
2.0–0.14 / 1.4 (last 2 days)	s.c. / orally	3rd–15th day		↑					(112)
		15th–17th day		↑					(112)
		7th, 17th, 24th week		0					(112)
		40 hr						↓	(123)
		40 hr						↓	(123)
		40 hr						↓	(123)
		40 hr						↓	(123)

ACTH during the preaddiction period, but there was no difference between the two conditions when the results were expressed as percents of pretreatment level for the rats receiving 15 mg/kg/day. At the 25 mg/kg/day-morphine dose the response was about one-half that seen during the preaddiction period. Urinary aldosterone levels were not changed by ACTH stimulation under any condition.[111]

E. Hydrocortisone Load Excretion

In man the rate of disappearance of hydrocortisone from plasma as well as the amount excreted in the urine during 24 hr after an infusion of hydrocortisone was the same during the addiction period as during the control period.[101]

F. *In Vitro* Steroid Synthesis

Paroli[130] found that while adrenal slices from rats treated with morphine in doses of 20 mg/kg/day for 5 days synthesized steroids as well or better than controls, adrenal slices from rats treated with this dose of morphine for periods of 15 or 30 days had an impaired ability to synthesize steroids.

Morphine added to whole adrenal and testical homogenates at concentrations of 50 and 100 $\mu g/g$ tissue produced a dose-related inhibition of 3-β-ol-dehydrogenase, indicating that morphine is interfering with the biosynthesis of steroids at the pregnenolone stage.[130]

Chronic morphine treatment also was shown to decrease the rate of conjugation of steriods by liver slices;[130] however, it is not known if these changes occur *in vivo*.

G. Hormonal Control of Metabolism

Strong evidence indicates that hydrocortisone and adrenaline are mutually antagonistic in their *in vitro* effects on glucose uptake and metabolism in diaphragm tissue from morphine-addicted rats. It has been demonstrated that both compounds retard glucose uptake in control tissue. In tissue from morphine-treated rats, adrenaline was without effect and hydrocortisone accelerated the uptake of glucose. When both compounds were added together, glucose uptake was accelerated in normal tissue while it was retarded in morphinized tissue. Differences also were observed in the way adrenaline and hydrocortisone influenced various stages of intracellular carbohydrate metabolism and respiration in tissue from morphine-treated rats. However, it was only in their effects on the glucose uptake mechanism that significant differences between control and chronically morphinized tissue were observed.[131]

The authors concluded from their experiments that morphine exerts its major influence at the cell membrane and that this is the site where the effects of hormonal influences are altered.

V. MORPHINE AND THE ADRENAL CORTICAL RESPONSE TO VARIOUS STIMULI

Although morphine in single doses stimulates the adrenal cortical response in some species it also antagonizes the effects of certain stimuli on the adrenal cortical response. In the guinea pig morphine in doses of 10 mg/kg reduced the increase in urinary 17-OHCS excretion following exposure to cold but did not inhibit this response to pitressin.[99] In rats morphine in single doses of 30–40 mg/kg did not block the increase in urinary corticosterone excretion following cold exposure or carbon tetrachloride administration[93,95]; however, repeated doses either partially (2×30 mg/kg) or completely (3×30 mg/kg) abolished the response. Furukawa[95] found that administration of morphine for 2 days (3×50 mg/kg) blocked the response to carbon tetrachloride. Similarly, Briggs and Munson[7] found that the administration of morphine for 4 days (20 mg/kg/day) blocked the depleting effects of histamine on adrenal ascorbic acid levels. In human subjects morphine in doses of 0.2 mg/kg antagonized the vasopressin-induced rise in plasma 17-OHCS.[103]

Several experiments suggest that tolerance develops to the morphine blocking effect on the stress-induced adrenal cortical response. Sobel[99] found that there was an enhanced excretion of corticosteroids in response to cold in guinea pigs that had received morphine in doses of 10 mg/kg/day for 12 days. In rats treated with morphine in doses of 15–25 mg/kg/day for 25 days Paroli[111] also found an enhanced excretion of corticosteroids and 17-KS in cold-stressed animals.

Morphine does not deplete adrenal ascorbic acid levels in the pentobarbital-anesthetized rat.[7] Pentobarbital anesthesia is known to block the adrenal cortical response to other stimuli such as minor surgical procedures[132] and exposure to cold but does not block the effects of more severe stimuli such as hemorrhage.[133,134]

When the stimulative effects of morphine on the adrenal cortical response is antagonized by anesthetic doses of pentobarbital, the ability of morphine to antagonize the adrenal cortical response to a variety of stimuli can be demonstrated. Thus morphine antagonized the adrenal cortical-stimulating effects of histamine,[7,135,136] epinephrine,[7,137] surgery,[7,137] exsanguination,[94] and vasopressin.[7,103] Furthermore, McDonald[103] demonstrated that morphine depressed the early morning rise in plasma 17-OHCS in patients who also had received 200 mg of pentobarbital. This effect of morphine can be antagonized by nalorphine.[103,135,136] It is important to regulate carefully the intensity of the stimulus since the blocking effect of morphine can be surmounted if the stimulus is sufficiently intense.[135,137,138] Leeman[138] demonstrated that there was no significant difference in responsiveness to ACTH between intact rats given pentobarbital alone and intact rats receiving pentobarbital and morphine and hypophysectomized rats. Similar results were demonstrated *in vitro*. Hypophysectomized rats and pentobarbital–morphine-treated rats were stimulated with graded doses of ACTH, and subsequently their excised adrenal glands were compared in their ability to synthesize steroids. The results indicated that there was no difference in the ability of the two preparations to produce corticosteroids following ACTH stimulation. These findings show that the blocking effect of morphine on the adrenal cortical response must be central to the adrenal cortex.

Morphine does not block adrenal cortical response to either hypothalamic or neurohypophyseal extracts.[136,138,139] From all these data, it can be concluded that the site of morphine's blocking action on the stress-evoked adrenal cortical response must be proximal to the CRF.

VI. SUMMARY

Single doses of morphine and other narcotic analgesics increase the adrenal cortical response in the unanesthetized rat, dog, and mouse, and this effect is probably antagonized by nalorphine. In man and the guinea pig single doses of morphine are either without effect or produce a liminal depression of 17-OHCS excretion, whereas the chronic administration of morphine and normorphine suppresses 17-OHCS excretion in man. In the rat the effects of morphine early in the course of chronic administration are similar to those of single doses. Continuing treatment, however, produces a depression of corticosterone and aldosterone excretion although the adrenal gland is hypertrophied at this time. In the rat nalorphine antagonizes the effects of morphine on corticosteroid excretion when it is chronically administered with equal doses of morphine. Both the rat and man develop some tolerance to the depressant effect of morphine on steroid excretion.

In contrast to the effects of morphine on the excretion of glucocorticoids and aldosterone, the initial and continued effect of morphine on urinary 17-KS excretion is a depressant one in both man and the rat. Some tolerance appears to develop to this effect with chronic administration of morphine.

The primary or early abstinence syndrome in man and the rat is associated with signs of adrenal cortical stimulation, whereas during protracted abstinence the urinary levels of corticosteroids are normal.

In man chronic morphine treatment does not impair the ability of the adrenal gland to respond to exogenously administered ACTH; on the contrary, the response appears to be increased. In addition human subjects appear to dispose of a hydrocortisone load normally during addiction. These experiments indicate that in man the depressed excretion of corticosteroids during morphine dependence is mainly due to a decreased release of ACTH rather than to a major alteration in biosynthetic and biotransformation pathways. In the rat, on the other hand, chronic morphine treatment either does not change the responsivity of the adrenal gland to ACTH stimulation or produces a decreased response, depending upon the dose. Furthermore, *in vitro* experiments with rat tissue show that morphine inhibits corticosteroid biosynthesis in adrenal slices and interferes with the steroid binding capacity of liver slices. In addition chronic morphine treatment antagonizes the *in vitro* stimulating effect of added hydrocortisone and epinephrine on glucose uptake in rat diaphragm tissue. The *in vivo* significance of these changes is not known.

Although morphine in single doses stimulates the adrenal cortical response in some species it also antagonizes the adrenal cortical-stimulating effect of stimuli such as cold exposure and carbon tetrachloride. Tolerance appears to develop to this blocking effect of morphine. In the pentobarbital-anesthetized animal morphine no longer stimulates the adrenal cortical response. Under these conditions morphine blocks the adrenal cortical-stimulating effect of a variety of stimuli including histamine, epinephrine, surgery, exsanguination, and vasopressin but does not inhibit the response to hypothalamic and neurohypophyseal extracts. In addition exogenous ACTH is equally as effective in stimulating the adrenal cortical response in hypophysectomized and pentobarbital–morphine-treated rats as in untreated controls.

Taken together these experiments indicate that morphine can both stimulate and depress the basal level of corticosteroid excretion. In addition they show that morphine can inhibit the corticosteroid response to many diverse stimuli that are thought to reach the central nervous system through different afferent pathways. The site (or sites) at which morphine exerts its effects are unknown but appears to be central to the adrenal gland and the anterior pituitary. Wherever this site(s) is located, the important effect of morphine on corticosteroid production is probably mediated through the stimulation and inhibition of the release of CRF(s).

VII. REFERENCES

1. E. M. MacKay and L. L. MacKay, *Proc. Soc. Exp. Biol. Med. 24*, 128 (1926).
2. C. Y. Sung, E. L. Way, and K. G. Scott, *J. Pharmacol. Exp. Ther. 107*, 12–23 (1953).
3. T. Tanabe and E. J. Cafruny, *J. Pharmacol. Exp. Ther. 22*, 148–153 (1958).
4. J. W. Sloan, J. W. Brooks, A. J. Eisenman, and W. R. Martin, *Psychopharmacologia 4*, 261–270 (1963).
5. J. W. Sloan and A. J. Eisenman, *Res. Publ. Ass. Res. Nerv. Ment. Dis. 46*, 96–105 (1968).
6. H. Selye, *Brit. J. Exp. Pathol. 17*, 234–248 (1936).
7. F. N. Briggs and P. L. Munson, *Endocrinology 57*, 205–219 (1955).
8. H. Isbell, *Bull. N.Y. Acad. Med. 31*, 886–901 (1955).
9. V. Puharich and F. R. Goetzi, *Perma. Found. Med. Bull. 5*, 19–22 (1947).
10. F. J. Friend and S. C. Harris, *J. Pharmacol. Exp. Ther. 93*, 161–167 (1948).
11. C. A. Winter and L. Flataker, *J. Pharmacol. Exp. Ther. 103*, 93–105 (1951).
12. E. Paroli and A. deArchangelis, *Arch. Ital. Sci. Farmac. 9*, 3–4 (1957).
13. C. R. Craig, *J. Pharmacol. Exp. Ther. 164*, 371–379 (1968).
14. R. K. Callow and F. G. Young, *Proc. R. Soc. London 157*, 194 (1936).
15. C. Jones, "The Adrenal Cortex," Cambridge University Press, London (1957).
16. R. I. Dorfman and F. Unger, "Metabolism of Steroid Hormones," Academic Press, New York (1965).
17. K. W. McKerns, "Steroid Hormones and Metabolism," Meredith Corporation, New York (1969).
18. J. Ashmore and D. Morgan, *in* "The Adrenal Cortex" (Albert B. Eisenstein, ed.), pp. 249–268, Little, Brown, Boston (1967).
19. L. J. Soffer and J. L. Gabrilove, *Metabolism 1*, 504–510 (1952).
20. P. J. Mulrow, *in* "The Adrenal Cortex" (Albert B. Eisenstein, ed.), pp. 293–313, Little, Brown, Boston (1967).
21. J. Tepperman, "Metabolic and Endocrine Physiology," Year Book Medical Publishers, Chicago (1968).
22. C. A. Gemzell, *Acta Endocrinol. 11*, 221–228 (1952).
23. E. D. Bransome, Jr., *Ann. Rev. Physiol. 30*, 171–212 (1968).
24. H. Selye, J. B. Collip, and D. L. Thompson, *Proc. Soc. Exp. Biol. Med. 32*, 1377–1381 (1935).
25. R. Gaunt, C. H. Tuthill, N. Antonchak, and J. H. Leathem, *Endocrinology 52*, 407–423 (1953).
26. M. Saffran, A. V. Schally, and B. G. Benfrey, *Endocrinology 57*, 439–444 (1955).
27. R. Guillemin and B. Rosenberg, *Endocrinology 57*, 599–607 (1955).
28. R. Guillemin, *Ann. Rev. Physiol. 29*, 313–348 (1967).
29. E. H. Venning, *Ann. Rev. Physiol. 27*, 107–132 (1965).
30. S. M. McCann, P. S. Dhariwal, and J. C. Porter, *Ann. Rev. Physiol. 30*, 589–640 (1968).
31. D. DeWeid, *Acta Endocrinol. 37*, 288–297 (1961).
32. K. B. Eik-Nes, *Endocrinology 69*, 411–421 (1961).
33. J. DeGroot and G. W. Harris, *J. Physiol. 111*, 335–346 (1950).

34. D. M. Hume and G. J. Wittenstein, *in* "Proceedings of the First Clinical ACTH Conference" (John R. Mote, ed.), pp. 134–147, Blakiston Division, McGraw-Hill, New York (1950).

35. A. Grollman, "Clinical Endocrinology and Its Physiological Basis," J. P. Lippincott Company, Philadelphia (1964).

36. D. I. Ingle and E. C. Kendall, *Science 86*, 245 (1937).

37. D. J. Ingle, *Amer. J. Physiol. 124*, 369–371 (1938).

38. J. B. Richards and R. L. Pruitt, *Endocrinology 60*, 99–104 (1957).

39. D. Das Gupta and C. J. Giroud, *Proc. Soc. Exp. Biol. Med. 98*, 334–339 (1958).

40. R. H. Egdahl, *J. Clin. Invest. 43*, 2178–2184 (1964).

41. F. G. Peron and R. I. Dorfman, *Endocrinology 64*, 431–437 (1959).

42. L. Martini, M. Fochi, G. Gavazzi, and A. Pecile, *Arch. Int. Pharmacodyn. 140*, 156–163 (1962).

43. S. Feldman, J. C. Todt, and R. W. Porter, *Neurology 11*, 109–115 (1961).

44. J. M. Davidson and S. Feldman, *Endocrinology 72*, 936–946 (1963).

45. I. Chowers, S. Feldman, and J. M. Davidson, *Amer. J. Physiol. 205*, 671–673 (1963).

46. S. Feldman, N. Conforti, and J. M. Davidson, *Neuroendocrinology 1*, 228–239 (1965–1966).

47. A. Corbin, G. Mangili, M. Motta, and L. Martini, *Endocrinology 76*, 811–818 (1965).

48. K. Matsuda, C. Duyck, J. W. Kendall, Jr., and M. A. Greer, *Endocrinology 74*, 981–985 (1964).

49. J. W. Kendall, Jr., K. Matsuda, C. Duyck, and M. A. Greer, *Endocrinology 74*, 279–283 (1964).

50. B. E. Eleftheriou, A. J. Zalovick, and R. Pearse, *Proc. Soc. Exp. Biol. Med. 122*, 1259–1262 (1966).

51. P. R. McHugh and G. P. Smith, *Amer. J. Physiol. 213*, 1445–1450 (1960).

52. M. A. Slusher and J. E. Hyde, *Amer. J. Physiol. 210*, 103–108 (1966).

53. S. A. D'Angelo and R. Young, *Amer. J. Physiol. 210*, 795–800 (1966).

54. P. R. McHugh and G. P. Smith, *Amer. J. Physiol. 212*, 619–622 (1967).

55. J. R. Blair-West, J. P. Coghland, D. A. Denton, J. R. Goding, M. Wintour, and R. D. Wright, *Recent Progr. Horm. Res. 19*, 311–363 (1963).

56. M. A. Newton and J. H. Laragh, *J. Clin. Endocrinol. Metab. 28*, 1006–1013 (1968).

57. J. O. Davis, J. Urquhart, and J. T. Higgins, Jr., *J. Clin. Invest. 42*, 597–609 (1963).

58. J. R. Blair-West, J. P. Coghlan, D. A. Denton, J. R. Goding, M. Wintour, and R. D. Wright, *Aust. J. Exp. Biol. Med. Sci. 44*, 455–474 (1966).

59. P. Vecesei, D. Lommer, H. G. Steinacker, and H. G. Wolff, *Eur. J. Steroids 1*, 91–93 (1966).

60. E. T. Marusic and P. J. Mulrow, Fifty-Seventh Meeting of the Americal Society of Clinical Investigation, Atlantic City, (May 1967).

61. P. J. Cannon, R. P. Ames, and J. H. Laragh, *J. Clin. Invest. 45*, 865–879 (1966).

62. B. H. Lamberg, T. Pettersson, A. Gordin, and R. Karlsson, *Acta Endocrinol. 54*, 428–438 (1967).

63. R. E. Miller, A. J. Vander, R. S. Kowalczyk, and G. W. Geelhoeld, *Amer. J. Physiol. 214*, 228–231 (1968).

64. S. L. Skinner, *Circ. Res. 15*, 64–76 (1964).

65. P. F. Binnion, J. O. Davis, T. C. Brown, and M. J. Olichney, *Amer. J. Physiol. 208*, 655–661 (1965).

66. J. R. Blair-West, G. W. Boyd, J. G. Coghlan, D. A. Denton, J. R. Goding, M. Wintour, and R. D. Wright, *Proc. Int. Union Physiol. Sci. 4*, 207–221 (1965).

67. J. O. Davis, *in* "The Adrenal Cortex" (Albert B. Eisenstein, ed.), pp. 203–247, Little, Brown, Boston (1967).

68. F. C. Bartter, I. H. Mills, E. G. Biglieri, and C. Delea, *Recent Progr. Horm. Res. 15*, 311–344 (1959).

69. J. Tepperman, F. L. Engel, and C. N. H. Long, *Endocrinology 32*, 373–402 (1943).

70. H. Selye, "Textbook of Endocrinology," Acta, Montreal (1949).
71. A. G. Hills, P. H. Forsham, and C. A. Finch, *Blood 3*, 755–768 (1948).
72. R. D. T. Carruthers, H. S. Robinson, R. A. Palmer, and H. W. McIntosh, *Lancet 2*, 111–112 (1952).
73. A. Aschkenasy, A. Bussard, P. Corvazier, and P. Grabar, *Rev. Haematol. 5*, 107–129 (1950).
74. R. W. Schayer, R. L. Smiley, and J. K. Davis, *Proc. Soc. Exp. Biol. Med. 87*, 590–592 (1954).
75. R. Hicks and G. B. West, *Nature 181*, 1342–1343 (1958).
76. B. N. Halpern, *Ciba Found. Symp. Histamine.* pp. 92–128 Churchill Ltd., London (1956).
77. C. F. Code, R. G. Mitchell, and J. C. Kennedy, *Mayo Clin. Proc. 29*, 200–204 (1954).
78. C. F. Code and R. G. Mitchell, *J. Physiol. 136*, 449–468 (1957).
79. E. Kelemen and G. Bikich, *Acta Haematol. 15*, 202–206 (1956).
80. R. K. Archer, "The Eosinophil Leucocytes," Davis, Philadelphia (1963).
81. G. Sayers, M. A. Sayers, H. L. Lewis, and C. N. H. Long, *Proc. Soc. Exp. Biol. Med. 55*, 238–239 (1944).
82. A. Brodish and C. N. H. Long, *Endocrinology, 66*, 149–159 (1960).
83. S. B. Koritz, *Biochim. Biophys. Acta 59*, 326–335 (1962).
84. V. Stollar, V. Buonassisi, and G. Sato, *Exp. Cell Res. 35*, 608–616 (1964).
85. D. Y. Cooper and O. Rosenthal, *Arch. Biochem. 96*, 331–335 (1962).
86. A. E. Kitabchi, *Fed. Proc. Fed. Amer. Soc. Exp. Biol. 26*, 484 (1967).
87. M. L. Sweat and M. J. Bryson, *Endocrinology 76*, 772–775 (1965).
88. R. E. Peterson, J. B. Wyngaarden, S. L. Guerra, B. B. Brodie, and J. J. Bunim, *J. Clin. Invest. 34*, 1779–1794 (1955).
89. R. H. Silber and C. C. Porter, *J. Biol. Chem. 210*, 923–932 (1954).
90. C. J. O. R. Morris and D. C. Williams, *Ciba Found Colloq. Endocrinol. 7*, 261 (1953).
91. I. J. Drekter, A. Heisler, G. R. Schism, S. Stern, S. Pearson, and T. H. McGavack, *J. Clin. Endocrinol. Metab. 12*, 55–65 (1952).
92. A. F. Holtorff and F. C. Koch, *J. Biol. Chem. 135*, 377–392 (1940).
93. O. Nikodijevic and R. P. Maickel, *Biochem. Pharmacol. 16*, 2137–2142 (1967).
94. J. T. Oliver and R. C. Troop, *Steroids 1*, 670–677 (1963).
95. T. Furkukawa, *Jap. J. Pharmacol. 16*, 131–137 (1966).
96. E. Paroli and P. Melchiorri, *Arch. Ital. Sci. Farmacol. 9*, 186–187 (1959).
97. M. A. Slusher and B. Browning, *Amer. J. Physiol. 200*, 1032–1034 (1961).
98. T. Suzuki, K. Yamashita, S. Zinnouchi, and T. Mitamura, *Nature 183*, 825 (1959).
99. H. Sobel, S. Schapiro, and J. Marmorston, *Amer. J. Physiol. 195*, 147–149 (1958).
100. A. J. Eisenman, H. F. Fraser, J. W. Sloan, and H. Isbell, *J. Pharmacol. Exp. Ther. 124*, 305–311 (1958).
101. A. J. Eisenman, H. F. Fraser, and J. W. Brooks, *J. Pharmacol. Exp. Ther. 132*, 226–231 (1961).
102. H. F. Fraser, A. J. Eisenman, and J. W. Brooks, *Fed. Proc. Fed. Amer. Soc. Exp. Biol. 16*, 298 (1957).
103. R. K. McDonald, F. T. Evans, V. K. Weise, and R. W. Patrick, *J. Pharmacol. Exp. Ther. 125*, 241–247 (1959).
104. R. George and E. L. Way, *J. Pharmacol. Exp. Ther. 125*, 111–115 (1959).
105. P. F. Dirk Van Peenen and E. L. Way, *J. Pharmacol. Exp. Ther. 124*, 261–267 (1957).
106. M. Holzbauer and M. Vogt, *Acta Endocrinol. 29*, 231–237 (1958).
107. R. George and E. L. Way, *Brit. J. Pharmacol. 10*, 260–264 (1955).
108. P. A. Nasmyth, *Brit. J. Pharmacol. 9*, 95–99 (1954).
109. F. N. Briggs and P. L. Munson, *J. Pharmacol. Exp. Ther. 110.* 7–8 (1954).
110. J. C. Szerb, *Can. J. Med. Sci. 31*, 8–17 (1953).
111. E. Paroli and P. Melchiorri, *Biochem. Pharmacol. 6*, 1–17 (1961).

112. A. J. Eisenman, J. W. Sloan, W. R. Martin, D. R. Jasinski, and J. W. Brooks, *J. Psychiat. Res. 7*, 19–28 (1969).
113. H. F. Fraser, A. Wikler, G. D. Van Horn, A. J. Eisenman, and H. Isbell, *J. Pharmacol. Exp. Ther. 122*, 359–369 (1958).
114. L. M. Gunne, *Nature (London) 184*, 1950–1951 (1959).
115. L. M. Gunne, *Psychopharmacologia 2*, 214–220 (1961).
116. W. R. Martin, A. Wikler, C. G. Eades, and E. T. Pescor, *Psychopharmacologia 4*, 247–260 (1963).
117. L. M. Gunne, *Acta Physiol. Scand. 58* (Suppl. 204), 1–91 (1963).
118. W. R. Martin and J. W. Sloan, *Pharmakopsychiat. Neuropharmakol. 1*, 260–270 (1968).
119. C. K. Himmelsbach, *Arch. Int. Med. 69*, 766–772 (1942).
120. H. Isbell and W. M. White, *Amer. J. Med. 14*, 558–563 (1953).
121. W. R. Martin, C. W. Gorodetzky, and T. K. McClane, *Clin. Pharmacol. Ther. 7*, 455–465 (1966).
122. H. Isbell and H. F. Fraser, *J. Pharmacol. Exp. Ther. 99*, Part II, 355–397 (1950).
123. H. F. Fraser and H. Isbell, *Ann. Int. Med. 38*, 234–238 (1953).
124. A. Wikler, H. F. Fraser, and H. Isbell, *J. Pharmacol. Exp. Ther. 109*, 8–20 (1953).
125. S. Kaymakcalan and L. A. Woods, *J. Pharmacol. Exp. Ther. 117*, 112–116 (1956).
126. W. R. Martin and C. G. Eades, *J. Pharmacol. Exp. Ther. 146*, 385–394 (1964).
127. A. Wikler and R. L. Carter, *J. Pharmacol. Exp. Ther. 109*, 92–101 (1953).
128. S. Irwin and M. H. Seevers, *J. Pharmacol. Exp. Ther. 106*, 397 (1952).
129. W. R. Martin, *Pharmacol. Rev. 19*, 463–521 (1967).
130. E. Paroli and P. Melchiorri, *Biochem. Pharmacol. 6*, 18–20 (1961b).
131. M. L. Ng and E. O'F. Walsh, *Biochem. Pharmacol. 12*, 1003–1009 (1965).
132. C. Rerup and P. Hedner, *Acta Endocrinol. 39*, 518–526 (1962).
133. E. Ronzoni, *Amer. J. Physiol. 160*, 499–505 (1950).
134. R. Gaunt, J. J. Chart, and A. A. Renzi, *Rev. Physiol. Biochem. Exp. Pharmacol. 56*, 7–172 (1965).
135. B. H. Burdette, S. Leeman, and P. L. Munson, *J. Pharmacol. Exp. Ther. 132*, 323–328 (1961).
136. S. Epstein, H. Burdette, and P. L. Munson, *Fed. Proc. Fed. Amer. Soc. Exp. Biol. 16*, 294 (1957).
137. E. A. Ohler and R. W. Sevy, *Endocrinology 59*, 347–355 (1956).
138. S. E. Leeman, "The problems of neurohormonal stimulation of the secretion of adrenocorticotropic Hormone," unpublished doctoral dissertation, Radcliffe College, Cambridge, Massachusetts (1958).
139. R. Guillemin, W. E. Dear, B. Nichols, Jr., and H. S. Lipscomb, *Proc. Soc. Exp. Biol. Med. 101*, 107–111 (1959).

Chapter 13

HYPOTHALAMUS: ANTERIOR PITUITARY GLAND*

Robert George

Department of Pharmacology and Brain Research Institute
University of California
The Center for the Health Sciences
Los Angeles, California

I. INTRODUCTION

Prior to 1940 very few definitive studies were done concerning the effects of narcotic analgesics on endocrine function. In general most of the early reports dealt with either the toxicity of morphine following extirpation or administration of endocrine tissue or morphological changes in endocrine organs after administration of the drug. These data are thoroughly reviewed by Krueger *et al.*[1]

At the time of these early reports very little was known about the part played by the central nervous system in the regulation of pituitary function.

II. NEURAL REGULATION OF THE ADENOHYPOPHYSIS

Within the past two decades the field of neuroendocrinology has expanded very rapidly, and numerous books[2-10] and reviews[11-17] have been published covering all aspects of this relatively new discipline. Thus it is the aim here to cite only briefly some evidence for the existence of neural control of the anterior pituitary gland. It is generally agreed that the hypothalamus acts as a 'final common path' or 'nodal point' for the integration of all neural activity effecting anterior pituitary function. In the discussion that follows emphasis will be placed on the hypothalamus, rather than the entire central nervous system, as the pituitary regulator.

As long ago as 1936 Marshall[18] had drawn attention to the role of the

* The author expresses his thanks to Dr. Peter Lomax for his kind comments and criticisms of the manuscript. Part of the work reported in this chapter was supported by research grants from the American Medical Association Education and Research Foundation and USPHS, NB-04499 and MH-17691.

nervous system in the control of gonadotropic hormone secretion, and it was not until several years later that studies were undertaken to challenge this concept of pituitary control. Numerous investigations using classical neurophysiological techniques revealed that electrical stimulation of specific hypothalamic areas could selectively increase or decrease secretion of one or more pituitary trophic hormones and that lesions in these hypothalamic areas produced effects opposite to those of electrical stimulation. Concurrently with these studies attempts were made to determine if this influence of the hypothalamus was mediated by secretomotor fibers ending in the anterior lobe of the pituitary, although the available evidence pointed to an absence of such innervation. These findings added a great deal of significance to the functional role of the hypophyseal portal system first described in 1930 by Popa and Fielding.[19] This system of vessels arises as capillaries in the median eminence of the tuber cinereum, which then form vascular trunks on the pituitary stalk and finally break up into sinusoids within the anterior lobe. The direction of the blood flow is from the hypothalamus to the pituitary. All of these observations led Green and Harris[20] to propose the existence of a neurohumoral mechanism for the control of anterior pituitary function. It was found that pituitary stalk section or pituitary transplantation procedures that eliminate the contiguity of the portal vessels tend to reduce anterior pituitary secretions. Revascularization of the gland by the portal system after stalk sections[21] returned pituitary function to normal. Similarly, transplantation of the pituitary gland to a distant site, such as the temporal lobe[22] or kidney,[23] reduced its activity, except in the case of prolactin secretion. Retransplantation of the gland to the sella turcica, and subsequent revascularization by the portal vessels, restored normal pituitary function.

The above findings soon resulted in intensive research efforts by numerous investigators to identify specific hypothalamic transmitter substances that might regulate the secretion of anterior pituitary hormones. The presence of transmitters in the hypothalamus was shown for the first time by Saffran and Schally,[24] who demonstrated the existence of a corticotropin-releasing factor (CRF). Since then hypothalamic extracts have been shown to contain several substances that alter anterior pituitary secretions. It is presumed that these substances that have been extracted from the median eminence and pituitary stalk are liberated either from nerve endings of hypothalamic nerve tracts or neurosecretory cells into the capillaries of the portal vessels in the median eminence. The releasing factors are denoted by the pituitary hormones they regulate. Five of these factors are concerned with stimulation of the release of anterior pituitary hormones and one with inhibition of release of a pituitary hormone (prolactin). Thus the releasing factor for regulating secretion of corticotropin (ACTH) is designated as CRF; for thyrotropin (TSH), as TRF; for somatotropin or growth hormone, as SRF or GH-RF; for follicle-stimulating hormone (FSH), as FSH-RF; and for luteinizing hormone (LH), as LH-RF. The remaining factor, mentioned above, appears to be involved with an inhibitory process concerned with prolactin secretion and is denoted as prolactin-inhibiting factor (PIF).

Although the presence of a releasing factor was described 15 years ago, only recently has the chemical structure of one been identified. Guillemin and co-workers[25] have identified TRF as a tripeptide with the sequence pyroglutamyl (2-pyrrolidone-5-carboxylic acid)-histidyl-proline-NH_2, abbreviated PCA-His-Pro-NH_2. The synthesis of this compound as well as its methylester derivative, which has an $-OMe$ group replacing $-NH_2$ on the C terminal, proline has been achieved. Both compounds are active in releasing TSH when administered orally to mice as well as being active *in vitro*. The remaining releasing factors found in the hypothalamus thus far appear to be polypeptides with molecular weights ranging from 1000 to 2500 and appear to lose some of their activity when exposed to proteolytic enzymes.[26-28]

III. EFFECTS OF NARCOTIC ANALGESICS ON ANTERIOR PITUITARY ACTIVITY

Although a number of reports in the literature indicate that narcotic analgesics may alter pituitary activity, no studies have been done in which plasma levels of anterior pituitary hormones have been measured. The possibility of future work with radioimmunoassay, a sensitive method now used for measuring protein and polypeptide hormones, undoubtedly will yield more definitive data regarding the effects of and mechanisms by which narcotic analgesics influence pituitary function. In the sections to follow all of the results and conclusions are based upon indirect indices of pituitary activity; cytological changes in the pituitary gland; morphological changes of target glands; biochemical changes in target glands; and secretory activity of target glands, as reflected in either plasma or urinary levels of the hormones or their metabolites.

A. Adrenocorticotropin (ACTH)

Morphine is known to have a dual action on the central nervous system, that is, excitation and depression, and its effect on ACTH secretion is no exception. Acute administration of morphine to rats has been reported by several workers[29-34] to produce a marked reduction in adrenal ascorbic acid, a rather specific index for measuring ACTH secretion, that is, the lower the adrenal ascorbic acid level, the higher the rate of ACTH secretion. This effect is mediated through the pituitary since hypophysectomy abolishes the response.[29]

Since a variety of stresses and pharmacologic compounds can evoke adrenal ascorbic acid depletion, George and Way[29] made an effort to determine the relative specificity of this response to morphine by comparing its action with *d*- and *l*-methadone in terms of their potency ratios, LD_{50}/ED_{50}. They found that one-fifth of the LD_{50} dose of both morphine and *l*-methadone produced a marked decrease in adrenal ascorbic acid levels, whereas d-methadone even at one-half of the LD_{50} dose was not as effective. The specificity of the adrenal response to these analgesics was further examined by pretreating the animals with nalorphine, using a dose that did not affect adrenal ascorbic acid levels. Nalorphine significantly blocked the adrenal response to morphine and l-methadone. These data suggested that different receptors or sites are concerned with mediating the pituitary–adrenal effect of different pharmacologic agents.

A single injection of morphine (doses ranging from 3.75 to 30 mg/kg) produces a marked elevation in plasma corticosterone in the rat[33–36] and an increase in plasma 17-hydroxycorticosterone in the dog.[37] Additionally, morphine has been reported to decrease the content of pituitary ACTH in rats. Nikodijevic and Maickel[33] found that a single dose of morphine (30 mg/kg), although enough to deplete adrenal ascorbic acid and to increase plasma corticosterone significantly, did not affect pituitary ACTH content. However, three doses of morphine (30 mg/kg each) within 24 hr caused a reduction of pituitary ACTH by 64% and blocked the pituitary response to cold stress. Similar findings have been noted in a preliminary study by Westerman et al.[38] in which they say, "Morphine in rats produces such an outpouring of ACTH that the pituitary is almost depleted in a few hours and the animals no longer respond to stressful stimuli." They also observed the similarity of this morphine effect to that of chlorpromazine and reserpine.

More data confirming an initial stimulatory effect of morphine come from the findings of Paroli and Melchiorri,[39] who were investigating the effects of chronic administration of morphine on the urinary levels of hydroxysteroids and 17-ketosteroids in rats. For a period of 10 days hydroxysteroid excretion increased, and this was associated with a persistence of the morphine analgesic effect. In the periods following (10–40 and 40–140 days) the urinary level of hydroxysteroids decreased as well as the analgesia.

Finally, it has been shown that chronic administration of morphine to rats results in adrenal cortical hypertrophy.[40–42] The hypertrophy is dependent upon an intact pituitary[42] and suggests that during the course of morphine administration there is a continuous increase of ACTH secretion. This effect of morphine is paradoxical since, on the one hand, the drug produces adrenal cortical hypertrophy while, on the other hand, it inhibits the pituitary–adrenal response to numerous stresses that normally increase ACTH output (see the discussion below). Intermittent nalorphine administration does not enhance the hypertrophy, indicating that this phenomenon is not associated with drug withdrawal.[42]

Based upon these data one can only conclude that morphine, when administered acutely, activates the pituitary–adrenal axis. Since morphine is a centrally acting agent, it generally has been assumed that this effect is mediated via the hypothalamus. However, several other possibilities exist.

It is well documented that morphine is capable of stimulating release of epinephrine from the adrenal medulla and of antidiuretic hormone (ADH) from the posterior lobe of the pituitary. Both of these hormones are released during most of the conditions of stress that enhance secretion of ACTH. In fact, antidiuretic hormone[43] and epinephrine[44] at one time were considered to be the transmitter substances responsible for the discharge of ACTH. It also is possible that morphine has a direct effect on the anterior pituitary.

These possibilities were investigated by George and Way by challenging with morphine rats that had been adrenal medullectomized[29] and rats with electrolytic lesions placed bilaterally in the hypothalamus.[45] Removal of the adrenal medulla did not impair the adrenal cortical response to morphine. In fact significant ascorbic acid depletion occurred with much smaller doses of morphine than those required in normal animals. Lesions in which at least one-half of the anterior median eminence was destroyed abolished the adrenal cortical response to morphine; diabetes insipidus was observed in 9 of these 11 animals. In addition to these findings it was noted that a dose of morphine as small as 0.1 mg/kg was effective in producing an antidiuretic effect in normal rats, and this dose was found to be as little as 1/200 of the usual effective dose (20–30 mg/kg) required to deplete adrenal ascorbic in normal animals. Although the effective lesions involved the median eminence, the hypophyseal portal vessels were found to be intact on microscopic examination and adrenal and testicular weights were normal, indicating that the pituitary function was normal.

These results clearly showed that morphine activates the pituitary to secrete ACTH through a central mechanism that is independent of prior release of epinephrine from the adrenal medulla or ADH from the posterior pituitary and suggests that the hypothalamus is the site of this action of morphine. However, morphine exerts its effects by acting on many structures in the central nervous system (CNS) and, since the hypothalamus contains numerous connections from many rostral and caudal regions of the CNS, it is conceivable that morphine produces corticotropin release by acting on some extrahypothalamic area.

In order to elucidate this possibility Lotti et al.[34] injected microquantities of morphine (5–50 μg) into various regions of the hypothalamus of conscious rats. They localized an area in the hypothalamus at which injection of morphine produced both a marked reduction of adrenal ascorbic acid and an elevation of plasma corticosterone. The area involved the middle region of the hypothalamus including, in whole or in part, the anterior, paraventricular, ventromedial, and dorsomedial hypothalamic nuclei. This region of the hypothalamus closely approximates the area of the hypothalamus where lesions abolish the adrenal ascorbic acid-depleting effect of morphine. The intrahypothalamic doses of morphine used in this study were between 1/100 and 1/1000 of the systemic dose necessary to stimulate ACTH secretion, thereby making it unlikely that entry of the drug into the systemic circulation was responsible for this effect. It is also improbable that this action of morphine was the result of diffusion of the drug into the ventricular system since, in an earlier finding,[46] injection of morphine (50 μg) into the anterior hypothalamus produced hypothermia but the same dose failed to lower body temperature when injected intraventricularly.

In conclusion, it can be stated that acute administration of morphine produces an increase in the rate of ACTH secretion. This action of morphine is mediated via the midventral hypothalamus and involves the release of CRF in the median eminence of the tuber cinereum. The mechanism responsible for the release of CRF is as yet unknown, although there is some evidence that it may be linked to the turnover of one or more of the biogenic amines found in the hypothalamus (see the discussion below).

As previously mentioned, morphine has a dual action on the pituitary–adrenal system. In general the effect of acute administration of morphine on ACTH secretion is stimulatory, whereas chronic administration of the narcotic results in an inhibition of both the basal level of secretion and the increased output in response to stress. There is, however, some evidence that under specific conditions the acute administration of morphine may inhibit ACTH secretion. Munson and Briggs[47] observed that preadministration of an anesthetic dose of pentobarbital abolished the stimulating action of morphine on the pituitary–adrenal axis in the rat. In addition, they noted that the combination of pentobarbital followed soon by a single injection of morphine prevented adrenal ascorbic acid depletion to a variety of stresses (histamine, vasopressin, and laparotomy). These results have been confirmed by others measuring either adrenal ascorbic acid[48] or plasma corticosterone[35,49] levels in rats. This blocking action of morphine can be overcome by increasing the intensity of the stress.[47]

Meperidine, in conjunction with nitrous oxide, also has been reported[50] to prevent a rise in the plasma level of 17-hydroxycorticosteroids (17-OHCS) during anesthesia and surgery: In four of six surgical patients previously anesthetized with meperidine–nitrous oxide the plasma 17-OHCS decreased gradually when meperidine was continued in intermittent doses of 50–100 mg/hr. During surgery three of the six patients showed a slight decrease in plasma 17-OHCS. It is interesting to note that in one patient an episode of severe hypoxia and carbon dioxide retention occurred as a result of airway obstruction, probably a condition of severe stress, and this produced a marked elevation in the 17-OHCS level. Although these results, in general, support the findings of the combined effect of pentobarbital and morphine, it should be pointed out that the sample was small and not statistically analyzed. Recently Oyama *et al.*[51] in studying a group of 19 patients were unable to show a blocking action of meperidine on adrenocortical stimulation induced by surgical stress.

Morphine is also capable of blocking the diurnal rise of corticosteroid secretion in the rat and in man. Slusher and Browning[36] found that injection of a single dose of morphine via chronic indwelling venous catheters into conscious rats in the morning produced a fivefold increase in plasma corticosterone within an hour but blocked the diurnal rise that normally occurs at 5 PM. Exposure to cold stress after 5 PM markedly increased adrenal secretion. Experiments in man[52] have shown that morphine injected at 3 AM after premedication with sodium pentobarbital prevented the normally occurring early morning diurnal rise of plasma hydrocortisone levels. In the same group of subjects injection of morphine alone reduced the ACTH release in response to vasopressin.

The effect of chronic administration of morphine on ACTH secretion has been studied in the rat, guinea pig, and man. Injection of morphine for 5 days in the rat inhibited ascorbic acid depletion in response to a variety of stresses.[47] The inhibition was not due to refractive adrenals since exogenous ACTH produced a normal response in the morphinized rats. This blocking action of morphine can be antagonized by nalorphine with approximately one-eighth of the morphine dose used.[53] The injection of morphine into rats for periods of 10–40 days and 40–100 days decreased urinary levels of hydroxysteroids, and treatment for 25 days blocked the increased urinary steroid response to cold stress.[54] In the latter study the experimenters also noted a partial decrease in adrenal cortical response to exogenous ACTH. A subsequent *in vitro* study[55] showed that adrenal slices from rats that were injected with morphine for 15–30 days synthesized corticoids at a reduced rate. The slices were less sensitive to ACTH added to the incubate; however, the reliability of *in vitro* studies of this nature are open to criticism.

A reduction in the basal level of urinary 17-hydroxycorticoids has been demonstrated in guinea pigs injected with morphine for a 12-day period. The normal increase in urinary corticoid excretion in response to cold stress and Pitressin also was markedly reduced by morphine.[56]

Studies in the human have shown that during two- to four-month cycles of morphine addiction there is a reduction in both plasma and urinary 17-OHCS levels. The reduction in these two parameters of adrenal function is due to a decrease in production of 17-OHCS rather than to an increase in their destruction. Furthermore, injection of ACTH caused a pronounced increase in the concentration of plasma 17-OHCS during control and addiction periods[57] (see Chapter 12).

It appears to be rather clear from these data that when morphine is administered chronically there is a reduction in adrenal cortical activity, and this effect is not due to a direct suppressive action of morphine on the adrenal cortex. On the basis of data reviewed regarding the excitatory mechanism and site of action of acutely administered morphine, it seems probable that morphine also exerts its inhibitory effect on ACTH secretion via the hypothalamic–pituitary system. The mechanism of morphine inhibition might be due to one of the following possibilities: There may be inhibition of synthesis and/or release of CRF or there may be an increase in the rate of CRF turnover to such a degree that pituitary ACTH concentration becomes subnormal throughout the period of morphine administration. Of these two possibilities the latter seems more probable for the following reasons: (1) repeated administration of morphine reduces pituitary ACTH content to such an extent that the gland becomes refractory to most stresses; (2) a very severe stress is still capable of releasing ACTH from a morphinized animal that presumably has reduced pituitary ACTH stores; (3) the adrenal cortex responds normally to exogenous ACTH, indicating continued adrenal stimulation by endogenous ACTH; and (4) chronic administration of morphine produces adrenal cortical hypertrophy, an effect that can only be due to excessive ACTH secretion.

B. Thyrotropin (TSH)

Narcotic drugs appear to have a dual action on thyrotropin secretion, which seems to be similar to their effect on ACTH secretion. However, the evidence for an excitatory effect on TSH secretion has been reported in only one publication, that of Redding et al.[58] These authors studied the effects of single injections and chronic administration (5 to 7 days) of morphine, codeine, levorphan, dextrorphan, meperidine, and dihydromorphinone on the uptake of ^{131}I in mice. Uptake and release of thyroidal ^{131}I were used as indexes of TSH secretion by the anterior pituitary: An increase in uptake and release represented increased TSH release, whereas depression of both indicated an inhibition of TSH secretion. A single subcutaneous injection of a small dose (100 μg/mouse) of each of the above drugs resulted in an increased ^{131}I uptake only with dihydromorphinone. Injection of the drugs subcutaneously with a higher dose (500 μg/mouse) produced a significant increase in the ^{131}I uptake only with levorphan and dihydromorphinone. Except for codeine, single intravenous injections of small doses of the drugs produced a significant increase of ^{131}I release. Administration of the larger dose intravenously resulted in a very rapid release of ^{131}I by each of the narcotics. Contrary results were observed when these analgesics were injected for 5 days, that is, there was a significant reduction in ^{131}I uptake. Furthermore, studies in hypophysectomized mice and in normal mice injected with radiothyroxine indicate that the stimulating effect of these narcotic agents is mediated through the pituitary and that the inhibitory effect from chronic administration is not due to an alteration in turnover rate of thyroxine. The results of these studies have led to the establishment of an *in vivo* assay method for TRF that consists of administration of a combination of codeine (1 mg) and *l*-thyroxine (1 μg) to ^{131}I-labeled mice.[59] The inhibitory effect of chronic administration of codeine has been confirmed by Schreiber *et al.*[60] who found that daily feeding of codeine (approximately 5 mg/kg) to rats for 14 days significantly reduced the uptake of [131]I. However, codeine was found to be ineffective in blocking methylthiouracil (MTU) induced thyroid and pituitary hypertrophy; the usual reduction in TSH content of the pituitary produced by MTU was not affected by simultaneous codeine administration. The investigators concluded that codeine is capable of lowering resting (basal) thyroid function but does not influence hyperfunction (hypersecretion of TSH) in the presence of thyroid hormone deficiency.

Inhibition of TSH secretion in rats after chronic administration of morphine has been reported from several laboratories. Injection of morphine twice daily (30 and 60 mg/kg) for 3 weeks was observed to decrease pituitary TSH content and to prevent the appearance of thyroidectomy cells in the pituitary.[61] Morphine, when administered for a 5-day period, has been shown to significantly inhibit TSH secretion as judged by the reduction in pituitary and thyroid weights, in the rate of ^{131}I uptake by the thyroid, in serum $PB^{131}I$ level and in ^{131}I release from the thyroid.[62] The latter findings were confirmed and extended in a later study by George and Lomax,[63] who found that release of thyroid hormone prelabeled with ^{131}I was markedly inhibited, in 90% of the animals tested, within 24 hr following the first morphine injection. The inhibition persisted throughout the 66-hr injection period, during which time the drug was injected twice daily (5 mg/kg), and continued for 24–30 hr following the last morphine injection. Additional studies were done to consider the possibility that depression of thyroid activity is due to a direct action of morphine on the thyroid gland. The increase in thyroid release rates in response to TSH was found to be the same in morphinized animals as in the controls. Since several reports had indicated that some stresses or administration of adrenal corticosteroids or ACTH, may inhibit thyroid activity, and since morphine stimulates ACTH secretion, the effect of morphine on the thyroid could be mediated through this mechanism. This proved not to be the case, because the inhibitory effect of morphine was found to be enhanced by adrenalectomy.

Although these data have clearly shown that morphine is capable of inhibiting thyroid hormonal secretion by an action on the pituitary or hypothalamic–pituitary axis, the mechanism is not clear. It is well known that anterior hypothalamic lesions prevent propylthiouracil-induced goiter[64] while electrical stimulation of this region promotes the release of thyroid hormones.[65] In view of this information, Lomax and George[66] investigated the effect of morphine on the release rate of ^{131}I-labeled thyroid hormones in rats having hypothalamic lesions placed in various regions of the hypothalamus. They found that bilateral electrolytic lesions placed in the medial mammillary nuclei abolished the thyroid inhibitory effect of morphine whereas animals with lesions in the region of the anterior hypothalamic/ventromedial nuclei showed marked inhibition of thyroid activity with release rates identical with those of controls receiving morphine. These investigators suggested that morphine exerts its thyroid-inhibiting effect via the caudal hypothalamus, possibly by activation of inhibitory neurons in the region of the medial mammillary nuclei rather than by a depression of the TSH facilitatory region of the anterior hypothalamus. Arguments in favor of such a hypothesis are based upon the findings that electrical stimulation of the posterior hypothalamus inhibits thyroid activity,[67] and the dose of morphine (5 mg/kg) used in these studies has been reported, from the same laboratory, to produce a rise in body temperature in the rat in contrast to higher doses that cause marked hypothermia. These temperature changes were shown to be due to a direct effect of morphine on the thermoregulatory centers in the anterior hypothalamus, and evidence for the stimulating effect of small morphine doses on these centers has been discussed in other reports by these experimenters.[68–69]

In a recent study Lomax et al.[70] designed experiments to determine the site(s) involved in the inhibition of TSH release by morphine by the focal injection of microquantities of the drug intrahypothalamically in rats. By using this experimental procedure they felt that such localization would also aid in resolving the question regarding the existence of two TSH regulating hypothalamic areas, one stimulatory and the other inhibitory. They found that bilateral injections of morphine (5 μg) into both rostral (supraoptic region) and caudal (posterior hypothalamic and supramammillary nuclei) sites depressed thyroid release rates of ^{131}I-labeled hormone. However, it was noted that in the case of injections into the supraoptic region there was a reversion to the normal thyroid release rate during morphine administration. In contrast to this the animals that were injected with morphine into the caudal hypothalamus showed inhibited release rates throughout the period of morphine administration; thus these data imply that tolerance to morphine developed only in the rostral site. Since it is generally accepted that tolerance develops to the depressive but not stimulative action of the narcotics[71] the experimenters propose that morphine exerts an inhibitory effect on TSH secretion through a dual action on the hypothalamus. There appear to be two target sites in the hypothalamus for morphine inhibition— the caudal site, an inhibitory one which is activated, and the rostral site, which is stimulatory and depressed. The results from the experiments regarding the mechanism of action of morphine on temperature regulation add supporting evidence for this hypothesis. Lotti et al.[46] found that microinjection of morphine into the rostral hypothalamus produced hypothermia, catatonia, respiratory depression, and an increase in pain threshold, effects that are seen following the systemic administration of a large dose of morphine. Conversely, microinjection of morphine with the same dose into the caudal hypothalamus resulted in hyperthermia and marked excitation, associated with increased motor activity and aggressive behavior.

In summary, it would appear that narcotic drugs exert a dual effect on the secretion of TSH, which is similar to their effect on ACTH secretion, via a hypothalamic mechanism involving the synthesis and release of TRF. Although acute administration of morphine may enhance TSH secretion, most of the evidence points to an inhibitory action of morphine. The inhibitory effect, in contrast to the one seen for ACTH, appears to be due primarily to the excitation of inhibitory neurons that lie in the caudal hypothalamus. It is conceivable that such neurons may activate the release of an inhibitory transmitter, thyrotropin inhibitory factor (TIF). Although such an inhibitory substance has not yet been demonstrated in the hypothalamus, three others with inhibitory activity have been found in hypothalamic extracts and are concerned with the regulation of secretion of prolactin (prolactin inhibiting factor, PIF), growth hormone (growth hormone-inhibiting factor, GIF), and melanocyte-stimulating hormone (melanocyte stimulating hormone-inhibiting factor, MIF).

C. Gonadotropins

One of the earliest suggestions that morphine can alter pituitary gonadotropic hormone secretion was in a survey in which it was found that human female addicts exhibited decreased libido, amenorrhea, and sterility during a period of morphine addiction. In many cases, even though the menstrual cycles returned to normal after withdrawal from the drug, sterility persisted.[72] Data from experimental studies concerning the effect of morphine on gonadotropin secretion, prior to 1955, have been conflicting and difficult to interpret. The discovery by Everett and Sawyer[73] that "spontaneous" ovulation in the rat involves a neurogenic timing factor with a 24-hr rhythmicity provided a new approach and possible solution to this problem. They found that the ovulatory discharge of LH depends upon a neurogenic stimulation of the hypophysis and that the critical period at which this stimulus occurs is between 2 and 4 PM on the day of proestrus. In subsequent studies Sawyer and his colleagues[74] found that acute administration of morphine (50 mg/kg) between 12 and 2 PM on the day of proestrus prevented ovulation from occurring that night in 81% of the injected animals. If treatment was delayed until 4 PM, 100% of the animals ovulated. Chronic administration, daily injections between 12 and 2 PM, for periods varying from 13 to 52 days inhibited ovulation only up to 25 days, and this was probably due to tolerance development. However, all animals showed irregular estrus cycles throughout the 52-day period. Although the animals lost weight during the addiction cycle their pair-fed controls, who also lost weight, continued to cycle and ovulate normally, thereby eliminating ination as a factor responsible for this effect of morphine. In another study Sawyer et al.[75] observed that in addition to morphine, atropine and the barbiturates (in doses that block release of pituitary gonadotropin) also depress the reticular activating system; the threshold of arousal produced by direct electrical stimulation of the midbrain reticular formation was markedly elevated by the three drugs. They believed that the blockade of ovulation was due to depression of the reticular activating system. However, in later investigations it was found that morphine, atropine, and pentobarbital could block ovulation in response to electrical stimulation of the posterior tuberal region in rabbits. The only animals to ovulate were those that had electrodes impinging on the median eminence. Sawyer concludes,[76] "These results place the limiting site of blockade by these agents just proximal to the median eminence."

Confirmation of these data comes from the report that morphine administration to male rats for 3 weeks produced a reduction in pituitary gonadotropin that was associated with atrophic changes in the seminal vesicles and prostates.[61] However, contrary results have been reported from two other laboratories. Rennels[77] determined the effect of morphine administration for a period of 2 weeks on pituitary cytology and gonadotropin content. In both male and female rats there was a threefold increase in the pituitary content of FSH but no change in the level of pituitary LH. Morphine produced persistent diestrous in one-half of the treated females. The cytological changes in the pituitary of males with increased FSH content consisted of a marked increase in the number, size, and staining intensity of the centrally placed PAS-purple gonadotrophic cells that are considered to be elaborators of FSH. The effects of morphine and nalorphine on pituitary LH secretion have been determined in a study by Giri,[78] who assayed LH secretion on the basis of comb growth, body weight, and testicular weight in cockrels. Injections of morphine (60 mg/kg) and nalorphine (10 mg/kg) for 26 days produced a significant increase in testicular weight, whereas only nalorphine increased comb growth and body weight. Although these data suggest that nalorphine facilitates LH secretion, it is possible that the drug is influencing testosterone secretion directly, rather than via the pituitary gland.

In general it appears that the primary action of morphine on pituitary gonadotropin secretion is one of inhibition, with the greatest effect on LH secretion. The evidence for this effect is favored by the relatively specific block of spontaneous ovulation by morphine in the rat and of artificially induced ovulation in the rabbit, a species that normally ovulates only after coitus. The inhibitory mechanism by which morphine blocks the release of LH is via the hypothalamus and appears to be due to a reduction in LHRF synthesis or release.

D. Growth Hormone (Somatotropin)

Detailed studies regarding the effect of morphine on pituitary growth hormone secretion are lacking. Tanabe and Cafruny[42] reported that chronic administration of increasing doses of morphine to rats for 21 days reduced their body weights to approximately 30% below controls. Body weights of hypophysectomized rats treated with morphine for 3 weeks were only 8% below the weights of their hypophysectomized controls. The authors suggest that morphine depressed growth hormone secretion. It is unfortunate that the controls were not pair-fed with the morphine treated group in order to eliminate this crucial variable. In their study concerning the effect of chronic morphine administration on ovulation, Barraclough and Sawyer[74] also found that rats injected with morphine for 25 days rapidly lost body weight but not to any greater extent than their pair-fed controls. In a recent study Muller and Pecile,[79] using insulin-induced hypoglycemia in the rat as a test for the ability of the pituitary to secrete growth hormone, found that acute administration of morphine slightly reduced the pituitary content of growth hormone. However, morphine was unable to block growth hormone release in response to insulin. Presumably this insulin-induced growth hormone response is mediated via the central nervous system since hypothalamic lesions abolish it. The growth hormone response to insulin was blocked by chlorpromazine and reserpine but not by pentobarbital.

E. Prolactin

The secretion of prolactin generally varies inversely with the secretion of FSH and LH. In addition, prolactin secretion increases in many stress conditions that increase ACTH secretion. Meites[80] has reported that morphine is capable of stimulating lactation, an index of prolactin secretion, and points out that its other effects on the pituitary are stimulation of ACTH and inhibition of LH secretion.

IV. SUMMARY AND CONCLUSIONS

At the present time it is evident that the narcotic analgesics, particularly morphine, influence pituitary function by their action on the central nervous system. The hypothalamus acts as a "final common path" for the integration of all neural activity that affects anterior pituitary activity, and it is the principal central structure through which analgesics exert their effects on the pituitary. Evidence for this has been presented in the case of ACTH, TSH, and gonadotrophic hormones. In general, acute administration of these drugs may stimulate the pituitary, whereas in nearly all cases chronic administration of these drugs inhibits pituitary function.

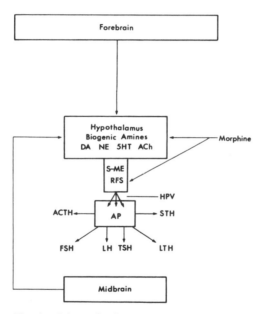

Fig. 1. Schematic diagram of possible site(s) and mechanism(s) of action of morphine on the hypothalamic–pituitary system. Abbreviations: ACh, acetylcholine; ACTH, adrenocorticotropin; AP, anterior pituitary; DA, dopamine; FSH, follicle-stimulating hormone; HPV, hypophyseal portal vessels; 5-HT, 5-hydroxytryptamine; LH, luteinizing hormone; LTH, prolactin; NE, norepinephrine; RFS, releasing factors; S-ME, pituitary stalk–median eminence; STH, somatotropin or growth hormone; and TSH, thyrotropin.

Although the mechanism(s) by which narcotic analgesics influence hypothalamic–pituitary activity have not been clearly defined, two possibilities arise (Fig. 1). The first concerns the presence of neurosecretory cells in the hypothalamus that elaborate humoral substances (releasing factors, RFS, and inhibiting factors, IFS) responsible for the control of the anterior pituitary. It is possible that the analgesics exert their effect on the pituitary through direct action on the neurosecretory cells by interfering with the synthesis and/or release of their products.

The second possibility focuses on the presence and role of cholinergic and monoaminergic pathways in the CNS[81,82] and the transmitters that are released at their terminals. The transmitters (norepinephrine, NE; dopamine, DA; 5-hydroxytryptamine, 5-HT; and acetylcholine, ACh) have been found in varying concentrations in the hypothalamus. If one assumes that the neurosecretory cells, which contain the RFS and IFS mentioned above, are in a postsynaptic position, then any alteration in the synthesis, release, or degradation of transmitter substances in the presynaptic neurons innervating these cells could influence the synthesis and/or release of the contents of the cells. Changes in turnover rates and concentrations of the transmitter substances in brain have been associated with alterations in anterior pituitary function, with particular emphasis on ACTH and LH secretion.

There is evidence of an increased turnover rate in rat brain of both NE and 5-HT by a variety of stresses.[83–86] Administration of DA systemically in rats[87] and 5-HT intrahypothalamically in guinea pigs[88] increase ACTH secretion in rats. Intrahypothalamic injection of carbachol increases 17-OHCS secretion in cats.[89,90] Systemic administration of atropine blocks circadian periodicity in cats,[91] and intrahypothalamic implantation of atropine inhibits ACTH release in rats to different stresses.[92]

The biogenic amines have been implicated also in the mechanism of secretion of LH. Elevated 5-HT levels in brain have been associated with blockade of ovulation while increased levels of catecholamines are related to the release of LH.[93–94] *In vitro* studies have shown that dopamine increases LH secretion of pituitaries incubated with hypothalamic tissue.[95–96]

Since morphine has been shown to raise brain acetylcholine[97] and 5-HT[98] levels and to decrease the concentration of norepinephrine in the brain[97] (see Chapters 10 and 11), it seems very likely that narcotic analgesics alter anterior pituitary activity via the latter-postulated mechanism.

V. REFERENCES

1. H. Krueger, N. B. Eddy, and M. Sumwalt, "Pharmacology of the Opium Alkaloids," U.S. Public Health Service, Public Health Report Supplement 165, Parts 1 and 2 (1941).
2. G. W. Harris, "Neural Control of the Pituitary Gland," Edward Arnold Ltd., London (1955).
3. A. V. Nalbandov (ed.), "Advances in Neuroendocrinology," University of Illinois Press, Urbana (1963).

4. V. Schreiber, "The Hypothalamo–Hypophyseal System," Publishing House of the Czechoslovak Academy of Sciences, Prague (1963).
5. L. Martini and W. F. Ganong (eds.), "Neuroendocrinology," Vols. 1 and 2, Academic Press, New York (1966 and 1967).
6. G. W. Harris and B. T. Donovan (eds.), "The Pituitary Gland," Vols. 1, 2, and 3, University of California Press, Berkeley (1966).
7. V. H. T. James (ed.), "Recent Advances in Endocrinology," Churchill Ltd., London (1968).
8. J. Szentagothai, B. Flerko, B. Mess, and B. Halasz, "Hypothalamic Control of the Anterior Pituitary," Akademiai Kiado, Budapest (1968).
9. W. F. Ganong and L. Martini (eds.), "Frontiers in Neuroendocrinology," Oxford University Press, New York (1969).
10. W. Haymaker, E. Anderson, and W. J. H. Nauta (eds.), "The Hypothalamus," C. C. Thomas (1969).
11. W. F. Ganong and P. H. Forsham, *Ann. Rev. Physiol. 22*, 579–614 (1960).
12. F. E. Yates and J. Urquhart, *Physiol. Rev. 42*, 359–443 (1962).
13. P. G. Smelik and C. H. Sawyer, *Ann. Rev. Pharmacol. 2*, 313–340 (1962).
14. J. Vernikos-Danellis, *Vitam. Horm. 23*, 97–152 (1965).
15. K. Brown-Grant and B. A. Cross (eds.), *Brit. Med. Bull. 22* (3), 195–277 (1966).
16. A. Brodish, *Yale J. Biol. Med. 41*, 143–198 (1968).
17. S. M. McCann, A. P. S. Dhariwal, and J. C. Porter, *Ann. Rev. Physiol. 30*, 589–640 (1968).
18. F. H. A. Marshall, *Phil. Trans. Roy. Soc. London Ser. B 226*, 423–456 (1936).
19. G. T. Popa and U. Fielding, *J. Anat. 65*, 88–91 (1930).
20. J. D. Green and G. W. Harris, *J. Endocrinol. 5*, 136–146 (1947).
21. G. W. Harris, *J. Physiol. London III*, 347–360 (1950).
22. G. W. Harris and D. Jacobsohn, *Proc. Roy. Soc. London Ser. B 139*, 263–276 (1952).
23. M. Nikitovitch-Winer and J. W. Everett, *Endocrinology 63*, 916–930 (1958).
24. M. Saffran and A. V. Schally, *Can. J. Biochem. 33*, 408–415 (1955).
25. W. Vale, R. Burgus, T. F. Dunn, and R. Guillemin, *J. Clin. Endocrinol. 30*, 148–150 (1970).
26. R. Guillemin, *Ann. Rev. Physiol. 29*, 313–348 (1967).
27. A. V. Schally, A. Arimura, C. Y. Bowers, A. J. Kastin, S. Sawano, and T. W. Redding, *Recent Progr. Horm. Res. 24*, 497–588 (1968).
28. I. I. Geschwind, *in* "Frontiers in Neuroendocrinology" (W. F. Ganong and L. Martini, eds.), pp. 389–431, Oxford University Press, New York (1969).
29. R. George and E. L. Way, *Brit. J. Pharmacol. 10*, 260–264 (1955).
30. P. A. Nasmyth, *Brit. J. Pharmacol. 9*, 95–99 (1954).
31. P. F. D. Van Peenen and E. L. Way, *J. Pharmacol. 120*, 261–267 (1957).
32. F. N. Briggs and P. L. Munson, *Endocrinology 57*, 205–219 (1955).
33. O. Nikodijevic and R. P. Maickel, *Biochem. Pharmacol. 16*, 2137–2142 (1967).
34. V. J. Lotti, N. Kokka, and R. George, *Neuroendocrinology 4*, 326–332 (1969).
35. J. T. Oliver and R. C. Troop, *Steroids 1*, 670–677 (1963).
36. M. A. Slusher and B. Browning, *Amer. J. Physiol. 200*, 1032–1034 (1961).
37. T. Suzuki, K. Yamashita, S. Zinnouchi, and T. Mitamura, *Nature 183*, 825 (1959).
38. E. O. Westermann, R. P. Maickel, and B. B. Brodie, *J. Pharmacol. 138*, 208–217 (1962).
39. E. Paroli and P. Melchiorri, *Biochem. Pharmacol. 6*, 1–17 (1961).
40. E. M. MacKay and L. L. MacKay, *Proc. Soc. Exp. Biol. Med. 24*, 129 (1926).
41. H. Selye, *Brit. J. Exp. Pathol. 17*, 234–248 (1936).
42. T. Tanabe and E. J. Cafruny, *J. Pharmacol. 122*, 148–153 (1958).
43. S. M. McCann, *Endocrinology 60*, 664–676 (1957).
44. C. N. H. Long, *Recent Progr. Horm. Res. 7*, 75–97 (1952).
45. R. George and E. L. Way, *J. Pharmacol. 125*, 111–115 (1959).
46. V. J. Lotti, P. Lomax, and R. George, *J. Pharmacol. 150*, 135–139 (1965).
47. P. L. Munson and F. N. Briggs, *Recent Progr. Horm. Res. 11*, 83–117 (1955).

48. E. A. Ohler and R. W. Sevy, *Endocrinology 59*, 347–355 (1956).
49. R. Guillemin, *Recent Progr. Horm. Res. 20*, 89–130 (1964).
50. Y. H. Han and E. S. Brown, *Anesthesiology 22*, 909–914 (1961).
51. T. Oyama, M. Takiguchi, T. Takazawa, and K. Kimura, *Can. Anaesth. Soc. J. 16*, 282–291 (1969).
52. R. K. McDonald, F. T. Evans, V. K. Weise, and R. W. Patrick, *J. Pharmacol. 125*, 241–247 (1959).
53. B. H. Burdette, S. Leeman, and P. L. Munson, *J. Pharmacol. 132*, 323–328 (1961).
54. E. Paroli and P. Melchiorri, *Biochem. Pharmacol. 6*, 1–17 (1961).
55. E. Paroli and P. Melchiorri, *Biochem. Pharmacol. 6*, 18–20 (1961).
56. H. Sobel, S. Schapiro, and J. Marmorston, *Amer. J. Physiol. 195*, 147–149 (1958).
57. A. J. Eisenman, H. F. Fraser, and J. W. Brooks, *J. Pharmacol. 132*, 226–231 (1961).
58. T. W. Redding, C. Y. Bowers, and A. V. Schally, *Acta Endocrinol. 51*, 391–399 (1966).
59. T. W. Redding, C. Y. Bowers, and A. V. Schally, *Endocrinology 79*, 229–236 (1966).
60. V. Schreiber, V. Zbuzek, and V. Zbuzková-Kmentova, *Physiol. Bohem. 17*, 253–258 (1968).
61. W. Hohlweg, G. Knappe, and G. Dörner, *Endokrinologie 40*, 152–159 (1961).
62. M. Sámel, *Arch. Int. Pharmacodyn. 117*, 151–157 (1958).
63. R. George and P. Lomax, *J. Pharmacol. 150*, 129–134 (1965).
64. M. A. Greer, *J. Clin. Endocrinol. 73*, 1259–1268 (1952).
65. G. W. Harris and J. W. Woods, *J. Physiol. (London) 143*, 246–274 (1958).
66. P. Lomax and R. George, *Brain Res. 2*, 361–367 (1966).
67. M. Vertes, Z. Vertes, and S. Kovacs, *Acta Physiol. Acad. Sci. Hung. 27*, 229–235 (1965).
68. V. J. Lotti, P. Lomax, and R. George, *J. Pharmacol. 150*, 420–425 (1965).
69. V. J. Lotti, P. Lomax, and R. George, *Int. J. Neuropharmacol. 5*, 35–42 (1966).
70. P. Lomax, N. Kokka, and R. George, *Neuroendocrinology 6*, 146–152 (1970).
71. A. K. Reynolds and L. O. Randall, "Morphine and Allied Drugs," University of Toronto Press, Toronto (1957).
72. E. Menninger-Lerchenthal, *Zentralbl. Gynaekol. 58*, 1044–1051 (1934).
73. J. W. Everett and C. H. Sawyer, *Endocrinology 47*, 198–218 (1950).
74. C. A. Barraclough and C. H. Sawyer, *Endocrinology 57*, 329–337 (1955).
75. C. H. Sawyer, B. V. Critchlow, and C. H. Barraclough, *Endocrinology 57*, 345–354 (1955).
76. C. H. Sawyer, in "Advances in Neuroendocrinology" (A. V. Nalbandov, ed.), pp. 444–457, University of Illinois Press, Urbana (1963).
77. E. G. Rennels, *Tex. Rep. Biol. Med. 19*, 646–657 (1961).
78. S. N. Giri, *Life Sci. 7*, 1183–1187 (1968).
79. E. E. Müller and A. Pecile, in "Growth Hormone" (A. Pecile and E. E. Müller, eds.), pp. 253–266, Excerpta Medical Foundation Milan (1968).
80. J. Meites, in "Neuroendocrinology" (L. Martini and W. F. Ganong, eds.), Vol. 1, pp. 669–707, Academic Press, New York (1966).
81. C. C. D. Shute and P. R. Lewis, *Brit. Med. Bull. 22*, 221–226 (1966).
82. K. Fuxe and T. Hökfelt, in "Frontiers in Neuroendocrinology" (W. F. Ganong and L. Martini, eds.), pp. 47–96, Oxford University Press (1969).
83. E. L. Bliss, J. Ailion, and J. Zwanziger, *J. Pharmacol. 164*, 122–134 (1968).
84. H. Corrodi, K. Fuxe, and T. Hökfelt, *Life Sci. 7*, 107–112 (1968).
85. A. M. Thierry, M. Fekete, and J. Glowinski, *Eur. J. Pharmacol. 4*, 384–389 (1968).
86. A. De Schaepdryver, P. Preziosi, and U. Scapagnini, *Brit. J. Pharmacol. 35*, 460–467 (1969).
87. A. B. King, *Proc. Soc. Exp. Biol. Med. 130*, 445–447 (1969).
88. E. V. Naumenko, *Brain Res. 11*, 1–10 (1968).
89. D. T. Krieger and H. P. Krieger, "Proceedings of the Second International Congress on Endocrinology," Part 1, pp. 640–645, Excerpta Medica Foundation, London (1964).

90. R. Eisenberg, N. Kokka, and R. George, unpublished data.
91. D. T. Krieger and H. P. Krieger, *Science 155*, 1421–1422 (1967).
92. G. A. Hedge and P. G. Smelik, *Science 159*, 891–892 (1968).
93. C. Kordon and J. Glowinski, *Endocrinology 85*, 924–931 (1969).
94. W. Lippman, R. Leonardi, J. Ball, and J. A. Coppola, *J. Pharmacol. 156*, 258–266 (1967).
95. H. P. G. Schneider and S. M. McCann, *Endocrinology 85*, 121–132 (1969).
96. I. A. Kamberi, R. S. Mical, and J. C. Porter, *Science 166*, 388–389 (1969).
97. E. W. Maynert, *Arch. Biol. Med. Exp. 4*, 36–41 (1967).
98. E. L. Way, H. H. Loh, and F. Shen, *Science 162*, 1290–1292 (1968).

Chapter 14

INORGANIC IONS: THE ROLE OF CALCIUM

Hiroshi Kaneto

Department of Pharmacology
Faculty of Pharmaceutical Sciences
Nagasaki University
Nagasaki, Japan

I. INTRODUCTION

The significant role of inorganic ions, above all, cations, in the various biological phenomena, especially in nervous activities, has been studied intensively, and it is well understood that the metabolism of the cells is regulated through the fluxes of ions at the cell membranes.

Calcium ion plays an essential role in the maintenance of membrane structure, not only as a constituent but also as a membrane stabilizer, and the fluxes of calcium ion across the cell membrane induce the various functional changes in the nervous tissues. Calcium ion also is known to be indispensable for muscle contraction and blood coagulation and, moreover, it is established that the release of biogenic amines such as catecholamines, acetylcholine, serotonin, and histamine all involve calcium ions.

The significant role of calcium ion in the mechanism of action of opiates has stimulated interest and studies along this line have appeared in the past several years.

This chapter describes the present status of the investigation of the role of inorganic cations, especially calcium ion, in the mechanism of action of opiates.

II. ANTAGONISTIC EFFECT OF CALCIUM ION ON MORPHINE ANALGESIA

Kakunaga *et al.*[1] reported that the intracisternal administration of calcium ion markedly suppressed the analgesic effect of intracisternally injected morphine, meperidine, and ohton (dimethylthiambutene). The antagonistic effect of calcium ion is proportional to the dose, and the dose–response curves are shown in Fig. 1.

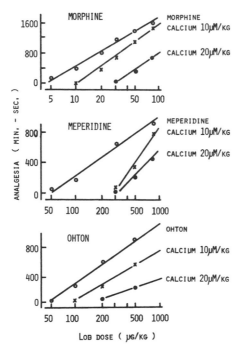

Fig. 1. Antagonistic effects of calcium ion on the analgesic response of morphine, meperidine, and ohton injected intracisternally into mice. The analgesics and calcium were injected simultaneously into the intracisternal space of mice. The analgesic effect was expressed as area under time–response curve in min-sec.

Similar results were obtained when these analgesics were administered subcutaneously. Subcutaneous or intraperitoneal administration of calcium ion also antagonized when it was injected repeatedly prior to the administration of the analgesics, although the antagonistic effect was not so remarkable as it was introduced intracisternally. Calcium ion also antagonized the established analgesic effect of these analgesic drugs.

On the other hand, decalcifying agents such as EDTA-2Na, cyclohexanediamine tetraacetic acid-2Na, sodium citrate, and sodium oxalate injected intracisternally produced little analgesic effect themselves and enhanced the analgesic effect of morphine, meperidine, and ohton and antagonized the antagonistic effect of calcium ion on the analgesic response. These facts seem to suggest that the chelating agents act through the removal of calcium ion with respect to the analgesic effect produced.

The antagonistic property is unique to calcium ions because other cations such as magnesium, barium, strontium, zinc, ferrous iron, aluminum, potassium, and sodium had no analgesic effect themselves and failed to produce the antagonistic effect on morphine analgesia or on the antagonistic effect of calcium ion to the analgesia produced by morphine. Thus the analgesic response to morphine-like analgesics may be associated with the level of calcium ions within the central nervous system.

Recently, however, in systematic experiments Kaneto et al.[2] examined the effect of various inorganic ions on morphine analgesia in mice and found enhanced effects by $CuCl_2$ and decreased effects by $SrCl_2$, in addition to $CaCl_2$, after large subcutaneous doses. High plasma concentration of copper ion in morphine-addicted patients was reported by Choi[3] and Stern,[4] also reported that intraperitoneal injection of $CuCl_2$ enhanced the analgesic effect of morphine. Recently, Watanabe et al.[5] studied the relationship between copper ion and morphine analgesia because sodium diethyldithiocarbamate, a dopamine–hydroxylase inhibitor, potentiated the morphine analgesia with a dose that did not induce a change of brain noradrenaline or dopamine level, suggesting that it may be acting as a chelating agent of the copper ion.

It is interesting to determine whether the antagonism by calcium ion is specific to the analgesic effect of morphine and, on the other hand, whether calcium ion acts antagonistically to other central and peripheral actions of morphine. Nutt[6] confirmed the antagonistic effect of calcium ion on morphine analgesia and extended the observation to the effect on guinea pig ileum and found that calcium ion noncompetitively antagonizes the inhibitory effect of morphine on the contraction of gut produced by coaxial stimulation. However, he also stated that the inhibitory effect of morphine on the action of acetylcholine or serotonin in guinea pig ileum is not reversed by calcium ion.

Intracisternal, intracerebral, or intraventricular injection of calcium ion produces torpor in rabbits, guinea pigs[7] and dogs[8]; typical and reversible signs of sleep in cats[9,10]; typical conditions of sleep in rats, rabbits, dogs,[11,12] and cats[11–14]; and tachypnea, muscular weakness, and ataxia followed by an anesthetic-like condition in mice.[1] These symptoms resemble somewhat the effect of morphine. Indeed, both agents act synergistically in their effect on EEG in rats.[15]

Thus the specificity of the antagonistic effect of calcium ion to the analgesic effect of morphine requires further investigation.

III. CHANGES OF CALCIUM ION LEVEL AFTER MORPHINE

In an experiment to support the hypothesis that the analgesic effect of morphine may be closely related to the concentration of calcium ion in the brain, Shikimi et al.[16] demonstrated that a single large dose of morphine, 100 mg/kg, lowered the brain calcium ion level of mice after a temporary increase, although smaller doses were without effect. The maximum decrease was obtained 30 min after the injection of morphine, and the level gradually rose to the control level within 24 hr. This effect of morphine of lowering

TABLE I

Brain Content of Calcium (μg/g) 30 min After the Administration of a Single Subcutaneous Dose of 100 mg Morphine/kg in Normal and Chronically Morphinized Mice

	Weeks of Morphinization			
	0	1	2	4
Control	75 ± 5^a	76 ± 6	77 ± 7	80 ± 8
After morphine injection	56 ± 3^b	61 ± 7	67 ± 6	77 ± 7

^{a} mean $\pm$ S.E.
^{b} $P < 0.05$

the calcium level in brain is obtained only in normal animals. The effect is gradually decreased during the chronic administration of morphine, and in tolerant animals the changes of brain calcium content were no longer obtained (Table I). However, the changes in calcium content in brain were not parallel to the analgesic effect, and the dose necessary to produce the decrease in level of calcium ions was higher than the usual analgesic dose.

Reciprocal changes of magnesium ion concentration in the brain were observed after an acute dose of morphine.[17] This finding may add to the explanation of the antagonistic effect of calcium on magnesium-induced anesthesia, although magnesium ion failed to reverse the antagonistic effect of calcium ion on morphine analgesia. Recently Vachon and Marchand[18] reported hypermagnesemia after intravenous administration of morphine. This effect was proportional to the dose and antagonized by nalorphine.

Functional activities bring about the changes of the distribution and content of inorganic ions in various parts of the central nervous system. Koeda[19] found a marked accumulation of calcium ion in the mes- and diencephalon of guinea pig during the time of excitation or sleep induced by various central stimulants and depressants. He stated that the accumulation of calcium ion is proportional to the degree of excitation or sleep and that the level returned to normal at the same time that the condition disappeared.

An acute dose of morphine produces an antidiuretic effect due to the liberation of antidiuretic hormone. However, tolerance to the antidiuretic effect develops during chronic treatment of rats with gradually increasing doses of morphine. Marchand and Fujimoto[20] reported an increase of urinary excretion of sodium and potassium at the beginning of treatment that returned to the control level as animals became tolerant to morphine. No changes in the level of these ions were observed in muscle or plasma. Recently Marchand and Denis[21] reported that the diuresis in morphine-tolerant rats is accompanied by an increase of urinary sodium, potassium, phosphate, calcium, and urea excretion. They also reported that the hypercalciurea and urea diuresis was partly reversed by nalorphine. These findings also support the significance of calcium ion in the action of morphine.

IV. CALCIUM CONCENTRATION AND THE INHIBITORY EFFECT OF MORPHINE ON THE RESPIRATION OF BRAIN SLICES *IN VITRO*

Takemori reported that morphine inhibited the K^+-stimulated oxygen consumption of cerebral cortical slices from nontolerant rats but not from rats chronically tolerant to morphine when the slices were incubated in Ringer's solution low in calcium[22] or in normal medium.[23] In carefully controlled experiments Elliott *et al.*[24] concluded that the inhibitory effect of morphine was only demonstrated in a calcium-free medium and not in conventional Ringer's solution. Kakunaga *et al.*[25] essentially confirmed these results using media containing various concentrations of calcium ion and concluded that the inhibitory effect of morphine depends upon the concentration of calcium ion in the incubating medium and that the inhibition is apparent only in medium containing a low level of calcium ion (Fig. 2).

It is widely accepted that the K^+-stimulated oxygen uptake can be demonstrated only in brain slices but not in brain homogenates and that it is a phenomenon closely linked with the function of brain cell membrane. Other drugs such as pentobarbital, ouabain, and cocaine also depress the oxygen consumption of brain slices. However, in the case of pentobarbital it inhibits both K^+-stimulated and unstimulated respiration of brain slices regardless of the concentration of calcium ion in the medium. Ouabain or cocaine, both agents closely concerned with membrane function, also depress the K^+-stimulated respiration of brain slices but differ from morphine in that they show their inhibitory effects even on unstimulated respiration when calcium concentration is low in the medium.

These facts seem to suggest that the respiratory depressant effect of

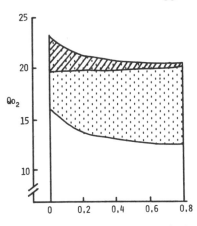

CONCENTRATION OF CALCIUM IN MEDIUM (mM)

Fig. 2. The relationship between the concentration of calcium in medium and the inhibitory effect of $10^{-3}M$ morphine on potassium-stimulated respiration ($0.1M$ KCl) in rat brain cortical slices. The lower line represents unstimulated oxygen uptake; the upper line, K^+-stimulated uptake. The hatched area represents the effect of morphine on K^+-stimulated oxygen uptake.

morphine is characteristic of the opiates and is closely connected with the specific morphological and functional state of brain cell membrane in the low calcium and high K^+ state.

Takemori[22] reported that the depressive effect of morphine on K^+-stimulated oxygen consumption was obtained in the cerebral cortical slices of nontolerant rats but not of rats tolerant to morphine. He also stated that the cellular adaptation of brain slices to morphine was partially reversed by nalorphine.[23] Kakunaga *et al.*[25] have confirmed the experiments with slices in a medium low in calcium and stressed the importance of the calcium ion in this effect.

V. FLUX OF CATIONS ACROSS THE CELL MEMBRANE

One of the mechanisms by which cell metabolism is regulated is through the transport of various cations across the cell membrane. Kakunaga[26] studied cation fluxes across cell membranes of rat brain slices in the presence of KCl (a membrane depolarizer), ouabain (an inhibitor of active cation transport as the result of specific action on $Na^+ + K^+$-activated ATPase), or EDTA (a decalcifying agent that induces depolarization as a secondary effect) and compared the effect of morphine and cocaine on cation fluxes in the presence or absence of calcium ion in the medium. The content of sodium, potassium, and calcium in rat brain cortex slices changed markedly on the addition of potassium or ouabain, and the changes were inhibited by cocaine but not influenced significantly by morphine.

^{45}Ca influx into brain slices was induced by potassium or ouabain regardless of the calcium ion concentration in the medium. This influx of ^{45}Ca was inhibited by cocaine in the presence or absence of ^{40}Ca. On the other hand, morphine inhibited the influx of ^{45}Ca induced by KCl only in the low calcium medium and had no influence on the influx induced by ouabain in both usual and low calcium medium. The efflux of ^{45}Ca from brain slices into calcium-free medium was stimulated by the addition of KCl, ouabain, EDTA, $CaCl_2$, or $CaCl_2 + KCl$. This stimulation of ^{45}Ca efflux by KCl or EDTA was inhibited by morphine while the stimulated ^{45}Ca efflux by these agents except EDTA was inhibited by cocaine. Thus morphine affects only the action of depolarizing agents but not the action of other agents. Apparently a remarkable difference exists between the mechanisms of action of morphine and cocaine. An important difference is that the effect of morphine is altered by the concentration of calcium ion in the incubating medium.

The inhibitory effect of morphine on the stimulated influx or efflux of ^{45}Ca was partially reversed by the simultaneous addition of nalorphine. Kakunaga[26] concluded that the analgesic effect of morphine may be associated with the inhibitory action on calcium ion fluxes at cell membranes in the nervous system.

VI. LIPID-FACILITATED TRANSPORT OF IONS

Lipids, especially phospholipids, are abundant in the central nervous system and are constituents of cell membranes. Calcium ions stabilize cell membranes and are closely related to the movement of other ions through cell membranes.

Since Wooley and Campbell have demonstrated the transport of calcium by lipids extracted from nerve and muscle, the important role of phospholipids as ion carriers in the transport at the cell membrane have been widely studied. Hano *et al.*[27] have reported the lipid-facilitated transport of ^{45}Ca and the replacement of lipid-bound calcium ion by morphine and nalorphine. Recently Mulé studied extensively the effect of various narcotic analgesics on phospholipid-facilitated ion transport and suggested that the binding of ions to phospholipids within the neuronal membrane may be involved in the action of the narcotic analgesics. This is discussed in Chapter 8.

VII. EFFECT OF CALCIUM ION ON THE RELEASE OF BIOGENIC AMINES

The fact that morphine inhibits cholinesterase activity *in vivo* and *in vitro* and that it increases the total acetylcholine content in mice brain[28] led to the hypothesis that the cholinergic mechanism may be involved in the effect of morphine on the central nervous system. On the other hand, morphine is known as an inhibitor of acetylcholine release from guinea pig ileum[29] and brains of anesthetized cats[30] and rat brain slices.[31]

Calcium ions play an essential role in the release of biogenic amines from their storage sites. Shikimi *et al.*[32] showed that morphine inhibits the

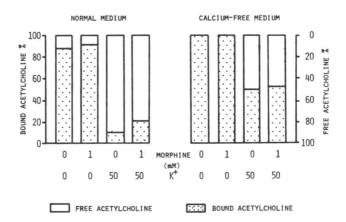

Fig. 3.   Effect of morphine on the spontaneous and potassium-stimulated liberation of acetylcholine from mouse cerebral cortical slices.

K$^+$-stimulated liberation of acetylcholine from cerebral cortical slices of rats suspended in usual medium but that morphine failed to exert its inhibitory effect in calcium-free medium (Fig. 3). They concluded from these results that calcium ion is indispensable for the exhibition of the inhibitory effect of morphine on the liberation of acetylcholine from cerebral cortical slices *in vitro*.

It is well known that the release of catecholamines from the adrenal gland by the stimulation of splanchnic nerve is mediated through acetylcholine. Douglas and Poisner[33] reported that the stimulatory effect of acetylcholine is accompanied by an increase in calcium ion uptake, and this facilitated the release of catecholamine. Recently Ray *et al.*[34] examined the catecholamine content of adrenal gland histochemically during morphine treatment of rats and stated that acute morphine administration to rats reduced the total catecholamine content of adrenal gland accompanied by an increased ATPase activity and an increased calcium ion content of the medullary cells. In chronically treated rats noradrenaline increased markedly accompanied by a decrease of ATPase activity and a decreased calcium ion content. In nalorphine-induced abstinence a reduction of total catecholamines and an increased ATPase activity and calcium level were observed.

Thus the changes of catecholamine content in adrenal gland after morphine treatment are accompanied by the changes in ATPase activity controlled by the level of calcium ion.

VIII. EFFECT OF CALCIUM ION ON THE DEVELOPMENT OF TOLERANCE

Many factors have been reported to influence the development of tolerance to morphine and related drugs. Weger and Amsler[35] stated that the development of tolerance to morphine is delayed in the guinea pig when the animals are fed a high-calcium diet, whereas the recovery from tolerance is accelerated by this diet. On the other hand, Detrick and Thienes[36] described a tissue hydration in morphine-addicted rats and stated that a high-calcium diet plus parathyroid hormone injection produced a greater loss of tissue water during withdrawal and weakened the withdrawal symptoms of these animals compared to rats on a low-calcium diet.

The fact that calcium ion antagonizes the analgesic effect of opiates[1] and that nalorphine delays the development of tolerance to morphine when administered chronically with morphine[37] stimulated an investigation of the effect of calcium ion on the development of tolerance to morphine. Kakunaga and Kaneto[38] found that the mice given repeated intracisternal injections of morphine plus calcium ion were less tolerant to the analgesic effect of morphine than were the group receiving morphine alone. On the other hand, disappearance of the acquired tolerance to morphine was not affected significantly by the intracisternal injection of sodium, potassium, calcium, magnesium, or EDTA.

IX. CONCLUSION

All the data presented in this chapter suggest that the action of morphine is closely linked with the metabolism of calcium ion. However, the participation of other ions cannot be excluded. It must be determined if the action of calcium ion is specific or if other ions also act similarly or reciprocally and whether the action of calcium ion is unique to morphine and related analgesic drugs or calcium ions respond nonspecifically to other classes of drugs. Moreover, it is important to establish whether the action of calcium ion is specific for the analgesic effect of morphine or if calcium ions affect other central and peripheral actions of morphine.

X. REFERENCES

1. T. Kakunaga, H. Kaneto, and K. Hano, *J. Pharmacol. Exp. Ther. 153*, 134–141 (1966).
2. H. Kaneto, H. Nakanishi, and M. Fujii, Reports at Thirty-fifth Regional Meeting (Kinki Area at Tokushima) and Twenty-second Regional Meeting (Seinan Area at Fukuoka) of Japanese Pharmacological Society (1969).
3. Shin Hai Choi, *New Med. J. (Korea) 3*(11), 83–98 (1960).
4. P. Stern, *Wien. Klin. Wochenschr. 80*(10), 181–185 (1968).
5. K. Watanabe, Y. Matsui, and H. Iwata, Report at Thirty-fifth Regional Meeting (Kinki Area at Tokushima) of Japanese Pharmacological Society (1969).
6. J. G. Nutt, *Fed. Proc., 27*, 753 (1968).
7. L. Stern and E. Rothlin, *Schweiz. Arch. Neurol. Psychiat. 3*, 234–254 (1918).
8. L. Stern and G.-J. Chvoles, *C. Séanc. Soc. Biol. 112*, 568–570 (1933).
9. V. Demole, *Naunyn-Schmiedebergs Arch. Exp. Pathol. Pharmakol. 120*, 229–258 (1927).
10. G. Marinesco, O. Sager, and A. Kreindler, *Z. Gesamte Neurol. Psychiat. 119*, 277–306 (1929).
11. M. Cloetta and H. Fisher, *Naunyn-Schmiedebergs Arch. Exp. Pathol. Pharmakol. 158*, 254–281 (1930).
12. M. Cloetta, H. Fisher, and M. R. van der Loeff, *Naunyn-Schmiedebergs Arch. Exp. Pathol. Pharmakol. 174*, 589–675 (1934).
13. W. Feldberg, *in* "Abstracts of the Twentieth International Physiology Congress" (C. P. Leake, ed.), p. 18, International Pharmacology Congress, Brussels (1956).
14. W. Feldberg and S. L. Sherwood, *J. Physiol. (London) 139*, 408–416 (1957).
15. H. Kaneto, unpublished.
16. T. Shikimi, H. Kaneto, and K. Hano, *Jap. J. Pharmacol. 17*, 135–136 (1967).
17. T. Shikimi and H. Kaneto, unpublished.
18. M. Vachon and C. Marchand, *Fed. Proc. 28*, 735 (1969).
19. T. Koeda, *Folia Pharmacol. Jap. 51*, 79–83 (1955).
20. C. R. Marchand and J. M. Fujimoto, *Proc. Soc. Exp. Biol. Med. 123*, 600–603 (1966).
21. C. Marchand and G. Denis, *J. Pharmacol. Exp. Ther. 162*, 331–337 (1968).
22. A. E. Takemori, *Science 133*, 1018–1019 (1961).
23. A. E. Takemori, *J. Pharmacol. Exp. Ther. 135*, 89–93 (1962).
24. H. W. Elliot, N. Kokka, and E. L. Way, *Proc. Soc. Exp. Biol. Med. 113*, 1049–1052 (1963).
25. T. Kakunaga, H. Kaneto, and K. Hano, *Folia Pharmacol. Jap. 62*, 31–39 (1966).
26. T. Kakunaga, *Folia Pharmacol. Jap. 62*, 40–50 (1966).
27. K. Hano, H. Kaneto, T. Kakunaga, A. Oshiro, and T. Shimomura, Report at Regional Meeting (Kinki Area at Osaka) of Japanese Pharmacological Society (1964).

28. K. Hano, H. Kaneto, T. Kakunaga, and N. Moribayashi, *Biochem. Pharmacol. 13*, 441–447 (1964).
29. W Schaumann, *Brit. J. Pharmacol. 12*, 115–118 (1957).
30. D. Beleslin and R. L. Polak, *J. Physiol. (London) 177*, 411–419 (1965).
31. M. Sharkawi and M. P. Schulman, *J. Pharm. Pharmacol. 21*, 548–549 (1969).
32. T. Shikimi, H. Kaneto, and K. Hano, *Jap. J. Pharmacol. 17*, 136–137 (1967).
33. W. W. Douglas and A. M. Poisner, *J. Physiol. (London) 162*, 385–392 (1962).
34. A. K. Ray, M. Mukherji, and J. J. Ghosh, *Neurochem. 15*, 875–881 (1968).
35. P. Weger and C. Amsler, *Naunyn-Schmiedebergs Arch. Exp. Pathol. Pharmakol. 181*, 489–493 (1936).
36. L. Detrick and C. H. Thienes, *Arch. Int. Pharmacodyn 66*, 130–137 (1941).
37. P. D. Orahovats, C. A. Winter, and E. G. Lehman, *J. Pharmacol. Exp. Ther. 109*, 413–416 (1953).
38. T. Kakunaga and H. Kaneto, *in* "Abstracts of the Twenty-third International Congress of Physiological Sciences," p. 526, Tokyo (1965).

Chapter 15

SINGLE CELLS

Eric J. Simon

Department of Medicine
New York University Medical School
New York, New York

I. INTRODUCTION

The use of single cells or unicellular organisms as model systems for the study of biochemical pathways has had wide application for a very long time. It has proved exceedingly useful for the elucidation of a variety of phenomena, including complex metabolic pathways and cycles and, more recently, mechanisms by which cells regulate the rate of metabolic processes. This approach suffers from the drawback of any model, namely, the uncertainty about the extent to which it exemplifies the behavior of the more complex systems, for which it serves as a model. Thus while much of the metabolic machinery of *Escherichia coli* has been found to be present in identical or similar form in the cells of higher plants, animals, and man, it is clear that differentiated cells of a complex organism carry on many functions not present in unicellular organisms or indeed in other cell types of the same organism.

This realization has been a major concern that has discouraged investigators from chosing such an approach to study the biochemistry of narcotic analgesic drugs. It is, however, a fact that the study of the pharmacological effects of narcotics is one of the oldest areas of research in biology and yet we have no understanding of the biochemical mechanisms that underlie the relief of pain, the production of tolerance to drug action, the development of physiological dependence on drugs, or their toxic manifestations such as respiratory depression.

A number of investigators, including the author of this chapter, therefore have decided to study the biochemical effects of narcotic analgesics in single cell systems. While it is clearly impossible to study pain relief or psychological dependence in anything less than an intact animal, the question has been legitimately raised whether or not the phenomena of tolerance and physiological dependence might be reproduced in single cells.

These effects bear, after all, some resemblance to the antibiotic resistance and dependence phenomena seen in bacteria.

Even if it should prove difficult to study tolerance and dependence in single cells, it should be possible to gain information about the effects of narcotics on metabolic pathways. Such systems may be useful in furnishing the answer to the important question of whether actions of narcotics are restricted to cells of the central nervous system (CNS) or whether similar effects can be seen in other types of cells. In the intact organism such effects could easily be obscured by the exquisite sensitivity of the CNS. It is, thus, the rationale of this approach to delineate the types of clear-cut, specific biochemical and physiological alterations that can be produced in single cell systems by narcotic analgesics in the hope that some of this knowledge may ultimately prove relevant to narcotic action in the CNS.

The present chapter reviews reports that have appeared on the study of morphine and related compounds in single cell systems, including mammalian cells in culture, microorganisms, and isolated single nerve fibers.

II. DEVELOPMENT OF TOLERANCE AND PHYSICAL DEPENDENCE IN TISSUE CULTURE

Evidence that phenomena resembling tolerance and physiological dependence might be produced in tissue explants appeared as early as the 1930s in a series of reports from the University of Kyoto, Japan[1-5]. These investigators used explants from organs (heart or iris) of 8- to 9-day-old chick embryos. These explants were cultivated by the cover glass procedure in plasma clots. The culture medium was composed of one drop of embryonic tissue juice and one drop of a mixture of four parts plasma and one part Ringer's solution. The drug added was dissolved in the Ringer's solution. The cultured tissue explants were transferred to fresh medium every 48 hr. The effect of the drug was determined by measuring the increase in explant size (a combination of cell division and migration outward of fibroblasts) and by comparing the area of outgrowth of treated explants to those of controls (the growth index is defined as the percent increase in area of outgrowth of a drug-treated culture divided by that of the control). Histological examination was carried out by the use of hematoxylin and Sudan III stains.

Morphine hydrochloride was found to suppress growth of fibroblast cultures at a concentration of $3.3 \times 10^{-5}M$ and higher. Complete inhibition was seen at $3.3 \times 10^{-3}M$. Histological changes, including rough and irregular arrangement of cells and an increase in fat granules, were observed and became more striking with the increasing concentration of morphine. At $3.3 \times 10^{-3}M$ morphine, the cells were circular with picnotic nuclei scattered around the mother tissue.

The "growth indexes" of cultures grown in media containing increasing concentrations of morphine are shown in Fig. 1. The cells appear to over-

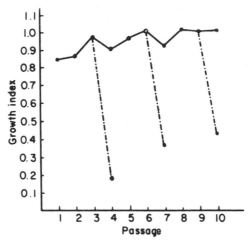

Fig. 1. Morphine tolerance in fibroblasts. Behavior of cultures pregrown in morphine towards normally lethal morphine concentration. ●——● = medium containing $3.3 \times 10^{-5}M$ morphine (first to third passages), $1.0 \times 10^{-4}M$ (fourth to sixth passages) and $1.7 \times 10^{-4}M$ (seventh to tenth passages). At passages 3, 6, and 9 cells were transferred to medium containing $3.3 \times 10^{-3}M$ morphine, ●—·—·—●. (From Sasaki,[4] Fig. 1.)

come the inhibitory effect of the drug. Thus during passages 7–10 essentially no decrease of growth was observed in the presence of $1.6 \times 10^{-4}M$ morphine, a concentration that is normally strongly inhibitory. Even more striking were the results obtained when, after three, six, and nine passages, the cultures were transferred to media containing $3.3 \times 10^{-3}M$ morphine, a dose that normally arrests growth completely. These "tolerant" cultures showed a persistence of growth in this medium that ranged from 15% of control after three passages to as high as 40% of control after nine passages in morphine-containing medium. Histological changes also decreased or failed to appear in these cells.

The authors further reported a type of abstinence phenomenon in these cultures. When explants, cultivated in the presence of increasing concentrations of morphine as described, were transferred to normal medium lacking morphine, their rate of growth was decreased abruptly. This effect became more striking the longer the tissues had been cultured in the presence of morphine (Fig. 2). Severe degeneration of the cells was observed on histological examination. In cultures pregrown for three passages in medium containing morphine, these effects, due to the withdrawal of morphine, disappeared spontaneously after two to three passages in normal medium; those previously exposed to morphine during six passages required

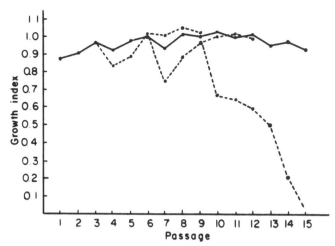

Fig. 2. Behavior of the cultures "addicted" to morphine on sudden interruption of morphine supply. ●——● = medium containing $3.3 \times 10^{-5}M$ (first to third passages), $1.0 \times 10^{-4}M$ (fourth to sixth passages), or $1.7 \times 10^{-4}M$ morphine (seventh to fifteenth passages). ●······● normal medium. (From Sasaki, [4] Fig. 2.)

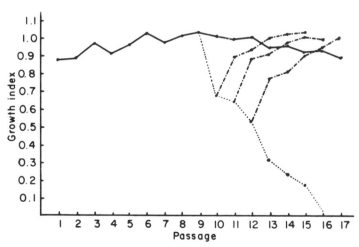

Fig. 3. Reversal of "abstinence syndrome" in fibroblast cultures by low concentration of morphine ($3.3 \times 10^{-5}M$). ●——● = medium containing $3.3 \times 10^{-5}M$ (first to third passages), $1.0 \times 10^{-4}M$ (fourth to sixth passages), or $1.7 \times 10^{-4}M$ morphine (seventh to seventeenth passages). ●······● = normal medium. ●—•—● = medium containing $3.3 \times 10^{-5}M$ morphine. (From Sasaki,[4] Fig. 4.)

somewhat more time to restore them to their normal state. However, in the explants cultivated in the presence of morphine for nine passages or longer, these reactions due to deprivation of drug no longer reversed spontaneously; in fact the growth rate continued to decrease and generally ceased entirely after five to six passages.

Figure 3 shows evidence that morphine, in quite small concentration, has a curative effect on the abstinence symptoms shown by morphine-habituated cultures in morphine-free medium. Thus when such cultures were placed in a medium containing $3.3 \times 10^{-5} M$ morphine, they gradually regained their normal rate of growth and appearance. A similar curative effect was seen when other opium alkaloids were substituted for morphine. In order of decreasing effectiveness, heroin, codeine, dionine, and papaverine could substitute for morphine. It is now known that papaverine is not related structurally to the morphine series and is not a narcotic. Its very weak activity may have been the result of contamination by small amounts of morphine or one of its derivatives.

It was not until 1952 that a similar type of study appeared in the literature. Heubner et al.[6] published investigations carried out in explants from the leg muscle of 8- to 9-day-old chick embryos. Except for the use

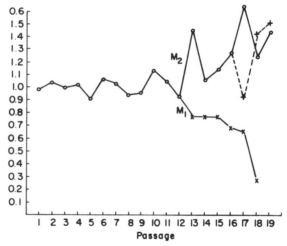

Fig. 4. Behavior of two different chick embryo fibroblast cultures (leg muscle) in the presence of morphine. ○——○ = medium containing $1.0 \times 10^{-4} M$ morphine (first to third passages), $2.0 \times 10^{-4} M$ (fourth to sixth passages), $1.5 \times 10^{-4} M$ (seventh to tenth passages), $3.0 \times 10^{-4} M$ (eleventh to thirteenth passages), and $3.45 \times 10^{-4} M$ morphine (fourteenth to nineteenth passages). $\times$——$\times$ = culture M_1 in medium containing $3.45 \times 10^{-4} M$ morphine (thirteenth to eighteenth passages). +——+ = culture M_2 in drug-free medium. (From Heubner et al.,[6] Fig. 1.)

of a different tissue, the experimental procedures of these authors were similar to those used by the Japanese workers. Evidence was obtained for the development of tolerance to morphine, primarily through observation of the disappearance of pathological symptoms (fat globules, vacuoles, and plump, wide, and rounded cell forms). At the same time, however, from the tenth passage on, there was a gradual decrease in growth rate that could not be reversed (Fig. 4). Of 20 cultures one culture grew very well at concentrations of morphine at which the others slowed their growth. This culture actually increased its growth rate above that of control cultures in the presence of $3.4 \times 10^{-4} M$ morphine. The authors felt that this culture resulted from a preexisting mutation exhibiting increased resistance to morphine and that this provided evidence for genetic heterogeneity of the cells with respect to their sensitivity to drugs.

Attempts to confirm the withdrawal phenomenon reported by the earlier investigators were unsuccessful. Even in those cases where growth appeared to slow down when morphine was abruptly withdrawn, the cells looked better morphologically than those of the cultures that received the drug. Unfortunately, the authors' study had to be terminated when all the cultures treated with morphine died of unknown causes after the nineteenth passage. It is noteworthy that the ten cultures from which morphine had been withdrawn earlier continued to grow. Experiments also were carried out with the synthetic narcotics meperidine (Dolantin) and methadone (Polamidon). These led to the development of a slight tolerance that the authors considered as possibly nonspecific.

A slightly different approach to the problem was taken by McCormick and Kniker.[7] Instead of attempting to produce the phenomena of tolerance and physiological dependence in cells in culture, these authors raised the question of whether such properties might be exhibited by tissues explanted from animals that had previously been habituated to large concentrations of narcotics. For this purpose albino rats were injected daily with increasing doses of morphine sulfate (up to 630 mg/kg), levorphanol (Dromoran) at one-third of the levels used for morphine, and sodium amytal (up to 190 mg/kg). Tissue explants were grown on cover slips in roller tubes. The plasma clot consisted of equal volumes of drug solution (four times the final concentration), embryonic extract, plasma containing penicillin (4000 units/ml), and rabbit serum. After the clot was formed, 2 ml of nutrient fluid (equal volumes of Gey's solution containing two times the final drug concentration and human ascitic fluid containing 5% embryonic extract) were added. Tubes were rotated eight times per hour at 37°C. The cultures were checked for amount of outgrowth every 2 days. Tissues were fixed in methyl alcohol and stained by Jacobson's method after 10 days.

None of the tissue explants showed any physiological dependence on, or requirement for, drugs. As shown in Fig. 5, the authors did find, however, that tissues taken from drug-habituated animals were able to thrive in concentrations of morphine or Dromoran that were lethal to the tissues explanted from normal animals. This increased resistance to the toxic

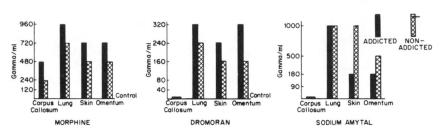

Fig. 5. Evidence for tolerance in tissue explants from drug-habituated rats. Height of bars represents highest drug concentration at which cellular outgrowth occurred *in vitro*. (From McCormick and Kniker,[7] Fig. 1.)

action of narcotics may represent a cellular tolerance similar to that reported by the Japanese investigators.

The most recently published report on the production of tolerance and dependence in cell culture appeared in 1964 from the laboratory of Corssen and Skora.[8] These workers studied epithelial-like cells isolated from the cervix of a patient with cervical carcinoma. Phase-contrast photomicrography and time-lapse microcinematography were employed. For these studies cells grown on a cover slip were introduced into a Rose perfusion chamber filled with nutrient medium. The chamber was placed on the stage of a microscope encased by an incubator in which the temperature was thermostatically controlled. Continuous phase-contrast, time-lapse cinematographic records of the cultures were taken at 15–30-sec intervals for periods of 4–12 hr each day through several days. Films were analyzed at projection speeds ranging from 8 to 24 frames/sec. At the end of the experiment some of the cultures were stained (Sudans III and IV) and photomicrographs were prepared. During the study cells were subcultured every 3–5 days depending on their rate of growth.

Prolonged exposure of cells to morphine sulfate (MS) resulted in marked degenerative changes, including irregular cell shapes, fat droplets and vacuoles in the cytoplasm, and picnotic nuclei. However, at MS concentrations of 5×10^{-4} molal or less, no retardation of growth was observed. At 1×10^{-3} molal MS or higher concentrations rapid cell deterioration resulted, with cell death ensuing in 2–6 days. Even 7×10^{-4} molal MS proved incompatible with prolonged maintenance of cultures. However, most cell cultures previously cultivated in lower concentrations of the narcotic for prolonged periods seemed to have developed marked tolerance to MS. As shown in Table I, such cultures, when transferred to growth-inhibitory concentrations of the drug, can generally be maintained for longer periods.

The effects of sudden cessation of MS administration in cultures previously grown in media containing various concentrations of the drug are shown in Table II. Time-lapse microcinematographic records of such cultures always indicated an abrupt decrease in growth rate. Pinocytosis rapidly diminished, and the cells took on irregular forms with pointed protoplasmic projections. The intensity of these changes, produced by

TABLE I
**Effect of Extremely High Morphine Sulfate Concentrations on
Cells Previously Exposed**

| | Previous Exposure of MS[a] | | Transferred to MS | | Cell condition at Termination of Experiment | | |
Cell Culture	Concentration (Molality)	Passages (Number)	Concentration (Molality)	Passages (Number)	Growing Well	Growth Retarded or Arrested	Dead
1	5.0×10^{-4}	15	1.0×10^{-3}	4	+		
2	2.5×10^{-4}	15	1.0×10^{-3}	2	+		
3	1.0×10^{-4}	5	6.7×10^{-4}	1	+		
	5.0×10^{-4}	4					
4	5.0×10^{-4}	19	1.0×10^{-3}	4	+		
5	5.0×10^{-4}	10	1.0×10^{-3}	1		+	
6	2.0×10^{-4}	1	1.0×10^{-3}	2			+

[a] MS denotes morphine sulfate. (From Corssen and Skora,[8] Table 3.)

withdrawal of MS, varied directly with the length of the period of previous exposure and the concentration of drug to which the cells were exposed. After relatively short exposure to low drug concentrations (for example, 18 days to 1×10^{-4} molal MS) the rate of growth and pinocytotic activity was impaired only temporarily. However, in virtually all cell cultures maintained for long periods in the presence of MS, withdrawal of the narcotic resulted in arrest of cell growth or rapid deterioration of the cells followed by death. In most cultures readministration of MS not later than 2 or 3 days after withdrawal resulted in rapid restoration of cytoplasmic

TABLE II
Effect of Withdrawal of Morphine Sulfate on Cell Growth

| | Previous Exposure to MS[a] | | Effect of Withdrawal | | |
Cultures (Number)	Concentration (Molality)	Passages (Number)	No Effect	Growth Retarded or Arrested	Dead
7	5.0×10^{-4}	10–19	1 culture	5 cultures	1 culture
3	2.5×10^{-4}	13–20		2 cultures	1 culture
1	2.0×10^{-4}	15		1 culture	
1	1.0×10^{-4}	5	1 culture		
1	5.0×10^{-4}	7			1 culture
	1.0×10^{-4}	9			

[a] MS denotes morphine sulfate. (From Corssen and Skora,[8] Table 4.)

TABLE III
Effect of Reintroduction of Morphine Sulfate on Cell Growth
Arrested by Withdrawal

| Previous Exposure to MS[a] | | | Effect of Reintroduction of MS | | |
| | | | | Restoration of Growth | |
Cultures (Number)	Concentration (Molality)	Passages (Number)	No Effect	Incomplete	Complete
3	5.0×10^{-4}	5–15		2 cultures	1 culture
2	2.5×10^{-4}	10–15			2 cultures
1	2.0×10^{-4}	15			1 culture
1	1.0×10^{-4}	5			1 culture
1	5.0×10^{-4}	7	1 culture		

[a] MS denotes morphine sulfate. (From Corssen and Skora[8], Table 5.)

function and growth (Table III). The irregular appearance of these cells also was usually reversed, but an actual increase of cytoplasmic fat droplets was often seen to accompany return to normal growth and cytoplasmic activity.

The authors indicate that, in their view, their system fulfills the criteria for "addiction." They suggest that tolerance and physiological dependence are cellular phenomena that are not restricted to tissues of the central nervous system, in confirmation of the findings of the Kyoto group.

III. TOXICITY OF NARCOTICS FOR SINGLE CELLS

A. Mammalian Cells in Culture

A number of laboratories have published results on the more modest approach of studying the toxic effects of narcotic drugs on cells or tissues in culture. I shall briefly summarize some of these findings.

In 1949 Painter et al.[9] published a report in which they surveyed the effect on spinal cord explants from 9-day-old chick embryos of agents known to affect the activity of the nervous system. Each explant was placed in a hanging drop made of one drop of cockerel plasma and one drop of equal parts of embryo extract and Tyrode solution containing the drug. The area and density of outgrowth and cell migration were measured every 24 hr. Staining was done with Bronin's solution by the Bodian technique. The minimum dose of drug that caused complete inhibition of fiber outgrowth (MID) and the minimum dose that caused detectable damage (LID) were determined. A large series of drugs was tested and divided into six groups according to their toxicities. Morphine, with an MID of 80

μg/ml and an LID of 8 μg/ml, fell into the most toxic group, which also contained strychnine, atropine, and lobeline.

An even more extensive report by the same group was published in the *Annals of the New York Academy of Sciences*[10]. In this study the toxicity of 110 compounds on a number of tissues in culture was reported. It is of interest that spinal cord tissue was found to be considerably more sensitive to narcotics than other tissues. Thus the LID of levorphanol was 62–125 μg/ml for heart, 15–30 μg/ml for spleen, but 4–8 μg/ml for cord.

In our laboratory (unpublished results) studies of the relative toxicities of morphine and related drugs in HeLa cell cultures were carried out by the cloning procedure developed by Marcus and Puck. Approximately 200 cells in Eagle's medium were placed in small plastic petri dishes. Cells were allowed to reproduce for 10–12 days, during which time they formed visible clones. The clones were fixed with formaldehyde, stained with hematoxylin, and counted. All drug concentrations and controls were carried out in triplicate dishes. Control cultures showed cloning efficiencies of 80–100%. Drugs were added to the medium in increasing concentrations up to a level that virtually completely suppressed formation of clones.

From a probit plot of the number of clones formed versus drug concentration, the dose that reduced the clone number to 50% of control (LD_{50}) was calculated. These LD_{50} values for morphine and a number of related drugs are listed in Table IV. It can be seen that the toxicities of these drugs show considerable variation, levallorphan being about 12 times as toxic as morphine. For comparison, LD_{50} values obtained in mice by intraperitoneal injection also are shown for some of the drugs. It is interesting to note that the toxicity of these compounds for HeLa cells varies generally in the same direction as that for mice.

Recently Noteboom and Mueller[15] measured the relative ability of 15 morphine alkaloids and morphinans to inhibit the growth of HeLa cells (Table V).

TABLE IV
Relative Toxicities of Morphine and Related Drugs in Tissue Culture and in Mice

Drug	LD_{50}, Mice (IV) (mg/kg)	LD_{50}, HeLa (M)	Relative Toxicity, Mice	Relative Toxicity, HeLa
Levallorphan		4×10^{-5}		12
Levorphanol	45[11], 41[12] 41.5[13]	5×10^{-5}	5–6	10
Dextrorphan	65[13], 75[11]	9×10^{-5}	3–4	6
Demerol	32[14]	$2–3 \times 10^{-4}$	5	2–3
Nalorphine	123[14]	4×10^{-4}	1.4	1.2
Morphine	225[11], 230[12]	5×10^{-4}	1	1
Dihydromorphine	180[14]	5×10^{-4}		1

TABLE V[a]
Inhibition of HeLa Cell Growth After 48 hr of Drug Treatment

	Compound		Percentage Inhibition of Growth at Molar Concentration		
			1×10^{-3}	1×10^{-4}	1×10^{-5}
I	N-allyl normorphine		57	0	0
II	N-allyl-3-hydroxy morphinan (levallorphan)		>100	30	0
III	N-methyl-3-hydroxy morphinan (levorphanol)		>100	16	0
IV	N-methyl-3-methoxy morphinan		l >100	13	0
V			d >100	13	0
VI	N-allyl-3-methoxy morphinan		>100	35	8
VII	phenazocine		>100	100	10
VIII	d,l-methadone		>100	40	0

[a] HeLa monolayers were treated with drugs and incubated in 5% CO_2 and air at 37°C. After 48 hr the cells were trypsinized and counted. Thebaine methiodide was not tested for inhibition of cell growth. In calculating the percentage inhibition of cell growth, the baseline was taken as the number of cells per monolayer at the time of addition of the compound under study. The percentage inhibition at 48 hr is then equal to

TABLE V (Continued)

	Compound		Percentage Inhibition of Growth at Molar Concentration		
			1×10^{-3}	1×10^{-4}	1×10^{-5}
IX	morphine		13	0	0
X	codeine		5	5	0
XI	dihydrocodeine		5	0	0
XII	dihydrocodeinone		5	5	0
XIII	thebaine		100	20	0
XIV	dihydrothebaine		20	0	0
XV	thebaine methiodide		—	—	—

$$\frac{\text{Control cells} - \text{treated cells}}{\text{Control cells} - \text{baseline cells}} \times 100$$

Thus, when the cells were killed, the percentage inhibition was greater than 100 and is so indicated in the table. An almost complete kill of cells occurred with drugs II–VIII at 1 mM by 48 hr. (From Noteboom and Mueller,[15] Table 1.)

These studies were carried out in monolayer cultures. While the relative inhibitory effectiveness of compounds that were tested in both studies agree well, the absolute inhibitory effectiveness was lower in the monolayer cultures. Thus levorphanol, which caused 50% inhibition of clone formation at $5 \times 10^{-5}M$, required a concentration about tenfold higher to produce a 50% inhibition of growth in monolayers. This discrepancy has been observed in our laboratory and has been ascribed to an "inoculum effect." For some reason, presently obscure, it requires less drug to inhibit the growth of 200 cells than of 2×10^5 cells. It is possible that the large numbers of cells present in the monolayer produce a substance that competes with or inactivates the drug, while in 200 cells the quantity formed of such a substance may be very small. This subject requires further investigation.

An interesting recent, and as yet unpublished, study comes from the Hadassah Medical Center in Jerusalem. E. Heller and A. Melinek (private communication) studied the effect of levorphanol on primary chick embryo cell cultures and on the production of viruses and interferon in such cultures. They found that marked inhibition of cell growth became manifest about 12 hr after the addition of 100–200 μg/ml of levorphanol to the culture medium. Spontaneous recoveries were frequently seen 36 hr after drug administration. These were thought to be the result of the drop in pH of the medium. Microscopic examination of levorphanol-treated cells showed cytological abnormalities, the most pronounced of which was cytoplasmic vacuolization. These effects were seen earlier than inhibition of growth and persisted after recovery of growth potential. Interferon production was found to be inhibited by levorphanol under conditions that did not inhibit cell growth when the drug was added shortly before or after an inducing virus.

The replication of a DNA virus (vaccinia) as well as an RNA virus (sinbis) was inhibited by levorphanol under conditions that did not affect host cell growth. The inhibition occurred within a one-step growth cycle of the virus.

The authors have not yet carried out biochemical investigations in their system. They speculate that the inhibition of interferon synthesis, virus replication, and cell reproduction may be the result of an effect of levorphanol on protein synthesis at the ribosomal level.

B. Microorganisms

Early reports in the literature indicated that single cell organisms, including bacteria and protozoa, are relatively insensitive even to very high levels of morphine.[16] Our findings in cell culture, described earlier, indicating that there are drugs considerably more toxic than morphine itself, led us to reexamine this question.

We found that E. coli was indeed little affected by morphine but that levorphanol inhibited bacterial growth at a concentration of about 3 mM in a minimal growth medium.[17] This growth inhibition is reversible as shown by viable cell counts. Only at concentrations considerably higher than those required for growth inhibition does one begin to see significant reduction in the number of viable cells.

The relative potencies of a variety of morphine congeners for bacteriostasis were in the same order as those shown for HeLa cells in Table IV. The allyl analogs, nalorphine and levallorphan, were again slightly more toxic than the corresponding parent compounds, and it has not been possible to demonstrate antagonism between them.

A study of optimal conditions for growth inhibition of *E. coli* by levorphanol showed that effectiveness increased as the pH of the growth medium was raised and as the concentration of divalent cations, in particular Mg^{2+} and Ca^{2+}, was kept as low as possible. Thus at pH 8.2 and $10^{-4}M$ magnesium complete arrest of growth was seen at about 1 mM levorphanol, while retardation of growth could be observed at 0.1 mM.

Other bacteria including *Bacillus subtilis*, *Bacillus megaterium*, *Micrococcus leysodeikticus*, and *Diplococcus pneumoniae* were all found to be quite sensitive to growth inhibition by levorphanol. More recently Gale[18] reported growth inhibition of *Staphylococcus aureus* by levorphanol, levallorphan, and heroin.

Protozoa, too, are sensitive to the presence of morphine-like compounds. Zimmerman[19] showed that 3–5 mM morphine was highly toxic to *Amoeba proteus*. Recent observations in our laboratory (E. J. Simon and A. Drexler, unpublished results) show that the growth of *Tetrahymena pyriformis* is blocked by 1–1.5 mM levorphanol in Difco tetrahymena broth at pH 7.0.

IV. METABOLIC STUDIES

The most extensive studies of metabolic effects by drugs of the morphine type in single cells have been carried out with bacteria. The studies in other systems are, in fact, largely based on the bacterial studies. Therefore, we shall begin the discussion on these biochemical studies in an evolutionary order.

A. In Bacteria

Metabolic studies were initiated in *E. coli* several years ago in an attempt to delineate the biochemical effects of morphine analogs that result in bacteriostasis. Levorphanol was used in most of these studies but other agonistic and antagonistic congeners of morphine (including morphine itself) also have been tested.

1. Effect on Macromolecular Synthesis

An effect of levorphanol observed promptly after its addition to a logarithmically growing culture of *E. coli* was a marked reduction in the incorporation of ^{32}P-labeled orthophosphate into the acid precipitable fraction.[20] Double labeling experiments of the type depicted in Fig. 6 indicated that the inhibition was specific for the labeling of RNA, while

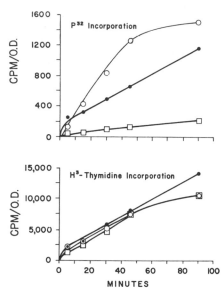

Fig. 6. Simultaneous determination of incorporation of ^{32}P into RNA nucleotides and of ^{3}H-thymidine into the acid-insoluble fraction. *E. coli* 15 T$^-$A$^-$U$^-$ was grown in low phosphate medium at pH 8.0, supplemented with casamino acids (0.2%), uracil (25 μg/ml), and thymidine (5 μg/ml). The doubling time was 45 min. Immediately after treatment with levorphanol or chloramphenicol, where indicated, tritiated thymidine (the final concentration was 20 μg/ml, 2 μc/ml) and ^{32}P (the final concentration was 0.015 M and 0.5 μc/ml) were added to the cultures. Upper curves show incorporation of ^{32}P into RNA nucleotides. Lower curves show incorporation of ^{3}H-thymidine into acid-insoluble fraction. ● = Growing culture; ⊙ = 100 μg/ml chloramphenicol; and □ = 1.8 × 10^{-3} M levorphanol. (From Simon and Van Praag,[20] Fig. 2.)

incorporation of thymidine into DNA continued undiminished for at least one generation. The culture treated with chloramphenicol is included as a nongrowing control in which nucleic acid synthesis is not inhibited.

Amino acid incorporation into proteins showed only a slight decrease in the presence of concentrations of levorphanol which reduced RNA synthesis by 85–90%. Chemical analysis showed that net accumulation of RNA was effectively blocked by levorphanol while protein and DNA continued to accumulate. This observation rules out the possibility that the observed decrease in isotope incorporation into RNA is the result of reduced uptake by the cells of labeled RNA precursors. This is particularly important since, as will be discussed later, levorphanol has been found to alter the permeability properties of membranes.

Treatment with levorphanol of nongrowing cultures of *E. coli* able to synthesize RNA (cultures treated with chloramphenicol or methionine-starved cultures of the "relaxed" strain 58–161, described below) demonstrated that growth and protein synthesis are not required in order for the effect of the drug on RNA synthesis to become manifest.

A similar selective effect on RNA synthesis by levorphanol has been observed in the gram-positive bacterium, *B. subtilis*,[21] while a recent abstract by Gale[18] indicates that treatment of *S. aureus* with morphine analogs causes an inhibition of both RNA and protein synthesis.

Greene and Magasanik[22] reported that levorphanol and levallorphan at much higher concentrations than those used by Simon *et al.* (5–10 mM instead of 1–2 mM) cause effective inhibition of all macromolecular syntheses in *E. coli*.

2. Effect of Levorphanol on the Synthesis of Different Classes of RNA

The lack of effective inhibition of protein synthesis at concentrations at which RNA synthesis was reduced to a low level led us to examine the effect of levorphanol on the synthesis of the different classes of RNA.[23]

Since ribosomal RNA (*r*RNA) comprises 85% of the total cellular RNA of *E. coli* the extent of reduction of RNA synthesis (85–90%) left little doubt as to the inhibition of the synthesis of this class of RNA. This was confirmed by showing strong inhibition of ^{14}C-uracil incorporation into the RNA of partially purified ribosomes.

To study the effect of levorphanol on the synthesis of messenger RNA (*m*RNA) a number of procedures were utilized that are thought to reflect primarily the production of *m*RNA. These included induction of the enzyme β-galactosidase (whose prompt turning on and off led Jacob and Monod to postulate the existence of a messenger), labeling with RNA precursors for very brief periods (pulse labeling) or after transfer of the culture from a rich to a minimal growth medium (shiftdown), as well as the reproduction of DNA-containing bacteriophages. The latter must induce production of a phage-specific *m*RNA in order to reproduce. All of these processes were only slightly inhibited by levorphanol (20–50%) at concentrations that reduced *r*RNA synthesis by 90% or more.

The synthesis of transfer RNA (*t*RNA) was measured by two techniques. The first involved incorporation of ^{14}C-uracil into the *t*RNA (4S) peak separated by sucrose density gradient centrifugation. The second method involved the use of the "relaxed" *E. coli* strain 58–161.[24] This strain appears to have lost a control mechanism whereby the synthesis of RNA is regulated by the availability of essential amino acids. In the absence of the required amino acid methionine, it cannot grow but, unlike stringent strains, con-

TABLE VI
Effect of Levorphanol on Methylation of "Starved" sRNA

Levorphanol (M)	Uracil incorporation (cpm)	Percent of Control	Methylation of sRNA (μmoles)[a]	Percent of Control
0	20,600	100	0.241	100
1.1×10^{-3}	9,800	47	0.204	85
1.8×10^{-3}	3,100	15	0.148	62

[a] mμmoles of methyl-^{14}C incorporated in 60 min when sRNA (500 m μmoles of nucleotides) was incubated with ^{14}C-methyl-S-adenosylmethionine and the enzyme that converts uracil to thymine in sRNA. (From Simon and Van Praag[23], Table 3.)

tinues to synthesize RNA for at least 2–3 hr. The RNA made during this period is devoid of methylated bases. The amount of tRNA made during starvation, in the presence and absence of levorphanol, can be measured by the extent of introduction of labeled methyl groups from ^{14}C-S-adenosylmethionine into the soluble RNA fraction, using one of the highly specific tRNA methylases isolated by Gold and Hurwitz.[25] The results of such experiments showed that tRNA synthesis was inhibited only relatively slightly as compared to rRNA synthesis (Table VI).

The results of all this work demonstrated a selective inhibition of rRNA synthesis by levorphanol. The synthesis of mRNA and tRNA is decreased only 20–50% when rRNA synthesis is inhibited 90–95%.

More recently in the laboratory of Dr. P. Fromageot in Saclay, France,[26] we showed that levallorphan exerts the same selective effect on rRNA synthesis as levorphanol. This study utilized a totally different approach. Control and treated cultures were exposed to radioactive uridine labeled with different isotopes (^{14}C and ^{3}H, respectively). The cultures were mixed and extracted together. Different classes of RNA were separated on a methylated albumin–Kieselguhr (MAK) column by the method of Mandell and Hershey.[27] The degree of inhibition was determined by measuring the

TABLE VII
The Relative Synthesis of RNA Classes at Various Concentrations of Levallorphan

Concentration of Levallorphan (mM)	4-5 RNA	16-S and 23-S RNA Synthesis[a] (%)	5-S RNA
1.0	87	55	34
1.1	65	30	28
1.2	40	17	10
1.5	26	12	10

[a] Calculated from ^{3}H/^{14}C ratios. The isotope ratio in DNA was taken as 100%. (From Roschenthaler et al.,[26] Table II.)

^{3}H/^{14}C ratio under the various peaks. This study not only confirmed, with levallorphan, our earlier results obtained by different methods with levorphanol but showed that the synthesis of 5S RNA is inhibited to about the same extent as the synthesis of the high molecular weight 16S and 23S ribosomal RNA's (Table VII). This is of considerable interest since it provides evidence that 5S RNA behaves like a ribosomal RNA in the manner in which the rate of its synthesis is controlled rather than like the 4S RNA, which it resembles closely in size.

3. Effect of Levorphanol on Membrane-Related Processes

Evidence has now come from several laboratories indicating that levorphanol and related drugs produce alterations in processes involving the cell membrane such as transport of substances in and out of cells and phospholipid metabolism.

In a study by Simon, Cohen, and Raina[28] of the possible relation between polyamines and levorphanol (see below) the levels of putrescine

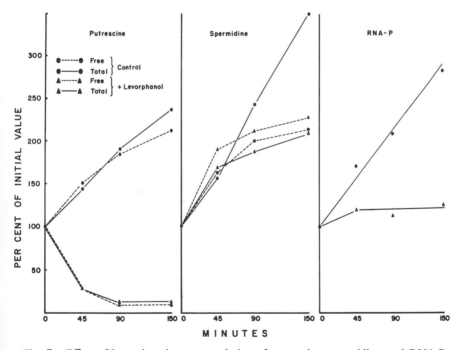

Fig. 7. Effect of levorphanol on accumulation of putrescine, spermidine, and RNA-P in *E. coli* 15 T$^-$A$^-$U$^-$. Results are expressed as percent of the initial content. Aliquots (150 ml) of the cultures were removed at the indicated intervals and analyzed for polyamines and RNA, as described by Raina and Cohen[31]. The actual zero time values for cellular polyamines and RNA-P in millimicromoles were putrescine-free, 593; total, 641; spermidine-free, 203; total, 282; and RNA-P, 8100. (From Simon, Cohen, and Raina,[28] Fig. 4.)

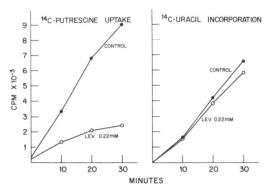

Fig. 8. Effect of a low level of levorphanol on putrescine uptake and RNA synthesis. A log phase culture of *E. coli* B leu⁻ was grown in a minimal medium containing $1.2 \times 10^{-2}M$ phosphate buffer, pH 7.3, $10^{-4}M$ MgCl₂, 2% NH₄Cl, 100 μg/ml leucine, and 0.1% sodium lactate. To a portion of the culture ¹⁴C-putrescine ($8 \times 10^{-7}M$, specific activity 5 mc/mmole) was added. Aliquots (2 ml) were taken at intervals, filtered through millipore membranes, and washed with cold medium. To another portion of the culture ¹⁴C-uracil (0.02 μc and 1 μg/ml) was added. Aliquots were collected in equal volumes of 10% TCA, filtered through millipore membranes, and washed with 5% TCA. All millipore membranes were placed in counting vials containing scintillation fluid and counted in a Packard Tricarb liquid scintillation counter.

and spermidine were determined in levorphanol-treated *E. coli*, as shown in Fig. 7. A striking and unexpected finding was the disappearance within 45–60 min of 90% of the cellular putrescine pool. More recent studies[29] have demonstrated that this is the result of putrescine efflux into the growth medium. Evidence has been obtained for two independent effects of levorphanol that result in rapid excretion of putrescine. Putrescine efflux is greatly stimulated, while the uptake of putrescine by *E. coli* cells is markedly inhibited. This latter effect is the most sensitive so far observed. As shown in Fig. 8, putrescine uptake is decreased by 80% at 0.2 m*M* levorphanol, a concentration that is virtually without effect on RNA synthesis. Alterations of transport also have been observed in *E. coli* for other small molecules and ions that include K⁺, spermidine, and some amino acids.

Greene and Magasanik[22] reported that levorphanol and levallorphan caused a rapid disappearance of cellular ATP and GTP pools from *E. coli*. The triphosphates were hydrolyzed, and the bulk of the hydrolysis products appeared in the growth medium (Table VIII).

These investigators also reported that levallorphan treated cells were unable to concentrate ^{14}C-methylthio-β-D-galactoside against a concentration gradient.

Gale[18] found that levorphanol and heroin in *S. aureus* caused a decreased uptake of certain amino acids, such as lysine and proline, and an increased uptake of other amino acids, such as aspartate and alanine. These results suggested alteration in membrane properties and led him to investigate the effect of the drugs on phospholipid metabolism. Glycerol incorporation into the phospholipids of *S. aureus* was found to be increased by heroin. Chromatographic examination of the phospholipids showed the presence of a new component. Qualitatively similar results were obtained with levorphanol and levallorphan, but the new components obtained differed in chromatographic properties from that obtained in the presence of heroin.

Similar experiments have been carried out in *E. coli* (Wurster, Elsbach, and Simon, unpublished observations). Overall ^{14}C-glycerol incorporation into phospholipids in the presence of levorphanol (1.3 mM) was not appreciably altered, but there was a marked decrease in the amount of label

TABLE VIII^a

The Effect of Inhibitors on Adenine Nucleotides of *E. coli* K12 During Aerobic Growth on Glucose

Experiment	Treatment	ATP	ADP	AMP	Adenosine or Adenine
1. Cells	None	100	13	24	13
Medium		10	0	0	0
2. Cells	5 mM levallorphan, 10 min	0	19	48	0
Medium		20	0	72	0
3. Cells	5 mM levallorphan, 20 min	0	0	10	0
Medium		8	30	132	0
4. Cells	5 mM levorphanol, 10 min	5	33	25	0
Medium		26	25	75	0
5. Cells	100 μg/ml chloramphenicol, 10 min	88	38	27	0
Medium		4	0	0	0
6. Cells	10 mM Na azide, 10 min	90	37	32	0
Medium		0	0	0	0

^{a} Cultures of *E. coli* K12 were prelabeled with adenine-^{14}C (10 μc/21.6 μg/mg) for one generation, filtered free of label, and resuspended in medium without radioactive adenine. They were then treated as indicated in the table, after which the cells and medium were separated by filtration and the nucleotides of each fraction were extracted and separated by thin-layer chromatography. The thin-layer chromatograms were counted in a strip counter. The radioactive peaks obtained were then cut out and weighed. The intracellular ATP peak of the control culture was assigned an arbitrary value of 100, and all other peaks were assigned a value relative to this peak. (From Greene and Magasanik,[22] Table 2.)

present in PE and to a lesser degree in PG. A new labeled component was formed, the radioactivity of which increased with time of incubation in the presence of drug.

It should be pointed out that the observed effects are subtle and readily reversible. Thus Greene and Magasanik find no effect on the rate of hydrolysis of *o*-nitro phenylgalactoside, a process for which transport is the rate-limiting step. Treatment of *E. coli* with levorphanol does not render them permeable to actinomycin D (Wurster and Simon, unpublished), a result obtained by treating cells with EDTA.[30] Moreover, removal of drug results in rapid resumption of growth, RNA synthesis, and return to normal membrane function.

It is not yet known whether these drug effects on membrane processes are exerted directly on the cell membrane or whether the primary site of action is elsewhere resulting in the observed apparent changes in membrane function. Another question as yet unanswered is the relationship, if any, between these effects and the inhibition of macromolecular synthesis discussed earlier.

4. The Mechanism of Action of Levorphanol in Bacteria

Studies on the action of narcotic analgesic drugs in bacteria have clearly uncovered a number of interesting and profound biochemical effects. The primary site of action that results in growth inhibition has not yet been defined, nor do we understand the biochemical mechanism by which the drugs elicit the observed metabolic alterations.

A number of hypotheses can be advanced to explain how levorphanol at growth-inhibitory levels can selectively inhibit the synthesis of *r*RNA in *E. coli*. Three hypotheses, on which some work has been done, are discussed below: (a) direct inhibition of RNA synthesis, (b) inhibition of the formation of ribosomal particles, and (c) effects on other metabolic pathways resulting indirectly in a reduction of *r*RNA synthesis.

a. Direct Inhibition of RNA Synthesis. The inhibitory effect of levorphanol and related drugs can be demonstrated on the incorporation of a variety of radioactive precursors as well as on the net accumulation of RNA. It is therefore attractive to postulate that the effect is exerted directly on RNA synthesis. This is made unlikely, but not ruled out, by repeated lack of success in producing inhibition in a cell-free system. Numerous experiments have been performed using purified RNA polymerase (Hurwitz, and Simon, unpublished observations) and crude *E. coli* extracts (Simon, unpublished observations), but no inhibition was seen even at very high levorphanol concentrations. Such negative findings cannot be regarded as conclusive. Very little ribosomal RNA is synthesized in cell-free systems, and the possibility has not been ruled out that levorphanol may require metabolism to an "active form," a reaction that may be lost during cell breakage. A direct effect, however, is thought to be unlikely on theoretical grounds since it would imply a recognition by a small molecule of the specific nucleotide sequence of the cistrons that code for the *r*RNA's.

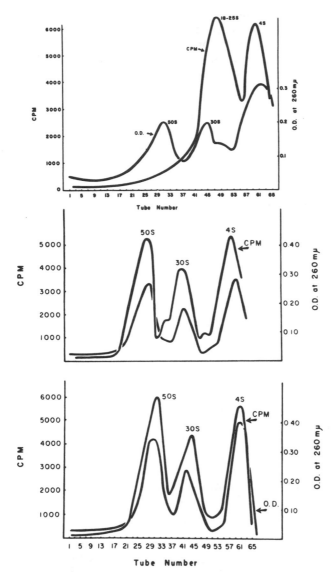

Fig. 9. Sucrose density gradient profile of extracts of *E. coli* 58-161 after methionine starvation (top) and after methionine replenishment in the absence (middle) and presence (bottom) of levorphanol. Culture was incubated for 1 hr in the absence of methionine and in the presence of ^{14}C-uracil. Cells were washed with isotope-free medium and resuspended. A portion of the culture was removed and chilled. The remaining culture was divided and incubated for 30 min in the presence of methionine, with and without levorphanol. Crude extracts were prepared from the three portions of the culture. These were centrifuged on a sucrose density gradient as described[23].

b. Effect on Formation of Ribosomal Particles. A selective effect could be explained readily if the inhibition were exerted on a reaction unique for the formation of ribosomes, for example, the "packaging" of *r*RNA and protein to form mature ribosome particles. Of course, one would have to postulate that such a block results in the feedback inhibition of *r*RNA synthesis.

An attempt to examine this possibility experimentally has been carried out (Simon and Wurster, unpublished). The relaxed strain of *E. coli* 58–161, when incubated in the absence of methionine, incorporates labeled RNA precursors into immature particles, known as "relaxed particles," but not into mature ribosomes. These particles, which sediment at 18-25S, contain the normal complement of 5S, 16S, and 23S ribosomal RNA's but only a fraction of the ribosomal proteins. When a culture containing prelabeled relaxed particles is reincubated in the presence of methionine the particles are rapidly converted to mature ribosomes. It can be seen from Fig. 9 that levorphanol does not interfere with this maturation process. Not only are 50S and 30S ribosomes formed, but their specific activity is the same whether formed in the presence or absence of drug. It is concluded that levorphanol does not interfere with the complexing of the "late" ribosomal proteins to the particles. An effect of the drug on an earlier step in ribosome formation is not ruled out, though perhaps it is made more unlikely.

c. Effects on Other Metabolic Pathways Resulting Indirectly in the Inhibition of rRNA synthesis. The hypothesis that the primary site of action of levorphanol is on another metabolic pathway that secondarily results in inhibition of *r*RNA synthesis has received support from the results of several laboratories.

Raina and Cohen[31] have reported evidence for the existence of a regulatory mechanism for the synthesis of *r*RNA that utilizes effectors of the polyamine type. This finding led to an examination of the effect of polyamines on the inhibition of RNA synthesis by levorphanol. Simon, Cohen, and Raina[28] showed that inhibition of RNA synthesis by levorphanol can be largely prevented (or reversed) by the addition of spermidine, as shown in Fig. 10. Putrescine at comparable concentrations did not exhibit this protective effect. Further studies are in progress to determine whether levorphanol, indeed, may exert its inhibitory effect by interfering with the normal stimulatory action of spermidine for RNA synthesis.

On the basis of their observation that levorphanol causes a rapid hydrolysis of cellular ATP, Greene and Magasanik[28] have suggested that the primary action of levorphanol may be to reduce the amount of available metabolic energy. The concentration of drug used by these investigators (10 mM at pH 7.0) was considerably higher than that used by Simon *et al.* (1.3 mM at pH 7.8). In order to explain the selective action of lower levels of levorphanol Greene and Magasanik postulated that a relatively slight decrease in ATP level may result in a "shiftdown."[32] This is a temporary arrest of stable RNA synthesis seen when bacteria are shifted from a medium containing an effective energy source to one with a poor one that

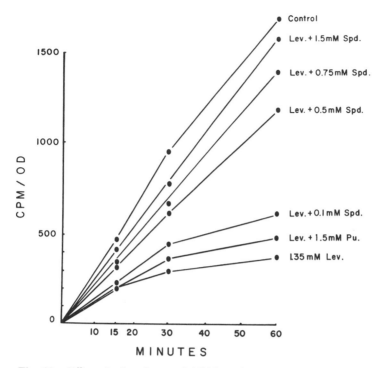

Fig. 10.   Effect of polyamines on inhibition of RNA synthesis by levorphanol. ^{14}C-uracil (7.2 μg/ml, 0.06 μc/ml) was added to a culture of *E. coli* K-13 in exponential growth 5 min after addition of levorphanol and polyamine. Samples were removed into 1 *M* perchloric acid at indicated intervals, filtered through millipore membranes, and counted as described previously.[20] The optical density at 550 mμ was measured in a Lumetron colorimeter. Spd = spermidine, Pu = putrescine, and Lev = levorphanol. (From Simon, Cohen, and Raina,[28] Fig. 1.)

will support a much slower growth rate (for example, glucose to succinate). This is thought to be a mechanism that adjusts the cellular ribosome content to that needed for a given growth rate.

This hypothesis still lacks experimental proof. Moreover, it is not consistent with some of the observed behavior of levorphanol. Thus Gale[18] found an increase in phospholipid turnover that is difficult to reconcile with a lack of available ATP. The efflux of putrescine has been found to require energy (Simon, Schapira, and Wurster, unpublished results). Levorphanol markedly accelerates this efflux, and this stimulation is virtually abolished by depriving the cells of a carbon source or by the addition of metabolic inhibitors such as 2,4-dinitrophenol or sodium arsenite. Figure 11 shows an experiment in which there was considerable spontaneous putrescine efflux that was further accelerated by levorphanol. Absence of a carbon source virtually prevented both the spontaneous efflux

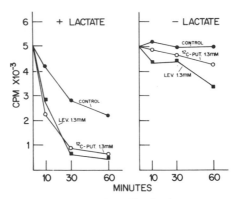

Fig. 11. Putrescine efflux in the presence
and absence of a carbon source. A culture
of *E. coli* K-13 was grown in a triethano-
lamine-buffered minimal medium.[20] The
culture was prelabeled by incubation with
14C-putrescine for 30 min. Cells were washed and resuspended in the same medium
and incubated with additions indicated on
graph. Aliquots were processed as described in the legend of Fig. 8.

and the stimulation by levorphanol. It is clear that in this instance the
action of levorphanol is opposite to effects of treatments designed to
decrease available energy.

Finally, the suggestion may be put forth that the effect of levorphanol
on cell–membrane related processes may be the primary site of action.
This hypothesis suggests that the ATP breakdown observed by Greene and
Magasanik is the result of ATPase stimulation due to membrane damage.
It is not yet clear whether or not levorphanol acts directly on some portion
of the cell membrane, nor is it known what connection, if any, exists between the membrane alterations and the inhibition of *r*RNA synthesis
produced by the drug.

B. In Other Systems

1. Protozoa

A study by Zimmerman[19] on the effect of morphine on the protozoan
Amoeba proteus is of particular interest since it is the only instance in which
antagonism of a morphine effect by nalorphine was observed in a micro-
organism. This investigator utilized high pressure to study pseudopodial
stability. Increasing pressure tends to weaken the plasmagel structure of
amoebas by shifting sol–gel equilibria toward the sol state. When sufficiently
high hydrostatic pressure is applied the amoeba loses its pseudopodia and
becomes spherical due to surface tension. Morphine showed a distinct
stabilizing effect on the pseudopodia of *A. proteus*, as indicated by a

decrease in the percentage of cells that became spherical at a given pressure. Figure 12 shows the stabilizing effect of $2 \times 10^{-3}M$ morphine at 5000 lb/in.^{2} pressure. It also depicts the reversal of the morphine effect when morphine-treated cells were transferred to a medium containing nalorphine (nalline). It is suggested that morphine alters the sol–gel equilibrium within the cell and thus may affect the integrity of proteins associated with pseudopodial stability. Nalorphine appears to antagonize the action of morphine, perhaps by competing for a common site.

2. Mammalian Cells in Culture

Noteboom and Mueller[34] studied the metabolic effects of levorphanol and levallorphan in cultures of HeLa cells. They found that levorphanol and levallorphan inhibited both RNA and protein synthesis about equally at concentrations (1–2 mM) at which they had little effect on DNA and phospholipid synthesis. Transfer of the cells to a drug-free medium caused an immediate and rapid acceleration of RNA and protein synthesis. The resumption of protein synthesis was found to occur even when RNA

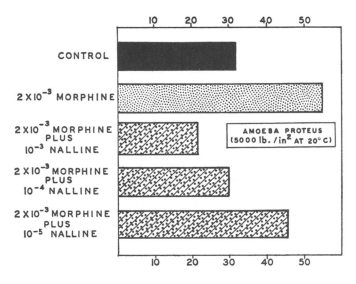

PSEUDOPODIAL STABILITY:

PERCENTAGE OF AMOEBA WITH SOME
PERSISTING PSEUDOPODIA

Fig. 12. Effects of varying concentrations of nalline (*N*-allylnormorphine) on pseudopodial stability of morphine-treated amoebae. Cells were treated for 1 hr in $2 \times 10^{-3}M$ morphine and then transferred to varying concentrations of nalline for 1 hr. Each value illustrates the percent of nonrounded specimens with some persisting pseudopodia after a 20-min exposure to a pressure of 5000 lb/in.^{2} at 20°C. (From Zimmerman,[19] Fig. 6.)

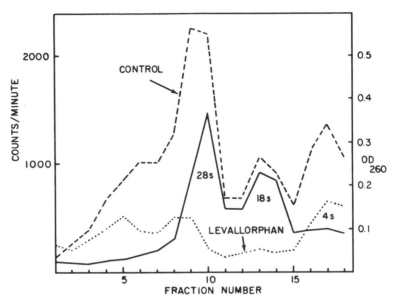

Fig. 13. Effect of levallorphan on the synthesis of various types of RNA. Cells were treated for 60 min with 0.001 M levallorphan. Guanine-8[14]C (0.1 μc/ml medium, 10 mc/mmole) added for 90 min. RNA extracted with phenol (2.5 optical density units, O.D.$_{260}$) layered on gradient —— = O.D.$_{260}$ of treated and control cells; - - - - = radioactivity of control cells and ...... = radioactivity of levallorphan-treated cells. (From Noteboom and Mueller,[33] Fig. 2.)

synthesis was prevented by the addition of actinomycin D. The sedimentation characteristics, in a sucrose density gradient, of the RNA made in levallorphan treated and control cultures are shown in Fig. 13. It can be seen that the incorporation of guanine-8-^{14}C into all classes of RNA is depressed by the drug (although the synthesis of 4S RNA appears to be inhibited less strongly). RNA polymerase activity *in vitro* was shown to be unaffected by the presence of the drugs, but the level of polymerase activity in isolated cell nuclei was found to be reduced in cells pretreated with 0.001 M levallorphan (Fig. 14). The authors also report that the level of polysomes is reduced in such cells and that ribosomes isolated from drug-treated cells are virtually unable to incorporate amino acids into proteins. They seem to lack a usable messenger since upon addition of poly-U they incorporate phenylalanine into polyphenylalanine as effectively as control ribosomes.

The authors feel that their results are more compatible with a primary action of the drugs on protein synthesis of which the reduction of the level of RNA polymerase may be a consequence. They point out, however, that they cannot exclude the possibility that the inhibition of RNA and protein synthesis may be the result of a direct effect on the metabolism of RNA, much like that seen in bacteria. The decrease in protein synthesis could be due to a superimposed effect, as has been suggested by Gale[18] for *S. aureus*.

A more recent report by Noteboom and Mueller,[15] already cited briefly, shows that the growth inhibition of HeLa cells by a group of morphine alkaloids and morphinans correlates well with their ability to inhibit protein and RNA synthesis. In an attempt to relate structure to activity the authors point out that substitution of the C ring as well as conversion of the tertiary nitrogen of the D ring to a quarternary state results in a drastic loss of inhibitory effectiveness. They suggest as one possible interpretation that a receptor may exist in HeLa cell membranes that recognizes the C and D ring regions of morphinans and related compounds. They do stress, however, that the possibility cannot be ruled out that the structural changes produce drugs that penetrate the cells less readily.

An interesting finding is the breakdown of polysomes to free ribosomes in cells treated with such narcotic analgesics as levorphanol, phenazocine, and methadone (Fig. 15). Although similar results were obtained with both *d*- and *l*-methorphan, treatment with thebaine, a nonnarcotic morphine congener, did not lead to polysome dissolution, although it is an effective inhibitor of protein synthesis. This suggests a different mode of action for

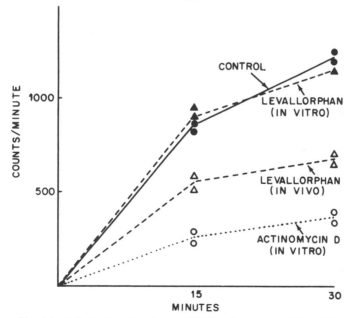

Fig. 14. Effect of levallorphan on RNA polymerase activity. Cells incubated for 60 min with 0.001 *M* levallorphan *in vivo* or 0.001 *M* levallorphan added *in vitro*. Each point represents the incorporation of CTP-^{3}H (1 μc, 1.25 c/mmole) by nuclei from 2×10^6 cells. (From Noteboom and Mueller,[33] Fig. 3.)

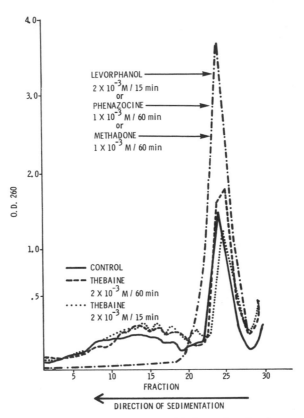

Fig. 15. Effect of treatment with levorphanol, phenazocine, methadone, and thebaine on polysome distribution
in HeLa cells. Cultures were treated with the drugs tested
for 15 or 60 min as indicated. Cells were harvested, and
polysomes were prepared as described. Polysomes from
40×10^6 cells were centrifuged through 10–40% linear
sucrose density gradients. The ordinate represents optical
density at 260 mμ. (From Noteboom and Mueller,[15]
Fig. 4.)

thebaine from that of the other drugs studied. Further confirmation for
this suggestion comes from the finding that microsomes isolated from
thebaine-treated cells incorporated ^{14}C-leucine into protein at a near-normal
rate while the protein-synthesizing ability of microsomes from levorphanol-
treated cells was greatly reduced (Table IX).

3. The Giant Axon of the Squid

At the Marine Biological Laboratory in Woods Hole, Massachusetts,
it was possible to examine the effects of narcotic analgesics on the giant
axon of the squid.[34] This single nerve fiber has been extensively studied

TABLE IX[a]

Effect of Thebaine and Levorphanol on Incorporation of ^{14}C-Leucine into Protein by Intact Cells and by a Microsomal System Derived from Treated Cells

Prior Treatment	Incorporation of ^{14}C-Leucine	
	Intact Cells[b] cpm/10 min	Microsomal System[c] cpm/60 min
Control	817	5837
Thebaine, 2 mM	82	5258
Levorphanol, 2 mM	82	993

[a] Cultures containing 350,000 cells/ml were treated with 2 mM levorphanol or thebaine for 15 min. To determine protein synthesis in intact cells, 5-ml aliquots were removed and incubated with 0.5 μc of DL-leucine-1-^{14}C for 10 min. The remaining cells were harvested, and a 15,000g supernatant fraction (microsomal system) was prepared which contained the cytoplasm equivalent to that of 40 × 10^6 cells/ml. Aliquots (0.1 ml) of this fraction were incubated for 60 min with supporting constituents anf 1.0 μc of L-leucine-^{14}C (240 μc/μmole) in a volume of 1.0 ml.
[b] Radioactivity incorporated by 1 × 10^6 cells.
[c] Radioactivity incorporated by supernatant fraction equivalent to 1 × 10^6 cells. (From Noteboom and Mueller,[15] Table 3.)

with respect to its electrical activity and transport of ions across the membrane.

Levorphanol at $10^{-3}M$ was found to block completely the evoked action potential when the axon was immersed in seawater at pH 8.0. The effect was readily reversible by placing the axon in drug-free seawater. The drug was much less effective in blocking conduction at lower pH. The experiment shown in Fig. 16 demonstrates inhibition of the action potential by levorphanol and virtually complete reversal of this effect by decreasing the pH to 6.0. The action potential returned to its original value when the axon was briefly placed in drug-free seawater. Other drugs also were tested, and

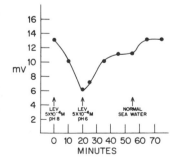

Fig. 16. Effect of pH on ability of levorphanol (LEV) to decrease the action potential of the giant axon of the squid. The ordinate represents the height of the action potential recorded with external electrodes.

once again the order of effectiveness was similar to that found for growth inhibition of HeLa cells and bacteria: Levallorphan > levorphanol = dextrorphan > nalorphine > morphine.

When the axon is placed in artificial seawater containing low concentrations of Mg^{2+} and Ca^{2+} (25% of normal) it exhibits a hyperexcitability known as spontaneous, repetitive firing. This electrical activity is blocked by $5 \times 10^{-5} M$ levorphanol.

While the concentration of drugs required to block conduction of the axon, or even to block spontaneous firing of the hyperexcitable axon, are inordinately high compared to the concentrations assumed to be pharmacologically active in the CNS of animals, the axon experiments provide further evidence for membrane alterations produced by narcotic analgesics and related drugs.

V. COMMENTS

The foregoing review of studies of biochemical sites of action of narcotic analgesics in single cells is far from exhaustive. Most of the more recent reports have been included with sufficient detail to permit the reader to form an idea of the types of techniques utilized and the kinds of results obtained.

A number of the studies reviewed have suggested that phenomena akin to tolerance and physiological dependence can be observed in mammalian cells in culture. From some of the conflicting reports it is apparent, however, that it is too early to be certain that single cells are useful for the study of these phenomena. Even if the findings of increased resistance to drugs and of "withdrawal symptoms" of cells are confirmed, it remains to be determined whether they represent adaptive changes similar to those seen in the CNS of animals and man.

On the other hand, the use of single-cell systems has permitted the detection of several interesting biochemical actions of drugs related to morphine. In bacteria, in which the most detailed studies have been carried out, effects on macromolecular synthesis on ATP levels, and on membrane-related processes have been observed.

One may well ask about the relevance of these findings in single cells to the mode of action of narcotic analgesics in man. The observed effects require relatively high drug concentrations, exhibit little stereospecificity and, in most instances, are not antagonized by compounds known to be effective antagonists of analgesic and narcotic effects of morphine and its analogs in animals and man. It is, therefore, tempting to dismiss these findings as totally irrelevant. However, it should be pointed out that stereospecificity and agonist–antagonist interaction may be the properties of highly specific receptors present only in the CNS of animals and man. Once the drug is attached to a receptor its biochemical action, which produces pharmacological results, may resemble that seen in a model system. In any case, the observation of striking, relatively specific metabolic alterations allows us to gain insight into the biochemical capabilities of narcotic analgesics. It is likely that this type of knowledge will ultimately prove useful for an understanding of the mode of action of these interesting and important compounds in the CNS of man.

VI. REFERENCES

1. S. Semura, *Folia Pharmacol. Jap. 17*, 34–45 (1933).
2. K. Sanjo, *Folia Pharmacol. Jap. 17*, 219–229 (1934).
3. K. Sanjo, *Jap. J. Med. Sci. IV, Pharmacol. 9*, 13 (1936).
4. M. Sasaki, *Arch. Exp. Zellforsch. 21*, 289–307 (1938).
5. T. Kubo, *Arch. Exp. Zellforsch. 23*, 269–277 (1939).
6. M. A. Heubner, E. Barocke, and H. Kewitz, *J. Mt. Sinai Hosp. 19*, 47–52 (1952).
7. W. C. McCormick and W. T. Kniker, *Tex. Rep. Biol. Med. 11*, 274–282 (1953).
8. G. Corssen and I. A. Skora, *J. Amer. Med. Ass. 187*, 328–332 (1964).
9. J. J. Painter, C. M. Pomerat, and D. Ezell, *Tex. Rep. Biol. Med. 7*, 417–455 (1949).
10. C. M. Pomerat and C. D. Leake, *in* "Tissue Culture Techniques in Pharmacology," *Ann. N.Y. Acad. Sci. 58*, 1110–1128 (1954).
11. K. Fromherz, *Arch. Int. Pharmacodyn. 85*, 387–398 (1951).
12. L. O. Randall and G. J. Lehmann, *J. Pharmacol. Exp. Ther. 99*, 163–170 (1950).
13. W. M. Benson, P. I. Stefko, and L. O. Randall, *J. Pharmacol., Exp. Ther. 109*, 189–200 (1953).
14. I. Shemano and H. Wendel, *Toxicol. Appl. Pharmacol. 6*, 334–339 (1964).
15. W. D. Noteboom and G. C. Mueller, *Mol. Pharmacol 5*, 38–48 (1969).
16. H. N. Krueger, N. B. Eddy, and M. Sumwalt, "The Pharmacology of the Opium Alkaloids," Part I, Suppl. 165, pp. 629–631, U.S. Public Health Reports (1941).
17. E. J. Simon, *Science 144*, 543–544 (1964).
18. E. F. Gale, *J. Gen. Microbiol. 55*, VIII–IX (1969).
19. A. M. Zimmerman, *J. Protozool. 14*, 451–455 (1967).
20. E. J. Simon and D. Van Praag, *Proc. Nat. Acad. Sci. U.S. 51*, 877–883 (1964).
21. E. J. Simon, N. Burton, J. Lyons, and L. Lyons, *Fed. Proc. 25*, 276 (1966).
22. R. Greene and B. Magasanik, *Mol. Pharmacol. 3*, 453–472 (1967).
23. E. J. Simon and D. Van Praag, *Proc. Nat. Acad. Sci. U.S. 51*, 1151–1158 (1964).
24. E. Borek, A. Ryan, and J. Rockenback, *J. Bacteriol. 69*, 460–467 (1955).
25. M. Gold, J. Hurwitz, and M. Anders, *Biochem. Biophys. Res. Commun. 11*, 107–114 (1963).
26. R. Roschenthaler, M. A. Devinck, P. Fromageot, and E. J. Simon, *Biochim. Biophys. Acta 182*, 481–490 (1969).
27. J. D. Mandell and A. D. Hershey, *Anal. Biochem. 1*, 66–77 (1960).
28. E. J. Simon, S. S. Cohen, and A. Raina, *Biochem. Biophys. Res. Commun. 24*, 482–488 (1966).
29. E. J. Simon, L. Schapira, and N. Wurster, *Bull. N.Y. Acad. Med. 45*, 500 (1968).
30. L. Leive, *Biochem. Biophys. Res. Commun. 18*, 13–17 (1965).
31. A. Raina and S. S. Cohen, *Proc. Nat. Acad. Sci. U.S. 55*, 1587–1593 (1966).
32. F. Neidhardt, *Progr. Nuc. Acid Res. 3*, 145–179 (1964).
33. W. D. Noteboom and G. C. Mueller, *Mol. Pharmacol. 2*, 534–542 (1966).
34. E. J. Simon and P. Rosenberg, *J. Neurochem. 17*, 881–889 (1970).

Chapter 16

THE NERVOUS SYSTEM*

Herbert L. Borison

Department of Pharmacology and Toxicology
Dartmouth Medical School
Hanover, New Hampshire

I. INTRODUCTION

Our purposes in this chapter are (1) to catalog actions of morphine on the nervous system, (2) to assign sites of action, and (3) to examine modes of action. Insofar as possible, the nervous system will be viewed as a communications device that receives, stores, and processes input signals and responds in measurable ways to the kind and quantity of information delivered to it.

The major clinical use of morphine for relief of pain depends on the least understood of its actions. Since pain, like love, is a personal and intangible feeling, we will concern ourselves only with the behavioral manifestations of responses to stimuli that may or may not be interpreted subjectively as painful. The difficulty of assessing pain outside the realm of human intercourse is well illustrated by Woods' description of the chronic decerebrate rat.[1] In this experimental preparation the midbrain is the highest functioning level of nervous organization which, by analogy in the human, means the absence of all conscious awareness. Indeed, the decerebrate state is accepted in Great Britain (at least for purposes of animal experimentation) as the legal equivalent of death. Woods states: "The rats nibbled at objects, edible and inedible, held against their snouts. They grasped pipettes with their forepaws and drank from them. The rats took care of their coats–a response which is lost in the cat following far less extensive brain removal. Finally, nociceptive stimulation elicited typical rodent defensive behavior (vocalization, attempts to escape, and use of claws and teeth) with accurate localization of the stimulus." Did the rat

---

* Rosaline Borison shared in large measure the burden of preparing this chapter. Previously unpublished experimental observations made in the author's laboratory were supported by PHS Grant NB 04456.

sans cerebrum feel pain? It behaved as if in pain. We have only to ponder what makes us different from the rat to appreciate the agony of pain (see Beecher[2] for further discussion).

Development of drug tolerance implies two kinds of adaptation that are not always differentiated. Simple elevation in dose threshold, without associated behavioral compensation, does not necessarily invoke partici-pation of a neural factor in the tolerance development.[3] On the other hand, if any modification of degree and/or quality of responsiveness accompanies the change in drug requirement, then a suitable mechanism must invoke participation of a neurologic process. We will assume through-out this chapter that compensatory neural behavior contributes to the development of tolerance in all of the forms described.

Mechanisms proposed to explain tolerance and physical dependence, whether at the biochemical or physiological level, invariably lead to a dynamic representation involving a slow negative feedback process that operates to counteract a fast initiating or modifying action of the addictive agent.[4] This concept is generalized using systems language in Fig. 1. If morphine (M) activates an input to a regulatory system or alters the ongo-ing level of activity by affecting its setpoint, the detected displacement (error signal) is translated rapidly into effector action. However, restoration of initial conditions (development of tolerance) occurs slowly as the negative feedback level rises in the capacitive link between output and detector. When the direct action of morphine is discontinued, stored feedback reverses the error signal until the capacitor is discharged (physical depen-dence). Morphine may be reinstituted at any time to balance the negative feedback overshoot through the rapid path to the effector. This exercise fails, of course, to attach biological elements to the system components. Nevertheless, it points up the uselessness of particularizing a mechanism for tolerance and physical dependence when specific working details are missing.

Synaptic transmitter substances are for obvious reasons highly attrac-tive fishes in the pool of addiction hypotheses. But only acetylcholine has withstood all accepted tests for a neurotransmitter role, and this is limited to junctions in the central recurrent paths of motor axon collaterals.[5] Indeed, insufficient evidence is available to support unequivocally the roles

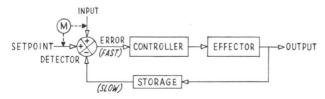

Fig. 1. Control systems scheme for development of tolerance and physical dependence. M points to sites of initiating or modi-fying actions of morphine. See the text for further explanation.

of acetylcholine or any of a host of amines and amino acids found in the central nervous system, either as excitatory or inhibitory neurotransmitters, at primary afferent and interneuronal synaptic connections. Accordingly, we cannot profitably concern ourselves within the purview of this chapter with presumed transmitter involvement in the discrete actions of morphine on reflex activities.

Electrophysiological monitoring of neural impulse traffic and associated phenomena constitutes a most powerful means of studying functions of the nervous system. We must not, however, lose sight of the limitations of such observations, especially as they relate to the more complex expressions of neural organization. It is necessary to take into account the increasing divergence of signal distribution with ascending progression in the neuraxis and to recognize the hazards of functional extrapolation in the absence of behavioral concomitants. Unfortunately, criteria for valid physiological interpretation of electrical events in the nervous system are all too often ignored.

Actions of a drug on the nervous system can be identified in three ways: (1) through alteration of a spontaneous function; (2) through alteration of an evoked activity; and (3) through elicitation of a novel response. The first two kinds of actions may be termed "modifying actions" and suggest forms of intervention in normally active circuits. The third type is conveniently labeled an "initiating action" and suggests activation of a normally silent circuit. This chapter will scrutinize the modifying and initiating actions of morphine in successively more complex spheres of central neural organization starting with the spinal cord segment.

II. THE ISOLATED SPINAL CORD

A. Basic Neural Circuitry

Examples of reflex circuits known to operate at the segmental level—the lowest order of central control—are illustrated in Fig. 2. According to Eccles,[5] the monosynaptic pathway (two-neuron arc) that effects the muscle stretch reflex response operates solely through excitatory transmission. Central postsynaptic inhibition requires the interposition of an inhibitory neuron. The elemental reflex functions of the spinal cord segment consist of (1) the fast local excitatory monosynaptic motor unit contractile response to spindle stretch, (2) inhibition of the motoneuron through a disynaptic pathway, such as originates in the muscle tendon when stretch tension becomes excessive, (3) ipsilateral limb withdrawal in response to appropriate skin stimulation transmitted through a polysynaptic pathway, which results in excitation of flexor muscles and simultaneous inhibition of extensor muscles. Intensification of the skin stimulus with recruitment of interneurons and spread of motor activity results successively in contralateral limb extension (crossed extensor reflex), rhythmic kicking movements, and ultimately a convulsive discharge (mass reflex) of the

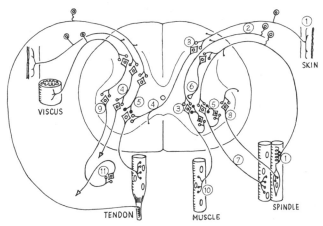

Fig. 2. Possible sites of drug influence at the level of the spinal cord segment. ①. Sensitivity of the input receptor. Examples given are free nerve endings in the skin, the muscle spindle stretch receptor, the golgi tendon organ, and visceral nerve endings. ②. Afferent nerve conduction into the spinal cord. The associated dorsal root ganglion cell is considered to have a trophic function only. ③. Primary excitatory synaptic transmission resulting in chemical generation of initial depolarizing excitatory postsynaptic potential (EPSP) and activation of monosynaptic stretch reflex response. ④. Excitatory interneuronal relay transmission responsible for chemical generation of depolarizing EPSP in alpha-motoneurons. ⑤. Inhibitory interneuronal relay transmission responsible for chemical generation of hyperpolarizing postsynaptic potential (IPSP); includes efferent recurrent inhibition of alpha-motoneuron by Renshaw cell. Filled end knobs indicate presence of inhibitory transmitter. ⑥. Presynaptic inhibition. Sustained partial depolarization of primary afferent nerve endings (through excitatory interneurons) reduces action potential mobilization of excitatory transmitter from those nerve endings, thereby diminishing resultant EPSP on motoneuron. ⑦. Efferent nerve conduction in ventral root leaving the spinal cord. ⑧. Small motor nerve (gamma-motoneuron efferent) activity responsible for intrafusal (spindle) muscle contraction and adjustment of stretch receptor sensitivity. ⑨. Sympathetic preganglionic nerve activity responsible for visceral control. ⑩. Skeletal neuromuscular transmission. ⑪. Autonomic ganglion transmission to postganglionic neuron that conducts impulses to visceral effectors.

spinal segments involving autonomic as well as somatic effectors. The degree of responsiveness depends in good measure on the chronic condition of the isolated spinal cord.

A more complex reaction related to the use of the leg as a supporting structure is the "extensor thrust" that results in simultaneous excitation of extensor and flexor muscles in response to the spread of the toes. This reflex effect as well as the stepping movements that are observed in the spinal animal represent complicated patterned forms of behavior that are laid down at the segmental level and that generally emerge only after longstanding cord separation from the brain. This chronic augmentation of spinal cord excitability has been attributed to development of "denervation supersensitivity," but this is no more than a restatement of the problem.

In addition to the basic reflex motor patterns built into the spinal segment, three sensitivity controls are available for the adjustment of responsiveness. Intersegmental and suprasegmental influences figure importantly in the modulation of these sensitivity controls. Certainly this is the case for the small motor nerve (gamma efferent) mechanism that adjusts the level of spindle stretch receptor sensitivity through contraction of the series-connected spindle muscle fiber. The Renshaw cell serves to complete an internal negative feedback loop in the large (alpha) motoneuron outflow. Blockade of this inhibitory pathway, as with strychnine, results in convulsions of spinal origin. The most intriguing segmental sensitivity control mechanism, whose significance has yet to be appreciated, is the input negative feedback process known as primary afferent depolarization (PAD), which operates presynaptically through dorsal horn recurrent *excitatory* interneurons to reduce the availability of transmitter for the generation of EPSP. PAD can be detected electrically on the dorsum of the spinal cord or as the so-called dorsal root reflex potential that appears in neighboring rootlets when one is stimulated. It is believed that the recurrent modulation of input gain may have an important and widespread influence on functions of the nervous system.

A presynaptic gating theory for the control of pain input at the segmental level has been proposed by Melzack and Wall[6] to account for the apparent modulation of a pain-inducing stimulus through behavioral conditioning or by application of concurrent nonpainful stimuli. Thus modality-specific inputs via large afferent fibers (touch, position, and some temperature impulses) would tend to close the pain gate through local as well as long-distance connections, whereas nonspecific inputs via small afferent fibers (nociceptive impulses) would tend to open the gate. This gating theory has been challenged by Viklicky et al.,[7] who have been unable to obtain the kind of dorsal root potential interaction predicted by the theory. Furthermore, the arguments advanced by Melzack and Wall do not lead exclusively to localization of the pain "gate" to the primary input synapse. The idea, however, that drugs may be able to modify presynaptic gating processes[8] offers an exciting prospect for the pursuit of chemical effects on the transmission of sensory information at all levels of nervous integration.

A variety of autonomic reflex functions and viscerosomatic interactions is maintained in elemental form at the segmental level. Urogenital and anorectal responses including appropriate posturing, vascular effects (both visceral and in skin), sweating, and piloerection (gooseflesh), and even involvement of liver glycogenolysis have been described in the spinal animal.[9] So it is evident that the functional substrates for certain direct actions of the narcotic analgesics, and particularly for the expression of signs of narcotic withdrawal, are contained within the basic neural circuitry of the spinal segment.

B. Alterations of Reflex Behavior

The modifying actions of morphine on functions of the isolated spinal cord in the nontolerant animal, without regard to level of transection, duration of cordotomy, or species examined, are summarized as follows:[10-17] (1) Spontaneous activity is depressed; (2) the monosynaptic reflex (stretch, and knee jerk) is usually unchanged or slightly facilitated, but it can be depressed by a large dose; (3) simple somatic polysynaptic reflexes (flexor withdrawal, tail flick, crossed extensor, Philippson's tendon inhibition and reciprocal motor effects, and skin twitch) are depressed as are those of visceral origin; (4) the extensor thrust reflex, involving more complex polysynaptic organization, is usually enhanced to a considerable extent.

A large body of evidence obtained from both *in vivo* and *in vitro* studies dismisses any significant direct peripheral depressant action of morphine on physiological receptor excitability, on conduction characteristics of any of the fiber types, on autonomic transmission, or on motor and secretory functions (smooth muscle activation may be an exception).[18-23] Possible influence of morphine upon high frequency nerve conduction failure deserves further exploration.[24]

A number of further negative inferences may be drawn from the ineffectiveness of morphine to modify the stretch reflex. Thus morphine does not interfere with spindle receptor excitation, conduction in large afferents, primary afferent depolarization (presynaptic inhibition), monosynaptic EPSP, recurrent (Renshaw cell) motoneuron inhibition, large efferent conduction, neuromuscular transmission, and muscle contraction. Kruglov,[25] however, has described variable depression by morphine of recurrent monosynaptic inhibition that could account for the occasional enhancement of the knee jerk reflex reported by others.

The ability of morphine to modify segmental excitatory polysynaptic reflex behavior introduces the first major clue on the central action of the drug. Depression of a wide variety of multineuronal responses of nociceptive origin—ipsilateral, contralateral, and intersegmental—whether from skin or viscera, points to a selective action on an interneuronal element (see Number 4 in Fig. 2).[26,27] The polysynaptic depressant effect of morphine is, understandably, most evident in the chronic high spinal preparation in which interneuronal excitability is expected to be greatest by comparison with low transections.

Of considerable importance is the fact that the depressant action of morphine extends as well to inhibitory processes.[26] The finding of Renshaw cell inactivation by Kruglov[25] has already been mentioned. Further evidence is provided by abolition of postural reciprocal inhibition, elimination of nociceptive (exteroceptive and interoceptive) inhibition of the stretch reflex, and interference with the tendon inhibitory reflex (see Number 5 in Fig. 2). Since all forms of central postsynaptic inhibition are thought to require an inhibitory interneuron and only polysynaptic reflexes are depressed by morphine, the common site of action is logically narrowed down to the input junction of the ubiquitous interneuron, whether excitatory or inhibitory.

An apparent inconsistency with the postulated interneuronal blocking action of morphine is the reported facilitation of the extensor thrust reflex. This response is unique, however, in that it is best explained as resulting from a generalized disinhibition of the stretch reflex initiated by large proprioceptive afferents from the toe joints. Thus further loss of inhibition through the action of morphine serves to enhance the reflex response. Variability of effect can be attributed to nonuniformity of technique in eliciting the extensor thrust response by variable contamination of the proprioceptive input with nociceptive stimulation.

While morphine initiates certain effects in the intact animal, it has essentially no direct initiating actions on the isolated spinal cord. Thus systemic doses of morphine that produce cephalic convulsions leave the separated portion of the spinal cord unaffected.[28] It is evident, therefore, that morphine does not share the segmental anti-inhibitory action of strychnine as an antagonist of transmitter-induced IPSP, but it acts instead to promote descending facilitation of supraspinal origin. We have observed, however, in the cat with a tight ligature placed around the spinal cord and meninges above the splanchnic outflow, that injection of morphine into the caudal subarachnoid space evoked movements of the hindquarters but did not produce hyperglycemia.[29] It has not been established whether the effect of topically applied morphine in the cerebrospinal fluid is due to a direct action of the drug or results from an indirect action as, for example, the release of histamine. The failure of morphine to elicit hyperglycemia despite the availability of a complete segmental arc to the adrenal medulla indicates that splanchnic activity is largely dependent on descending control. Intrathecal injection at the same level in an intact cat elicited behavioral excitement as well as hyperglycemia effected through adrenomedullary secretion.

Valdman and Arushanyan[26] have pointed out similarities between effects of morphine and ischemia on spinal reflex patterns. The possibility that morphine may be able to produce selective effects on the nervous system through local alteration of blood supply should be given serious consideration.

C. Tolerance and Withdrawal

Development of tolerance is well manifested at the segmental level by the progressive restoration of predrug polysynaptic reflex behavior with increasing resistance to depression from elevated doses of morphine. As an exception, the extensor thrust response continues to be hyperactive in the otherwise tolerant spinal dog. Abrupt withdrawal of morphine or the administration of an antagonist results in hyperactive reflexes where initially they were depressed, while the extensor thrust and even the knee jerk become depressed as the inverse of their initial modification by treatment with morphine. The appearance of spontaneous stepping movements implies a measure of overshoot in gross central excitatory state. Possible influence of circulating humoral and metabolic factors upon the disconnected spinal cord itself or on its effectors greatly complicates the interpretation of withdrawal effects at the segmental level. For example, the presence of muscle tremors below as well as above the level of transection suggests the involvement of a systemically distributed chemical agent. But Wikler[10] points out that the isolated spinal reflex alterations follow a different time course from that of the generalized abstinence syndrome. It appears safe to conclude, in any event, that spinal cord transection below the midthoracic level does not influence the course of opiate tolerance and abstinence above that level.

The visceral consequences of withdrawal attributable to spinal cord isolation are especially difficult to evaluate because of the diffuse nature of autonomic innervation as well as the continuing reactivity of visceral effectors to bloodborne mediators. So it is not surprising that isolated spinal autonomic participation both in the development of tolerance and in physical dependence has not been thoroughly investigated.

III. THE DECEREBRATED NEURAXIS

A. Cranial Nerves and Vital Centers

If we think of cranial nerves I (olfactory) and II (optic) in the more exact sense as extensions of the brain itself rather than as connecting nerve

TABLE I

Functions of Cranial Nerves Utilizable in the Midbrain Decerebrate Animal Preparation that Spares the Edinger–Westphal Nucleus

III.	Eye movement; pupil constriction and lens focus (parasympathetic)
IV.	Eye movement
V.	Sensations from face, jaw, and head; and chewing movement
VI.	Eye movement
VII.	Taste; face movement; and salivation and other secretions (parasympathetic)
VIII.	Hearing and orientation in space
IX.	Taste; blood pressure; gas tension and H^+ reception; swallowing movement; and salivation (parasympathetic)
X.	Visceral input; swallowing movement; vocalization; and visceral output (parasympathetic)
XI.	Neck and shoulder movement
XII.	Tongue movement

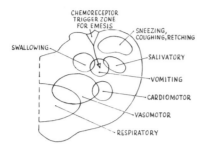

Fig. 3. Unilateral topographic relationship of various control centers in the medulla oblongata localized stereotaxically by means of electrical stimulation. The chemoreceptor trigger zone in the area postrema is not electrically excitable but is activated instead by emetic substances applied on the surface or delivered to it through the bloodstream; the trigger zone in turn elicits emesis by exciting the deeper lying vomiting center.

trunks, it becomes obvious that *all* sensory and motor nerve connections between neuraxis and other organs are made below the cerebrum and, indeed, are utilizable in the midbrain decerebrate animal preparation. Suprasegmental lines of communication are summarized in Table I. The special ingredient that has been added to the spinal cord, however, comprises the "control centers" for the vital functions situated in the reticular formation of the lower brainstem. These are illustrated schematically in Fig. 3. It cannot be overemphasized that the reticular interneuronal conglomerates that exercise control over immediate life-sustaining functions, as the respiration and the circulation, are not autonomic centers *per se* but serve to coordinate all appropriate effector machinery, somatic as well as visceral, required for execution of the complex processes.

Another quantity that has entered the picture is a larger apparatus for general modulation of excitatory tone–sensory and motor, descending and ascending—that appears to be localized mainly at the pontine level. Fundamental questions that have arisen in connection with these tonic facilitatory and inhibitory centers are their roles in sleep and arousal, control of muscle tension, and setpoint adjustment for homeostatic processes. The cerebellum, too, remains an enigma. Its special structure, sparse content of neurohumors,[30] and apparent lack of influence upon sensorium and vital processes leaves a large void of understanding yet to be filled. At best the cerebellum is known to serve for subconscious conditioning of skeletal muscular activity, and it participates in reflexes of spatial orientation, as in motion sickness, for example.

In short, the midbrain decerebrate animal possesses the basic wherewithal for short-term survival and even shows acyclic sleep-wakefulness swings of behavior, but it is missing the more complex regulatory mechanisms required in climatic adaptation and in fulfillment of nutritional and reproductive drives.

B. Alterations of Lower Brainstem Functions

1. Muscle Tone and Movement

Depending on the precise level of brain transection, duration of transection, and species under investigation, the decerebrate state is characterized by more or less extensor rigidity, more or less spontaneous activity, and more or less reactivity to sensory stimulation. The major facts that emerge with respect to the actions of morphine on the musculoskeletal system in the midbrain decerebrate preparation are that the drug does surprisingly little to modify descending facilitation of gamma-motoneurons, and of alpha-motoneurons when these are activated by decerebellation, but it tends to amplify those alterations of segmental reflexes that are already evident in the spinal animal.[11,31] Except for episodic enhancement of extensor tonus and running movements, which reflect a reduction in descending inhibition from the lower brain stem, primary initiation of frank convulsions by morphine is questionable.[10] Indeed the incidental bursts of central hyperactivity are likely to be occasioned by respiratory and circulatory disturbances. The influence of morphine on the irregular alternations of primitive sleep and arousal behavior manifested in decerebrate animals has not been adequately assessed.[32] The evidence thus far accumulated appears to point to actions of morphine on mechanisms of muscular control located rostral to the midbrain as mainly accounting for the postural and excitant effects of the drug in the intact animal.

2. Circulation

For a classic pharmacodynamic analysis of the hypotensive action of morphine, the reader is referred to the work of Schmidt and Livingston,[33] to which little has since been added. According to them initial intravenous injection of morphine in nonrodent species elicits a fall in blood pressure through an action independent of the nervous system, the heart, and even of the release of histamine. Acute tolerance develops rapidly. Work in the human, however, demonstrating narcotic obtundation of the vascular compensatory response to tilt,[34] suggests that morphine also causes an alteration in reflex vasomotor behavior that is consistent with the more general influence of the drug on the brain stem reticular formation. It is significant that direct electrical stimulation of the medullary vasomotor[35,36] (and respiratory) centers revealed no depressant action of morphine on these output-integrating mechanisms; this indicates that interference with reflex behavior must occur in a modulator component of the signal-processing system. Midbrain decerebration only partly eliminates the further complication of sustained sympathetic discharge, best elicited by morphine in unanesthetized animals (see Fig. 4 for associated hyperglycemic effect). Thus the presence or absence of a hypotensive effect of morphine depends on a variety of factors including species, anesthesia, dosage schedule, use of artificial ventilation, and neurological status.[37] It is favored by conditions that depress central reactivity.

The mechanism of cardiac slowing in response to morphine remains obscure,[38] although it is known to be elicitable in the decerebrate animal and to depend on the vagus nerve for its expression. It has not yet been established, however, whether the action is central or reflex, direct or indirect (as through release of a humoral intermediate).

The cardiovascular effects of morphine in the unanesthetized decere-

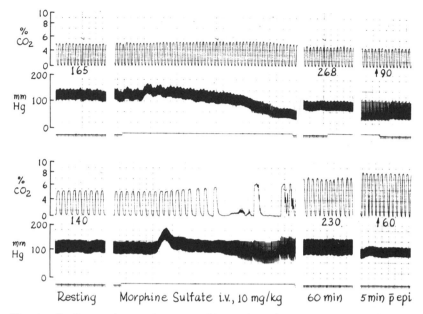

Fig. 4. Cardiovascular, respiratory, and hyperglycemic responses to morphine in unanesthetized decerebrate cats. Morphine (10 mg/kg) was infused intravenously over a period of 2 min. Recorded parameters are end tidal percent CO_2 and arterial blood pressure. Numbers between channels represent blood glucose concentration in milligram percent; values after injection of epinephrine (in far right column) indicate step rise. Upper row of panels shows responses in a cat with ventilation assisted artificially to prevent morphine-induced changes in end tidal CO_2 tension. Lower set shows effects during unassisted spontaneous ventilation. Note the marked hypotensive response to morphine with only slight bradycardia evident in the upper set of tracings, in contrast to the marked bradycardia with minimal hypotensive effect (following a brief rise in blood pressure) that accompanied the elevation of CO_2 in the lower set. Hyperglycemia (*ca.* 50% increase) occurred in both cases. (Borison and McCarthy, unpublished observations.)

brate cat are illustrated in Fig. 4. The hypotensive action was best manifested during assisted ventilation that prevented an elevation in alveolar carbon dioxide tension, whereas cardiac slowing was most evident during the onset of hypercarbia (and hypoxia) in the absence of ventilatory assistance; reflex vasoconstriction undoubtedly helped to reduce the hypotensive effect in the latter case. Development of hyperglycemia (relatively mild by comparison with the intact cat owing to extirpation of the posterior hypothalamus) was uninfluenced by the use of artificial ventilation to forestall the occurrence of hypercarbia.

3. Respiration

The central respiratory control machinery can be subdivided into at least seven working parts, each potentially susceptible to an action of morphine, as follows: (1) CO_2 detector, (2) tidal volume modulator, (3) frequency modulator, (4) panting controller, (5) oscillator, (6) cough controller, and (7) output integrator (latter two components shown in Fig. 3). The oscillator may be incorporated in the output integrator. Because of the multiplicity of targets, differential capability of tolerance development, and interactions with general anesthetics and antagonists, experimental studies have yielded a bewildering fund of information concerning the actions of morphine on the respiration.[39,40] Indeed, under certain acute experimental conditions, it is practically impossible to kill cats and dogs[33] with morphine even though respiratory failure is the essential cause of death in clinical cases of narcotic intoxication.

Since general anesthesia indiscriminately depresses all of the central components and appears to interfere with the development of acute tolerance, morphine produces a different picture in the anesthetized animal than in the unanesthetized. For example, respiratory periodicity tends to be more profoundly and more longlastingly slowed in the presence of anesthesia.[41] Table II gives data on the successive respiratory effects of two doses of intravenous morphine in decerebrate unanesthetized cats.[42] The tidal volume control mechanism (which is related to CO_2 detection) shows less propensity for acute tolerance development than does respiratory frequency control. At the hypothalamic level, morphine facilitates the respiration in contrast to its depressant actions on the lower brain stem.[43] Figure 5 compares the time course of effects on spontaneous respiration of single doses of morphine injected at various sites in the cerebrospinal fluid spaces as well as by the intravenous route in anesthetized cats.[43] Injection into the third ventricle, in proximity to the hypothalamus, elicited a unique stimulant response by comparison with those obtained by the other means.

TABLE II
Alterations in Respiratory Parameters Produced by Intravenous Morphine[a]

	N	f	V_{Tsp}[b] (ml)	V_E (ml/min)	F_{ACO_2}[c] (% CO_2)	Apneic Threshold (% CO_2)	$\dfrac{\Delta V_T}{\Delta \log}$[d] [CO₂]	$\dfrac{\Delta f}{\Delta \log}$[d] [CO₂]
Control	8	38.8±5.0	19.7±0.8	754±89	3.8±0.1	2.7±0.1	47.0±4.1	24.3±3.7
1 mg/kg		31.1±3.4	18.0±0.7	552±61	4.7±0.2	3.6±0.2	40.2±3.3	22.8±2.3
3 mg/kg		36.5±3.2	16.5±0.8	550±61	5.1±0.2	4.1±0.2	38.2±2.5	28.2±2.0

[a] Results are given as mean ± S.E.M. N = number of cats; f = frequency in breaths per min; V_{Tsp} = tidal volume of spontaneous breathing; V_E = resting pulmonary ventilation; F_{ACO_2} = resting end-expiratory concentration of CO_2; apneic threshold = concentration of CO_2 at respiratory cutoff (zero tidal volume) achieved by hyperventilation; $\Delta V_T/\Delta\log$ [CO₂] = tidal volume gain function related to concentration of CO_2; $\Delta f/\Delta\log$ [CO₂] = frequency gain function related to concentration of CO_2. (From Flórez and Borison.[42])

[b] BTPS.

[c] Mean bar. pressure, 745 mm Hg.

[d] Arbitrary units.

This is consistent with the initiation of thermoregulatory panting by morphine in the intact unanesthetized animal, but the concurrent increase in tidal volume obtained after third ventricular injection suggests that a more general form of stimulation has been activated in the upper brain stem.

It is quite meaningful that morphine in the midbrain decerebrate cat did not depress the respiratory output integrator, as judged by its undiminished responsiveness to electrical stimulation, at the same time that the drug elevated the CO_2 apneic threshold (see Fig. 6).[42] This is best interpreted as resulting from a selective action on the CO_2 detector mechanism to shift the setpoint upward. In the experiment illustrated, no depression (rather an increase) of CO_2 gain (slope of V_T-log $[CO_2]$) is evident, although a statistically significant reduction in slope was obtained for the entire series of experiments (see Table II). Thus it is possible to obtain an independent assortment of effects of morphine upon CO_2 detection (setpoint), tidal volume modulation, frequency modulation, and integrator activity. Ngai[44] showed in the *pontine* decerebrate cat that morphine prevents vagotomy-induced apneustic breathing, which lends support to a postulated depressant action on the depth modulator. In the midbrain decerebrate cat morphine eliminated the respiratory acceleratory response to electrical stimulation of the central end of the vagus nerve, which is indicative of a depressant action on the frequency modulator. The basic oscillator itself, like the integrator, would appear to be more resistant than its respective modulator to the influence of morphine.

The final selective respiratory action of morphine to be considered is that on the cough controller. This antitussive effect, which apparently displays no tolerance, does not depend on the narcotic property of the opiates since nonaddicting isomers are also effective cough suppressants.[45,46]

In reviewing the relative influences of morphine on input, output, and modulating functions of the central respiratory control system, we may conclude that the major actions of morphine are exercised on modulator processes with relatively minor direct actions on the final integrative mechanisms. Insofar as input influence is concerned, it may be added that the respiratory effects of morphine are not modified qualitatively by interruption of feedback sources from the lungs and the arterial chemo- and pressoreceptors,[42] again emphasizing the more important influence on modulating functions.

Fig. 5. Time course of the respiratory effects produced by morphine administered i.v., into the third ventricle, into the fourth ventricle, and into the bulbar subarachnoid space. Values for frequency, tidal volume, and end expiratory percent CO_2 are represented as the percent change in relation to the control period. Bracketed lines indicate mean $\pm$ S.E.; n indicates number of cats. Standard errors of control values serve to compare degree of variation in terms of percent of the mean prior to the conversion to 100%. Sites of injection into the CSF spaces are shown in the diagram of the cat brain. [Dotted line indicates level of midbrain decerebration employed in *other* experiments described in the text (see Fig. 6 and Table II)]. (From Flórez *et al.*[43])

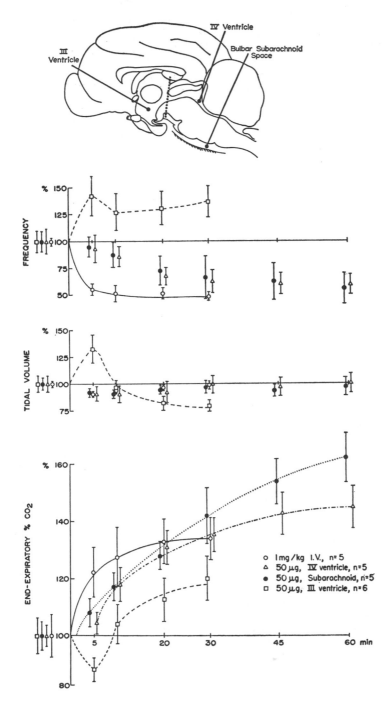

o 1mg/kg I.V., n=5
△ 50 μg, IV ventricle, n=5
● 50 μg, Subarachnoid, n=5
□ 50 μg, III ventricle, n=6

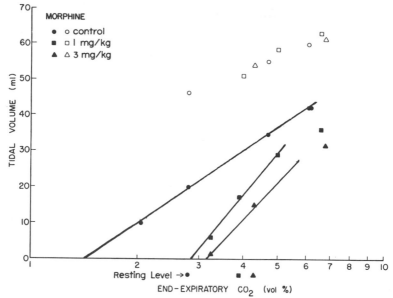

Fig. 6. Effects of morphine on the steady-state relationships between CO_2 and spontaneous tidal volume (closed symbols) and the inspiratory response to direct electrical stimulation of the medulla oblongata (open symbols) in decerebrate cats. Point of intersection of response line with abscissa is the CO_2 apneic threshold. Note lack of influence of morphine on the CO_2– electrical response relationship. (From Flórez and Borison.[42])

4. Vomiting

The discovery by Wang and Borison[47,48] that destruction of a unique entity in area postrema of the medulla oblongata resulted in permanent refractoriness to the emetic action of apomorphine, without disturbing the vomiting response to stimuli presented through other inputs, established that the opiates stimulate a specialized receptor to initiate a specific behavioral effect. This finding was later extended to morphine by Wang and Glaviano.[49] The emetic dose–effect relationship of morphine characteristically shows a median range of effectiveness. Large doses are not only less effective in evoking vomiting but they produce a nonspecific antiemetic effect attributable to depression of the vomiting center itself.[50] The morphological relationship of the emetic receptor site (chemoreceptor trigger zone on the surface of the medulla) and the deeper lying vomiting center is shown schematically in Fig. 3. Thus morphine stimulates a particular input to initiate a given effect and it depresses the neural coordinating mechanism to counteract that very effect.

The two obvious alternatives that might account for the development of tolerance to the emetic action of morphine are (1) receptor inactivation or (2) threshold elevation for excitation of the vomiting center. Receptor blockade is demonstrable by local application of the drug,[48] but this acute phenomenon may not be applicable to long-term tolerance. On the other hand, the direction of tolerance development in a depressed emetic center is theoretically toward the acquisition of greater excitability. Similar difficulties arise in trying to explain how abstinence rebound vomiting is effected, whether by receptor activation or through facilitation of the emetic center. Receptor participation should be excluded surgically to test the question, but an educated guess favors the likelihood that withdrawal-induced vomiting would still occur in the receptor-ablated animal. This is because tolerance and associated compensatory behavior seem to require mobilization of reiterative neuronal circuitry. The complexity of systems interaction in the development of tolerance and physical dependence is underscored by such remote effects as the enhancement of ACTH secretion that is said to accompany morphine-induced nausea[52] and the claim that frontal lobotomy in humans prevents withdrawal-induced vomiting;[10] the appearance of abstinence vomiting, however, has been reported in decorticated dogs.[53]

Attempts have been made to relate drug alteration of vestibular function with emetic responsivity owing to the increased incidence of vomiting and dizziness reported in ambulatory patients by comparison with bed patients treated with morphine.[51] Vestibular involvement in morphine-induced vomiting has not, however, been demonstrated objectively, and it is more likely that associated hypotension in upright persons is responsible for the augmented discomfort.

C. Implications of Decerebration for Addiction

The retention of all final common pathways and integrative machinery for motor expression in the midbrain decerebrate animal raises certain important implications for the understanding of addiction, yet to our knowledge no careful long-term study has been performed on the effects of morphine after decerebration. In the chronic spinal animal, the forebrain still maintains humoral communication with the neuronally isolated portion of the body, and in the decorticate condition the basal ganglia are still connected to the lower brain stem. The provocative point repeatedly crops up, however, that morphine does not much affect the output integrative machinery but exercises its greatest influence upon modulating processes. Because of the progressive encephalization of behavioral control in evolutionary advancement, the concentration of subconscious modulating functions in the hypothalamus and basal ganglia may well hold a key to mechanisms of addiction in modern species. The capability of the midbrain decerebrate animal to develop chronic tolerance and physical dependence to the narcotic analgesics remains to be determined.

IV. THE HYPOTHALAMUS AND BASAL GANGLIA

A. Adaptive Processes

The input scope of the decorticate animal is increased over that of the decerebrate by the availability of the optic nerve (cranial nerve II) for pupilloconstrictor activation and for certain optokinetic reflexes that do not require conscious perception of visual experience. The higher order of functional representation in the hypothalamus and basal ganglia adds a new and greatly expanded dimension to the fixed reflex patterns of activity displayed by the decerebrate preparation (including the rat). But it is doubtful, even with the further capability in affective expression, climatic adaptation, and reproductive behavior, that distinctive qualities of "personality" are recognizable in the decorticated animal. If the paleocortex (rhinencephalon) with its olfactory input (cranial nerve I) is retained with the subcortical brain, we have all the essential neural machinery that operates in lower forms for survival of the species in a hostile environment. It may be assumed from many reported descriptions of the decorticate preparation that, owing to problems of neurosurgical technique, variable amounts of rhinencephalic cortical structures were left behind—hence accounting for a residual sense of smell and related behavioral manifestations.

"Sham rage" behavior in response to nociceptive stimuli is well manifested in the decorticate dog (which in certain respects may be likened to the decerebrate rat), but expressions of anticipation through recognition of "painful" stimuli have never made themselves evident. This affective rage response, except for the vocalization component, is readily antagonized by morphine.[53] The suprasegmental origin of morphine-induced seizures also is not fully accounted for by the response characteristics of the decorticate animal. According to Wikler,[10] convulsive twitches but not well-developed seizures were produced by large doses of morphine in chronic decorticate dogs, which is in apparent conflict with older literature.[28]

Long-term maintenance of the decorticate preparation is greatly facilitated by its incorporation of the hypothalamo–hypophyseal endocrine control machinery. Indeed the range of responsivity of the neuroendocrine system to a wide variety of input and feedback signals,[54] both neural and humoral, makes a pharmacodynamic analysis of the actions of morphine on this system vulnerable to many pitfalls. One isolation approach that eliminates neural inputs to the controlling apparatus is the technique of the "hypothalamic island" (see Woods[1]). The endocrinologic effects of morphine are examined in other sections of this book.

B. Alterations of Subcortical Functions

1. Pupillary Diameter

Other effects of morphine on the eye, as on ocular movement (nystagmus), lens transparency, and accommodation are not considered because of limited information and the relatively minor changes that occur by comparison with alteration of pupil size. The miotic effect of morphine, as occurs in man and other species especially in relation to the occurrence of sedation, is herein treated at the subcortical level because pupilloconstriction is still elicitable after decortication but not decerebration (presumably above the oculomotor nucleus).[28] According to McCrea et al.[55] the miotic action of morphine in the dog is mainly an exaggeration of the light reflex which of course is dependent on the integrity of the optic nerve. Unfortunately, only one of the optic nerves was cut by these investigators so that it is impossible to know because of the consensual light reflex whether the action of morphine absolutely requires an input from the retina. And so it has not yet been established whether the pupilloconstrictor effect is due to an initiating or modifying action or both. Nevertheless, the known facts that morphine-induced miosis is considerably reduced in the dark and that photophobia is a concomitant effect in the light strongly suggest that the drug action is largely a modifying one on light input activity. Furthermore, these facts are entirely compatible with, and even favor, the possibility that the drug interaction with light input takes place at the retina itself.

Curiously, no mention is made in reviews of this subject, at least in those that have come to our attention, of the pupillary effect of morphine in humans with bilateral optic nerve blindness. If, in fact, morphine no longer produced miosis after interruption of both optic nerves, this would definitely exclude a central initiating action, and a modifying action on the light-to-impulse translation process would appear to be the most likely mechanism for the miotic effect of morphine. The neural organization of the retina is sufficiently complex to accommodate a modulating or a gating action for morphine-induced miosis. Exclusion of a central modifying action at the Edinger–Westphal nucleus, for example, would require the use of an artificial input signal by electrical stimulation of the optic nerve after separating it from its normal impulse source.

Mydriasis, unlike miosis, is elicitable in the decerebrate preparation and is simply one part of the more generalized capability for morphine-induced discharge of the sympathetic nervous system. However, it must be reemphasized that other expressions of sympathetic activation, such as hyperglycemia, may be produced independently of the mydriatic response. The photophobic effect of morphine undoubtedly also occurs in species that show adrenergic pupillodilatation. Indeed it is evident that light aversion contributes to the excitement observed in intact cats because darkening the environment produces an abatement of the characteristic morphine-induced cries of distress.[56] Mydriasis with photophobia is a regenerative feedback process since the resulting excitement leads to greater pupillodilatation resulting in greater light input resulting in greater excitement and so on. Morphine-induced mydriasis is readily blocked by general anesthesia, which interferes with all expressions of sympathetic activation including the hyperglycemic effect, discussed next.

2. Blood Glucose Concentration

The elicitation of hyperglycemia by morphine depends to a major extent on the integrity of the posterior hypothalamus[57] and to a lesser extent on the lower brain stem (see Fig. 4). Execution of the response requires an intact pathway from the brain stem to the sympathetic splanchnic outflow leading primarily to the adrenal medulla and secondarily to the liver. The hyperglycemia may occur in the absence of behavioral excitation and, conversely, morphine-induced excitement is not necessarily accompanied by hyperglycemia.[58] Pupil size is also unrelated to the presence or absence of hyperglycemia, which suggests that pupillodilatation, when it occurs, is due mostly to direct sympathetic activation of the radial muscle of the iris rather than to an indirect effect of circulating epinephrine and that splanchnic sympathetic discharge is separately controllable by the central nervous system. Selective depletion of norepinephrine from the hypothalamus was invariably correlated with the development of hyperglycemia in response to optimal doses of morphine administered intravenously or locally into the third ventricle.[58] This is in contrast to earlier reports of nonselective catecholamine depletion from the brain after intraperitioneal administration of large doses of morphine. The hypothalamic content of norepinephrine was not reduced when hyperglycemia was not evoked, even in the presence of behavioral excitement obtained after the intracisternal injection of morphine. Thus it is apparent that hypothalamic catecholamine turnover may be concerned more with central descending influence than with ascending. Neither the dopamine content of the caudate nucleus[58] nor the overall 5-hydroxytryptamine content of the brain[59] was altered concomitantly with the elicitation of hyperglycemia. Suggestive evidence has been obtained by Borison et al.[60] to involve the ependymal subfornical organ in the hyperglycemic action of morphine, at least for the response to injection into the third ventricle. Nevertheless, it has still not been established whether the blood sugar response to morphine is due to an initiating action on a receptor (as is the case with vomiting) or is the result of a modifying action on a homoestatic feedback mechanism. Moreover, the massive release of adrenal cetecholamines evoked by morphine in the decorticate animal must not only be considered responsible for the hyperglycemic response but also must be implicated to an extent that has not yet been adequately assessed in the initiation and/or modification of the neuroendocrine, thermoregulatory, and locomotor behavioral effects controlled above the level of the midbrain.

3. Body Temperature

The dynamics of temperature regulation are determined by a variety of processes related to heat production, storage, distribution, and loss. Consequently, the end effect of morphine on body temperature depends on the balance of influences on those processes.[61] In instances where morphine can be expected to turn up heat production, as through generalized somatic excitation along with increased sympathetic discharge, body temperature is, appropriately, observed to rise; whereas when morphine produces a more restricted action on the hypothalamic temperature controller in association with its sedative effect, the end result is a coordinated response to rid the body of heat as if the drug causes a lowering of the thermoregulatory setpoint.[62] Thus, depending on environmental conditions, species, dose, route of administration, neurologic status, etc., morphine may cause a rise, fall, or biphasic change in body temperature that is perhaps the least predictable index of narcotic effect.[63] It is evident, however, that morphine in accord with its other actions does not depress the neural output or effector components of the thermoregulatory system.

C. Implications of Decortication for Addiction

With due regard to the uncertainty of judging the extent of decortication, it may be concluded that dogs chronically deprived of large parts of the cerebral cortex—but certainly all of the neocortex—outwardly respond much like normal humans to acute and prolonged treatment with morphine as well as to withdrawal of the drug. Thus that part of the brain concerned with ideational behavior seems not to be required for any part of the physiological sequence manifested in the clinical addiction cycle. The role of the paleocortex, on the other hand, still remains a great mystery. Interestingly enough, this biologically ancient "olfactory" cortex (rhinencephalon, limbic system) is generally described inclusively with structures in the basal ganglia and hypothalamus because of the rich and redundant interconnections between these closely related evolutionary transitional elements. The prominent central representation of visceromotor functions in the olfactory cortex adds further to the heuristic nature of the problem.

Just as an exacting study of chronic tolerance and physical dependence to morphine is needed in the midbrain decerebrate preparation in order to define the minimal neural apparatus required, so a thorough study is needed at the rhinencephalic level to determine the upper limits of neural sufficiency for the effectuation of addiction.

V. THE WHOLE NERVOUS SYSTEM

A role of the neocortex in morphine-induced modulation of the defense reaction to nociceptive stimulation is made evident in connection with a surprising influence on vocal behavior. Wikler[53] found in decorticate dogs that morphine effectively blocked the gross motor expression of "sham rage" but it spared the vocal component, whereas in normal dogs the drug suppressed the entire pain reaction including vocalization. This additional influence on the intact brain suggests that the opiates activate cortical inhibitory processes rather than depress the cortex, which is consistent with the weak general anesthetic effect of morphine as contrasted with the barbiturates, for example. On the other hand, evidence of cortical facilitation (or disinhibition) is provided by the fuller display of morphine-induced seizures in intact animals. More engaging perhaps are questions relating to those influences of morphine on learning, motivation, mood, and appreciation of pain that are effected through actions involving the cerebral cortex, but these problems have greater bearing on the sociologic consequences of drug abuse than on the physiologic mechanisms of addiction.

Evaluation of studies on evoked electrical activity in central sensory circuits has been postponed to this point because such works are concerned ultimately with explaining the effects of opiates on pain perception.[64,65] The relationship of analgesia to addiction is not, however, self-evident. Major sources of difficulty in the interpretation of electrophysiological findings are as follows: (1) Input stimulation is not pain specific, (2) conduction pathways in the central nervous system are not pain specific,[66] (3) lability of evoked electrical activity increases directly with complexity of connections in the conduction pathway, (4) pattern of evoked activity varies with experimental technique, and (5) expression of drug effect varies with dosage and experimental conditions. These considerations having been given, we will now summarize the reported effects of morphine upon electrically evoked activity in afferent pathways and central neuron pools, without regard to species or experimental conditions: (1) Primary afferent neurons are not depressed,[67] (2) secondary and higher order neurons activated from visceral inputs are more readily depressed than somatic,[68,69] (3) nociceptive pathways are more readily depressed than the nonnociceptive,[67,70] (4) special-sense pathways are not depressed,[67,70] (5) thalamic and cortical evoked potentials delivered via direct pathways are not influenced, whereas activity conducted over diffuse pathways may be either depressed or facilitated,[69-72] (6) activity in paleocortex is more readily depressed than in neocortex,[73-75] and (7) high-frequency discharges are more readily depressed than low-frequency ones.[69,76]

Although much capital has been made of species differences in the effects of morphine, Krueger et al.[28] comment on the remarkable fact that Claude Bernard made no mention of the commonly described excitement occurring in his observations on cats. Indeed, they concluded that the overall effect of morphine is qualitatively similar in all species, due allowance being made for the sensibility of the animal, including man. On the other hand, even for so primitive a function as vomiting, quite marked differences in emetic responsiveness to the opiates are evident among species.

VI. CONCLUDING REMARKS

It is estimated that the nervous system contains about 10^{10} neurons. We have looked at some of the better but still poorly understood ways in which small samplings of this vast complex of microcomputers receive and emit information. In the main we have relied for our present purposes on the logic of exclusion imposed by the available time-worn methods of fractionating the functions of the nervous system. To state the obvious, this has provided an oversimplified view of the neural actions of morphine. So long as we lack more effective means of studying the communication mechanisms of the brain, we shall have to face the dilemma as to how the

neurochemical approach to addiction can yield meaningful answers if they
do not rest upon sound neurophysiological footings.

VII. REFERENCES

1. J. W. Woods, *J. Neurophysiol. 27*, 635–644 (1964).
2. H. K. Beecher, "Quantitative Effects of Drugs. Measurement of Subjective Responses," Oxford University Press, New York (1959).
3. A. Goldstein, L. Aronow, and S. M. Kalman, "Principles of Drug Actions," Hoeber Medical Division, Harper & Row, New York (1968).
4. A. Wikler (ed.), "The Addictive States," Research Publication of the Association for Nervous and Mental Disorders, Vol. 46, Williams & Wilkins, Baltimore (1968).
5. J. W. Eccles, "The Physiology of Synapses," Springer-Verlag, Berlin (1964).
6. R. Melzack and P. D. Wall, *Science 150*, 971–979 (1965).
7. L. Vyklicky, P. Rudomin, F. E. Zajac, III, and R. E. Burke, *Science 165*, 184–186 (1969).
8. W. A. Krivoy and R. A. Huggins, *J. Pharmacol. Exp. Ther. 134*, 210–213 (1961).
9. W. R. Ingram, *in* "Handbook of Physiology, Section 1, Neurophysiology" (J. Field, H. W. Magoun, and V. E. Hall, eds.), Vol. II, pp. 951–978, American Physiological Society, Washington, D.C. (1960).
10. A. Wikler, *Pharmacol. Rev. 2*, 435–506 (1950).
11. W. R. Martin, *in* "Physiological Pharmacology" (W. S. Root and F. G. Hofmann, eds.), Vol. 1, pp. 275–312, Academic Press, New York (1963).
12. A. Wikler and K. Frank, *J. Pharmacol. Exp. Ther. 94*, 382–400 (1948).
13. W. R. Martin and C. G. Eades, *J. Pharmacol. Exp. Ther. 146*, 385–395 (1964).
14. H. Takagi, M. Matsumura, A. Yanai, and K. Ogiu, *Jap. J. Pharmacol. 4*, 176–187 (1955).
15. W. Koll, J. Haase, G. Block, and B. Muhlberg, *Int. J. Neuropharmacol. 2*, 57–65 (1963).
16. B. Silvestrini and G. Maffii, *J. Pharm. & Pharmacol. 11*, 224–233 (1959).
17. S. Irwin, R. W. Houde, D. R. Bennett, L. C. Hendershot and M. H. Seevers, *J. Pharmacol. Exp. Ther. 132*, 132–143 (1951).
18. P. W. Wagers and C. H. Smith, *J. Pharmacol. Exp. Ther. 130*, 89–105 (1960).
19. A. B. Cairnie and H. W. Kosterlitz, *Int. J. Neuropharmacol. 1*, 133–136 (1962).
20. H. W. Kosterlitz and D. I. Wallis, *Brit. J. Pharmacol. 22*, 499–510 (1964).
21. H. W. Kosterlitz and D. I. Wallis, *Brit. J. Pharmacol. 26*, 334–344 (1966).
22. U. Trendelenberg, *Brit. J. Pharmacol. 12*, 79–85 (1957).
23. R. K. S. Lim, *in* "Pain," Henry Ford Hospital International Symposium (R. S. Knighton and P. R. Dumke, eds.), pp. 117–154, Little Brown and Company, Boston (1966).
24. W. A. Krivoy, *J. Pharmacol. Exp. Ther. 129*, 186–190 (1960).
25. N. A. Kruglov, *Int. J. Neuropharmacol. 3*, 197–203 (1964).
26. A. V. Valdman and E. B. Arushanyan, *in* "Progress in Brain Research, Pharmacology and Physiology of the Reticular Formation" (A. V. Valdman, ed.), Vol. 20, pp. 223–242, Elsevier Amsterdam (1967).
27. S. J. De Salva and Y. T. Oester, *Arch. Int. Pharmacodyn. 124*, 255–262 (1960).
28. H. Krueger, N. B. Eddy, and M. Sumwalt, "The Pharmacology of the Opium Alkaloids," Public Health Report 56, Suppl. 165, U.S. Government Printing Office, Washington, D.C. (1941).
29. H. L. Borison, B. R. Fishburn, N. K. Bhide, and L. E. McCarthy, *J. Pharmacol. Exp. Ther. 138*, 229–235 (1962).
30. H. McLennan, "Synaptic Transmission," Saunders, Philadelphia (1963).
31. I. Jurna, *Int. J. Neuropharmacol. 4*, 177–184 (1965).
32. E. K. Killam, *Pharmacol. Rev. 14*, 175–223 (1962).
33. C. F. Schmidt and A. E. Livingston, *J. Pharmacol. Exp. Ther. 47*, 411–441 (1933).

34. J. E. Eckenhoff and S. R. Oech, *Clin. Pharmacol. Ther. 1*, 483–524 (1960).
35. S. H. Ngai, *J. Pharmacol. Exp. Ther. 153*, 495–504 (1966).
36. G. V. Kovalyov, *in* "Progress in Brain Research, Pharmacology and Physiology of the Reticular Formation" (A. V. Valdman, ed.), Vol. 20, pp. 187–209, Elsevier, Amsterdam (1967).
37. A. G. J. Evans, P. A. Nasmyth, and H. C. Stewart, *Brit. J. Pharmacol. 7*, 542–552 (1952).
38. J. E. Walker, H. E. Hoff, R. A. Huggins, and S. Deavers, *Arch. Int. Pharmacodyn. 170*, 216–228 (1967).
39. C. G. Breckenridge and H. E. Hoff, *J. Neurophysiol. 15*, 57–74 (1952).
40. Ma Chuang Gen and A. V. Valdman, *in* "Progress in Brain Research, Pharmacology and Physiology of the Reticular Formation" (A. V. Valdman, ed.), Vol. 20, pp. 148–170, Elsevier, Amsterdam, (1967).
41. P. Pentiah, F. Reilly, and H. L. Borison, *J. Pharmacol. Exp. Ther. 154*, 110–118 (1966).
42. J. Flórez and H. L. Borison, *Resp. Physiol. 6*, 318–329 (1969).
43. J. Flórez, L. E. McCarthy, and H. L. Borison, *J. Pharmacol. Exp. Ther. 163*, 448–455 (1968).
44. S. H. Ngai, *J. Pharmacol. Exp. Ther. 131*, 91–99 (1961).
45. H. A. Bickerman, *Clin. Pharmacol. Ther. 3*, 353–368 (1962).
46. N. K. Chakravarty, A. Matallana, R. Jensen, and H. L. Borison, *J. Pharmacol. Exp. Ther. 117*, 127–135 (1956).
47. S. C. Wang and H. L. Borison, *Arch. Neurol. Psychiat. 63*, 928–941 (1950).
48. H. L. Borison and S. C. Wang, *Pharmacol. Rev. 5*, 193–230 (1953).
49. S. C. Wang and V. V. Glaviano, *J. Pharmacol. Exp. Ther. 111*, 329–334 (1954).
50. S. C. Wang, *in* "Physiological Pharmacology" (W. S. Root and F. G. Hofmann, eds.), Vol. 1, pp. 255–328, Academic Press, New York (1963).
51. L. B. Gutner, W. J. Gould, and R. C. Batterman, *J. Clin. Invest. 31*, 259–266 (1952).
52. L. S. Goodman and A. Gilman, "The Pharmacological Basis of Therapeutics," Macmillan, New York (1970).
53. A. Wikler, *Amer. J. Psychiat. 105*, 329–338 (1948).
54. E. Mills and S. C. Wang, *Amer. J. Physiol. 207*, 1405–1410 (1964).
55. F. D. McCrea, G. S. Eadie, and J. E. Morgan, *J. Pharmacol. Exp. Ther. 74*, 239–246 (1942).
56. H. L. Borison, unpublished observations.
57. C. M. Brooks, R. A. Goodwin, and H. N. Willard, *Amer. J. Physiol. 133*, 226–227 (1941).
58. K. E. Moore, L. E. McCarthy, and H. L. Borison, *J. Pharmacol. Exp. Ther. 148*, 169–175 (1965).
59. E. W. Maynert, G. I. Klingman, and H. K. Kaji, *J. Pharmacol. Exp. Ther. 135*, 296–299 (1962).
60. H. L. Borison, B. R. Fishburn, and L. E. McCarthy, *Neurology 14*, 1409–1053 (1964).
61. H. L. Borison and W. Clark, *Advan. Pharmacol. 5*, 129–212 (1967).
62. A. Hemingway, *J. Pharmacol. Exp. Ther. 63*, 414–420 (1938).
63. U. Banerjee, W. Feldberg, and V. J. Lotti, *Brit. J. Pharmacol. 32*, 523–538 (1968).
64. E. F. Domino, *in* "The Addictive States," Nervous and Mental Disorders (A. Wikler, ed.), Vol. 46, pp. 117–149, Williams & Wilkins, Baltimore (1968).
65. A. V. Valdman *in* "Progress in Brain Research, Pharmacology and Physiology of the Reticular Formation" (A. V. Valdman, ed.), Vol. 20, pp. 19–35, Elsevier, Amsterdam (1967).
66. L. Kruger *in* "Pain, Henry Ford Hospital International Symposium" (R. S. Knighton and P. R. Dumke, eds.), pp. 67–81, Little, Brown, Boston (1966).
67. L. N. Sinitsin, *Int. J. Neuropharmacol. 3*, 321–326 (1964).
68. S. Fujita, M. Yasuhara, and K. Ogiu, *Jap. J. Pharmacol. 3*, 27–38 (1953).

69. S. Fujita, M. Yasuhara, S. Yamamoto, and K. Ogiu, *Jap. J. Pharmacol. 4*, 41–51 (1954).
70. S. Silvestrini and V. G. Longo, *Experientia 12*, 436 (1956).
71. H. Gangloff and M. Monnier, *J. Pharmacol. Exp. Ther. 121*, 78–95 (1957).
72. J. Heng Chin and E. F. Domino, *J. Pharmacol. Exp. Ther. 132*, 74–86 (1961).
73. J. S. McKenzie and N. R. Beechey, *Electrencephalog. Clin. Neurophysiol. 14*, 501–519 (1962).
74. J. S. McKenzie, *Electroencephalog. Clin. Neurophysiol. 17*, 428–431 (1964).
75. A. Soulairac, C. Gottesman, and J. Charpentier, *Int. J. Neuropharmacol. 6*, 71–81 (1967).
76. M. Matsumura, S. Takaori, and R. Inoki, *Jap. J. Pharmacol. 9*, 67–74 (1959).

Chapter 17

THE KIDNEY

James M. Fujimoto

Department of Pharmacology
Marquette School of Medicine
Milwaukee, Wisconsin

I. INTRODUCTION

The kidney as an organ system plays an important role in the biochemical pharmacology of morphine and its surrogates because it exhibits definite responses to narcotic analgesics and at the same time excretes these drugs and their metabolites. Studies are described in which investigators have attempted to relate mechanisms of development of tolerance to the effects of narcotics on renal function. If it is assumed for heuristic purposes that the antidiuretic effect of morphine is due to the release of antidiuretic hormone, it would be reasonable to infer that decreases in the antidiuretic effect from repeated doses of morphine involve changes in this antidiuretic hormone–renal response system. The remainder of the chapter considers mechanisms of transport of certain narcotic analgesics and their metabolites. It is evident how widely the general concepts developed for drug transfer have been applied to excretion of narcotics and it also is demonstrated how studies with narcotic analgesics have in turn contributed in a small but important way to furthering understanding of renal tubular transport of drugs.

II. ANTIDIURETIC EFFECT

It is well established that morphine and its surrogates produce an antidiuretic effect in certain animals.[1-4] The bulk of evidence[1-4] favors the concept first propounded by de Bodo[5] that the antidiuretic effect of morphine in the dog is due to release of antidiuretic hormone.

However well favored this mechanism is, it has been far from satisfactory as a unitary hypothesis to explain all actions of narcotics on formation of urine by the kidney. In man, Papper and Papper[4] indicated that renal hemodynamic changes and the antidiuretic effect of morphine occur without activation of the release of antidiuretic hormone. Also in man, Schnieden and Blackmore[6] and Becker and Moeller[7] could not attribute the antidiuretic action of morphine entirely to antidiuretic hormone release. Under somewhat different circumstances, Handley and Keller[8] have shown that both antidiuretic hormone release and renal hemodynamic changes may be involved in the antidiuretic response to morphine in dogs. George and Way[9] indicated that, in the rat, the antidiuretic hormone release mechanism appeared to be possibly more sensitive to morphine than was the direct effect on the kidney. Several additional mechanisms were implicated by Lipschitz and Stokey[10] in the rat. Morphine-induced antidiuresis was in part mediated through the renal nerves since one component of the action of morphine was not obtained in rats with decapsulated, denervated kidneys. A most interesting finding of Mills and Wang[11] was that morphine and meperidine *blocked* release of antidiuretic hormone usually obtained by stimulation of the ulnar nerve in the dog.

Morphine administration does not invariably produce antidiuresis. Studies in man indicate that morphine is useful in treating patients with acute mountain sickness.[12] Although the therapeutic effect was related to improved pulmonary function, morphine augmented the furosemide-induced diuresis in cases where oliguria was associated with the hypoxic stress of mountain sickness. The diuretic effect appeared to be related to ACTH release in that improvement of adrenal cortical function or administration of betamethasone also produced diuresis. In patients with circulatory disturbances, Hopmann[13] found that morphine produced a marked nocturnal diuresis. Boyd and Scherf[14] stated that in patients with cardiac asthma and pulmonary edema, morphine produced diuresis whereas in normal patients loaded with water, antidiuresis was obtained. Thus the effects of morphine were complex.

The same situation held for certain animals, as indicated by the work of Marchand.[15] Morphine produced antidiuresis in the water-loaded rat but produced diuresis in the nonhydrated rat. (These experiments are discussed later.) In rabbits Eckhard[16] reported polyuria in a paper entitled "Uber den Morphiumdiabetes." In the chicken morphine infusion elicited a diuretic response.[17] The diuresis appeared to be mediated by an extrarenal mechanism since diuresis was bilateral even though infusion of morphine into the saphenous vein led to greater delivery of morphine to the corresponding kidney. In a case of ingestion of a fatal dose of dextropropoxyphene, a marked polyuria was observed.[18] Bower *et al.* performed studies with a toad bladder preparation *in vitro* and demonstrated an antagonism of the action of antidiuretic hormone by dextro and levo propoxyphene.[19] Thus it appeared that the polyuria from dextro propoxyphene might be due to nephrogenic diabetes insipidus produced by inhibition of antidiuretic hormone action at the renal tubular level.

III. DEVELOPMENT OF TOLERANCE TO THE ANTIDIURETIC EFFECT

A. Explanation of Terms

Because no thorough discussion exists on the topic of development of tolerance to the antidiuretic effect of narcotic analgesics, a detailed presentation is given. In this chapter the word "tolerance" is used in its broadest context, and no attempt is made to differentiate between acute tolerance, tachyphylaxis, and chronic tolerance, because these phenomena cannot be delimited and defined precisely in regard to the kidney and also because definitions are highly dependent on the technique used to manifest the phenomena. The first technique involves the use of water-loaded animals. The antidiuretic effect of the narcotic is manifested as a prolongation of the time necessary to excrete a specified portion of the water load or as a decrease of the percentage of water excreted during a specified interval. Tolerance to the antidiuretic effect then means that the drug produces significantly less antidiuretic effect. Reduction in antidiuretic effect also may be synonymous with return to control levels of diuresis. It is possible for the excretion time to be significantly shorter than the control time (more rapid diuresis than controls). This is tolerance also. Logically it would not be possible to show tolerance unless the basic antidiuretic response has been established in using this technique. In the orally water-loaded rat, minor difficulties may arise in assessing the antidiuretic effect. Inhibition of water absorption from the gastrointestinal tract may occur with morphine[6] and levorphanol,[20] an effect to which tolerance develops. Retention of urine in the bladder may ensue, but this effect can be overcome by massaging the area over the bladder[21] or pulling the tail.[10] The retention can be circumvented by exteriorizing the bladder.[15,22]

Another technique involves *ad libitum* water drinking and measurement of urine production. The animal may show minimal or no reduction in 24-hr drinking or urination as manifestation of an antidiuretic effect. But continued treatment with the drug leads to increased water intake and urine excretion, an indication of tolerance. In contrast to rats and mice, the first dose of morphine induces an increase in urine flow in the chicken. Diuresis being the usual response in the chicken, reduction in diuretic response then would be a manifestation of tolerance.

Hypotheses on the mechanism of the antidiuretic effect dictate particular approaches to studying the mechanisms of tolerance with an initial problem of demonstrating that tolerance does indeed occur. In turn, the method used to demonstrate tolerance has imposed restrictions on the breadth of studies possible. On the evidence that antidiuretic hormone release is one of the main mechanisms for the antidiuretic effect of narcotics, studies are described that deal with changes in this system as tolerance to the antidiuretic effect develops upon repeated treatment with these agents.

B. Studies in Rats

1. Water-Loading and Urinary Excretion

Newsome et al.[21] used the measurement of excretion times of water load in rats to demonstrate development of tolerance to the antidiuretic effect of levorphanol. Control rats given oral water loads equal to 5% of their body weight had excretion times of about 100 min. A single dose of levorphanol administered subcutaneously or intraperitoneally at the same time as the water load resulted in the expected antidiuretic effect of levorphanol being manifested by prolongation of the excretion times. Tolerance to levorphanol was demonstrated in several ways: (1) The same dose of levorphanol given twice a day did in time lead to reduction of the excretion times toward control values, (2) in order to maintain the same effectiveness of levorphanol as the initial dose, the dose of levorphanol had to be raised, and (3) in spite of larger repeated doses, the excretion time diminished. In Fig. 1 are illustrated the results obtained on one group of animals. This figure shows that not only did the excretion time return to control values but on days 18 and 19 the excretion time was shorter than for controls. The times of 81 ± 6 (S.D.) and 73 ± 11 on days 18 and 19 differed significantly from the respective control values of 96 ± 9 and 93 ± 11 min, showing that rats on large doses of levorphanol could excrete the water load faster than control animals. This last response was reported to occur to large doses of morphine by Shimai et al.[23] They found a 50% increase in the rate of excretion of the water load when the dose of morphine became 100 mg/kg. Marchand and Fujimoto[24] also demonstrated tolerance to morphine by using water loading. Thus the results unequivocally show that tolerance develops to the initial antidiuretic response to morphine or levorphanol.

Experiments have been performed that implicated changes in the antidiuretic hormone renal response system as a basis for development of

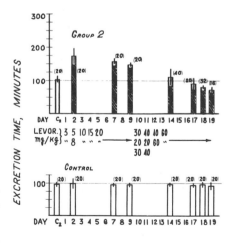

Fig. 1. Effect on the excretion time of a standard water load in rats treated with levorphanol (Group 2 above) and controls (below). Levorphanol and distilled water were both administered subcutaneously twice daily except on days 10 and 11, where three injections were given. The abscissa gives the days of treatment and the ordinate gives the excretion time for the water load in the rats. Vertical lines indicate the standard deviation, and the number in parentheses is the number of rats tested for each determination. C is the control response before the experiment was begun. (From Newsome et al.[21] by courtesy of the *Journal of Pharmacology and Experimental Therapeutics.*)

this tolerance. Shimai *et al.*[23] used Gomori's chrome alum–hematoxylin–phloxin stain to study the histological appearance of neurosecretory granules under the conditions where the 100 mg/kg dose of morphine was producing diuresis. Compared to control animals, the supraoptic and paraventricular nuclei of the hypothalamus from the morphinized rats showed exceedingly fine and apparently diminished quantities of neurosecretory material. Similarly, the content of granules in the neurohypophysis was reduced. Thus the authors felt that the repeated administration of morphine inhibited the neurosecretory activity of the hypothalamo-neurohypophyseal system, leading to the increased rate of urine excretion. They also performed experiments in which rats were examined at the shorter time interval of 3, 7, 9, and 14 days of treatment. No difference from controls was observed at 3, 7, and 9 days. On day 14 decreases in granules were seen in both the paraventricular and supraoptic nuclei, whereas no effect was seen on the neurohypophysis.

Rodeck and Braukmann[25] investigated the same problem. Rats were treated twice daily with 2 mg morphine sulfate, and the neurohypophysis was examined by the same histochemical stain as above. Thirty minutes after the first dose of morphine, no loss in granules was seen. Rats treated 3–5 days showed without exception a considerable reduction in "Herring bodies," and only finely dispersed granules were diffusely located in the posterior pituitary gland. In animals treated for 5–10 days, a return toward a normal picture already was evident. By 15–20 days, the picture appeared normal and perhaps even an increase in neurosecretory granules tended to occur in the posterior pituitary. Essentially parallel changes occurred in the paraventricular and supraoptic nuclei. Another group of animals was treated with 10 times the dose, that is, twice a day with 20 mg of morphine sulfate. The posterior pituitary and the paraventricular and supraoptic nuclei were normal on the twentieth day. Injection of twice the dose of morphine on the twentieth day (40 mg) caused no reduction in the staining. This same dose of morphine in previously untreated rats produced a marked depletion of neurosecretory granules. Thus this last experiment indicated that tolerance had developed. The tolerance was manifested by the lack of effect on the Gomori-staining granules. The authors explained that this tolerance was nonspecific in that other agents such as nicotinic acid produce the same phenomenon; they felt that these tolerance effects might be related to the general adaptation syndrome of Selye. However, they stressed the fact that these manifestations of tolerance were not seen in the neurohypophyseal system to the specific stimulus of increased osmotic pressure of the blood with salt solution.

If the work of Shimai *et al.* and of Rodeck and Braukmann are considered together, some confusion results because it seems that in one case tolerance led to depletion of neurosecretory granules while in the other case tolerant animals had a normal level of granules that then did not respond to morphine. Perhaps these findings may reflect the differences in the design of the experiments. Unfortunately, histological studies without some parallel studies on the functional state of diuresis detract from the value of the work by Rodeck and Braukmann.

Two groups examined the concentration of antidiuretic hormone (ADH) in chronically treated rats. Shimai et al.[23] used the method of Birnie[26] to bioassay for ADH. In control rats the hypophysis contained 340 (mU)/100 g and the serum 5 mU/ml. In rats on continuing doses of morphine, 100 mg/kg, the ADH content decreased to 200 mU/100 g in the hypophysis and 0.8 mU/ml in serum. These findings along with the histological picture discussed above indicated that there was depletion of antidiuretic hormones in the chronically treated animal. Newsome et al.[21] used essentially the method of Jeffers et al.[27] to bioassay for ADH in the blood of rats completely tolerant (undergoing diuresis) to levorphanol. They found in contrast to Shimai et al. that the ADH-like activity was severalfold higher than in control rats. They did not, however, completely eliminate the possibility that the blood contained residual amounts of levorphanol that could have compromised the specificity of the bioassay. In any event,' both groups proposed that tolerance to the antidiuretic effect resulted in part from decreased sensitivity to the antidiuretic hormone. Experiments by Newsome's group on changes in sensitivity to injected vasopressin were not conclusive, but the tolerant rats were less sensitive to the antidiuretic effect of nicotine, a releaser of antidiuretic hormone. The excretion time for control rats of 116 ± 9 min was increased to 137 ± 7 min by 5 mg/kg of nicotine. In rats treated with large doses of levorphanol for 12 days, nicotine did not produce significant antidiuresis (excretion time was 87 ± 4 min). Shimai's group found decreased sensitivity to vasopressin in their morphinized rats. Vasopressin produced a 60% decrease in urinary excretion rate in controls but produced only a 30% decrease in morphine tolerant animals. The findings thus far are suggestive of the narcotic producing at least a functional diabetes insipidus. Ad libitum water drinking experiments appear to support such a concept.

This may be a logical point at which to discuss studies on the effects of narcotic antagonists. Schieden and Blackmore[6] found that the antidiuretic effect of morphine in the water-loaded rat was antagonized by nalorphine. Winter, Gaffney, and Flataker[28] suggested that nalorphine prevented the antidiuretic action of morphine by blocking the release of antidiuretic hormone. If, then, development of tolerance (as discussed above) is a matter of change in the ADH hormone release–response system, what would be the effect of narcotic antagonist on this tolerance phenomenon? Newsome et al.[21] tested control and chronically levorphanol-treated rats at various times with single doses of levallorphan. In Fig. 2 it may be seen that the dose of 10 mg/kg of levallorphan alone (lower panel) had no effect on the excretion time of the water load. In the rats maintained on levorphanol, a marked antagonism of the antidiuretic effect was obtained with levallorphan, as shown by the shortened excretion time. In fact, in many cases the antagonism went beyond control levels. For example, in Fig. 2, on day 7 the excretion time for the levorphanol plus levallorphan-treated rats was shorter

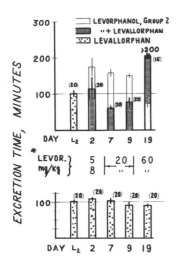

Fig. 2. Effect on the excretion time of a standard water load in rats treated daily with levorphanol but given levallorphan (10 mg/kg, subcutaneously) on day L_2, 2, 7, 9, and 19. The asterisk indicates only those doses of levorphanol which were given on the day of the test. The rats were, however, receiving levorphanol throughout this period according to the following schedule in which the numbers in parentheses indicate the day and the others indicate the dose of levorphanol in milligrams per kilogram; (1) 3, 3; (2) 5, 8; (3) 10, 10; (4) 15, 15; (5 through 9) 20, 20 each day; (10) 30, 20, 30; (11) 40, 20, 40; (12) 40, 60 and (13 through 19) 60, 60. (From Newsome et al.[21] by courtesy of the *Journal of Pharmacology and Experimental Therapeutics*.)

than the respective time for the levallorphan control ($P=0.05$). One further observation was that animals showing a diuretic effect to levorphanol (day 19) responded with a marked antidiuretic effect to the same dose of levallorphan. Thus antagonism led in this instance to reversal in the opposite direction, that is, antidiuresis was obtained by giving levallorphan to the chronically levorphanol treated rats.

2. Ad Libitum Water Drinking and Urine Excretion

Holmes, Carter, and Fujimoto[29] observed polydipsia and polyuria in rats maintained on high doses of levorphanol for longer than 7 days. The polydipsic response started usually 5–10 min after injection of the levorphanol with the rats drinking vigorously and consuming 50–75 ml of water in 4 hr. The tremendous increase in daily water intake and urine excretion is shown in Fig. 3. The results obtained by Shimai et al. on water intake and urine excretion in rats treated with morphine twice a day are shown in Fig. 4. Urine excretion gradually increased, and the increase was particularly noticeable after 2 weeks of morphine treatment. In this case daily water intake did not appear to increase. Increase in both water intake and urine excretion were later observed by Marchand and Denis[30] in morphinized rats. These results by the latter group agree with the earlier work of Flowers, Dunham, and Barbour,[31] who stated that morphine addiction in rats increased water intake 25–50% and urine output by at least 200%. Unfortunately, Flowers et al. did not state whether this response was observed early or late in the treatment cycle. Wikler et al.[32] observed that the 24-hr water intake in tolerant rats, maintained on a continuing single large dose of morphine in the morning, was not different from control rats. In the tolerant group water drinking occurred in the first 7-hr period with an abrupt drop occurring 13–14 hr after the morning injection of morphine, whereas the control rats consumed the largest part of their intake during the last 17-hr period of observation. Since the 13–14-hr period in the morphinized rats coincided with the development of signs of abstinence, it may be assumed that the lack of difference in 24-hr water intakes between morphinized and control animals was due to temporal overlap of drug effects and withdrawal effects in the morphinized group. The direct drug effect would appear to have been that of increasing water intake. In another study (Martin *et al.*)[33] it was mentioned that tolerant rats would "consume food and water voraciously if allowed." The duration of action of the drug and dose schedules therefore must play important roles, and it should be remembered that water intake–urine excretion measurements may frequently be made over longer time intervals than the duration of action of the drug.

It is not clear how these studies on water intake and excretion relate to some of the interesting early findings of changes in tissue hydration. Ac-

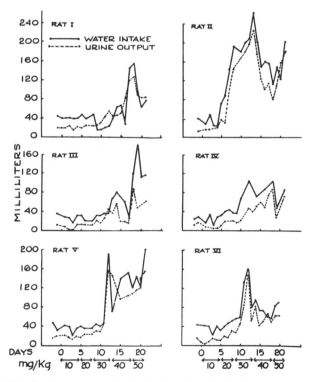

Fig. 3. Occurrence of polydipsia and polyuria in rats treated chronically with levorphanol. Days of treatment and dose of levorphanol are given in the abscissa. Levorphanol was given subcutaneously at the designated dose every 12 hr. (From Holmes *et al.*[29] by courtesy of the *Proceedings of the Society for Experimental Biology and Medicine.*)

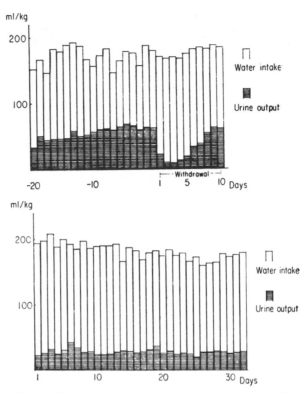

Fig. 4. Influence of morphine administration and withdrawal on water intake and urine excretion (top panel). Morphine and administered subcutaneously twice a day, increasing in dose from 20 to 100 mg/kg. The latter dose had been reached at the time of withdrawal of the drug. Water intake and urine volume were measured daily. The bottom panel is for control animals. (From Shimai *et al.*[23] by coutesy of *Okajimas Folia Anatomica Japonica*.)

cording to Flowers *et al.*[31] there was an "addiction edema" of tissues. Water content of the tissues was increased and was maximal at different times during continued treatment with morphine for each tissue: Brain and liver, day 10; muscle, 23; and skin, 3 and 14. These data agree in general with those of Detrick and Thienes.[34]

In *ad libitum* water drinking and urine excretion experiments, Shimai *et al.*[23] found that when rats were withdrawn from morphine the volume of urine decreased to minimal values on the day 3 with gradual recovery over the next several days (Fig. 4). At the same time Gomori-positive granules increased maximally in the hypothalamo-hypophyseal system at 3 days. Bioassay ascertained the histological picture in that on day 1, 610 mU/100 g of ADH were found in the hypophysis and day 4, 510 mU/100 g. The serum had increased concentrations of 7.8 and 6.3 mU/ml on the first and fourth days of withdrawal. Thus they concluded that on withdrawal of morphine, the activity of the hypothalamo-hypophyseal system increased up to and beyond control values and then gradually decreased down to control levels. The observations of Wikler et al.[32] were mentioned earlier where abruptly decreased water intake occurred with the onset of withdrawal signs. Martin et al.[32] observed that after the fourth day of abrupt withdrawal of morphine, the withdrawn group consumed more water than controls. This increased intake persisted for at least 63 days after the last dose of morphine. During this interval the depressed body weights were regained. Decreased drinking and decreased excretion had been observed in early stages of withdrawal by Flowers et al.[31] In this work water content of several tissues was measured and a "withdrawal edema" was found. The brain and liver became edematous rapidly on the first day after withdrawal of morphine. The skin seemed to partially compensate for this change by becoming drier. These observations would be compatible with the proposed changes in the hypothalamo–hypophyseal system postulated by Shimai et al.[23]

3. Exteriorized Bladder in Intact Rats

Inturrisi and Fujimoto[22] pursued the problem of changes in sensitivity to vasopressin (Pitressin) in morphinized rats. A continuous hydrating solution was infused intravenously into the rats, and urine excretion was measured as an outflow from the surgically exteriorized bladder.[35] In Fig. 5(a), the responses of a control rat to 8 mg/kg morphine sulfate administered subcutaneously and 10 or 20 μU of vasopressin are seen. These same doses of morphine and vasopressin were repeated in animals that had been treated with increasing doses of morphine for 1, 3, or 5 days. After one day of treatment, some tolerance has been developed to the antidiuretic effect of morphine. Definite tolerance was seen on days 3 and 5, while the response to vasopressin was unchanged. Therefore it was concluded that at least some tolerance developed to the antidiuretic effect of morphine in the absence of a change of sensitivity to vasopressin. It was suggested that the ability of morphine to release antidiuretic hormone had been decreased by the repeated administration of morphine. The effect of morphine treatment on adrenal ascorbic acid depletion was cited as a possible parallel situation. The acute administration of morphine stimulates the release of ACTH as measured by adrenal ascorbic acid depletion.[9] When morphine is given daily for 6 days, the release of ACTH to a challenge dose of morphine is not only absent but the release of ACTH by other drugs is blocked in morphine-treated animals. The same sort of mixed situation may be produced acutely by giving a combination of morphine and pentobarbital where blockade of ACTH release occurs.[36]

It was also observed that with the development of tolerance to the antidiuretic effect, a diuretic response component began to appear, as seen in Fig. 5D. This response component also was elicited in consistent fashion using 0.4 mg/kg intravenously of morphine sulfate in place of the 8 mg/kg

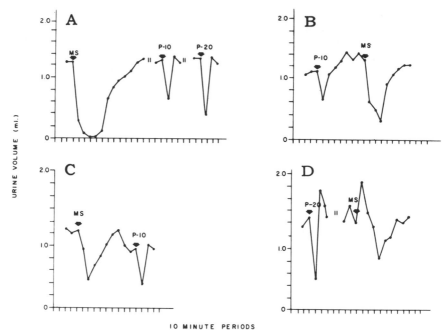

Fig. 5. The development of tolerance to the antidiuretic effect of morphine (MS) 8 mg/kg given subcutaneously. MS = morphine sulfate 8 mg/kg subcutaneously (s.c.); P–10 = Vasopressin (Pitressin) 10 μU (i.v.); P–20 = Vasopressin (Pitressin) 20 μU (i.v.). A. The response to MS and P–10 and P–20 of a previously untreated rat. B. The response to MS and P–10 of a rat treated for 1 day with MS 8 mg/kg s.c., b.i.d.. C. The response to MS and P–10 of a rat treated for 3 days with MS 8 mg/kg s.c., b.i.d., increased 8 mg/kg per day so that by day 3 the dose was MS 24 mg/kg s.c.. D. The response to MS and P–20 of a rat treated for 5 days. The first 3 days of morphine treatment were as in C. On days 4 and 5 the doses were 36 and 44 mg/kg, respectively. (From Inturrisi and Fujimoto[22] by courtesy of the *European Journal of Pharmacology*.)

subcutaneous challenge dose. This diuretic component became more prominent as the number of treatment days increased. This response was not sensitive to changes in the challenging dose of morphine in that going from 0.16 to 1.2 mg/kg intravenously did not enlarge or prolong the response. The antidiuretic component was much more dose dependent. It is not known whether this diuretic component in the action of morphine in the rat is the same action that produces diuresis in the chicken with the first dose of morphine.

Recently Marchand[15] used the same preparation to obtain results that provide some provocative thoughts on the action of morphine. He demonstrated dose-dependent antidiuretic responses to morphine in well-hydrated rats. But a single dose of 3 mg/kg of morphine sulfate produced *diuresis* in the *nonhydrated* rat. As the dose was increased to 12 mg/kg, the diuresis decreased. Thus Marchand demonstrated that morphine may have a diuretic or antidiuretic effect in the rat, depending upon whether the animals had been hydrated or not hydrated. He pointed out that most workers who have demonstrated the antidiuretic action of narcotic analgesics have used hydrating conditions. This work is important in providing a bridge between the various experiments done in a number of species and in man in which the experimental results have not all been explicable by the unitary hypothesis of antidiuretic hormone release and development of tolerance through changes in this system.

4. Inorganic Ion Excretion in Rats

There are changes in urinary constituents that may or may not relate to changes in antidiuretic hormone activity. Marchand and Fujimoto[24] found increases in urinary sodium ion excretion in rats treated with morphine. These initial increases were followed by what appeared to be the development of tolerance so that by day 14 of treatment differences were no longer observed between treated and control groups. Of course there may have been a limit as to how long such differences could be sustained. Although the urinary output of potassium ion was temporarily increased, the effects were not as consistent. It appears from the work of Tomizawa *et al.*[37] that once rats were in the diuretic phase from morphine treatment, sodium excretion dropped while tissues retained sodium. Subsequently Marchand and Denis[30] confirmed the effects on sodium and potassium excretion but also reported that calcium excretion was markedly affected. The initial dose of morphine had no effect, but as repeated doses were given the daily excretion of calcium rose about tenfold over a 12-day period (Fig. 6).

Vachon and Marchand[38] examined the dynamics of ^{45}Ca distribution after intravenous injection and simultaneous treatment with morphine. For up to 2 hr the specific activity of ^{45}Ca (expressed as ^{45}Ca/mg Ca) remained higher in the serum of the morphine-treated animals than the controls. Since the total calcium concentration in the serum did not change, these differences reflected changes in the rate of turnover of serum calcium. In relation to tissue calcium, it was concluded that morphine either decreased the exit of ^{45}Ca from serum to tissues or increased the entry of ^{45}Ca back into the

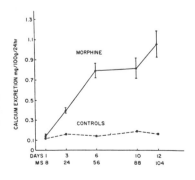

Fig. 6. The effect of chronic morphine treatment (MS) in milligrams per kilogram, three times a day, on urinary calcium excretion. Each value represents the mean of 18 control ($N = 6$) and 21 morphine-treated rats ($N = 7$). Except on the first day urinary calcium excretion was significantly higher in morphine-treated animals. (From Marchand and Denis[30] by courtesy of the *Journal of Pharmacology and Experimental Therapeutics.*)

serum from tissue. The calcium excretion in the 24-hr urine samples was increased in the morphine-treated animals. On the other hand, the excretion of ^{45}Ca was reduced. Thus the fate of the ^{45}Ca was hard to interpret in relation to the total calcium pool. Presumably similar changes would be found with other narcotics since Marchand[39] showed that dihydrocodeinone produced hypercalciuria as well as changes in sodium and potassium excretion. Comparatively, the hypercalciuria with morphine was greater than that with dihydrocodeinone treatment. With long-term treatment with this latter compound, Marchand noted that about 40% of his rats showed calcium deposits in the kidney. These calcium deposits in the medulla of the kidney were located in the lumen of tubules or the interstitium. Sometimes the tubular epithelium was encroached upon by the calcium deposits. On the other hand, Marchand and Denis[30] did not find such deposits in the kidneys of rats treated with morphine. In a later study[48] microscopic examination of the kidneys of chronically morphine-treated rats did show some pathologic changes. These inconstant, variable changes consisted of necrosis, fibrosis, congestion, cellular infiltration, casts, etc.

Certainly changes in calcium excretion lead to provocative implications. Kaneto (Chapter 14) discusses the relationship between morphine analgesia and calcium concentrations in the central nervous system. Takemori (Chapter 7) considers some biochemical aspects of the interaction of calcium and morphine. Interrelationships between calcium, phospholipids, and narcotics are discussed by Mulé (Chapter 8). The role of calcium in antidiuretic hormone release and changes in renal sensitivity to antidiuretic hormone[41,42] and the changes in phospholipid content that take place during diuresis[43] are topics that may be of heuristic value in considering actions of narcotic analgesics.

Because of the effects of morphine on calcium excretion, studies were undertaken to observe effects on magnesium metabolism.[44] An acute dose of morphine caused hypermagnesemia. Magnesium concentrations decreased in muscle, heart, liver, and brain. Magnesium in bone increased. The distribution of ^{28}Mg also changed under the influence of morphine. Studies using chronically morphine-treated animals have not yet been described.

5. Additional Mechanisms of Tolerance

Evidence exists to indicate that the diuresis occurring during chronic morphine treatment was an osmotic diuresis. Marchand and Denis[30] found that on the first day of morphine treatment, U_{OSM} V was low but that from day 3, solute excretion increased. Total urea osmolal clearance and nonurea osmolal clearance were calculated. In control rats 37% of the total solute clearance was urea, whereas in the chronically morphine-treated rats, urea accounted for 54%. The increase in urea excretion in the latter group was proportional to the increase in solute excretion. Blood urea concentrations also were elevated. The authors suggested that decreased sensitivity of the rat at this time to vasopressin was consonant with the reduction in sensitivity to antidiuretic hormone which took place during an osmotic diuresis. Both morphine and vasopressin were ineffective in depressing a mannitol induced diuresis.[22]

In spite of all the studies, one does not have definite evidence that it is the kidney itself or some mechanism outside the kidney that has changed during the development of tolerance. A direct renal mechanism might be involved, as proposed by Marchand, Cantin, and Côté,[40] who observed that in the morphinized rats there were changes in mitochondrial morphology and a thickening of the basement membrane. These changes were thought to be compatible with increased sodium, potassium, and calcium excretion and a decreased responsiveness to antidiuretic hormone. The problem was that the structural changes were not consistent enough to correlate with the physiological changes. Thickening and fibrillary changes of the basement membrane have been observed in water-loaded rats without morphine treatment.[45] Since degenerative changes occur in the aging rat kidney that lead to increased water intake, increased urine excretion, and decreased capacity to concentrate urine,[46] it may be speculated that chronic morphine treatment enhances such natural processes.

Malnutrition may be a factor to contend with in assessing mechanisms of development of tolerance to the antidiuretic effect. Flowers et al.[31] alluded to malnutrition as a possible factor in the production of "addiction edema." Also Detrick and Thienes[34] thought that the less abrupt development of "addiction edema" in their rats than in those of Flowers et al. was due to the difference in diets. Shimai et al.[23] and Marchand and Denis[30] found that chronic treatment with morphine depressed body weight gain; malnutrition could have contributed to changes in the responsiveness of these animals to morphine. Handling of the animals, accommodation to the new environment, age, and sex may be factors affecting the relationship between food and water ingestion in the rat,[47] and it is possible that morphine modifies such interactions.

C. Other Animals

There has been a paucity of work on the development of tolerance to the antidiuretic action of narcotics in animals other than the rat. A number of publications have appeared on studies of the mechanism of the antidiuretic effect especially in the dog (see the review by Papper and Papper),[4] but this advance has not been followed by attempts to examine mechanisms of tolerance in these same animals. Because substantially no progress has been made in this latter respect, only new findings in certain animals are discussed.

The effect on the chicken is most interesting because the response of this animal to a dose of morphine is diuresis[17] instead of antidiuresis. The presence of a renal portal circulation in the chicken allowed morphine to be introduced intravenously into the saphenous vein to perfuse first, and in highest concentration, the tubules of the kidney on the same side. Effects that predominate on this side were taken to be indicative of a tubular effect, while a bilateral effect would have suggested lack of a predominantly renal tubular effect. A bilateral diuretic effect was found to morphine, a response also reflected by bilateral increases in electrolyte excretion. The diuretic effect occurred 15–30 min after the beginning of morphine infusion in hydrated hens. In a companion experiment it was shown that the amount of radioactivity excreted after ^{14}C-N methyl morphine infusion was much greater on the infused than on the noninfused side. Thus the diuretic response was not directly related to the concentration of morphine that was presented to the tubules. The response was mediated by a systemic mechanism. Since the chicken possesses a neurohypophyseal system capable of elaborating antidiuretic hormone, the production of diuresis indicated that morphine did not stimulate this system to release the hormone. If morphine had both diuretic and antidiuretic properties, it may be that in the chicken only the diuretic action occurs. Such a dual effect of morphine might serve to explain the previously discussed findings of Inturrisi and Fujimoto[22] given above in the exteriorized bladder preparation. It would be interesting to see whether tolerance to the diuretic effect occurs in the chicken.

Tolerance was shown to develop to the antidiuretic action of morphine in the mouse.[48] This tolerance developed more slowly than tolerance to the analgesic action of morphine. On day 7 of treatment, sensitivity to the antidiuretic effect of vasopressin appeared unchanged even though tolerance had developed to the antidiuretic effect of morphine. Body weight decreased during morphine administration. Measurements of water and food consumption and urine excretion gave significantly different results depending upon whether these parameters were measured over 24 hr or between 8 A.M. to 4 P.M. or 4 P.M. to 8 A.M. Thus possible contributing factors based on changes in nutritional factors must be assessed further.

In the future the pellet implantation technique of administering narcotics[49,50] should facilitate studies on tolerance by making larger numbers of animals easily available. For these purposes other drug treatment schedules may be helpful. Dihydromorphinone has been administered to mice in milk by Shuster et al.,[51] and etonitazene may be administered to rats in drinking water in dilute concentrations to obtain opioid effects.[32,52] To implicate changes in the hypothalamo–hypophyseal system as the basis for tolerance, additional evidence must be gathered. Development of improved bioassay techniques for antidiuretic hormone may be one approach. Another would be to follow the rate of synthesis of antidiuretic hormone, perhaps by improvement of the *in vivo* labeling techniques of Sachs.[53] Some value may be derived by investigating the rate of degradation of antidiuretic hormone; work by Marchand and Fujimoto[54] gives some preliminary data. Availability of the Brattelboro strain of rats with hereditary diabetes insipidus[55] may provide an answer as to whether antidiuresis and subsequently tolerance to this effect can occur under conditions where the role of the antidiuretic hormone mechanisms has been minimized. Preliminary experiments indicate that antidiuretic effects are produced by morphine in these rats. A polydipsic polyuric syndrome has been induced in rabbits by food deprivation.[56] These rabbits retained sensitivity to vasopressin.[57] Studying the response of these rabbits to narcotic analgesics might be fruitful in elucidating the mechanism of polyuria induced by food deprivation and the development of tolerance to narcotic analgesics (as discussed in Section III B5).

IV. EXCRETION OF NARCOTICS AND METABOLITES IN URINE

Quantitatively, the urine is the most important route of excretion for the narcotics and many of their metabolites. In this section, some of the mechanisms that take part in transferring these compounds from the blood into urine are discussed. Most narcotics have not been thoroughly studied in this regard because their pharmacologic potency frequently precludes giving them in the doses necessary to obtain quantitative data in studying such mechanisms, as discussed by Baker and Woods[58] in their studies on the renal clearance of morphine. Availability of radioactively labeled narcotics have in part ameliorated this problem, but the number of radioactively labeled narcotics is still small. Therefore this section is written to illustrate some concepts based on studies of a few compounds rather than to list exhaustively the mechanisms involved in the excretion of every narcotic. No evidence exists to indicate that narcotic analgesics are handled any differently by the kidney than are other organic bases. Since not all mechanisms applicable to excretion of bases in general have been investigated for the narcotics, only certain renal mechanisms can be discussed. The metabolism of narcotics by the kidney will be included as it relates to the transport mechanisms. Excellent reviews have been published that serve as broad background material. Milne et al.[59] and Weiner and Mudge[60] discuss nonionic diffusion and excretion of weak acids and bases; Weiner[61] gives instructive examples of the excretory processes through which drugs are handled; Forster[62] examines the biochemical mechanisms; and Peters[63] describes some of the methodology and reviews the base transport system.

A. Meperidine and Levorphanol Excretion and pH Dependence

If a narcotic possesses high lipid solubility, it is likely that little of it will be excreted in the urine. This hydrophobic property leads to reabsorption of the drug through the renal tubules. If a pH gradient exists between the urine and the blood and the narcotic can be made to exist more in an ionized form in the fluid of the tubular lumen, then reabsorption can be reduced by ionic trapping of this ionized species. Ionization makes the narcotic more hydrophilic and less hydrophobic. Milne et al.[59] calculated that for organic bases to be excreted by this pH-dependent ionic trapping process, the pK_a of the bases (in our case, narcotic analgesics) should range between 6.5 and 10.0. This process applies to the excretion of meperidine and levorphanol.

Asatoor et al.[64] studied the excretion of meperidine and one of its metabolites, normeperidine, in humans. In patients who received sodium bicarbonate and were excreting highly alkaline urine, the total excretion of meperidine and normeperidine in 48 hr was less than 5% of the subcutaneously administered dose of meperidine. In patients whose urines were highly acidic due to ammonium chloride ingestion, the total excretion of meperidine and normeperidine in 48 hr accounted for about 50% of the administered dose of meperidine. Of this 50%, 22% was meperidine itself. Since the pK_a of meperidine was 8.63 and that of normeperidine was 9.68, relatively more of the un-ionized free base form was present at alkaline than at acid pH. Thus, with the pH of blood at 7.4 and the urine alkaline, more reabsorption of the bases occurred than when the urine was acidic. In fact more than a severalfold increase in excretion occurred with acidification of the urine. However, these experiments do not indicate whether processes other than pH-dependent nonionic diffusion were involved. Nor do they in themselves prove the nonionic diffusion hypothesis. However, the pH sensitivity and the evidence cited in the several review articles above provide assurances that nonionic diffusion plays an important role in meperidine excretion. For certain organic bases it has been possible to attribute excretion completely to the nonionic diffusion mechanism.[65,66] However, in other cases factors may militate against efficient operation of this mechanism.[60] Note that in Asatoor's experiments it was not necessary to know the status of glomerular filtration; in fact, meperidine may have reduced the glomerular filtration rate.[4] Asatoor et al.[64] have demonstrated that ethoheptazine excretion also is sensitive to pH changes.

Quantitative relationships between the excretion of parent drug and its metabolites may be affected by pH changes. Thus Asatoor et al. showed that meperidine and normeperidine predominated in acid urine while in alkaline urine meperidinic acid, normeperidinic acid and their conjugates were relatively greater. Their advice to consider pH-dependent excretion was followed subsequently by Beckett. Taylor, Casy, and Hasson[67] in a report on metabolism of methadone. The subjects' urine were kept acidic to enhance the excretion of bases. Beckett's group developed sophisticated analog computer similation applicable to controlled acidic and fluctuating urinary pH values (Beckett et al.[68] among others). The application of this approach to narcotic excretion may prove to be valuable. Mathematical approaches to estimating certain drug excretion parameters have been reviewed by Wagner.[69] Perhaps the rabbit will prove to be a useful animal in studying pH dependent excretion of narcotics since simple feeding and fasting produce large shifts in urinary pH in a circadian fashion, shifts which have been correlated with drug excretion and metabolism changes for phenobarbital.[70]

Excretion of levorphanol (pK_a of 9.2) was shown by Braun et al.[71] to be pH sensitive in the rat. After an intraperitoneal injection of 6 mg/kg levorphanol, the amount of free drug excreted in 6 hr was measured. In going from a pH range of 8.0 to 8.7 down to 5.5 to 5.8, the excretion of levorphanol doubled. Perhaps because the collection period was relatively short, this doubling only increased excretion from 1% at the higher pH to 2% of the administered dose at the lower pH.

B. Morphine and Dihydromorphine Excretion

1. Conventional Clearance

Baker and Woods[58] bravely tackled the problem of determining renal clearance for morphine in the dog. Handley and Keller[8] had shown that morphine at a dose of 2 mg/kg, administered intravenously, reduced renal blood flow, and Shideman and Johnson,[72] using the same dose of morphine, obtained hypotensive effects. Recognizing that these pharmacologic effects might interfere with a clear interpretation of clearance experiments, Baker and Woods devised a morphine treatment protocol that was in part successful in avoiding these effects. After three or four consecutive control periods for the clearance determinations for inulin and p-aminohippuric acid (PAH), 30 mg/kg of morphine was injected subcutaneously. This morphine treatment did not alter inulin or PAH clearances. Forty minutes following this treatment, intravenous infusion of 7.5 mg/kg/hr of morphine was instituted, and 20 min later clearances for inulin, PAH, morphine, and conjugated morphine were measured for four or five periods of 15 min each. This combination of subcutaneous and intravenous morphine produced no change in inulin clearance, indicating that the glomerular filtration rate was unaffected. At the same time an 11–45% decrease in PAH clearance occurred even though the Tm for PAH (determined in separate experiments) was shown to be unaffected by this combination treatment. Thus it appeared that renal blood flow had been reduced by morphine even though glomerular filtration rate was not affected. Under these conditions the clearance for free morphine and conjugated morphine (formed in the animal from morphine) equaled inulin clearance. Therefore the conclusion was that morphine was excreted by glomerular filtration. It was also conjectured that since about 25% of morphine in the plasma may be bound to proteins, correcting for this factor would raise the clearance value for morphine. This new value being greater than inulin clearance then suggested that some tubular secretion was occurring. Unequivocal experimental evidence for this latter suggestion was provided subsequently by others working with dihydromorphine.

2. Stop Flow

Hug, Mellett, and Cafruny[73] proved that dihydromorphine was secreted by the proximal tubules in the dog and monkey using stop-flow analysis. Specifically, the urine-to-plasma concentration ratios (u/p) were calculated for each constituent that was measured during free and stop-flow periods; these ratios were divided by the calculated urine-to-plasma ratio for inulin to correct for reabsorption of water in the formation of urine. Thus the point at which the minimum occurred for the corrected ratio "U/P" for sodium or chloride during the stop flow period identified urine from distal tubules. The maximum for the U/P of PAH identified urine from proximal tubules. Minima and maxima in U/P indicated reabsorption and secretion respectively. ^{3}H-dihydromorphine (DHM) was administered in 1 or 2 mg/kg doses subcutaneously, intravenously, or by infusion. DHM produced a stop flow pattern of U/P with a maximum that coincided with the location of the PAH peak. Thus DHM was being secreted by the proximal tubules in both the dog and monkey. In dogs these DHM maxima were 2.4–5.4 times the free flow values while in the monkey these were about 1.7 times. Thus the magnitude of the maxima differed in the two species. Since these experiments indicated active secretion, the transport system was characterized by responsiveness to inhibitors. Classically, drugs that are actively transported fall into two groups: organic anions transported by the acid transport system and organic cations transported by the base transport system.[59,60] Transport of drug by the acid system can be blocked by an acid, probenecid, and that by the base system by a variety of bases, cyanine 863, mepiperphenidol, and quinine, each to varying degrees. In the experiments of Hug et al.[73] both mepiperphenidol and cyanine 863 blocked transport of DHM, as indicated by the drop in U/P compared to control stop flow values. At the same time the transport of PAH was unaffected by these base transport inhibitors. Probenecid on the other hand had no effect on the stop flow pattern of DHM. Therefore DHM appeared to be transported only by the base system in these stop flow experiments. They investigated the transport of the metabolite of DHM presumed to be the glucuronide conjugate of DHM. This endogenously formed metabolite had U/P maxima in proximal tubular urine. The average value was 2.5 times the free flow values. Therefore there was an active proximal tubular secretion of this metabolite. Both mepiperphenidol and probenecid reduced free flow and stop flow U/P ratios for this metabolite. Because it appeared that this metabolite was the glucuronide conjugate, it was suggested that it was not unlikely for this conjugate to be transported by both the acid and base systems since DHM glucuronide can exist in the zwitterion form at physiologic pH. Fujimoto and Way[74] have demonstrated that morphine-3-glucuronide has a $pK_1 = 3.2$ and $pK_2 = 8.1$ and proposed a zwitterion structure for morphine-3-glucuronide.

The matter of ionization was considered by Hug's group in relation to pH-dependent excretion. Since pK_a of DHM was 8.55, nonionic diffusion was a possibility. Although their experiments were not designed to study this mechanism, it appeared from their limited data that nonionic diffusion played a minimal role in DHM excretion. Generally a decrease in urinary pH decreased DHM excretion so that the effect was opposite to what one would have expected by the nonionic diffusion mechanism. The metabolite excretion was somewhat sensitive to pH changes; decreasing pH tended to decrease U/P. Their data were not sufficient to make a firm conclusion, but they pointed out that conjugated drugs such as the DHM metabolite generally penetrated lipid barriers poorly. Sensitivity to pH changes was unexpected.

How closely this work with DHM relates to the work of Baker and

Woods with morphine is still an open issue. Hug *et al.* emphasized the qualitative similarity in pharmacologic action and metabolism of dihydromorphine to morphine. It would be interesting to perform clearance and stop flow experiments with labeled morphine in view of some of the difficulties encountered with large doses of unlabeled morphine.

3. Sperber Preparation

In the following discussion renal tubular transport and metabolism of morphine in the intact unanesthetized chicken are examined. For this examination it is necessary to describe the technique briefly because an intelligible interpretation of results requires this background. In contrast to the mammalian and rodent species commonly used in the laboratory, the chicken differs in possessing a renal portal system. Sperber[75,76] must be given full credit for developing the technique and recognizing its usefulness in studying renal tubular transport. Using the technique, he demonstrated the tubular excretion of strong organic bases.[63] The technique as it is used today is as follows. An unanesthetized hen is restrained in a stand and hydrated with water by crop tube. Under local anesthesia, collecting cannulas are sutured over each of the two ureteral openings in the urodeum. Urine is collected separately from each kidney. If an agent is infused into the saphenous vein of one leg, venous blood from that leg passes through the ipsilateral renal portal system without being filtered by the glomeruli. The agent therefore first comes into contact with the tubular cells of this kidney and has the opportunity to be secreted by the tubular cells before passing into the renal vein and general circulation. Any part of the infused material that is not excreted in one pass reaches the systemic circulation and then returns to both kidneys in equal concentrations. Thus an entity called the "apparent tubular excretion fraction" (ATEF) can be calculated where EXC_I is the amount excreted in the urine from the ipsilateral kidney, EXC_C is the amount from the contralateral kidney, and INF is the amount infused during the same period as the urine collection: $ATEF = (EXC_I - EXC_C)/INF$. The ATEF will be a fraction less than one, or it can be more conveniently multiplied by 100 to express it as a percentage. In effect, the ATEF is the percent of the infused dose that has been excreted by the infused kidney during a given period.

Since it is possible for the agent to reach the urine by diffusion through the tubular cell rather than by secretion, an estimate has been made of what the ATEF would be for a pure diffusion mechanism. This estimate was placed by Sperber at an ATEF of 0.10 or 10%. He derived this value as a maximum based on the diffusion volumes on one hand (based on the rate of delivery taken as the renal plasma flow) and on the other hand based on the maximal volume into which diffusion could occur given by the glomerular filtration rate. The ratio of glomerular filtration rate to renal plasma flow was 1/13. Therefore an ATEF of 0.10 is a safe estimate of maximal diffusion. Also, all the blood from the saphenous vein does not necessarily pass through the renal portal system. There it is shunt, controlled by a venous valve that enables some of the blood to bypass the portal system on its way to the heart. The ATEF value for PAH represents the percentage of saphenous blood passing through the renal portal system, assuming that all of the PAH delivered to the portal system is removed by the tubular cells. The part of the blood that bypasses the renal portal system by way of the shunt system returns to both kidneys and contributes directly to the value of EXCc. If most of the saphenous blood is shunted and flows directly into the renal vein, a low ATEF for PAH is found. On the other hand, a high ATEF for PAH indicates that most of the saphenous blood is now routed through the ipsilateral peritubular network. In studying a particular agent, that is morphine, an ATEF for the agent may be calculated relative to the ATEF for PAH. The ATEF value divided by the ATEF for PAH times 100 is an estimate of the relative efficiency of removal of morphine compared to PAH. In essence, what this last calculation does is to correct for quantitative difference in the degree of shunting occurring from chicken to chicken. Since a certain amount of the infused agent usually flows into the systemic circulation through the shunt system, this relative ATEF provides a better estimate of tubular secretion by the infused kidney.

The renal portal system of the chicken also can be used to study the metabolism of drugs by the ipsilateral kidney.[77-79] By the same reasoning as for the parent drug, if a metabolite of that drug were excreted in excess by the ipsilateral kidney, this excess must have been formed from the drug at sites between the point of infusion of the drug (the saphenous vein) and the kidney or urine. Unless there are very active enzymes in the blood or blood vessels, the transit of drug from the point of injection to the renal tubules is so rapid that metabolism would be unlikely. Also, it is highly unlikely for most drugs to be metabolized directly by the urine. Since metabolism by blood and urine can be checked experimentally by testing their enzyme activity, these can be eliminated easily as sources of the excess metabolite, leaving the metabolism to the kidney itself.

May, Fujimoto, and Inturrisi[79] found the ATEF for ^{14}C after infusion of morphine-N-$^{14}CH_3$ to be 29.5 ± 3.3 (S.E.) percent. The PAH ATEF was $58.5 \pm 4.6\%$ and the ^{14}C/PAH, relative ATEF, was 52.5 ± 6.0. Thus administration of ^{14}C morphine led to the appearance of large amounts of ^{14}C in the urine on the ipsilateral side; the large ATEF was indicative of tubular secretion. More than 50% of the radioactivity reaching the tubular cells of the ipsilateral kidney was secreted during the initial passage. These experiments were repeated using the same amount of radioactive morphine to which increasing quantities of cold morphine were added. Addition of 0.125, 0.35, and 0.5 mg/kg/min of morphine caused noticeable to almost complete decrease in ^{14}C ATEF in a dose-related manner. This effect of unlabeled on labeled morphine was taken as evidence against diffusion since, if diffusion were the main process, such an effect of morphine would not have occurred. The concentration gradient for the radioactive morphine had not been changed by adding the unlabeled morphine. The volume from which the radioactive morphine would have diffused was unchanged. Therefore no effect should have occurred. The observed effect could only be accounted for by a concentration-limited process such as active transport. These observations on the differences between tracer and carrier doses may possibly form the basis of explanation for the apparent difference in findings between Hug et al.[73] and Baker and Woods.[58] Thus Hug et al., using relatively small doses of dihydromorphine, found tubular secretion. These results seem to parallel the findings with tracer doses in the chicken. Baker and Woods found primarily glomerular filtration for their large dose of morphine. These results may correspond to conditions where low ATEF values were found in the chicken with the addition of unlabeled morphine. It seemed likely at least in the chicken that the transport system was easily saturated.

In the experiments by May et al.[79] the radioactivity in the urine samples was separated into morphine-N-$^{14}CH_3$ and a ^{14}C metabolite fraction. Both morphine and metabolite were excreted in excess by the ipsilateral kidney. The excess of metabolite indicated that it had been formed from the metabolism of morphine by the kidney. Cyanine 863 reduced the ATEF for both morphine and metabolite formed from morphine. Probenecid had no effect on either ATEF. These results were compatible with an interpretation that the cyanine 863 inhibited active transport of morphine into the tubular cell and therefore led to a parallel fall in formation of the metabolite. That this interpretation was correct was supported by subsequent work by May and Fujimoto[80] and Watrous, May, and Fujimoto.[81]

In the meantime, this metabolite of morphine was isolated by Fujimoto and Haarstad from the cat as well as the hen and was shown to be morphine-3-ethereal sulfate, MES.[82] The isolation was greatly facilitated by use of Amberlite XAD-2 resin, as was the isolation of two metabolites of nalorphine and two of naloxone.[83-85] The isolation of MES allowed this metabolite to be studied for the differential effects of transport inhibitors on infused and intracellularly formed MES.

Figure 7 is a diagram of the proposed scheme of the mechanism of renal tubular transport of morphine and MES in the chicken.[80,81] Evidence on the transport and metabolism of morphine as presented in the previous discussion was amplified, and new information was provided by studies on MES. MES that was administered by infusion (designated as MES* in the figure) was shown to be transported by the acid system. Infused MES* was excreted unchanged, and its high ATEF indicated tubular secretion. This ATEF was markedly reduced by the administration of probenecid. The administration of mepiperphenidol or quinine did not block MES* transport. Thus MES* was clearly transported by the acid system. Confirmatory experiments were repeated to show that base transport blocking agents, cyanine 863 and mepiperphenidol, blocked the transport of morphine. Also, MES that would have been formed in the tubular cell from morphine failed to appear in the urine. Probenecid used in place of the base transport blocking agents had no effect on the amount of morphine and MES. The important deduction was that probenecid must act on the peritubular side of the cell to block transport of MES* because the ATEF of MES formed from

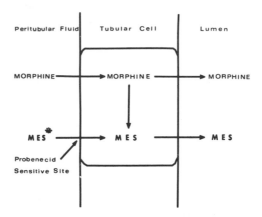

Fig. 7. Diagram of proposed scheme of the mechanism of renal tubular transport of morphine and morphine ethereal sulfate. (From Watrous *et al.*[81] by courtesy of the *Journal of Pharmacology and Experimental Therapeutics.*)

morphine within the cell was not reduced by probenecid. This work represented clear evidence for one site of action of probenecid that was consistent with the site of acid transport indicated by the work of Wedeen and Jernow[86] using autoradiography of nephrons of the rat. Other areas of argument and disagreement are described in the paper by Watrous *et al.*[81]

The concepts developed from the study of the morphine-MES system in the chicken should be applicable in a more general way to the study of other drugs and inhibitors. One such study has been that of Hakim and Fujimoto[87] on the serotonin-5-hydroxyindoleacetic acid (5-HT–5-HIAA) system. In summary, it may be stated that 5-HT was transported into the renal tubular cell and metabolized in part to 5-HIAA and both were excreted in the urine. 5-HT transport was blocked by base transport inhibitors. Excretion of 5-HIAA formed in the cell from 5-HT was not blocked by probenecid, while the transport of infused 5-HIAA was inhibited by probenecid. These results paralleled the findings discussed for morphine and MES and again placed the site of action of probenecid on the peritubular border of the cell. Because of interest in the interaction of the 5-HT–5HIAA system with morphine in the central nervous system (Way, Chapter 10) and the effects of probenecid thereon, the study of these drugs in the kidney may be useful as a model system to investigate some of the interaction between these drugs. Perhaps this model system would be useful in demonstrating a mechanism of interaction between narcotics and monoamine oxidase inhibitors since monoamine oxidase inhibitors potentiate the action of certain narcotics. Several mechanisms of interaction have been suggested.[88-90] Since inhibition of monoamine oxidase can be obtained without affecting transport of 5-HT,[87] the interaction of the narcotic with 5-HT transport may be studied as well as the interaction of the narcotic with monoamine oxidase function.

There are differences among results obtained in the several preparations that bear mentioning. The two agents, mepiperphenidol and probenecid, blocked the transport of the DHM conjugate in the dog whereas the transport of injected MES was blocked by probenecid but not by mepiperphenidol in the chicken. Since the DHM conjugate was not isolated and readministered, it is not clear whether the effect of mepiperphenidol was on the metabolite circulating in the blood or metabolite formed in the kidney. If the metabolite were formed in the kidney from DHM, then the block of DHM transport into the tubular cells could explain the blocking effect of mepiperphenidol on the metabolite excretion. Examination of the results published by Hug et al.[73] tend to support the view that mepiperphenidol blocked the transport of circulating metabolite. If this is true, the difference in effect of mepiperphenidol between the dog stop flow and the chicken ATEF experiments could be a species difference. Species differences in the metabolism of some of these compounds is evident. In the dog morphine was metabolized to morphine-3-glucuronide,[91] a metabolite that appeared not to be transported in the conventional clearance study by Baker and Woods.[58] The hen may show species peculiarity in that both morphine and nalorphine were conjugated as ethereal sulfate[82,83] but naloxone administration yielded a glucuronide in which the 6-keto group of naloxone was reduced to an alcohol group.[84] The cat was similar to the hen with respect to formation of morphine ethereal sulfate and nalorphine ethereal sulfate.[82,83] Most if not all of these metabolites exist in zwitterion form at physiologic pH. Thus the question of whether each form is transported by the acid (anionic) or the base (cationic) transport systems, or by both, is relevant to the matter of specificity of the transport systems.

Studies on the excretion of compounds other than narcotics suggest limitations on the value of transport inhibitors in defining the mechanism of transport. Creatinine transport was inhibited by both probenecid and cyanine 863 in the hen.[92] In the dog, similarly, probenecid and mepiperphenidol inhibited creatinine transport.[93] It was stated in this work that it seemed unusual for the acid transport inhibitor to be active since creatinine was a base. The exact form of the basic ionization was not known, according to Rennick.[92] Cyanine 863 and probenecid inhibited norepinephrine transport in the chicken.[94] Epinephrine transport that was blocked by probenecid and not by cyanine[95] was found more recently to be blocked by quinine.[96] Dopamine transport was blocked by both probenecid and cyanine[94,96] Riboflavin transport appeared to be inhibited by both base and acid blocking agents.[97] Catechol interacted with catecholamine transport, but catechol transport had its own peculiarities.[98] The specificity of such agents as cyanine 863 and probenecid as base and acid transport inhibitors, respectively, therefore, has become clouded, and conclusions about the mechanism of transport of narcotic analgesics and their metabolites must be tentative in view of these difficulties. The same caution should be applied to the use of probenecid as a transport blocking agent in the liver and central nervous system in view of studies showing that a new major effect of probenecid is to decrease the volume of distribution and indirectly increase the rate of metabolism of some antibiotics.[99]

4. In Vitro Kidney Slice Experiments

Results of studies using this approach for narcotics in general have supported expectations based on the *in vivo* studies. In two publications Hug[100] and Hug *et al.*[73] described experiments in which the uptake of tritium-labeled dihydromorphine (DHM) was determined for renal cortex slices from the rat and dog. The slice-to-media (S/M) concentration ratio for DHM reached peak values in 1–2 hr at the usual incubation temperature of 37°C. S/M was about four. That this accumulation of DHM in the slice was due to an energy-requiring or active uptake process was indicated by the fact that the ratio was markedly reduced by classical metabolic inhibitors such as iodoacetate, fluoroacetate, sodium cyanide, 3,4-dinitrophenol, and nitrogen. Reduction of the incubation temperature to 25°C reduced the ratio slightly while reduction to 0°C markedly reduced the uptake of drug. Under nitrogen and with a range of DHM concentrations from 10^{-3} to 10^{-7} molar, a constant, low S/M was obtained. This S/M was considered to indicate nonspecific tissue binding. The normally high S/M ratio under oxygen decreased rapidly toward this nonspecific S/M value when the concentration of DHM was increased from 10^{-7} to 10^{-3} molar. The uptake therefore was due in part to a saturable process. The uptake was stimulated by 5–10 mM acetate in the medium. Mepiperphenidol, N-methylnicotinamide, and nalorphine inhibited DHM uptake—results that suggested participation of the base transport system in producing high S/M ratios. Determinations of the S/M ratio for morphine, nalorphine, levorphanol, dextrorphan, and *l*-methorphan as separate substrates indicated that all of these bases were accumulated by the slice. In testing for the participation of the acid transport system, probenecid was included in the DHM incubation medium. Probenecid reduced the accumulation of DHM. In more limited experiments probenecid also inhibited the uptake of levorphanol and *l*-methorphan into the slice. These results with probenecid were not explicable in terms of the dissociable phenolic acid group of the narcotics since *l*-methorphan has a methoxy rather than a phenolic hydroxy group. Nonspecific action of probenecid appeared to be the more likely possibility. It should be recalled that Hug *et al.*[73] had not found inhibition of DHM transport by probenecid *in vivo* in their stop flow experiment. Also, the work by other workers on the chicken had shown no effect of probenecid on morphine ATEF. Thus accumulation in the slice may not be measuring the same process as the *in vivo* experiments. Furthermore, McIsaac[101] found that large species variation existed in the ability of kidney slices to accumulate certain organic bases. With some of these bases, uptake by the slices was not correlated with active secretion in the intact animals. With the morphine surrogates more studies must be done before such a general comparison can be made among kidney slices from different animals.

ACKNOWLEDGMENT

The author gratefully acknowledges the efficient secretarial help of Mrs. Rita Mongan in the preparation of this manuscript.

V. REFERENCES

1. H. Krueger, N. B. Eddy, and M. Sumwalt, "The Pharmacology of the Opium Alkaloids," Suppl. 165, Public Health Service Report, U.S. Government Printing Office, Washington, D.C. (Part I, 1941; Part 2, 1943).
2. A. K. Reynolds and L. O. Randall, "Morphine and Allied Drugs," University of Toronto Press, Toronto (1957).
3. O. Schauman, "Handbuch der Experimentellen Pharmakologie," Vol. 12, Springer-Verlag, Berlin (1957).
4. S. Papper and E. M. Papper, *Clin. Pharmacol. Ther. 5*, 205–215 (1964).
5. R. C. de Bodo, *J. Pharmacol. Exp. Ther. 82*, 74–85 (1944).
6. H. Schnieden and E. K. Blackmore, *Brit. J. Pharmacol. 10*, 45–50 (1955).
7. G. Becker and J. O. Moeller, *Arzeim.-Forsch. 10*, 239–243 (1960).
8. C. A. Handley and A. D. Keller, *J. Pharmacol. Exp. Ther. 99*, 33–37 (1950).
9. R. George and E. L. Way, *J. Pharmacol. Exp. Ther. 125*, 111–115 (1959).
10. W. L. Lipschitz and E. Stokey, *Amer. J. Physiol. 148*, 259–268 (1947).
11. E. Mills and S. C. Wang, *Amer. J. Physiol. 207*, 1405–1410 (1964).
12. I. Singh, P. K. Khanna, M. C. Srivatava, M. Lal, S. B. Roy, and C. S. V. Subramanyan, *New England J. Med. 280*, 175–184 (1969).
13. R. Hopmann, *Z. Klin. Med. 107*, 582–605 (1928).
14. L. J. Boyd and D. Scherf, *Med. Clin. N. Amer. 24*, 869–876 (1940).
15. C. Marchand, *Proc. Soc. Exp. Biol. Med. 133*, 1303–1306, 1970.
16. C. Eckhard, *Betrage Anat. Physiol. 8*, 79–99 (1879).
17. C. E. Inturrisi, D. G. May, and J. M. Fujimoto, *Eur. J. Pharmacol. 5*, 79–84 (1968).
18. W. H. McCarthy and R. L. Keenen, *J. Amer. Med. Ass. 187*, 460–461, (1964).
19. B. Bower, L. C. Wegienka, and P. H. Forsham, *Proc. Soc. Exp. Biol. Med. 120*, 155–157 (1965).
20. J. M. Fujimoto, R. J. French, J. K. Graham, and C. E. Inturrisi, *Proc. Soc. Exp. Biol. Med. 114*, 193–195 (1963).
21. H. H. Newsome, W. Tobin, and J. M. Fujimoto, *J. Pharmacol. Exp. Ther. 139*, 368–376 (1963).
22. C. E. Inturrisi and J. M. Fujimoto, *Eur. J. Pharmacol. 2*, 301–307 (1968).
23. K. Shimai, M. Akita, S. Tomizawa, and H. Kondo, *Okajima Folia Anat. Jap.*, 40, 911–933 (1965).
24. C. R. Marchand and J. M. Fujimoto, *Proc. Soc. Exp. Biol. Med. 123*, 600–603 (1966).
25. H. Rodeck and R. Braukmann, *Z. Zellforsch. Mikrosk. Anat. 75*, 517–526 (1966).
26. J. H. Birnie, *Endocrinology 54*, 33–38 (1953).
27. W. A. Jeffers, M. M. Livesey, and J. H. Austin, *Proc. Soc. Exp. Biol. Med. 50*, 184–188 (1942).
28. C. A. Winter, C. E. Gaffney, and L. Flataker, *J. Pharmacol. Exp. Ther. 111*, 360–364 (1954).
29. J. S. Holmes, M. K. Carter, and J. M. Fujimoto, *Proc. Soc. Exp. Biol. Med. 99*, 319–321 (1958).

30. C. Marchand and G. Denis, *J. Pharmacol. Exp. Ther. 162*, 331–337 (1968).
31. S. H. Flowers, E. S. Dunham, and H. G. Barbour, *Proc. Soc. Exp. Biol. Med. 26*, 572–574 (1929).
32. A. Wikler, W. R. Martin, F. T. Pescor, and C. G. Eades, *Psychopharmacologia 5*, 55–76 (1963).
33. W. R. Martin, A. Wikler, C. G. Eades, and F. T. Pescor, *Psychopharmacologia 4*, 247–260 (1963).
34. L. Detrick and C. H. Thienes, *Arch. Int. Pharmacodyn. Ther. 66*, 130–137 (1941).
35. J. W. Czaczkes, C. R. Kleeman, and M. Koenig, *J. Clin. Invest. 43*, 1625–1640 (1964).
36. F. N. Briggs and P. L. Munson, *Endocrinology 57*, 205–219 (1955).
37. S. Tomizawa, H. Kondo, K. Shimai, and M. Akita, *23rd Int. Congr. Physiol. Sci. Abstr. 1228*, 509 (1965).
38. M. Vachon and C. Marchand, *J. Pharmacol. Exp. Ther. 172*, 122–127, (1970).
39. C. Marchand, *Toxicol. Appl. Pharmacol. 15*, 385–392 (1969).
40. C. Marchand, M. Cantin, and M. Côté, *Can. J. Physiol. Pharmacol. 47*, 649–655 (1969).
41. G. Farrell, L. F. Fabre, and E. N. Rauschkolb, *Ann. Rev. Physiol. 30*, 557–588 (1968).
42. G. A. Robison, R. W. Butcher, and E. W. Sutherland, *Ann. Rev. Biochem. 37*, 149–174 (1968).
43. Ch. Philippson, J. Löbe, and S. Gursky, *Pfluegers Arch. 281*, 122–128 (1964).
44. M. Vachon and C. Marchand, *Fed. Proc. 28*, 735 (1969).
45. M. S. Sabour, M. K. MacDonald, A. T. Lambie, and J. S. Robson, *Quart. J. Exp. Physiol. 49*, 162–170 (1964).
46. W. A. Foley, D. C. L. Jones, G. K. Osborn, and D. J. Kimeldorf, *Lab. Invest. 13*, 439–450 (1964).
47. L. J. Cizek and M. R. Nocenti, *Amer. J. Physiol. 208*, 615–620 (1965).
48. C. E. Inturrisi and J. M. Fujimoto, *Toxicol. Appl. Pharmacol. 13*, 258–270 (1968).
49. F. Huidobro and C. Maggiolo, *Acta Physiol. Lat. Amer. 11*, 201–207 (1961).
50. C. Maggiolo and F. Huidobro, *Acta Physiol. Lat. Amer. 11*, 70–78 (1961).
51. L. Shuster, R. V. Hannam, and W. E. Boyle, *J. Pharmacol. Exp. Ther. 140*, 149–154 (1963).
52. A. Wikler, P. C. Green, H. D. Smith, and F. T. Pescor, *Fed. Proc. 19*, 22 (1960).
53. H. Sachs, *J. Neurochem. 5*, 297–303, 1960.
54. C. Marchand and J. M. Fujimoto, *Pharmacologist 7*, 163 (1965).
55. H. Valtin and H. A. Schroeder, *Amer. J. Physiol. 206*, 425–430 (1964).
56. L. J. Cizek, *Amer. J. Physiol. 201*, 557–566 (1961).
57. M. R. Nocenti and L. J. Cizek, *Proc. Soc. Exp. Biol. Med. 124*, 767–770 (1967).
58. W. P. Baker and L. A. Woods, *J. Pharmacol. Exp. Ther. 120*, 371–374 (1957).
59. M. D. Milne, B. H. Scribner, and M. A. Crawford, *Amer. J. Med. 24*, 709–729 (1958).
60. I. M. Weiner and G. H. Mudge, *Amer. J. Med. 36*, 743–762 (1964).
61. I. M. Weiner, *Ann. Rev. Pharmacol. 7*, 39–56 (1967).
62. R. P. Forster, *Fed. Proc. 26*, 1008–1019 (1967).
63. L. Peters, *Pharmacol. Rev. 12*, 1–35 (1960).
64. A. M. Asatoor, D. R. London, M. D. Milne, and M. L. Simenhoff, *Brit. J. Pharmacol. 20*, 285–298 (1963).
65. J. Orloff and R. W. Berliner, *J. Clin. Invest. 35*, 223–235 (1956).
66. B. H. Scribner, M. A. Crawford, and W. J. Dempster, *Amer. J. Physiol. 196*, 1135–1140 (1959).
67. A. H. Beckett, J. F. Taylor, A. F. Casy, and M. M. A. Hassan, *J. Pharm. Pharmacol. 20*, 754–762 (1968).
68. A. H. Beckett, J. A. Salmon, and M. Mitchard, *J. Pharm. Pharmacol. 21*, 251–258 (1969).
69. J. G. Wagner, *J. Clin. Pharmacol. J. New Drugs 7*, 89–92 (1967).

70. J. M. Fujimoto and R. A. Donnelly, *Clin. Toxicol. 1*, 297–307 (1968).
71. W. Braun, I. Hesse, and G. Malorney, G., *Arch. Exp. Pathol. Pharmakol. 245*, 457–470 (1963).
72. F. E. Shideman and H. T. Johnson, *J. Pharmacol. Exp. Ther. 92*, 414–420 (1948).
73. C. C. Hug, L. B. Mellett, and E. J. Cafruny, *J. Pharmacol. Exp. Ther 150*, 259–269 (1965).
74. J. M. Fujimoto and E. L. Way, *J. Amer. Pharm. Ass., 47*, 273–275 (1958).
75. I. Sperber, *Nature 158*, 131 (1946).
76. I. Sperber, *in* "Biology and Comparative Physiology of Birds" (A. J. Marshall, ed.), Vol. 1, pp. 469–492, Academic Press, New York (1960).
77. E. E. Owen and R. R. Robinson, *Amer. J. Physiol. 206*, 1321–1326 (1964).
78. B. R. Rennick, M. Z. Pryor, and B. G. Basch, *J. Pharmacol. Exp. Ther. 148*, 270–276 (1965).
79. D. G. May, J. M. Fujimoto, and C. E. Inturrisi, *J. Pharmacol. Exp. Ther. 157*, 626–635 (1967).
80. D. G. May and J. M. Fujimoto, *24th Int. Congr. Physiol. Sci. 7*, 285 (1968).
81. W. M. Watrous, D. G. May, and J. M. Fujimoto, *J. Pharmacol. Exp. Ther. 172*, 224–229 (1970).
82. J. M. Fujimoto and V. B. Haarstad, *J. Pharmacol. Exp. Ther. 165*, 45–51 (1969).
83. J. M. Fujimoto, W. M. Watrous, and V. B. Haarstad, *Proc. Soc. Exp. Biol. Med. 130*, 546–549 (1969).
84. J. M. Fujimoto, *J. Pharmacol. Exp. Ther. 168*, 180–186 (1969).
85. J. M. Fujimoto, *Proc. Soc. Exp. Biol. Med. 133*, 317–319 (1970).
86. R. P. Wedeen and H. I. Jernow, *Amer. J. Physiol. 214*, 776–785 (1968).
87. R. Hakim and J. M. Fujimoto, *Pharmacologist, 11*, 233 (1969).
88. I. M. Vigran, *J. Amer. Med. Ass. 187*, 953–954 (1964).
89. A. Carlsson and M. Lindquist, *J. Pharm. Pharmacol. 21*, 460–464 (1969).
90. K. J. Rogers and J. A. Thornton, *Brit. J. Pharmacol. 36*, 470–480 (1969).
91. L. A. Woods, *J. Pharmacol. Exp. Ther. 112*, 158–175 (1954).
92. B. R. Rennick, *Amer. J. Physiol. 212*, 1131–1134 (1967).
93. J. M. B. O'Connell, J. Romeo, and G. H. Mudge, *Amer. J. Physiol. 203*, 985–990 (1962).
94. B. Rennick and M. Z. Pryor, *J. Pharmacol. Exp. Ther. 148*, 262–269 (1965).
95. B. Rennick, M. Pryor, and N. Yoss, *J. Pharmacol. Exp. Ther. 143*, 42–46, (1964).
96. A. Quebbemann and B. R. Rennick, *J. Pharmacol. Exp. Ther. 166*, 52–62 (1969).
97. B. R. Rennick, *Proc. Soc. Exp. Biol. Med. 103*, 241–243 (1960).
98. A. J. Quebbemann and B. R. Rennick, *Amer. J. Physiol. 214*, 1201–1204 (1968).
99. M. Gibaldi and M. A. Schwartz, *Clin. Pharmacol. Ther. 9*, 345–349 (1968).
100. C. C. Hug, Jr., *Biochem. Pharmacol. 16*, 345–359 (1967).
101. R. J. McIssaac, *J. Pharmacol. Exp. Ther. 168*, 6–12 (1969).

Chapter 18

PERIPHERAL TISSUES

Marta Weinstock *

Department of Pharmacology
St. Mary's Hospital Medical School
London, W2, England

I. THE ACTION OF OPIATES ON THE GASTROINTESTINAL TRACT

The action of morphine on the gastrointestinal tract has been extensively studied for more than 70 years in various animal species including man. Evidence exists from several studies that the effect of morphine on the intestine of the intact conscious animal is largely due to a central action having a visceral efferent component. Vaughan-Williams and Streeten[1] found that much smaller doses of both morphine and methadone inhibited propulsion of the intestinal contents of dogs *in vivo* than in isolated gut preparations. The intracerebral injection of 1 μg of morphine in mice produced a 50% reduction in the passage of a charcoal meal through the intestine. Transection of the spinal cord between C4 and L4 had no effect on this constipating action, indicating that morphine acted at a site above the fourth cervical vertebra.[2]

A. Gastrointestinal Effects of Opiates in Whole Animals

The effect produced on the gastrointestinal tract by the administration of morphine is somewhat variable and depends upon a number of factors: the species, the dose, the time of injection in relation to food intake, and the method of assessment. The effects described below are those which have been reported to occur most often in the various portions of the gastrointestinal tract.

1. Stomach

In the dog, a dose of 0.5–1 mg/kg morphine produced an increase in the

---

* Present address: Department of Pharmacology, Tel Aviv University, Ramat Aviv, Tel Aviv, Israel.

amplitude of peristaltic waves and a diminution in the amplitude and frequency of large contractions. The increase in motility was abolished by atropine and potentiated by prostigmine.[3,4] A dose of 6 mg/kg produced spasm of the pyloric sphincter and a delay of several hours before food passed from the fundus to the pylorus.[5]

Morphine also delays emptying of the stomach in man, but Abbot and Prendergast suggested from their x-ray studies that this is not due to pylorospasm, as in the dog, but to a powerful contraction of the proximal half of the duodenum. The increased duodenal resistence prevents emptying of the stomach.[6]

In the rabbit and cat decreased motility of the stomach with pylorospasm and delayed emptying also have been observed after morphine.[7]

2. Small Intestine

The initial effect of small doses of morphine (0.1–5 mg/kg) in the dog intestine is to increase the muscle tone and the amplitude of contractions but not their frequency. This is usually followed by a decrease in frequency of peristaltic waves but no change in their amplitude. Plant and Miller claimed that this increased tone of the ileum was not antagonized by atropine or supradiaphragmatic vagotomy,[3] but Adler and Ivy found that atropine did antagonize the effects of morphine on the ileum in the dog.[8] This difference may have been due to the relative amounts of both drugs used in the two experiments. In a later detailed study in dogs, Vaughan Williams and Streeten found that morphine had an inhibitory action on intestinal propulsion and suggested that this was brought about by a prolonged contraction that closed the intestinal lumen and prevented the development of coordinated propulsive movements.[1]

In man the results obtained with various techniques indicate that morphine increases both the tonus and motility of the intestine but diminishes propulsive activity so that the passage of contents along the gut is slowed.

The guinea pig differs from dog and man in that morphine (1 mg/kg) depresses both the tone and the motility of the small intestine.[9]

3. Large Intestine

The effects of morphine on the large bowel have been less extensively studied than those on the small intestine. In general the effects are similar to those produced on the ileum in the different species. With the exception of the guinea pig there is an increase in tone of the muscle accompanied by a decrease in propulsive activity.[7]

4. Secretions

Morphine suppresses the reflex secretion of saliva caused by electrical stimulation or chemical irritation of the tongue but not that resulting from stimulation of the chorda tympani. These observations led Claude Bernard to conclude that this was a central action of morphine.[7]

Gastric secretion is initially depressed by morphine, but then it increases 1–2 hr after the injection. The increase in secretion is abolished by atropine and vagotomy.[7]

Schonduke and Lurman found that the secretion of bile is decreased by morphine and that this action is not affected by prior administration of atropine.[7] Roentgenological studies have shown that morphine produces spasm of the biliary tract and of the sphincter of Oddi. This spasm prevents emptying and causes intraductal pressure to rise.[10] There is also some evidence that morphine inhibits contractions of the gall bladder.

Secretion of pancreatic juice also is reduced by morphine. This may be a direct action on the pancreas, or it may occur indirectly through the initial suppression of gastric juice. Alternatively, it may result from closure of the pylorus, which would have the same effect as decreased secretion of hydrochloric acid on the formation of pancreatic juice.[7]

5. Summary

The constipating action of morphine in the intact animal results from a combination of several factors, the relative importance of which varies in different species and even within individuals of the same species. It has been shown that at least 50% of the total delay in the evacuation of the intestinal tract occurs in the stomach.[5] This is partly due to the initial failure of hydrochloric acid secretion and to pylorospasm, as well as to decreased gastric motility. Similarly, the delay occurring in both small and large bowels is due to a reduced secretion of digestive juices, closure of sphincters, and reduced propulsive activity. Since fecal material remains for a longer time in the large intestine, more reabsorption of water can take place, causing a drying of material and making it more difficult to pass through partially closed sphincters. In addition, a reduced awareness of the need to defecate occurs through the central depressant action of the drug.

6. Action of Opiates Other Than Morphine

In general all the other opiates have the same qualitative effects on the gastrointestinal tract as morphine.

Meperidine is said to be less constipating than morphine in equi-analgesic doses. Its spasmogenic effect is also less than that of morphine. Whereas the effects of physostigmine and morphine in the dog intestine are often synergistic, meperidine on the other hand reduces the stimulant effects of anticholinesterase drugs and pilocarpine because of its pronounced atropine-like action.[10] As might be expected, meperidine is much less effective than morphine in treating intractable diarrhea.

Methadone is also less constipating than morphine in equianalgesic doses.[1] Pentazocine delays gastric emptying time and causes a slight inhibition of propulsive activity in the rat.[11] In the dog it decreases tone in the duodenum, unlike morphine, and does not cause spasm of the pyloric sphincter.[12]

B. Effects of Opiates in Isolated Intestinal Preparations

Although most of the earlier work on the action of morphine on the gastrointestinal tract was carried out in whole animals, more recent studies have concentrated on the action in various isolated intestinal preparations. Most workers have used guinea pig ileum, but it should be emphasized that many of the effects of opiates seen on the gut of this species are not seen in isolated preparations from other animals.

In the guinea pig's gut the action of morphine usually seen is that of depression, so that almost all stimuli, whether they are electrical, mechanical, or chemical, are inhibited by morphine.

The preparations most widely used to study the action of opiates on the intestine have been that of Trendelenburg, used to elicit a peristaltic reflex,[13] and segments of gut in which contractions of the longitudinal muscles are produced either by electrical stimulation[14] or by chemical agents.[15] Contractions of the circular muscles also have been recorded.[16]

1. Peristaltic Reflex of Small Intestine

Intestinal movement can result from either stimulation of intramural nervous elements or from intrinsic rhythmic activity of smooth muscle cells. Trendelenburg showed that distention of the lumen of the ileum by a pressure of 1–2 cm of water first caused a contraction of the longitudinal muscles with shortening of the gut. This was followed by a contraction of the circular muscles in a wave, causing the gut contents to be expelled while the longitudinal muscles relaxed.[13]

The contraction of the longitudinal muscles increases with the degree of distention. Both phases of the peristaltic reflex are evoked by distention and are called by Kosterlitz "the graded reflex of the longitudinal muscle" and the "peristaltic reflex proper." The later phase has been shown to have at least one reflex with a cholinergic synapse, since it is inhibited by atropine and hexamethonium.[17]

Numerous workers have reported that morphine can inhibit both phases of the peristaltic reflex in the guinea pig ileum.[17–19] Morphine markedly reduced the second phase of the reflex in concentrations of 10^{-8} to $10^{-7}M$. Atropine also partially antagonized the peristaltic reflex proper in concentrations of $10^{-8}M$. However, neither of these drugs abolished the reflex entirely, even when concentrations were increased tenfold. This suggests that there must be another reflex pathway sensitive to stretching that is insensitive to morphine and atropine.

Unlike atropine, morphine does not significantly reduce the responses of the longitudinal muscles of the ileum to applied ACh in doses in which it inhibits the peristaltic reflex.[15] Schaumann has shown that morphine can inhibit spontaneous release of ACh from segments of guinea pig ileum incubated in Tyrode solution at 37°C. Therefore it can be assumed that morphine does not prevent the action of ACh on the muscle but it depresses the peristaltic reflex by preventing ACh release.

The actions of numerous other analgesic drugs on the peristaltic reflex have been shown to be similar to those of morphine.[20] Inhibition of peristalsis appeared to parallel analgesic potency. Furthermore, there is good evidence of stereospecificity for this action of opiates as for many other effects of these drugs. Thus only levorphan has any effect on the peristaltic reflex, dextrorphan being inactive.[20] Also the levo isomer of methadone is 15 times more potent than the dextro form.[21]

The inhibitory action of morphine and related opiates is antagonized by small doses of nalorphine and levallorphan. Larger doses of these antagonists act like morphine and inhibit the peristaltic reflex.[20]

The ileum of guinea pigs made tolerant to the sedative and analgesic actions of morphine by repeated injections are much less sensitive to the inhibitory actions of the drug. Atropine can still exert its full inhibitory effect in guinea pig ileum refractory to morphine, indicating that the production of tolerance in this preparation is specific for opiates.[22] Isolated ilea from guinea pigs given repeated injections of pethidine, however, still showed the full inhibitory effect of the drug.[23] This may have been due to the atropine-like action of pethidine.

2. Response of Longitudinal Muscle of Ileum to Electrical Stimulation

In 1955, Paton described a technique of stimulating pieces of ileum by passing a current from one electrode placed in the bath fluid to another within the lumen of the gut. This coaxial stimulation resulted in contractions of the longitudinal muscles due to the release of ACh.[14]

The contractions of the stimulated ileum were inhibited by morphine in concentrations of $10^{-8}-10^{-7}M$, which were the same range as those that depressed the peristaltic reflex.[14,28] The effect of morphine on the stimulated ileum was shown to be due to an inhibition of the release of ACh. A close correlation was demonstrated between the ability of several opiates to inhibit ACh release and their analgesic potency.[24] Nalorphine and levallorphan were found to be equipotent with morphine and levorphan in depressing the twitch of the stimulated ileum. However, when given in doses too low to depress the contractions of the ileum, these drugs were shown to antagonize the inhibitory effect of morphine and other analgesic drugs.[24] Both agonist and morphine antagonist activity also was shown by naloxone, cyclazocine, and pentazocine.[25]

When morphine was given in repeated doses its effect on the stimulated ileum declined. This tolerance occurred with all opiates tested, and cross tachyphylaxis also was observed. In general this tachyphylaxis developed much more rapidly to the narcotic antagonist drugs nalorphine and cyclazocine than to the opiates.[25]

3. Response of the Longitudinal Muscle to Chemical Agents

Morphine has been found to inhibit the contractions of the longitudinal muscles of guinea pig ileum elicited by 5-hydroxytryptamine,[15,17,26,27] barium chloride,[17,27] darmstoff,[28] angiotensin,[29] and arachidonic acid peroxide.[38] All these agents have been shown to exert their effect on the gut either wholly or partially by causing the release of ACh. The concentrations of morphine that antagonize all these agents are in the range of 1 in 10^7 to 1 in 10^8, as for inhibition of the peristaltic reflex. The response to all the above substances is not completely abolished by morphine, so that even when higher concentrations of the analgesic are employed a small residual contraction of the ileum usually remains.

The antagonistic action of several other opiates against 5-HT has been shown to parallel analgesic potency.[31] Nalorphine and levallorphan acted like morphine in inhibiting 5-HT-induced contractions. Neither drug significantly antagonized the effect of morphine on this preparation.[17] On the other hand, contractions elicited by nicotine were inhibited by morphine but not by levallorphan and nalorphine. Both compounds antagonized the inhibiting effect of morphine on nicotine-induced contractions.[17]

A similar relationship between inhibitory activity and analgesic potency was demonstrated for opiates when arachidonic acid peroxide was used to contract the ileum.[30] In this preparation nalorphine had no inhibitory activity alone. Its ability to antagonize the action of morphine was not tested in these experiments.

The ilea of morphine-tolerant guinea pigs are much less sensitive to the depressant action of morphine against contractions induced by 5-HT than are those of nontolerant animals.[32]

4. Response of Circular Muscle to Chemical Agents

The circular muscle strip prepared from guinea pig ileum is generally much less responsive to drugs than is longitudinal muscle. However, if the preparation is incubated with an anticholinesterase, mipafox, the sensitivity of the muscle to ACh is increased 4000 times. Only after such treatment will the muscle contract to 5-HT, nicotine, and histamine. The response to all three agents can be antagonized by morphine.[16] It should be noted that morphine does not significantly reduce the response of the longitudinal muscle to histamine.

5. Action of Morphine in Isolated Ilea of Other Species

The response of the ilea of other species to morphine differs markedly from that of the guinea pig. Morphine does not antagonize the effect of 5-HT on the rabbit jejunum[27] or on the rat ileum or stomach strip.[32] Even in concentrations of $10^{-5}M$ morphine shows only slight inhibitory activity on the peristaltic reflex elicited in rabbit gut.[27]

Burke and Long studied the effects of morphine on the intestine *in situ* in anesthetised dogs. They showed that morphine 20 μg by intraarterial injection caused the release of 5-HT from the intestine. Other analgesic drugs including nalorphine had the same effect.[33] They speculated from their studies that the stimulant action of morphine on the dog intestine could be initiated by 5-HT. This in turn could stimulate intestinal ganglia with the subsequent release of ACh. This hypothesis is substantiated by the enhancement of the responses to morphine in the dog by physostigmine and their antagonism by atropine.

Although morphine increases the tone of the intestine in the dog, propulsive activity is diminished. Burke and Long suggest that this is due to decreased intraluminal space and increased mechanical resistance to the movement of contents down the bowel. The action in the guinea pig, however, is quite different. Here morphine produces only depression of movement and tone. There is ample evidence that this effect is due to an inhibition of the release of ACh.

II. THE ACTION OF OPIATES ON THE MUSCLE OF THE GENITO-URINARY TRACT

A. Urinary Tract

Studies on the effect of morphine on the muscle of the genit-ourinary tract have been sparse compared with those on the gastrointestinal tract, and very few experiments have been performed in the past 20 years.

Morphine has been shown to increase the tone of the ureters and their frequency of contractions *in situ* in almost all animals studied.[34,35] This action also occurs in excised ureters of the guinea pig and rabbit and is not antagonized by either atropine or ergotoxine.[34,36] In the intact animal, however, atropine does abolish the effect of morphine on the ureter.

Urinary retention occurs with all species even with analgesic doses of morphine. The exact mechanism by which this is achieved, however, differs in different species. In the guinea pig, morphine produces spasm of the vesical sphincter that is unaffected by atropine or papaverine.[37] This action of morphine appears to be mediated via the sacral portion of the spinal cord, not on the muscle itself, since section at the sacral level but not in the lumbar region abolished its effect.[37]

In the dog morphine produces contractions of the bladder that are unaffected by destruction of the brain or spinal cord or by the administration of atropine.[38] The stimulant action of morphine on the detrusor muscle in the dog, however, is inhibited by Trasentin.[39]

Morphine in doses of 8–30 mg rapidly produces an increase in tone and an amplitude of contraction of the intact human ureter. This effect usually persists for about 3 hr and can be prevented by the simultaneous administration of atropine, 0.6 mg.[10] The tone of the detrusor muscle also is increased in man, giving rise to urinary urgency. However, morphine also contracts the vesical sphincter, making it more difficult to pass urine. The central depressant action of the drug makes the patient less attentive to impulses from the bladder, which also contributes to urinary retention.

Unlike morphine, methadone causes relaxation of the ureters in the anesthetised dog.[40] Meperidine also has a spasmolytic effect on the intact ureters of both humans and dogs, and this is particularly marked in subjects in which initial uretal tone is high.[41] The spasmolytic effect of pethidine has been shown by Lapides to result from the reduction in urine secretion, not from a direct depressant action on the ureters.[42]

B. Uterus

Studies of the action of morphine on the human uterus have produced conflicting results. Caldayro-Barcia and his collaborators conducted their investigations before and during labor and measured changes in amniotic pressure to determine the effect of morphine on uterine contractility. They found that doses of 10–20 mg exerted little or no effect on uterine contractions during labor. In the days preceding labor, morphine increased the tone and frequency of large contractions but reduced the tone of small contractions.[43] In contrast to these findings, Bickers reported that morphine may abolish the mobility of the uterus during labor.[44] Rucker also showed that doses greater than 10 mg diminished uterine contractions during labor.[45]

In the rabbit, cat, and guinea pig, morphine stimulates the uterus *in situ* in small doses but depresses it in larger ones. In the rabbit the stimulant effect of morphine can be antagonized by vagotomy, atropine, or lumbar anesthesia.[46] In the anesthetized nonpregnant dog, Slaughter and Gross observed a diminished tone and decreased height of contractions and concluded that these effects of morphine were chiefly central in origin.[47] The direct effect of morphine seen in the isolated uterus of nearly all species is to produce an increase in tone in concentrations of 0.5 mg/ml.[48]

Since meperidine is so often employed in obstetric analgesia its effects on the human uterus have been comparitively well studied. Most observers agree that this drug does not significantly influence rhythmic contractions nor does it delay labor.[49,50] The drug is said to relax the uterine cervix and shorten the duration of labor. It does, however, appear to increase the tone, frequency, and intensity of contraction of the uterus made hypersensitive by oxytocics.

III. THE EFFECT OF OPIATES ON THE RODENT LENS

A. The Nature of the Lens Opacity

In 1958, Weinstock and her associates reported that a reversible opacity of the lens could be produced in mice by the acute administration of morphine-like analgesic drugs.[52] With most of the compounds tested the effect developed within 20–30 min of parenteral administration and lasted $1\frac{1}{2}$ to $2\frac{1}{2}$ hr, depending upon the dose given.[53] Slit-lamp examination of the eyes revealed that the opacities appeared in the anterior subcapsular region. In some cases the cloudiness formed in a crescent shape in the periphery of the lower portion of the lens and then extended inward until the whole area of the anterior portion of the lens was obscured. In some mice, and always in rats, the opacity appeared first in close relation to the anterior sutures from which it radiated outward to the periphery. In both types, as the opacity increased in density it extended deeper into the lens cortex but never as far as the nucleus.[54] The effect could be produced in rats and guinea pigs as well, but the incidence was much lower than in mice when similar doses were given on a weight basis. Larger amounts of nearly all the opiates tested proved to be too toxic in the larger species. The majority of the studies concerning the mechanism of action of opiates on the lens were therefore carried out in mice.[53]

B. Production of Opacities in Isolated Lenses

Smith, Karmin, and Gavitt showed that the lens took up ^{3}H-labeled levorphanol after parenteral administration.[55] This suggested that opacity might result from a local interaction between the analgesic drug and some constituent of the lens capsule. However, when lenses, removed from untreated mice, were incubated in modified aqueous humor fluid containing either morphine, methadone, or pethidine in concentrations of 10^{-5} to $2 \times 10^{-4} M$, the lenses remained perfectly clear. It was possible to produce opacities *in vitro* by increasing the concentrations of glucose or sodium chloride, thereby raising the tonicity. Lenses also became opaque when the pH of the incubation fluid was reduced from 7.4 to 7.15. Both changes could be reversed by restoring the lenses to the original incubation fluid.[54]

These findings suggested that lenses might become opaque *in vivo* if the tonicity of the aqueous humor increased. The effect of a hypertonic aqueous humor on the lens could be to draw water out of the cells immediately below the capsule. This, in turn, could alter the solubility of a protein constituent of the lens, causing it to precipitate. On restoration of the normal tonicity of the aqueous humor, water would again be taken up into the subcapsular cells and the lens would regain its transparency.

C. Influence of Respiratory Depression on the Production of Lens Opacities

It is well known that all the morphine-like drugs that produce lens opacities are powerful respiratory depressants. Although no data is available for the influence of these drugs on the biochemistry of mouse aqueous humor, it has been shown that methadone, 15 mg/kg, reduced the blood pH to 7.17 and raised both blood lactic acid and pCO_2 by 50–60%. This dose of methadone produced an incidence of opacity of 85%.[54]

There is some experimental evidence that the production of reversible cataracts are accompanied by changes in the constituents of the aqueous humor in larger animals. The concentration of glucose increased markedly in the aqueous humor of rabbits in which reversible cataracts were produced by oxygen deprivation.[56] The lactic acid content also was raised more than threefold.[57]

The prominent role of respiratory depression in the action of opiates on the lens was demonstrated by the findings that both oxygen and respiratory stimulants markedly reduced the effect.[54] Further support for the role of hypoxia in the production of opacities was obtained from experiments in solitary mice. It was found that the incidence of cataracts in mice housed individually after drug injection was much lower than in those kept in groups of 2–10. The greater effect in grouped animals was shown to be associated with higher rates of movement than in solitary mice. On the other hand, mice caged singly and forcibly moved showed a high incidence of opacities. It was concluded from these findings that the increased oxygen needs from the muscular activity resulted in more severe hypoxia in mice in which the respiratory center was already depressed by the opiate. The effect of the hypoxia on the pH or tonicity of the aqueous humor was held responsible for the greater incidence of opacity in grouped mice.[58]

D. Role of the Sympathetic Nervous System in the Lenticular Effect of Opiates

Respiratory depression is not the only action of analgesic drugs that contributes to the lenticular effect. It has been shown that pretreatment with reserpine, which antagonized the analgesic effect of morphine, also prevented the formation of the lens opacity. Unlike the analgesic effect, however, the opacity was restored in reserpinized mice by the injection of epinephrine or norepinephrine with the opiate.[59] Phentolamine and guanethidine also prevented the action of methadone on the lens.[60] These findings suggested that this action of morphine was mediated by the sympathetic nervous system.

When opiates are administered to mice a characteristic proptosis develops within a few minutes, preceding the development of the opacity. Smaller doses of the drugs that do not cause eyelid retraction fail to produce opacities. Cataracts also fail to develop if the eyelids are taped shut after the injection of morphine.[61] In a study of the influence of ptosis on the degree of lens opacity it was shown that sympathetic nerve blocking agents also prevent its development by closing the eyes.[62]

The site of the sympathetic stimulant action of morphine was shown to be in the central nervous system by two independent experiments. First it was found that intracerebral levorphanol produced opacities in less than one-fifth of the dose required by parenteral administration.[63] Then Weinstock and Marshall showed that unilateral preganglionic cervical sympathectomy prevented the formation of an opacity only in the operated eye. The analgesic drug still produced its full effect in the eye with the intact sympathetic nerve supply.[62]

Stimulation of the sympathetic supply to the mouse eye does not appear to be a direct consequence of the anoxia produced by opiates, since it was found that respiratory stimulants do not reduce eyelid retraction, and yet they antagonize the effect on the lens.[62] This indicates that although morphine-like drugs cannot produce opacities unless the eyelids remain wide open, respiratory depression and its resultant biochemical changes also must play an important role in their formation. It therefore can be concluded that opiates produce lens opacities by increasing the tonicity of the aqueous humor by at least two independent actions. The first of these is to stimulate the sympathetic nervous system centrally and bring about a prolonged opening of the eyes. This allows water to evaporate through the exposed cornea, rendering the aqueous humor hypertonic. Morphine-like drugs also may increase the glucose concentration and lower the pH of the aqueous humor by causing some degree of hypoxia.

E. Use of Lenticular Action of Opiates as a Model for Studying Drug–Receptor Interactions

It was shown by Weinstock that all compounds that produced lens opacities in mice in sublethal doses had analgesic activity. A close correlation existed for the two actions. It also was shown that in a pair of optical enantiomorphs the isomer with the greater analgesic activity also was the more active in producing an opacity. Dextrorphan and dextromethorphan were completely devoid of either activity in the amounts used, whereas the levo isomers produced both effects in relatively small doses.[64]

It was found that prior administration of nalorphine only prevented the production of an opacity by those compounds the analgesic action of which it abolished. Some compounds such as papaverine and the dextro rotatory isomer of methadone had significant analgesic and lenticular actions, but neither was reduced by nalorphine.[64] This suggested that these drugs did not produce either effect by combining with the same receptors as morphine. As a result of these findings it was suggested that the action of drugs on the lens can be used as the basis of a method for screening potential morphine-like analgesic drugs. If the new drug produces an opacity, antagonism by nalorphine should be ascertained. This will give further evidence that the compound has similar actions to morphine and is therefore potentially addictive.

The effect of opiates on the lens also was used as a model for demonstrating the competitive nature of the antagonism by nalorphine in whole animals.[65] ED_{50}s were obtained for the incidence of opacity given by a number of drugs alone and in the presence of nalorphine. Dose ratios were computed for each compound from the ED_{50}s and were used to calculate the

TABLE I
"PA₂" Values for Opiates with Nalorphine in Opacity and Analgesic Tests[a]

Drug	"PA_2" Opacity	"PA_2" Analgesia
Etorphine	4.20	4.26
Methadone	3.99	4.20
Morphine	3.92	4.18
Dextromoramide	4.22	4.25
Levorphanol	4.19	—
Pethidine	3.90	—
Diacetylmorphine	4.25	—

[a] Some of these figures are reproduced from B. M. Cox and M. Weinstock, *Brit. J. Pharmacol.* **22**, 289 (1964).

equivalent of a PA_2 value. This value was based on the dose of nalorphine injected that was assumed to be proportional to the concentration of antagonist at the receptor site. "PA_2" values for antagonism of the opacity and analgesia in mice are shown in Table I. It can be seen that the PA_2 values for opacity are very similar for the different drugs tested and also almost identical to those for the antagonism of analgesia by nalorphine. These findings indicate that nalorphine antagonizes both effects of opiates by competing with them for essential receptor sites.

F. Tolerance to the Lenticular Effect of Opiates

It has been shown that the incidence of opacity obtained with levorphanol,[66] methadone, and morphine[67] diminishes rapidly on repeated injection. The second administration of levorphanol 50 mg/kg reduced the effect from 80 to 20%. This diminished response to one injection of levorphanol persisted up to 20 days. When a smaller dose was given initially, 12.5 mg/kg, the duration of tolerance was shorter. Cross tolerance also was demonstrated between different opiates. Preliminary studies also indicate that tolerance develops more rapidly to the lenticular effect of morphine, pethidine, and etorphine than to the analgesic action of these drugs in mice.[67]

Smith and his colleagues have shown that the development of tolerance to levorphanol can be prevented by the prior administration of the protein synthesis inhibitors actinomycin or puromycin.[66] These compounds and other protein synthesis inhibitors also reduce tolerance development to the analgesic action of morphine.[68,69] It therefore can be concluded from these findings that tolerance to both the lenticular and analgesic effects of morphine involves a change in the pattern of protein synthesis in the brain.

IV. REFERENCES

1. E. M. Vaughan Williams and D. H. P. Streeten, *Brit. J. Pharmacol. 5*, 584–603 (1950).
2. S. Margolin, *Proc. Soc. Exp. Biol. Med. 112*, 311–315 (1963).
3. O. H. Plant and G. H. Miller, *J. Pharmacol. Exp. Ther. 27*, 361–383 (1926).
4. D. Slaughter, A. B. Goddard, and W. M. Henderson, *J. Pharmacol. Exp. Ther. 76*, 301–308 (1942).
5. H. Krueger, *Physiol. Rev. 17*, 618–645 (1937).
6. W. O. Abbott and E. P. Prendergast, *Amer. J. Roentgenol. 35*, 289–299 (1936).
7. A. K. Reynolds and L. O. Randall, *in* "Morphine and Allied Drugs," pp. 66–84, University of Toronto Press, London (1957).
8. H. F. Adler and A. C. Ivy, *J. Pharmacol. Exp. Ther. 70*, 454–459 (1940).
9. E. Rentz, *Arch. Exp. Pathol. Pharmakol. 191*, 172–182 (1938).
10. L. S. Goodman and A. Gilman, *in* "The Pharmacological Basis of Therapeutics," pp. 216–280, Macmillan Company, New York (1956).
11. I. E. Danhof, W. P. Blackmore, and G. L. Upton, *Toxicol. Appl. Pharmacol. 2*, 356–362 (1966).
12. I. E. Danhof and W. P. Blackmore, *Toxicol. Appl. Pharmacol. 90*, 223–232 (1967).
13. P. Trendelenburg, *Arch. Exp. Pathol. Pharmakol. 81*, 55–129 (1917).
14. W. D. M. Paton, *Brit. J. Pharmacol. 12*, 112–127 (1957).
15. G. P. Lewis, *Brit. J. Pharmacol. 15*, 425–431 (1960).

16. J. Harry, *Brit. J. Pharmacol. 20*, 399–417 (1963).
17. H. W. Kosterlitz and J. A. Robinson, *J. Physiol. 136*, 249–262 (1957).
18. O. Schaumann, M. Giovanni, and K. Jochum, *Arch. Exp. Pathol. Pharmakol. 215*, 460–468 (1952).
19. W. Schaumann, *Brit. J. Pharmacol. 12*, 115–118 (1957).
20. E. A. Gyang, H. W. Kosterlitz, and G. M. Lees, *Arch. Exp. Pathol. Pharmakol. 248*, 231–246 (1964).
21. W. Schaumann, *Arch. Exp. Pathol. Pharmakol. 229*, 41–51 (1956).
22. E. Rentz and D. N. Kesarbani, *Arch. Exp. Pathol. Pharmakol. 148*, 107–113 (1941).
23. D. M. Kesarbani, *Arch. Exp. Pathol. Pharmakol. 198*, 114–120 (1941).
24. B. M. Cox and M. Weinstock, *Brit. J. Pharmacol. 27*, 81–92 (1966).
25. E. A. Gyang and H. W. Kosterlitz, *Brit. J. Pharmacol. 27*, 514–527 (1966).
26. J. H. Gaddum and Z. P. Picarelli, *Brit. J. Pharmacol. 12*, 323–328 (1957).
27. W. Schaumann, *Brit. J. Pharmacol. 10*, 456–461 (1955).
28. W. Vogt, *Arch. Exp. Pathol. Pharmakol. 235*, 550–558 (1959).
29. C. A. Ross, C. T. Ludden, and C. A. Stone, *Proc. Soc. Exp. Biol. Med. 105*, 558–559 (1960).
30. R. Jaques, *Helv. Physiol. Acta 23*, 156–162 (1965).
31. M. Medakovic, *Arch. Int. Pharmacodyn. 114*, 201–209 (1958).
32. M. Mattila, *Acta Pharmacol. Tox. 19*, 47–52 (1952).
33. T. F. Burks and J. P. Long, *J. Pharmacol. Exp. Ther. 156*, 267–276 (1967).
34. D. I. Macht, *J. Pharmacol. Exp. Ther. 11*, 394 (1918).
35. C. N. Gruber, *Proc. Soc. Exp. Biol. Med. 33*, 532–534 (1936).
36. W. F. Ockerblad, H. E. Carlson, and J. F. Simon, *J. Urol. 33*, 356–365 (1935).
37. T. Ikoma, *Arch. Exp. Pathol. Pharmakol. 102*, 145–166 (1924).
38. D. E. Jackson, *J. Lab. Clin. Med. 1*, 862–866 (1916).
39. R. E. Van Duzen, D. Slaughter, and I. C. Winter, *J. Urol. 44*, 667–676 (1940).
40. C. C. Scott, K. K. Chen, R. G. Kohlstaedt, E. B. Robbins, and F. W. Israel, *J. Pharmacol. Exp. Ther. 91*, 147–156 (1947).
41. R. Climenko and H. Berge, *J. Urol. 49*, 255–258 (1943).
42. J. Lapides, *J. Urol. 59*, 501–533 (1948).
43. R. Caldeyro-Barcia, H. Alvanez, and J. J. Posiero, *Arch. Int. Pharmacodyn. 101*, 171–188 (1955).
44. W. Bickers, quoted by A. K. Reynolds and L. O. Randall *in* "Morphine and Allied Drugs," p. 87, University of Toronto Press, London (1957).
45. M. P. Rucker, *Anesth. Analg. 5*, 235–246 (1926).
46. A. K. Reynolds and L. O. Randall, *in* "Morphine and Allied Drugs," p. 88, University of Toronto Press, London (1957).
47. D. Slaughter and E. C. Gross, *J. Pharmacol. Exp. Ther. 59*, 350–357 (1937).
48. C. N. Gruber, J. T. Brundage, A. De Note, and R. Heiligman, *J. Pharmacol. Exp. Ther. 55*, 430–434 (1935).
49. N. R. Schumann, *Amer. J. Obstet. Gynecol. 47*, 93–104 (1944).
50. B. Gallen and F. Prescott, *Brit. Med. J. 1*, 176–179 (1944).
51. J. Jaffe, *in* "Pharmacological Basis of Therapeutics," (L. S. Goodman and A. Gilman, eds.) pp. 266–270, MacMillan, New York, 1967.
52. M. Weinstock, H. C. Stewart, and K. R. Butterworth, *Nature (London) 182*, 1519–1520 (1958).
53. M. Weinstock and H. C. Stewart, *Brit. J. Ophthalmol. 45*, 408–414 (1961).
54. M. Weinstock and J. D. Scott, *Exp. Eye Res. 6*, 368–375 (1967).
55. A. A. Smith, M. Karmin, and J. Gavitt, *J. Pharmacol. Exp. Ther. 151*, 103–109 (1966).
56. G. Morone and M. Citroni, *Riv. Aeron. 14*, 464–479 (1951).
57. J. Bellows and D. Nelson, *Arch. Ophthalmol. 31*, 250–252 (1944).
58. M. Weinstock and A. S. Marshall, *J. Pharmacol. Exp. Ther.* (1969) (in press).
59. A. A. Smith, M. Kaplan, and J. Gavitt, *Recent Advan. Biol. Psychiat. 6*, 208–213 (1963).

60. M. Weinstock, *Arch. Exp. Pathol. Pharmakol. 259*, 201–202 (1968).
61. F. T. Fraunfelder and R. P. Burns, *Arch. Ophthalmol. 76*, 599–601 (1966).
62. M. Weinstock and A. S. Marshall, *J. Pharmacol. Exp. Ther. 166*, 8–13 (1969).
63. A. A. Smith, M. Karmin, and J. Gavitt, *J. Pharm. Pharmacol. 18*, 545–546 (1966).
64. M. Weinstock, *Brit. J. Pharmacol. 17*, 433–441 (1961).
65. B. M. Cox and M. Weinstock, *Brit. J. Pharmacol. 22*, 289–300 (1964).
66. A. A. Smith, M. Karmin, and J. Gavitt, *J. Pharmacol. Exp. Ther. 156*, 85–91 (1967).
67. M. Weinstock, unpublished observations.
68. M. Cohen, A. S. Keats, W. Krivoy, and G. Ungar, *Proc. Soc. Exp. Biol. Med. 119*, 381–384, (1965).
69. B. M. Cox, M. Ginsburg, and O. M. Osman, *Brit. J. Pharmacol. 33*, 245–256 (1968).

Chapter 19

TOLERANCE AND PHYSICAL DEPENDENCE

Louis Shuster

Department of Biochemistry
Tufts University School of Medicine
Boston, Massachusetts

I. THE RIDDLE OF PHYSICAL DEPENDENCE

> The continued administration of morphine exerts an entirely different effect on a morphinist from that exerted by a single medium dose of morphine injected into a healthy person. While this latter causes nausea, even vomiting, a feeling of faintness, pulse acceleration and lowering of the blood pressure, the former occasions just the opposite feelings, sensations and states—namely, pleasant feeling, euphoria, increased power, and, in the heart and vessels, strengthening of the contraction, invigoration of the pulse, and rise of the blood pressure. Since every morphinist has once had the first morphine injection, there arises the question: by what means and at what time of the continued abuse does this reversal of effect take place? It is brought about as follows: the morphine, originally *foreign* to the body, becomes an *intrinsic* part of the body as the union between it and the brain cells keeps growing stronger; it then acquires the significance and effectiveness of a heart tonic, of an indispensable element of nutrition and substinence, of a means of carrying on the business of the entire organism.[1]

The statement quoted here describes, but does not explain, the most striking feature of narcotic addiction—that is, the development of an apparent physiological requirement for a toxic foreign substance. As a result of this dependence the addicted animal seems to be "well" while intoxicated but becomes ill when the poison is removed. This illness is called the withdrawal syndrome. The poison can be removed from its receptors by stopping administration or by administering a competitive antagonist.

II. CRITERIA FOR TOLERANCE AND PHYSICAL DEPENDENCE

A. Tolerance

The degree of narcotic tolerance can be quantitated by measuring the response to a test dose of a standard narcotic such as morphine. Parameters that are commonly measured, such as analgesia, rate of respiration, or the regulation of body temperature, usually reflect depression of the central nervous system by the narcotic. The response of a tolerant animal can be expressed as a percentage of the response of a nontolerant control animal or of the response measured before the start of the treatment.

In order to produce analgesic tolerance many investigators have given a series of injections spaced one or more days apart and a schedule of increasing doses of the narcotic drug. It has become clear from recent studies that appreciable tolerance can be achieved within 1 day or less if fairly high levels of narcotic are given continuously. For example, Martin and Eades produced tolerance within 8 hr by infusing 3 mg/kg of morphine per hour into dogs.[2] When pellets of morphine base were implanted subcutaneously into mice analgesic tolerance could be measured within 1 day.[3] Another method that can be used is to feed mice dihydromorphinone in dilute evaporated milk. During the night a mouse can consume an amount of this mixture exceeding his body weight and by noon of the next day it will exhibit appreciable analgesic tolerance.[4] Cox et al. have reported that when rats were given morphine by continuous intravenous infusion at a rate of 7.5 mg/kg/hr, analgesic tolerance developed within 3 or 4 hr (Fig. 1).[5]

These investigations were not directed toward establishing a record for the attainment of narcotic tolerance. The reason for using shorter periods of narcotic administration is to allow greater flexibility in biochemical experiments, particularly those involving labeled precursors of proteins and

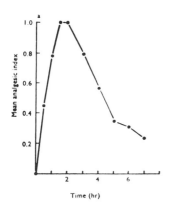

Fig. 1. Development of analgesic tolerance in rats given a continuous intravenous infusion of morphine. Morphine was infused at a rate of 7.5 mg/kg/hr. The analgesic response was measured by applying pressure to the tail. Each point was obtained from five animals. (Taken from Cox et al.,[5] by permission of Brit. J. Pharmacol.)

nucleic acids, and metabolic inhibitors. Where physical dependence has been looked for in short-term experiments, it seems to develop in parallel with analgesic tolerance.[3,4]

B. Physical Dependence

In measuring the extent of physical dependence it is necessary to quantitate the degree of illness that follows either termination of narcotic administration or treatment with a narcotic antagonist. One reliable indicator of illness in man and other animals is anorexia and weight loss. Kolb and Himmelsbach[6] studied addicts that had been receiving about 300 mg of morphine daily. Within 2 days after withdrawal there was a 70% decrease in caloric intake and an average weight loss of 2 kg. Akera and Brody[7] found that within 24 hr of withdrawal, 210 g rats that had been treated chronically with morphine lost 30 g of weight (Fig. 2). When food was withdrawn instead of morphine, weight loss was only 20 g. While the authors conclude from this observation that weight loss is not merely the result of anorexia, it should be pointed out that water was not withdrawn when the rats were deprived of food, and anorexia may have caused the animals to stop drinking. Diarrhea may also contribute to weight loss in withdrawn animals. Akera and Brody have concluded as follows: "In our opinion, the loss of body weight upon withdrawal is the best index of addiction in rats. It is objective, easy to recognize, and dose-dependent with both morphine and levorphanol."[7]

The weight loss in mice undergoing narcotic withdrawal is also quite

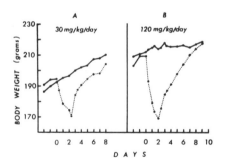

Fig. 2. Weight loss of rats during withdrawal from morphine. A. The rats were injected with increasing doses of morphine sulfate for 4 weeks up to a final level of 30 mg/kg/day. B. The dose was raised to 120 mg/kg/day over a period of 8 weeks. Morphine was injected every 8 hr. On day 0, morphine injections were continued for one group of rats (solid lines) and replaced by saline for another group (dotted lines). (Taken from Akera and Brody,[7] by permission of *Biochem. Pharmacol.*)

striking. It ranged from 1 g in 24 hr for mice that had been drinking dihydromorphinone for 1 day to 3 g for mice that had been drinking for 14 days. At the same time there was a marked decrease in the consumption of dilute evaporated milk.[4] Mice that were not given any evaporated milk or water for 24 hr also lost about 3 g weight.

These observations pose a serious dilemma for the biochemist who wishes to measure changes associated with narcotic tolerance and physical dependence. As Akera and Brody have pointed out, the effects of injected morphine are short-lived, and the drug must be administered to rats every 8 hr or the animals will go into withdrawal.[7] If experiments are carried out during these 8 hr they will be influenced by acute effects of the narcotic. Tolerance to all the acute effects does not develop uniformly. Therefore the biochemical changes observed in a tolerant animal shortly after an injection of morphine may or may not include: alteration of carbohydrate metabolism in liver, brain, and skeletal muscle; a breakdown of polysomes and decreased amino acid incorporation into brain proteins; increased incorporation of ^{32}P into brain phospholipids; and other metabolic actions attributable to the release of epinephrine and corticosteroids from the adrenal glands, the release of ACTH and antidiuretic hormone from the pituitary, and the release of histamine from mast cells. The biochemical responses of the nervous system and other tissues to the acute administration of narcotics have been summarized by Clouet in a recent review.[8]

If, on the other hand, the researcher waits until the narcotic has been eliminated from the body, then he will be working with a starving animal undergoing the stress of withdrawal. One of the consequences of this stress is a massive discharge of epinephrine from the adrenals as well as depletion of catecholamines from the central nervous system.[7,9] Both starvation and the release of epinephrine cause the breakdown of liver glycogen.

In early experiments with the mixed function oxidase system of the endoplasmic reticulum of liver, most investigators added NADP to the incubation mixture. They depended upon the soluble enzymes in their extracts to generate NADPH from endogenous substrates such as glucose-6-phosphate or isocitrate. Roth and Bukovsky have demonstrated that starvation for 16 hr causes a decrease in the activity of rat liver N-demethylase unless the incubation medium is supplemented with glucose-6-phosphate.[10] This finding might explain the apparent decrease in demethylase activity that is associated with narcotic tolerance. In the experiments of Axelrod[11] and Cochin and Axelrod[12] N-demethylase activity was measured 24 hr after the last injection of morphine. Under these conditions only the morphine-tolerant animals should show the effects of starvation.

Another simple method for determining physical dependence in mice involves measurement of the decrease in spontaneous activity that occurs upon withdrawal.[4] Because morphine and other narcotics increase the locomotor activity of mice, the first observations of this phenomenon were interpreted as the development of tolerance to the exciting effects of morphine. There was good correlation between analgesic tolerance and the

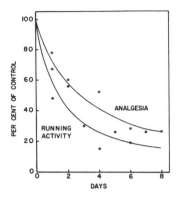

Fig. 3. Change in tolerance of mice drinking dihydromorphinone for various time periods. The average daily dose was 70 mg/kg. The test dose of dihydromorphinone HCl was 1 mg/kg for analgesia (tail-flick method) and 5 mg/kg for running activity (activity cage). Each point is the value for a group of eight mice. (Taken from Shuster, Hannam, and Boyle,[4] by permission of *J. Pharmacol. Exp. Ther.*)

running response to a test dose of dihydromorphinone (Fig. 3). With a smaller test dose of dihyromorphinone, the decline in motor activity after treatment for 1 day was actually a more sensitive measure of apparent tolerance than was the change in analgesic response (Fig. 4). Goldstein[13] also has described what appears to be tolerance to the exciting action of levorphanol in mice(Fig. 5). However, dihydromorphinone-tolerant mice were also "tolerant" to the exciting action of nonnarcotic stimulants such as cocaine, amphetamine, and methylphenidate. Furthermore, they were less active than controls when injected with saline or even when no injection was given. This decreased spontaneous activity persisted for several days after withdrawal of the narcotic. For example, the spontaneous activity of mice that had been consuming 60 mg/kg of dihydromorphinone for 6 days was 34% of the control level on the third day after withdrawal. Mice that were withdrawn by the injection of 10 mg/kg of nalorphine also showed a decrease in spontaneous activity. These experiments were carried out with C57 black mice. Decreased spontaneous activity also has been observed when nalorphine was injected into mice of the DBA strain that had been

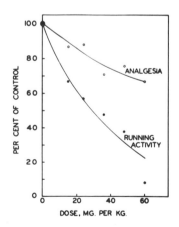

Fig. 4. Tolerance in mice drinking different amounts of dihydromorphinone for 1 day. Mice were given milk containing dihydromorphinone HCl in concentrations ranging from 0 to 12 mg/ 100 ml. The test dose of dihydromorphinone was 1 mg/kg for analgesia and 2 mg/kg for running activity. Each point represents eight mice. (Taken from Shuster, Hannam, and Boyle,[4] by permission of *J. Pharmacol. Exp. Ther.*)

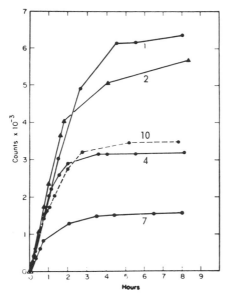

Fig. 5. Development of tolerance to the exciting action of levorphanol in mice. Levorphanol was injected subcutaneously every 8 hr in a dose of 20 mg/kg, and the mice were placed in an activity cage. The numbers under each curve refer to the injection sequence. For the tenth injection (broken line) the dosage was raised to 60 mg/kg. (Taken from Goldstein, Aronow, and Kalman,[13] by permission of the publishers.)

made tolerant by implantation of a pellet of morphine base.[14] On the other hand, the same authors found that this treatment increased the spontaneous activity in a different (Catholic University) strain. Way *et al.*[3] found that naloxone produced considerably less hyperactivity in morphine-tolerant Berkeley–Pacific Swiss mice than in animals of the same strain that were bred in the Hooper Foundation. Mice of different strains can differ markedly with respect to both brain neurohormone concentrations and response to drugs,[15] and these differences must be kept in mind in comparing results obtained by administering narcotics and narcotic antagonists.

The most effective way to induce a withdrawal syndrome in the early stages of the development of physical dependence is by giving a narcotic antagonist such as nalorphine or naloxone. Martin and Eades were able to produce a withdrawal syndrome by injecting nalorphine into dogs that had been infused with morphine for 8 hr.[2] Way *et al.*[3] have refined the technique of Maggiolo and Huidobro[14] to the point that physical dependence in mice can readily be detected within 3–6 hr of the subcutaneous implantation of a pellet of morphine base. One characteristic feature of the abstinence syndrome induced by injecting tolerant mice with a narcotic

TABLE I
Development of Tolerance and Physical Dependence in
Mice After Implantation of Morphine

Hours After Implantation	Morphine AD_{50} (mg/kg)	Naloxone ED_{50} (mg/kg)
0	10.7	>12.0
3	—	3.2
6	—	1.6
12	13	0.42
24	22	0.26
48	39	0.058
72	53	0.045

a The morphine AD_{50} is the median effective dose required to inhibit the tail-flick response to a thermal stimulus. The ED_{50} of naloxone is the median effective dose required to precipitate withdrawal jumping. (From Way et al.,[3] by permission of J. Pharmacol. Exp. Ther.)

antagonist is a tendency for the mice to jump. The degree of abstinence is measured by placing the mice on a platform and recording the percentage of animals that leap off within 15 min after the injection of naloxone. Withdrawal jumping occurs 4–10 hr after removal of the implanted morphine pellet or within a few minutes after the injection of naloxone even when the morphine implant is left in place. The amount of antagonist needed to produce jumping decreases with increasing tolerance (Table I).

Another measure that can be used as an index of withdrawal in narcotic-tolerant animals is the rate of lever pressing in monkeys or rats that have learned to inject themselves with narcotics.[16] This approach is discussed more fully in Chapter 22.

The experiments described in this discussion of tolerance and physical dependence lead to two conclusions that have an important bearing on attempts to define narcotic addiction in biochemical terms: (1) There are methods available to establishing and quantitating narcotic tolerance and physical dependence within 1 day or less. (2) There is a good correlation between the development of tolerance and physical dependence wherever this relationship has been examined carefully. This point is especially emphasized and clearly illustrated in the work of Way, Loh, and Shen.[3] These authors state that their evidence "argues strongly for an approach that seeks to link tolerance and physical dependence to a common underlying process."

III. THE THEORETICAL LINK BETWEEN NARCOTIC TOLERANCE AND PHYSICAL DEPENDENCE

The explanation of this sudden reversal of symptoms lies in the fact that morphine taken as a habit-forming drug upsets the normal equilibrium between the synpathetic and the autonomic systems, establish-

ing a new one. When this new equilibrium is destroyed by taking away
the drug, symptoms naturally result.[17]

This hypothesis by Kraus, while unproven, does attempt to express in
physiological terms the general theory linking narcotic tolerance with
physical dependence. That is, repeated exposure to the narcotic leads to a
new metabolic equilibrium in which the narcotic drug is an essential component. One result of this equilibrium is tolerance, so that a normally fatal
amount of drug produces few overt signs of intoxication. Removal of the
narcotic disturbs this new state of equilibrium, and the same metabolic
change that previously served to neutralize the toxic effects of the drug now
produces the withdrawal syndrome.

As pointed out in Chapters 4 and 6 of this volume, it seems quite clear
that narcotic tolerance cannot be attributed to a change in distribution or
an increased rate of metabolic inactivation. Therefore there is no basis for
theories such as those of Marmé, who claimed that tolerance resulted from
an increased rate of formation of a metabolite of morphine. Physical
dependence was explained by postulating that this metabolite was toxic and
that its toxicity could be counteracted by morphine.[18]

It seems clear that the biochemical basis for narcotic addiction has to
be sought not in the liver, where metabolic inactivation takes place, but in
the central nervous system, where tolerance and physical dependence are
most strongly expressed. Because so much is still unknown about the
chemistry and metabolism of the central nervous system, it is not surprising
that even modern theories are vague and unsatisfactory. The most to be
expected of such theories at the present stage is that they provide the
stimulus for meaningful new experiments.

A. The Analogy to Enzyme Induction

The only theories that are worth being considered are those that are
based on the chemistry of the morphinist.[19]

One way in which cells may respond chemically to the introduction of
a foreign substance is by an increase in the amount of an enzyme. Enzyme
induction is a reversible process, and enzyme levels return to normal when
the inducer is removed. The inducer need not be a substrate for the enzyme.
For example, barbital, which is not metabolized to any appreciable extent
in the liver, stimulates the formation of mixed function oxidases in the
smooth endoplasmic reticulum of liver cells.[20] While barbital does resemble known substrates of the mixed function oxidases, such as phenobarbital, even this resemblance is not necessary. A case in point is the
induction of hepatic δ-aminolevulinic acid synthetase by such chemically
different substances as allyl-isopropyl acetamide, griseofulvin, collidine
derivatives, and estrogens.[21] These compounds bear no obvious structural
similarity to either succinyl-Coenzyme A or glycine, the substrates of the
enzyme. The only structural requirement for an enzyme inducer is that it
possess the capacity for dissociating a specific repressor protein from the
appropriate cistron in the nuclear DNA of the affected cell.

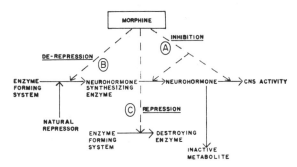

Fig. 6. Hypothetical scheme for explaining morphine tolerance and physical dependence in terms of repression and derepression of enzyme synthesis. (Taken from Shuster[22] by permission of *Nature*.)

The metabolic changes resulting from enzyme induction by a foreign substance are neither "good" nor "bad" and bear no obvious relationship to its toxicity.[20] They simply represent an intrusion into the regulatory machinery of the cell.

Narcotic addiction, in theory can be viewed as another example of such intrusion. There is a reversible response to the administration of a foreign substance, with several features that are comparable to enzyme induction.

One version of the enzyme induction hypothesis is shown in Fig. 6. It is assumed that a narcotic such as morphine may act at A to inhibit either neurohormone synthesis or function. At the same time repeated exposure to morphine causes a derepression (B) of the synthesis of an enzyme that makes the neurohormone. Alternatively there may be repression (C) of the synthesis of an enzyme that destroys the neurohormone. In either case the result is an increase in neurohormone level that counteracts the effect of morphine and produces tolerance. When the narcotic is withdrawn the

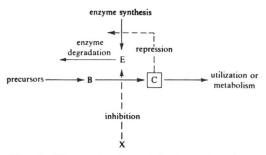

Fig. 7. "Depression theory" for the explanation of narcotic tolerance and physical dependence. (Taken from Goldstein, Aronow, and Kalman,[13] by permission of the publishers.)

persistence of unopposed high levels of neurohormone produces the abstinence syndrome.[22]

Goldstein and Goldstein[23] have proposed a somewhat similar hypothesis, and this is illustrated in Fig. 7. The narcotic is assumed to inhibit an enzyme E that catalyzes the formation of substance C, a reaction essential to a neuronal function or synaptic transmission. If C mediates an excitatory function, the acute response of the central nervous system to the narcotic is depression. If at the same time the level of C controls the synthesis and/or breakdown of enzyme E, the decrease in concentration or availability of C will increase the amount of enzyme E. After some hours or days a new steady state will be achieved in which the amount of C is increased, and tolerance results. Upon withdrawal the unbalanced excess of C leads to excitation and an abstinence syndrome.[13]

Induction of protein synthesis also is a central feature of the theory proposed by Collier.[24] According to this hypothesis, the amount of neurohumoral transmitter does not change in narcotic addiction. Instead there is an increased responsiveness due to the synthesis of new receptor proteins. As in the other induction theories, there is the suggestion that narcotics may inhibit the combination of a neurohormone with its receptors in the central nervous system. The result is a kind of "pharmacological denervation" that leads to denervation hypersensitivity. Axelsson and Thesleff[25] have shown that either surgical denervation or pharmacological denervation (produced by botulinum toxin[26]) increased the number of cholinergic receptors on the membrane of a striated muscle cell. The increase in the number of CNS receptors would be expected to lead to both narcotic tolerance and to the abstinence syndrome when the narcotic is withdrawn.

Pharmacological or "disuse" supersensitivity of the central nervous system has been described by Friedman et al.[27] These authors treated mice with scopolamine, a centrally acting anticholinergic drug. After several days or weeks the scopolamine was withdrawn and the mice were tested with pilocarpine, a cholinergic drug that produces a centrally mediated hypothermic response. The scopolamine pretreatment increased the degree and duration of the hypothermic response to pilocarpine. It is also worth noting that in these experiments the mice developed tolerance to the pilocarpine-blocking action of scopolamine. The development of tolerance followed the same time course as the development of supersensitivity. Both tolerance and supersensitivity were ascribed to an increase in the number of central cholinergic receptors.

An example of what may be disuse supersensitivity of central adrenergic synapses has been reported by Stolk and Rech.[28] They found that chronic treatment with reserpine made rats more sensitive to locomotor stimulation by d-amphetamine.

B. Consequences of the Induction Theory

The theories that attempt to explain narcotic addiction in terms of induced protein synthesis are at present not much more than a framework

TABLE II
Side Effects of Some Inhibitors of Protein and Nucleic
Acid Biosynthesis

Inhibitor	Side Effect	Reference
Puromycin	inhibition of cAMP-diesterase	(36)
	induction of intestinal phosphatase	(37)
	swelling of neural mitochondria	(38)
	lowered threshold for seizures	(39)
Actinomycin D	inhibition of phospholipid synthesis	(40)
	increased corticosterone secretion	(41)
	inhibition of respiration and glycolysis	(42)
	inhibition of protein synthesis unrelated to RNA synthesis	(43)
	disruption of polysomes	(44)
	stimulation of conversion of acetate to cholesterol	(45)
	decreased enzyme catabolism in vivo	(46)
Cycloheximide	decreased conversion of cholesterol to pregnenolone	(47)
	decreased enzyme catabolism in vivo	(46)

of ideas upon which experiments may be constructed. Their chief value is that they allow certain predictions to be made that can be tested in the laboratory. These predictions are:

1. *That the amount of some neurohormone in the central nervous system may increase during the development of narcotic tolerance.* As indicated in Chapter 10, this prediction has not been fulfilled. Such changes in norepinephrine levels that have been described seem to reflect either the release of catecholamines immediately after the injection of high doses of morphine or the stress of the withdrawal syndrome.[7,9] Morphine tolerance does not seem to affect the amount of 5-hydroxytryptamine, gamma-amino butyric acid,[29] or acetylcholine* in the brains of treated animals. These negative findings do not rule out changes in compartmentalization or turnover of neurohormones. Way *et al.* have reported an increased rate of serotonin turnover in the brains of morphine-tolerant mice.[30] It should be pointed out that the turnover of brain serotonin also increases with electroshock stress.[31]

2. *That the development of narcotic tolerance may be blocked by treatment with inhibitors of protein or nucleic acid synthesis.* Experiments relating to this prediction are described in Chapter 20. There have been claims that the development of analgesic tolerance to morphine can be blocked by treatment with actinomycin,[5] azaguanine[32] (inhibitors of nucleic acid synthesis), puromycin,[33] and cycloheximide[30,34] (inhibitors of protein

* L. Shuster and R. V. Hannam, unpublished results obtained with dihydromorphinone-tolerant mice.

biosynthesis). In such experiments it is essential to establish that the inhibitor is indeed preventing the *in vivo* incorporation of radioactive precursors into brain proteins or nucleic acids. The half-lives of at least some mammalian messenger RNAs are such that there is little change in protein synthesis for several days after an injection of actinomycin.[35] In general the amounts of these inhibitors that must be given in order to block biosynthesis *in vivo* are quite toxic and often produce unrelated side effects that must be taken into account. Some of these side effects are listed in Table II.

Even where there appears to be correlation between the inhibition of brain protein synthesis *in vivo* and the blockade of narcotic tolerance, interpretation of the results may be difficult. Studies with chlorpromazine provide a case in point. Clouet and Ratner found that chlorpromazine inhibited the development of narcotic tolerance in rats.[48] There also were reports by Piha *et al.*[49] and by Glasky[50] that chlorpromazine inhibited the incorporation of amino acids into rat brain proteins *in vivo*. These findings were confirmed in mice.[51] The inhibition of protein synthesis proved to be due to the hypothermia that results from interference with temperature regulation by chlorpromazine. It was not observed when chlorpromazine-treated mice were kept at 33°C.[51]

Mice that were injected with both chlorpromazine and dihydromorphinone and kept at 22° did not develop analgesic tolerance to dihydromorphine. Chlorpromazine treatment of mice kept at 33° did not block the development of tolerance (Table III). However, when these injections were continued beyond 3 days, the animals that were given chlorpromazine

TABLE III
Effect of Chlorpromazine Treatment on the Development of Narcotic Tolerance[a]

Treatment	Room Temperature	Analgesic Area	
		(min-sec)	(Percent of Control)
Saline	20°	532	100
Chlorpromazine		602	113
Dihydromorphinone		288	55
Dihydromorphinone and chlorpromazine		535	100
Saline	33°	501	100
Dihydromorphinone		318	63
Dihydromorphinone and chlorpromazine		306	61

[a] Each value represents the mean analgesic area for eight mice after a test dose of 1 mg/kg of dihydromorphinone.[4] Mice were injected twice daily for 3 days with 10 mg/kg of dihydromorphinone or chlorpromazine intraperitoneally. (Unpublished data of L. Shuster and R. V. Hannam.)

alone showed tolerance to the analgesic action of dihydromorphinone as well as tolerance to the depressant action of chlorpromazine. The chlorpromazine treatment also increased the activity of microsomal enzymes that demethylate both morphine and chlorpromazine.[52]

In the absence of other evidence, the recent experiments on the blockade of narcotic tolerance by treatment with cycloheximide provide the only experimental support for theories linking narcotic addiction with the induction of protein synthesis.[30] Even if this link were more firmly established, there still would be no clear indication of which particular proteins may be affected. Unless the protein can be identified as some enzyme that shows the expected changes in activity, this problem is not likely to be solved by the use of inhibitors of protein synthesis.

The amount of any given protein in a mammalian cell can increase in two ways—by an increased rate of synthesis or a decreased rate of breakdown—and these changes can be determined by labeling experiments.[53] It is highly unlikely that the changes produced by narcotic administration would be so extensive as to alter the rate of synthesis or breakdown of a major portion of the total brain protein. For this reason considerable fractionation would be necessary in order to detect such changes.

There is one distinguishing characteristic that may prove useful in looking for candidate proteins in the brain. Most brain proteins that have been examined turn over rather slowly. Half-lives for various fractions of rat brain proteins range from 12 to 22 days.[54,55] By contrast, the half-life of many liver proteins is a few days or, in some cases, a few hours.[53] Because narcotic tolerance and physical dependence can be induced within 1 day or less, brain proteins that may increase in this process should have a short half-life—perhaps a few days or less. It therefore would make sense to study only those brain proteins that are rapidly labeled *in vivo*.

The labeling characteristics of very few specific brain proteins have been studied carefully. The half-life of brain acetylcholinesterase, as determined with irreversible inhibitors, is about 1–2 weeks.[56] However, it has been reported that one isozyme of rat brain acetylcholinesterase has a half-life of about 3 hr.[57] The half-life of brain monoamine oxidase, as determined by recovery of activity after the administration of an irreversible inhibitor, is about 10 days.[58] Preliminary experiments have indicated that the choline acetyltransferase of mouse brain has a half-life of about 19 days.* The S-100 protein of brain is rapidly labeled, with a half-life of about 6 hr.[59] In general much more work needs to be done on the labeling of specific brain proteins, perhaps with the aid of microelectrophoretic techniques such as those developed by McEwen and Hydén.[59]

C. A Theory Based on Miscoding

The possibility remains that narcotic tolerance may be related to a change in the *quality*, rather than the quantity, of certain brain proteins.

* L. Shuster and C. O'Toole, unpublished data.

In this case narcotics need not alter the rate of turnover of brain proteins, but the development of tolerance still could be blocked by inhibitors of protein synthesis. Narcotic addiction bears some resemblance to streptomycin dependence in bacteria, although it must be emphasized that streptomycin dependence is not induced but is genetically determined. Certain bacterial mutants display not only resistance (tolerance) to streptomycin but also "physical dependence" in that they require the presence of the antibiotic for normal growth. The basis for streptomycin dependence is an alteration in the 30S subunit of the bacterial ribosome that leads to defective protein synthesis.[60] This defect is "corrected" by streptomycin, which causes misreading of the genetic code.[61] Other agents that cause similar misreading, such as aliphatic alcohols, can substitute for streptomycin in supporting the growth of streptomycin-dependent organisms.[62]

An analogous explanation for narcotic tolerance would imply that narcotic drugs cause misreading of the genetic code in the synthesis of some brain proteins. If the resulting altered proteins caused a central stimulation that could be antagonized by a narcotic drug, the result would be tolerance. Upon withdrawal of the narcotic the unopposed action of the altered proteins could give rise to a withdrawal syndrome.

Validation of this hypothesis would require evidence that there can be miscoding in a mammalian system. Ambiguity of the genetic code has been observed with a cell-free system from rabbit reticulocytes.[63] Miscoding is stimulated by high concentrations of Mg^{2+}. There was no increase in miscoding when streptomycin or aliphatic alcohols were added to a cell-free system from mouse spleen.[64]

If narcotic drugs did produce miscoding they also should act as inhibitors of protein synthesis. Morphine and related compounds, in fairly high concentrations, do inhibit protein synthesis in mammalian cells. The mechanism is not clear, but there is evidence for the breakdown of polysomes.[65] Unfortunately, there is no correlation between the narcotic activity of various morphinan derivatives and their potency as inhibitors of protein synthesis. Even if narcotic-induced miscoding in a cell-free system from brain cells could be demonstrated, it still would be necessary to establish that the functional defect in the altered protein is specifically corrected by narcotics. Here, too, there would be a requirement that the protein turn over rapidly.

IV. CONCLUDING REMARKS

It now should be quite obvious to the reader that theoretical castles are easy to build in the realm of narcotic addiction. A proper foundation for these edifices, that is, solid evidence of chemical differences between the brains of addicts and nonaddicts, is much more difficult to lay down. Whatever theory does survive the searching glare of experimental verification will have to explain the close association between narcotic tolerance and physical dependence. Improved techniques for the rapid induction and assay of both tolerance and physical dependence will ease the task of the researcher. However, the most formidable obstacle will continue to be the complexity of the central nervous system.

V. REFERENCES

1. A. Erlenmeyer, *Z. Gesamte Neurol. Psychiat. 103*, 705 (1926).
2. W. R. Martin and C. G. Eades, *J. Pharmacol. Exp. Ther. 133*, 262–270 (1961).
3. E. L. Way, H. H. Loh, and F. H. Shen, *J. Pharmacol. Exp. Ther. 167*, 1–8 (1969).
4. L. Shuster, R. V. Hannam, and W. E. Boyle, Jr., *J. Pharmacol. Exp. Ther. 140*, 149–154 (1963).
5. B. M. Cox, M. Ginsburg, and O. H. Osman, *Brit. J. Pharmacol. Chemother. 33*, 245–256 (1968).
6. L. Kolb and C. K. Himmelsbach, *Amer. J. Psychiat. 94*, 759–799 (1938).
7. T. Akera and T. M. Brody, *Biochem. Pharmacol. 17*, 675–688 (1968).
8. D. H. Clouet, *Int. Rev. Neurobiol. 11*, 99–128 (1968).
9. L. M. Gunne, *Acta Physiol. Scand. 58*, Suppl. 204 (1963).
10. J. S. Roth and J. Bukovsky, *J. Pharmacol. Exp. Ther. 131*, 275–286 (1961).
11. J. Axelrod, *Science 124*, 263–264 (1956).
12. J. Cochin and J. Axelrod, *J. Pharmacol. Exp. Ther. 125*, 105–110 (1959).
13. A. Goldstein, L. Aronow, and S. M. Kalman, "Principles of Drug Action," p. 597, Harper & Row, New York (1968).
14. C. Maggiolo and F. Huidobro, *Acta Physiol. Lat. Amer. 11*, 70–78 (1961).
15. A. G. Karczmar and C. L. Scudder, *Fed. Proc. 26*, 1186 (1967).
16. J. Weeks, *Science, 138*, 143–144 (1962).
17. W. M. Kraus, *J. Nerv. Ment. Dis. 48*, No. 1 (1918); quoted by C. E. Terry and M. Pellens, "The Opium Problem," p. 362, Committee on Drug Addictions and the Bureau of Social Hygiene, New York (1928).
18. W. Marmé, *Deutsche Med. Wochenschr. 9*, 197 (1883); quoted by C. E. Terry and M. Pellens, "The Opium Problem," p. 321, Committee on Drug Addictions and the Bureau of Social Hygiene, New York (1928).
19. P. Sollier, *Presse Med.*, April 23 and July 8 (1898); quoted by C. E. Terry and M. Pellens, "The Opium Problem," p. 325, Committee on Drug Addictions and the Bureau of Social Hygiene, New York (1928).
20. A. H. Conney, *Pharmacol. Rev. 19*, 317 (1967).
21. S. Granick, *J. Biol. Chem. 241*, 1359–1375 (1966).
22. L. Shuster, *Nature 189*, 314–315 (1961).
23. D. B. Goldstein and A. Goldstein, *Biochem. Pharmacol. 8*, 48–49 (1961).
24. H. O. J. Collier, *Nature 220*, 228–231 (1968).
25. J. Axelsson and S. Thesleff, *J. Physiol. 147*, 178–181 (1959).
26. S. Thesleff, *J. Physiol. 151*, 598–603 (1960).
27. M. J. Friedman, J. H. Jaffe, and S. K. Sharpless, *J. Pharmacol. Exp. Ther. 167*, 45–55 (1969).
28. J. M. Stolk and R. H. Rech, *J. Pharmacol. Exp. Ther. 163*, 75–83 (1968).
29. E. W. Maynert, G. I. Klingman, and H. K. Kaji, *J. Pharmacol. Exp. Ther. 135*, 296–299 (1962).
30. E. L. Way, H. H. Loh, and F. H. Shen, *Science, 162*, 1290–1292 (1968).
31. M. Fekete, A. M. Thierry, and J. Glowinski, *Abstr. 4th Int. Congr. Pharmacol. Basel 68*, (1969).
32. M. T. Spoerlein and J. Scrafani, *Life Sci. 6*, 1549–1564 (1967).
33. A. A. Smith, M. Karmin, and J. Gavitt, *J. Pharmacol. Exp. Ther. 156*, 85–91 (1967).
34. M. P. Feinberg and J. Cochin, *Pharmacologist 11*, 256 (1969).
35. M. Revel and H. H. Hiatt, *Proc. Nat. Acad. Sci. 51*, 810 (1964).
36. M. M. Appleman and R. G. Kemp, *Biochem. Biophys. Res. Commun. 24*, 564–568 (1966).

37. F. Moog, *Science 144,* 414–416 (1964).
38. P. Gambetti, N. K. Gonatas, and L. B. Flexner, *Science 161,* 900–902 (1968).
39. H. D. Cohen and S. H. Barondes, *Science 157,* 333–334 (1967).
40. I. Pastan and M. Friedman, *Science 160,* 316–317 (1968).
41. B. M. Lippe and C. M. Szego, *Nature 207,* 272–274 (1965).
42. J. Laszlo, D. S. Miller, K. S. McCarthy, and P. Hochstein, *Science 151,* 1007–1010 (1966).
43. G. R. Honig and M. Rabinovitz, *Science 149,* 1504–1506 (1965).
44. M. Revel, H. H. Hiatt, and J. P. Revel, *Science 146,* 1311–1313 (1964).
45. F. DeMatteis, *Biochem. J. 109,* 775–785 (1968).
46. F. T. Kenney, *Science 156,* 525–528 (1967).
47. W. W. Davis and L. D. Garren, *J. Biol. Chem. 243,* 5153–5157 (1968).
48. D. H. Clouet and M. Ratner, *J. Pharmacol. Exp. Ther. 144,* 362–372 (1964).
49. R. S. Piha, R. M. Bergström, L. Bergström, A. J. Uusitalo, and S. S. Oja, *Ann. Med. Exp. Fenn. 41,* 498 (1963).
50. A. J. Glasky, *Fed. Proc. 22,* 272 (1963).
51. L. Shuster and R. V. Hannam, *J. Biol. Chem., 239,* 3401–3406 (1964).
52. L. Shuster and R. V. Hannam, *Can. J. Biochem. 43,* 899–908 (1965).
53. R. T. Schimke, R. Ganschow, D. Doyle, and I. M. Arias, *Fed. Proc. 27,* 1223–1230 (1968).
54. K. von Hungen, H. R. Mahler, and W. J. Moore, *J. Biol. Chem. 243,* 1415–1423 (1968).
55. A. Lajtha and J. Toth, *Biochem. Biophys. Res. Commun. 23,* 294–298 (1966).
56. P. H. Glow and S. Rose, *Nature 202,* 422–424 (1964).
57. G. A. Davis and B. W. Agranoff, *Nature 220,* 277–280 (1968).
58. S. Barondes, *J. Neurochem. 13,* 721–727 (1966).
59. B. S. McEwen and H. Hydén, *J. Neurochem. 13,* 823–833 (1966).
60. L. Luzzatto, D. Apirion, and D. Schlessinger, *Proc. Nat. Acad. Sci. 60,* 873–880 (1968).
61. J. Davies, L. Gorini, and B. D. Davis, *Mol. Pharmacol. 1,* 93 (1965).
62. I. Gado and I. Horvath, *Life Sci. 2,* 741–748 (1963).
63. K. K. Bose, C. L. Woodley, N. K. Chatterjee, and N. K. Gupta, *Biochem. Biophys. Res. Commun. 37,* 179 (1969).
64. L. Stavy, *Proc. Nat. Acad. Sci. 61,* 347–353 (1968).
65. W. D. Noteboom and G. C. Mueller, *Mol. Pharmacol. 5,* 38–48 (1969).

Chapter 20

INHIBITORS OF TOLERANCE DEVELOPMENT*

Alfred A. Smith

Departments of Anesthesiology and Pharmacology
New York Medical College
New York, New York

I. INTRODUCTION

Tolerance to opioids is a complex phenomenon that depends in large measure on dose and duration of treatment. Recently the use of pharmacological agents to prevent tolerance has permitted a sharper view of some of the mechanisms responsible for this phenomenon. In order to examine more easily some of the experiments involving inhibition of tolerance to opioids, a brief review of several of the proposed mechanisms seems appropriate. This subject will be more thoroughly reviewed elsewhere in this volume.

In 1964, Cochin and Kornetsky[1] found that tolerance to morphine developed after a single dose. Tolerance to the analgesic effect was tested using the Eddy hotplate method. The speed of swimming also was tested and found less affected by morphine than in controls. After chronic treatment with morphine, tolerance persisted up to 1 year, suggesting the possibility that some fundamental change had occurred in the nervous system of the animal. For that reason the authors considered that tolerance might represent some sort of immune phenomenon. In their studies no relationship to the duration of tolerance was found in the recovery of N-demethylase activity in the livers of tolerant animals. The possibility was considered that the very fact of testing these animals produced an apparent tolerance, but this was ruled out by appropriate studies. However, others showed[2] that after repeated testing the normal response time fell slightly. In treated rats tolerance also developed more rapidly than it did in controls.

* Supported by a grant from the U.S. Public Health Service, MH-12988-04.

It therefore was concluded that experience seems to play some role in tolerance development.

The possible involvement of an immunological mechanism was considered by other investigators. Cohen et al.,[3] following this lead, reported that administration with morphine of actinomycin-D, an inhibitor of DNA-directed RNA synthesis, prevented the development of tolerance to the analgesic effects of the narcotic. Ungar and Cohen[4] then reported that they were able to transfer tolerance to mice by means of homogenates prepared from brains of animals made tolerant by chronic treatment with morphine. However, Smits et al.[5] and Tirri[6] were unable to repeat this experiment. More recently Ungar and Galvin[7] reported that a purified extract of brain also transferred tolerance. In rats treated up to 20 days with increasing doses of morphine, they found that a dialyzate of the brain homogenate, partitioned with phenol, produced tolerance when injected intraperitoneally into mice. Unfortunately there was some variation in effectiveness of the extract in regard to the degree of tolerance produced.

Smith and co-workers[8] have shown that tolerance to the cataractogenic effect of opioids developed in two stages: short-term stage lasting up to 8 hr and a second stage lasting at least 3 weeks. The existence of two stages in tolerance development somewhat resembles memory formation. This resemblance is made more striking by the finding that the long-term tolerance to opioids and long-term memory or the consolidation phase can be blocked by other inhibitors of protein synthesis such as puromycin[8] or acetoxy-cycloheximide[9] whereas the short-term phases of each phenomenon are unaffected by these drugs.

Biogenic amines have been implicated in the development of tolerance.[10] Tolerant animals have significantly higher levels of norepinephrine in brain and adrenals[11] and, furthermore, the onset of the abstinence syndrome was found to be associated with a discharge of catecholamines. Although dependency has been attributed[12,13] to a disuse phenomenon first postulated by Sharpless,[14] the mechanisms for tolerance development are not so readily explained. Goldstein[15] and Shuster[16] proposed the notion that enzyme inhibition and derepression may explain not only the phenomenon of tolerance but the development of the abstinence syndrome when the blocking agent is withdrawn. More recently Way[17] showed that tolerance to the analgesic effects of morphine can be inhibited by pretreatment of the animal with a serotonin depletor, p-chlorophenylalanine. Tolerance to opioids does not appear to be associated with significant changes in brain or plasma levels of the opioids.

It is apparent that many factors are involved in the development of tolerance to opioids. However, the ability of certain drugs to block protein synthesis or deplete biogenic amines or directly antagonize the effects of morphine may provide the key to unlock the larger question of the mechanism of action of opioids.

II. OPIOID ANTAGONISTS

In 1956, Axelrod[18] found that the enzyme N-demethylase was competitively blocked by nalorphine. This enzyme has been thought[19] to represent a model for the opioids receptor in brain since it possesses stereospecificity and is affected by chronic administration of narcotics. However, the recovery of the N-demethylating process proceeds more rapidly in the tolerant animal than does the return of the analgesic response.[20]

The administration of nalorphine, either systemically or directly into the anterior hypothalamus of the rat was found[21] to block development of tolerance to the hypothermic or to the analgesic effects of morphine, suggesting that tolerance may be initiated in the anterior hypothalamus. It also was shown[22] that parenteral nalorphine blocked the development of tolerance to the respiratory depressive effects of meperidine. Seevers and Deneau[23] found that the development of dependency and presumably tolerance to morphine could be prevented in morphine-treated monkeys by simultaneous administration of levallorphan. A complete block of dependency was obtained only when the dose of levallorphan equaled that of morphine. Levallorphan also was found[8] to prevent development of tolerance to the cataractogenic effect of levorphanol. But in order to completely block the action of levorphanol the dose of levallorphan had to be twice that of levorphanol. These findings indicate that blockade of the opioid receptor prevents not only the immediate opioid effects but also inhibits the later development of tolerance.

III. EFFECTS OF BIOGENIC AMINE DEPLETORS

That reserpine antagonizes the analgesic effect of morphine has long been recognized.[24] Experiments have since been undertaken to determine whether reserpine administration, by depleting catecholamines and serotonin, prevents development of tolerance. Smith et al.[25] showed that the administration of this drug to the rat in a dose of 2 mg/kg, 3 hr before injection of levorphanol, did not block the development of tolerance to the analgesic or respiratory depressive effects of the drug. Yet at the time of levorphanol injection, brain norepinephrine was only 26% of normal. Tolerance also developed to the cataractogenic effect of levorphanol[8] despite pretreatment with reserpine. In these studies only one to two doses of the opioid were required to produce long-lasting tolerance. Animals then were tested after they had recovered from severe depression induced by reserpine treatment.

Using the implanted morphine pellet technique,[26] Way[17] showed that tolerance to the analgesic effect of morphine, which developed within 3 or 4 days, was partially prevented by p-chlorophenylalanine, a depletor of serotonin. The turnover rate of serotonin in tolerant animals was then compared with controls. After administration of pargyline, a monoamine oxidase inhibitor, tolerant animals showed a higher rate of synthesis of serotonin in the brain both 30 and 60 min after injection of the inhibitor. The increase was approximately five times above control. However, tetrabenazine, which depletes norepinephrine and dopamine more specifically than serotonin, also blocked[27] tolerance development to morphine in mice. In this study treatment was continued for 9 days, and the mice were tested on the tenth day using the tail-clip method. When *l*-DOPA was injected at the time of treatment with tetrabenazine and morphine, the ability to develop tolerance could be restored. A dose of 100 mg/kg returned[28] the dopamine level to normal, the norepinephrine level rose to half of normal in the brain.

Maynert and Klingman[11] showed that the brain stems and adrenal medullas of morphine-tolerant dogs contained a higher concentration of norepinephrine than did controls. Furthermore, the administration of monoamineoxidase inhibitors to animals treated with morphine resulted in a more rapid accumulation of brain norepinephrine than in similarly treated controls. It was concluded[29] that a causal relationship existed between norepinephrine accumulation, tolerance development, and perhaps dependency, since the norepinephrine levels fell sharply during abstinence. However, these and other extensive studies[30] of the role of biogenic amines in tolerance development do not decisively establish that amines contribute fundamentally, although they indicate a strong association with the tolerance phenomenon.

In order to help answer the important question of how biogenic amines are involved in tolerance development, the problems of hypothermia, dehydration, and weakness in the animal induced by large doses of reserpine, or similarly acting substances, must be overcome. It therefore is difficult to critically evaluate data from experiments in which the duration of tolerance is brief or when the study requires repeated injections of both opioid and the toxic amine depletors.

IV. INHIBITORS OF PROTEIN SYNTHESIS

The possibility that tolerance results from an immune response was first suggested by Cochin and Kornetsky.[1] In subsequent studies, Cohen and co-workers[3] found that repeated administration of actinomycin-D with morphine partially prevented the development of tolerance to the analgesic effect of the narcotic. However, many mice were killed by the large dose of actinomycin-D that doubtlessly produced much morbidity. In other studies[8] in which actinomycin-D also was used, long-term tolerance to the cataractogenic effects of opioids was found to develop within 6 hr after administration of a single large dose of levorphanol. In this instance it was possible to administer actinomycin-D in a dose that ordinarily kills most mice within 1–2 days. However, during the 6-hr test period, no morbidity or mortality was observed; yet tolerance failed to develop in mice pretreated with actinomycin-D. Interestingly, mice treated with the antibiotic 1 hr after injection of the levorphanol were able to develop tolerance. Cox *et al.*,[31] using a similar acute tolerance technique, found that rats continuously infused with morphine showed a maximum analgesic effect after 2 hr but, despite continued infusion of morphine, the analgesic effect fell to roughly 20% of maximum at 6 hr. When actinomycin-D, 10 μg/kg/hr, was infused together with morphine, the peak analgesic effect did not fall. Studies with etorphine or meperidine gave similar results. Repeated testing of analgesic effect during the experiment did not produce conditioning. A determination of the *N*-demethylase activity in the livers of treated rats revealed no change in enzyme activities despite the development of tolerance. These investigators concluded that protein synthesis probably was causally related to the development of tolerance to morphine.

Inhibitors of protein synthesis, which act by mechanisms other than by blockade of RNA synthesis, also have been utilized in prevention of tolerance development to opioids. Smith and co-workers[8] in 1967 showed that puromycin, 40 mg/kg given intraperitoneally, could prevent the development of tolerance to the cataractogenic effects of levorphanol. Feinberg and Cochin[32] showed that cycophosphamide (Cytoxin), 15 mg/kg injected at frequent intervals over a 10-week period of treatment with morphine, significantly reduced the development of tolerance. Cycloheximide, another inhibitor of protein synthesis, virtually blocked[33] the development of tolerance to morphine. Interestingly, the abstinence syndrome ordinarily produced by administration of naloxone or nalorphine did not develop in cycloheximide-treated mice. Another inhibitor of RNA synthesis, an analog of a purine base, 8-azaguanine,[34] also was found to block tolerance development.

V. OTHER INHIBITORS OF TOLERANCE

Levorphanol administered in a single small dose was found[35] to sensitize mice to the cataractogenic effect of the second dose if given within 5 days. Under these conditions considerably greater tolerance developed than would be expected merely from the additive effects of the two doses. This phenomenon was attributed to an anamnestic effect much like the immune response to a second injection of an antigen such as tetanus toxoid. The existence of a chemical memory is suggested since tolerance to opioids appears to show recall. It may be possible, therefore, to block the retention of the chemical experience in much the way that amnesia can be produced by electroshock therapy or anesthesia. This hypothesis was tested using the cataractogenic effect and also the respiratory–depressive action of opioids. Chloroform or ethanol, in anesthetic doses, was found[36] to block tolerance development when given simultaneously with levorphanol. More recently it also was shown[22] that an anesthetic dose of ethanol or pentobarbital prevented tolerance to the respiratory depression induced by meperidine. In each of these studies the opioid produced tolerance after the administration of a single dose. It therefore was possible to narcotize the mice during the several hours of opioid activity.

VI. SUMMARY AND CONCLUSIONS

Several studies have shown that opioid antagonists block the development of tolerance and dependency. The site for initiating tolerance development seems to be in the anterior hypothalamus since injection of nalorphine into this area effectively blocks tolerance. Because dependency and tolerance can both be prevented by an antagonist it can be concluded that the two phenomena are really part of the same process. But there is an objection to the ready acceptance of this hypothesis: Blockade of the receptor by an antagonist also prevents the depressive effects of the opioid on respiration, sensory appreciation, temperature regulation, etc. According to Sharpless,[14] continued blockade of nervous pathways produces compensatory changes leading to more normal responses. This phenomenon then can account for the development of supersensitivity and concomitant dependency. Thus the failure to develop dependency when an opioid is effectively antagonized may merely reflect the absence of chronic neuronal depression, possibly quite remote from the antagonized opioid receptor. Studies obviously are needed that can discriminate between tolerance and dependency, if indeed they are separate phenomena. Perhaps injection of antibiotics into specific brain nuclei, or selective destruction of the nuclei, will yield the desired result. In that event dependency might be produced without the development of tolerance.

The role of the biogenic amines in tolerance development remains unclear, principally because of the conflicting reports on the effect of depleting agents. Recently serotonin, rather than catecholamines, has been suggested as the amine involved in tolerance development. There seems to be little doubt that increased synthesis of these amines takes place in chronically treated animals. However, stress induced by repeated insulin injections can also induce an increase in the concentration of epinephrine in the adrenal medulla. Furthermore, the fall in brain norepinephrine after morphine injection requires exceedingly high doses of morphine.

Virtually all investigators using antibiotics to study tolerance development appear to agree that protein synthesis is causally related to tolerance development. The antibiotics used block protein synthesis in different ways. Some prevent synthesis of RNA, while others interfere with ribosomal synthesis of protein chains. Several of these have limited toxicity in that they do not irreversibly affect production of essential proteins. For this reason they must be administered frequently during the development of tolerance. However, the cumulative effect then may produce sufficient morbidity to make difficult a valid comparison between experimental and control animals, Experiments in which animals show development of acute tolerance can obviate this problem. That acute tolerance is not necessarily short term is shown in studies of the cataractogenic effect.

The finding that tolerance can persist for extraordinary periods of time, as much as 1 year, fits well with the conclusion that protein synthesis is the basis of tolerance to opioids. The protein may be an enzyme that synthesizes a neuroactive substance. This hypothesis is in accord with several current views. These views generally hold that tolerance and dependency are interrelated. But in view of the long duration of opioid tolerance, it would appear unlikely that dependency can persist for periods up to a year. Furthermore, depressive drugs such as ethanol or barbiturates can cause remarkable dependency. In man these appear far more intense than corresponding opioid dependency. Yet tolerance to these drugs is relatively slight. It may be more reasonable to assume that tolerance to opioids represents a rise in the threshold to a chemical stimulus that acts to depress specific nervous pathways. Introduction of a protein-containing molecule somewhere in this pathway may reduce the sensitivity of the synapse or neuron involved in transmission of the impulse.

It is even possible that the protein-containing molecule is specific for opioids in regard to composition. Injection of homogenates from brains of tolerant animals ought then to confer tolerance on the recipient. But if the protein is common, as this author supposes, and only the site of its deposition is critical, then it would be overly optimistic to expect that the injected protein would reach the desired site.

Certain anesthetic agents have the ability to block tolerance development to the respiratory or cataractogenic effects of opioids. These agents are generally thought to act at synapses or neuronal membranes. Tolerance development therefore may involve at least one synapse. Possibly the synapse modulates the sensory impulses arising from the periphery. Sympathetic nerves seem to do just that as shown by the increase in the depressive effect of morphine with central sympathetic activity and its decrease when biogenic amines are depleted. Reducing the sensitivity of the modulating mechanism by means of an alteration in membrane potential could be one way to change the response to opioid or, for that matter, to facilitate or depress nervous pathways representing storage of any information.

VII. REFERENCES

1. J. Cochin and C. Kornetsky, *J. Pharmacol. Exp. Ther. 145*, 1–10 (1964).
2. S. Kayan, L. A. Woods, and C. L. Mitchell, *Eur. J. Pharmacol. 6*, 333–339 (1969).
3. M. Cohen, A. S. Keats, W. Krivoy, and G. Ungar, *Proc. Soc. Exp. Biol. Med. 119*, 381–384 (1965).
4. G. Ungar and M. Cohen, *Int. J. Neuropharmacol. 5*, 183–192 (1966).
5. J. Smits, D. Stephen, and A. E. Takemori, *Proc. Soc. Exp. Biol. Med. 127*, 1167–1171 (1968).
6. R. Tirri, *Experientia 23*, 278–279 (1967).
7. G. Ungar and G. Galvin, *Proc. Soc. Exp. Biol. Med. 13*, 287–291 (1969).
8. A. Smith, J. Gavitt, and M. Karmin, *J. Pharmacol. Exp. Ther. 156*, 85–91 (1967).
9. S. H. Barondes and H. D. Cohen, *Proc. Nat. Acad. Sci. 58*, 157–164 (1967).
10. L. Gunne, *Acta Physiol. Scand., Suppl. 204*, 1–91 (1963).
11. E. W. Maynert and G. I. Klingman, *J. Pharmacol. Exp. Ther. 135*, 285–295 (1962).
12. J. Jaffe and S. Sharpless, *in* "The Addictive States" (A. Wikler, ed.), Williams & Wilkins, Baltimore (1968).
13. H. Collier, *Nature (London), 220*, 228–231 (1968).
14. S. Sharpless, *Ann. Rev. Physiol. 26*, 357–388 (1964).

15. A. Goldstein and D. Goldstein, *Biochem. Pharmacol. 8*, 48 (1961).
16. L. Shuster, *Nature (London) 189*, 314–314 (1961).
17. E. Way, *Science 162*, 1290 (1968).
18. J. Axelrod, *Science 124*, 263–264 (1956).
19. J. Axelrod in "The Addictive States" (A. Wikler, ed.), 247–264, Williams & Wilkins, Baltimore (1968).
20. J. Cochin and S. Economon, *Fed. Proc. 18*, 377 (1959).
21. P. Lomax and W. E. Kirkpatrick, *Med. Pharmacol. Exp. 16*, 165–170 (1967).
22. P. Y. Huang and A. Smith, reported at annual meeting of the Committee on Drug Dependence, National Research Council, Washington, D.C. (1970).
23. M. Seevers and G. Deneau, *in* "The Addictive States" (A. Wikler, ed.), 199–205, Williams & Wilkins, Baltimore (1968).
24. E. Sigg, B. Caprio, and I. Schneider, *Proc. Soc. Exp. Biol. Med. 97*, 97–100 (1958).
25. A. Smith, K. Lisper, and K. Hayashida, *Fed. Proc. 28*, 261 (1969).
26. F. Huidobro, J. Huidobro, and C. Larrain, *Acta Physiol. Lat. Amer. 18*, 59–67 (1968).
27. H. Takagi and H. Kuriki, *Int. J. Neuropharmacol. 8*, 195–196 (1969).
28. H. Takagi and M. Nakama, *Jap. J. Pharmacol. 18*, 54–58 (1968).
29. E. Maynert, *in* "The Addictive States" (A. Wikler, ed.), pp. 89–95, Williams & Wilkins, Baltimore (1968).
30. L. Gunne, *Acta Physiol. Scand., Suppl. 204*, 1–91 (1963).
31. B. Cox, M. Ginsberg, and O. Osman, *Brit. J. Pharmacol. Chemother. 33*, 245–256 (1968).
32. M. Feinberg and J. Cochin, *Pharmacologist, 10*, 191 (1968).
33. H. Loh, F. Shen, and E. Way, *Pharmacologist, 10*, 192 (1968).
34. J. Yamamoto, R. Inoki, Y. Tammari, and K. Iwatsubo, *Jap. J. Pharmacol. 17*, 140–142 (1967).
35. A. Smith, M. Karmin, and J. Gavitt, *Pharmacologist, 9*, 231 (1967).
36. A. Smith, M. Karmin, and J. Gavitt, *Biochem. Pharmacol. 15*, 1877–1879 (1966).

Chapter 21

ROLE OF POSSIBLE IMMUNE MECHANISMS IN THE DEVELOPMENT OF TOLERANCE

Joseph Cochin

Department of Pharmacology
Boston University School of Medicine
Boston, Massachusetts

I. INTRODUCTION

Although a number of chapters in this book deal with various aspects of tolerance (see Chapters 19 and 20), it would be extremely difficult to discuss some possible mechanisms involved in the development of tolerance without briefly reviewing the major theories that have been proposed to explain this puzzling and intriguing phenomenon.

A. Theories of Tolerance

There have, of course, been a large number of hypotheses proposed over the years to explain tolerance to the effects of the narcotic analgesics. Among the most important of these are those that propose (1) altered metabolic disposition and/or differential distribution in the tolerant vis-à-vis the nontolerant animal, (2) prevention of access of drug to the site of action of the drug, (3) occupation and saturation of these receptor sites, thus preventing access secondarily, (4) cellular adaptation, be it biochemical or physiological, and finally (5) some sort of change, ill-defined and vague at the cellular level, that resembles an immune reaction or a reaction analogous to memory. This last hypothesis is really a subclassification under (4) since it also postulates a basic adaptation at a cellular level that is extremely broad in scope.

1. Interference with Access to Receptors

There is no evidence that prevention of access of drug to the site of action plays any significant role in the type of tolerance we are discussing, although it purportedly does explain the "tolerance" to substances such as arsenic, where changes in membranes by prior doses of the toxic metal prevent efficient absorption of subsequent small doses.

2. *Altered Metabolism and Distribution*

The concept that tolerance may be due to altered metabolism and/or distribution and excretion has been widely held for many years. Much of the previous research in this area has centered upon this type of explanation of the phenomenon. Although early work by Pierce and Plant,[1,2] by Gross and Thompson,[3] and by Oberst[4] was interpreted to demonstrate differences between tolerant and nontolerant animals insofar as the metabolism and excretory pattern of unconjugated and conjugated morphine was concerned, careful examination of the analytical and statistical methods employed by these investigators does not support this conclusion. More recently, studies using extremely sensitive chemical or radioactive assays have not been able to demonstrate any significant differences in the way narcotic drugs are metabolized and excreted in the body in the tolerant animal as compared to the nontolerant animal—certainly no differences that are great enough to explain the many puzzling and sometimes incredible manifestations of tolerance.[5-9]

3. *Receptor Occupation*

The receptor-occupation hypothesis first suggested by Schmidt and Livingston[10] was based on their classic observations of acute tolerance to the vascular effects of morphine. It assumes that drug molecules exert their action at the time of occupation of the receptor sites and that once the receptor sites are occupied the original molecules exert no further effect but do prevent the initiation of a response by receptor combination with additional molecules of the same or similar drugs. Although many aspects of morphine action are explained by this theory, other facts point to much more profound changes than simple receptor occupation. Among these are the long-term persistence of tolerance and the appearance of abstinence symptoms long after the drug has been completely eliminated from the body. It is extremely difficult to visualize occupation of receptor sites by drug for periods of several weeks, let alone the 9 months to a year that Cochin and Kornetsky have shown tolerance to persist,[11] especially in the face of a complete lack of evidence that any drug residue persists for more than a week at the very most.

4. *Cellular Adaptation*

The idea that cellular adaptation based on biochemical changes may be one of the mechanisms of tolerance development is a natural consequence of contemporary biology. Enzymatic adaptation is a well-recognized phenomenon in biology, and cytochemical studies have shown that, for example, the distribution and concentration of H^+, SH groups, nucleic acids, etc. in liver cells are affected by extrinsic factors such as diet, temperature, physiologic condition of the animals, etc. Such biochemical adaptation may take the form of an increase or decrease in existing mechanisms normally operating to carry out a function or a reaction or by substitution of alternative pathways or initiation of new mechanisms. The forms that such adaptations take differ from preparation to preparation and from phenomenon to phenomenon, but the term "adaptation" is broad enough to cover a wide range of biochemical and biophysical changes, often intracellular, that are responses to external stimuli and the external environment. These can include changes in levels of enzyme activity, in the respiration of cortical slices, in the reaction of offspring of mothers that were treated with drugs chronically before pregnancy, in evoking mechanisms to maintain homeostasis, and in the responses of animals and man to drugs with reactions that resemble those seen in immune reactions and memory.

The theories invoking adaptive responses are quite old, although the terminology and explanations for these responses have changed markedly over the years. In the latter part of the nineteenth century, Marmé[12] proposed a unitary hypothesis to explain tolerance and dependence based on the formation of a morphine derivative with stimulant properties whose titer rose with chronic morphine administration, thus neutralizing the depressant effects usually seen with acute morphine administration. Upon withdrawal of morphine, the balance of depressant and stimulant effects was disrupted and an abstinence syndrome, primarily stimulant in nature, resulted. No evidence of the existence of such a compound has ever been found, however.

In 1953, Seevers and Woods[13] published a hypothesis based on earlier work of Tatum et al.[14] in which they proposed a mechanism to explain both tolerance and physical dependence based on the fact that morphine and its congeners exert a dual action in both man and animals. They postulated that the action of morphine is either depressant or stimulant, depending on the location of the receptor-drug combinations. Certain sites on the neuronal surfaces result in CNS depression, while other intracellular sites result in CNS stimulation—one balancing the other and thus initiating tolerance. After abrupt termination of drug, however, the receptors responsible for depression on the cell surface are freed of drug very quickly, while the intracellular drug–receptor combinations are broken much more slowly and the stimulation so characteristic of the withdrawal illness is seen. Thus this theory is a variant of a number of homeostatic adjustment hypotheses and attempts to explain both tolerance and dependence with one mechanism.

Although this is a most plausible and intriguing hypothesis, there are a number of requirements that cannot be fulfilled. In the first place, morphine must be present in nerve tissue during the entire period of withdrawal, and this is not true. Second, if the dual-action concept were valid, one would expect the stimulant actions of morphine-like drugs and the abstinence syndrome to be qualitatively similar, and this is not so. In addition, Seevers and Deneau showed that the abstinence syndrome can be aborted by simultaneous administration of nalorphine, a morphine antagonist, with morphine.[15] These experiments demonstrated that the receptors responsible for the depressant action of morphine are also involved in the development of physical dependence, since no abstinence syndrome is seen if one antagonizes that depressant activity. If it were true that the abstinence syndrome was due to the unmasking of the stimulant activity of morphine, then abstinence in this experiment should have been marked because there was no significant morphine depression observed during the course of the experiment. More recently, both Goldstein et al.[16] and Shuster[17] have demonstrated tolerance to the stimulant effect of morphine, a finding that makes it very difficult to accept homeostatic theories that are based on "neutralization" or "masking" of effects to explain tolerance as well as "unmasking" or "overshoot" of stimulant properties or formation of a stimulant derivative to explain the abstinence syndrome.

5. Biochemical Homeostasis

Although it is not within the province of this chapter to discuss specific biochemical theories of tolerance, it is important to mention in passing that they too invoke homeostatic adaptation as a unitary explanation for tolerance and dependence, and several are based on theories of "derepression"[18-20] involving drug-induced repression and subsequent derepression of certain neurohumors and/or enzymes involved in CNS activity. Although there are some striking examples of repression of enzyme activity induced by chronic administration of narcotic drugs, such as the marked decrease in N-demethylase activity described by Axelrod,[21] by Cochin and Axelrod,[22] and by Mannering and Takemori,[23] there has been no solid evidence that these changes which accompany tolerance are in any way causally related to tolerance or to dependence. The theories of derepression all invoke inhibition of a hypothetical neurohormone that is responsible in some way for selected CNS activity. The inhibition of the action or synthesis of this neurohormone causes the depression characteristically seen after administration of narcotic drugs. This inhibition sets off a complex feedback mechanism resulting in the synthesis of more neurohormone, thus neutralizing the depressant effects of morphine and initiating tolerance. Upon withdrawal, the excess neurohormone, whose action has been inhibited by the morphine present, is disinhibited or, alternatively, the synthesis of the neurohormone is disinhibited and a great overproduction occurs. Thus the abstinence syndrome is a result of very high levels of a neurohormone with stimulant properties.

These hypotheses postulate the existence of a neurohormone whose concentrations are affected by chronic administration of the narcotic analgesics. Is there any evidence for such changes in concentration of the known neurohormones and related compounds in the CNS during chronic drug administration? Unfortunately, those changes that have been found by Vogt,[24] by Maynert,[25] by Gunne,[26] by Martin et al.,[27] and by Akera and Brody[28] have not been great enough or temporally related in a convincing manner or sufficiently independent of other changes such as stress to allow one to postulate any causal relationship between catecholamine or catecholamine–precursor levels and tolerance and dependence.

More recently, however, Smith[29] and Weinstock[30] have independently studied the formation of lenticular cataracts following the administration of opioids and found a close and rather interesting relationship between the rate of cataract induction and catecholamine administration. Smith was able to show a reciprocal quantitative relationship between opioid dose and the dose of catecholamines. The higher the catecholamine dose, the lower the opioid dose necessary to produce lenticular cataracts in mice. Thus catecholamines (epinephrine, in this case) were directly implicated in the formation of lenticular cataracts, and the development of tolerance to the opioid effect also affects the amount of epinephrine needed to potentiate cataract production by the opioids. The tolerance to the lenticular effects, which both investigators were able to demonstrate, developed very rapidly. Smith described two forms of tolerance, one that persisted up to 8 hours and a second that was detectable for at least 3 weeks that he called "long-term" tolerance. The short-term tolerance did not develop when small amounts of levorphanol were given, suggesting the receptor sites need to be saturated with agonist in order to produce this form of tolerance. On the other hand, the long-term tolerance was dose related and could be detected at small doses.

Within the past few years, Way and his associates have investigated changes in the rate of brain serotonin turnover during chronic administration and after abrupt withdrawal of morphine in mice. Way[31] determined the rate of serotonin synthesis by blocking the conversion of serotonin to 5-hydroxyindole acetic acid (5-HIAA) with a monoamine oxidase inhibitor, pargyline. The rate of serotonin synthesis then can be calculated from the initial increase in serotonin levels if one makes the assumption that brain serotonin is converted solely to 5-HIAA. He found that serotonin turnover is increased markedly when the animals become tolerant after chronic

TABLE I

Effects of Chronic Administration of Morphine Sulfate to the Rat on a Subsequent Single Injection of Morphine (20 mg/kg) in the Rat

Time After Withdrawal (months)	Percentage of Initial Response to Morphine as Measured by Hot-Plate Procedure[a]
0	8
4	30
6	35
8	44
11	40
15	42

[a] Adapted from Cochin and Kornetsky.[11]

morphine administration and that initial serotonin levels return to normal values 2 weeks after discontinuation of morphine. He also found that inhibition of serotonin synthesis with p-chlorophenylalanine markedly affected the development of tolerance in animals receiving morphine chronically and inhibited the development of physical dependence. At this time the relationship between changes in serotonin turnover, tolerance, and physical dependence is not at all clear, but we can certainly agree with Way's assessment that they may all be "parts of closely related phenomena."[32] In addition to the implication of a biogenic amine in tolerance, Way's work is an additional piece of evidence pointing to the plausibility of those theories which view tolerance and dependence as part and parcel of the same homeostatic process.

II. CHARACTERISTICS OF THE DEVELOPMENT AND LOSS OF TOLERANCE

Some years ago, Kornetsky and I initiated a series of experiments designed to elucidate the characteristics of tolerance development and loss.[11] In the course of those studies we found evidence of residual tolerance in rats that had received morphine twice daily for 68 days and were then withdrawn for periods ranging from 3 to 15 months (Table I). We also were able to show residual effects 11 months after a single previous injection of morphine (Table II). The results clearly showed that what is important in maintaining or developing tolerance to the effect of morphine is not the length of the previous period of chronic morphine administration or the amounts given (within certain limits) but rather a history of several (not necessarily many) previous injections of drug, spaced regularly or irregularly at various intervals, long or short. Our results were consistent with those of Winter,[33] who reported the same degree of tolerance to morphine induced by daily doses of drug for 3 weeks or by weekly injections of the same amount of drug for the same period. Green and co-workers[34] reported

TABLE II
Effects of a Single Injection of Morphine Sulfate (20 mg/kg) on a Subsequent Injection of the Same Dose

Time After Initial Injection (months)	Percentage of Initial Response to Morphine as Measured by the Hot-Plate Procedure[a]
3	60
4	60
6	60
8	72
11	60
15	100

[a] Adapted from Cochin and Kornetsky.[11]

similar findings in the mouse using the tail-flick response to heat as a measure of drug effect and several weeks as the elapsed time. Fraser and Isbell also reported tolerance to the nauseant and emetic effects and to the effects of morphine on body temperature in man 6 months after termination of chronic morphine administration.[35]

In 1953 Eddy[36] noted that a second injection of morphine given to mice 24 hr following an initial dose does not demonstrate the initiation of any measurable tolerance by the first injection to the analgesic effect of the drug. However, if there is a 72- or 96-hr interval between the first and second injections, the effect of the second injection is much reduced. I also have found in the rat (unpublished observations) that the prolonged effect of one injection—the protracted tolerance induced by a single injection—becomes apparent under certain conditions only if the interval between injections is longer than 3 weeks. Recent work in our laboratory[37] also indicates quite clearly that the interval between injections is an extremely important factor in determining how much attenuation of effect is present in subsequent assays of drug effect, another indication that a process is taking place that requires a finite amount of time for completion and one that once instituted is amazingly persistent.

Findings such as these led me and Kornetsky[11,38] to question the widely accepted explanations of tolerance that involved receptor sites or changes in drug-metabolizing enzymes. In considering the types of changes that might account for the findings of many studies giving similar experimental results, we found the possibility of the induction of an immune mechanism or of a process resembling such a mechanism most inviting. It has been observed by many workers in the fields of immunochemistry and immunology that the effects of a single previous dose of an antigen is most persistent and might even last for the lifetime of an individual animal or man, and it has also been noted that these effects take a finite time to develop. Since tolerance can be extremely persistent and since it does take a finite time to develop, it was proposed that tolerance might well be such an immune phenomenon—a reaction to administration of a drug that was conditioned by the previous administration of that same drug or one closely related to it in structure or action.

III. POSSIBLE TRANSFERABLE FACTORS IN TOLERANT ANIMALS

A. Serum Factors

That some form of antigen–antibody reaction is initiated by the administration of morphine has been considered as a possibility for many years. Kreuger et al.[39] in their extensive and definitive review of the literature up to the 1940s, concluded that there was no evidence suggesting that serum from a morphine-tolerant animal transferred to a nontolerant test animal has any effect on the subsequent response of the test animal. However, these early experiments dealt with tolerance to the lethal effect of morphine and not with tolerance to the effect of the drug on behavior or on reactions to noxious stimuli "pain" and therefore may not be really germane to the latter. More recently, Kornetsky and Kiplinger[40] attempted to transfer serum from tolerant to nontolerant animals in order to see whether some factor was present in the serum of tolerant donor animals that would affect the response of nontolerant recipient animals to morphine. Dogs were given morphine chronically for several weeks, and serum was prepared from these animals and from comparable controls. Serum from the morphine-tolerant dogs was then injected into one group of rats and from control dogs into a second group of rats. Forty-eight hr after the injection of the serum into the recipient animals, the effect of a 20-mg/kg dose of morphine sulfate on swimming time was measured. The speed of swimming a measured course was then determined at 30-min intervals for 210 min. At 30 and 60 min after the injection of morphine the swim times of the rats that had received serum from tolerant dogs were significantly slower than the swim times of animals that had received serum from control animals 48 hr previously. These investigators also found that the lethal effect of morphine in the rat was not affected by prior injections of control and experimental serum. The unexpected finding that the serum from the tolerant dogs potentiated rather than attenuated the effect of morphine was, and is, inexplicable. Shortly after the publication of the paper cited above,[40] a paper by Kiplinger and Clift[41] extended the serum studies in dogs. In this study dogs and man were used as donors of serum and the mouse was the recipient species. Kiplinger and Clift confirmed the presence of a potentiating factor in the serum of tolerant animals and found that it appeared 27 days after the beginning of chronic administration of morphine and that it persisted for about 3 weeks postwithdrawal. They also observed that one could detect the effects of the potentiating factor for at least 96 hr after the serum was injected into mice, that it was not dialyzable, and that it was stable when refrigerated for 3–4 days. They concluded that the potentiating factor was a substance of large molecular weight, perhaps protein or polypeptide.

After the publication of the experiments described above, Kornetsky and Cochin proceeded independently to ascertain whether the potentiating factor was present in species other than the dog and whether it could be assayed by test procedures other than measurement of drug effect on swimming speed.[38] In one series of experiments, rabbits were used as the donor animal and the hot-plate procedure of Eddy and Leimbach[42] as the measure of drug effect. It was found that under certain conditions mice injected with serum taken from rabbits that had been given morphine twice daily for 157 days were less sensitive to the effects of a 10-mg/kg test dose of morphine than were comparable groups of mice injected with serum from untreated rabbits. Figure 1 shows representative results obtained in mice injected with rabbit serum collected 1 week after the termination of chronic morphine administration. The intervals shown are the elapsed times between

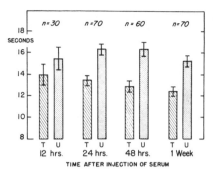

Fig. 1. The effect of serum obtained from
morphine-tolerant and nontolerant donor rabbits
on the hot-plate response of mice to a 10 mg/kg
dose of morphine sulfate. Both treated and untreated rabbits bled 1 week after the withdrawal
of the treated animals from morphine. Ordinate
= maximum increase in reaction time after injection of test dose to mice; abscissa = time of
injection of test dose after serum injection; T =
mice injected with serum from morphine-tolerant
rabbits; and U = mice injected with serum from
control rabbits. Shown are mean values ± SEM.
(From Cochin and Kornetsky.[38])

the injection of the serum into the mice and the analgesic assay. Significant
differences were found at 24 and 48 hr and at 1 week after passive transfer
of rabbit serum ($P < 0.005$). Serum also was collected from rabbits 24 hr
after termination of drug administration and injected in mice that were
subsequently tested at the intervals mentioned above following serum
transfer. Only those mice in which the transfer had been made 48 hr prior
to testing were significantly different from controls.

Because the results did not confirm the findings previously reported in
the dog,[40,41] a second series of investigations was undertaken using the rat
as the donor animal and the mouse as the recipient in order to confirm the
presence of an *attenuating* factor in a species other than the rabbit. Rats
were given 50 mg/kg of morphine twice daily for 3 weeks and then withdrawn
24 hr after abrupt termination of drug administration. Serum from these
animals and from control rats was then injected into groups of mice, and
these animals were tested 24 and 48 hr and 1 week later for their hot-plate
response to a test dose of morphine. Figure 2 shows the results of the experiments using serum taken from rats 24 hr after termination of drug administration. Only in the groups tested 48 hr after the serum injection were
there significant differences between mice receiving serum from morphine-
treated and control rats, and the differences were in the direction of attenuation of drug effect—results quite comparable to those observed in the
rabbit. Serum prepared from rats bled 1 week post withdrawal also at-

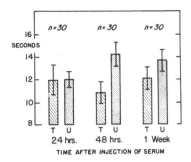

Fig. 2. The effect of serum obtained from morphine-tolerant and nontolerant donor rats on the hot-plate response of mice to a 10 mg/kg dose of morphine sulfate. Both treated and untreated rats bled 24 hr after the withdrawal of the treated animals from morphine. Ordinate = maximum increase in reaction time after injection of test dose to mice; abscissa = time of injection of test dose after serum injection; T = mice injected with serum from morphine-tolerant rats; and U = mice injected with serum from control rats. Shown are mean values ±SEM. (From Cochin and Kornetsky.[38])

tenuated the analgesic response of recipient mice to morphine, but there were significant differences between these mice and mice receiving serum from control rats at 24 as well as 48 hr after the injection of the serum. The results indicated that both the time of serum collection post withdrawal and the interval between serum injection and analgesic assay were important variables.

A comprehensive investigation was then undertaken in order to determine the effect of various conditions such as length of time of drug administration, the time post withdrawal when serum collections were made, the interval between injection of serum and testing of recipient animals, the size of the test dose, and so on. Despite a large number of experiments, further evidence that is statistically significant of a potentiating factor in the dog or an attenuating factor in the rabbit and rat has not been forthcoming. In all species all the results showed the same trends as previously reported—potentiation in dog and man and attenuation in rabbits and rats. Although the possibility exists that these results were artifacts (chance results due to errors of procedure, design, interpretation of results, or a combination of or all some of these), it is also possible that the findings are real and that the factor or factors described by Kornetsky, Kiplinger, Cochin, and others are extremely unstable and that this instability explains the difficulty that various investigators have had in duplicating previous findings.

B. Tissue Extracts

Ungar and Cohen[43] attempted to demonstrate in a somewhat different way that tissues of tolerant animals contain substances that can cause qualitative or quantitative alterations in the properties of the receptors subserving the analgesic narcotics. Extracts were prepared from brains of rats and dogs subjected to progressively increasing doses of morphine for a period of 2 weeks and then sacrificed and the brains removed, homogenized, treated in various ways, and injected intraperitoneally in mice. Brain extracts from untreated rats and dogs also were injected into mice. The recipient mice were then tested by a tail-pinch analgesic assay method. The results showed that mice that had received brain homogenates from tolerant rats were resistant to the effects of a test dose of 6 mg/kg of morphine sulfate—that is, the 6 mg/kg dose did not abolish the tail-pinch response in 64% of the animals, whereas the same dose in animals receiving brain homogenates from control rats abolished the response in all but 21–31% of the mice. Injections of homogenates of other tissues and serum produced no significant alteration of the morphine effect. Injection of a 26,000g supernatant had the same effect as the whole homogenate. Dialysis of the supernatant and injection of the dialysate also had the same effect. Comparison of rat brain and dog brain dialysates showed that rat brain was about four times more potent than dog brain. A single injection of rat brain preparation produced tolerance lasting at least 10 days. The injection of dialysate also conferred a degree of cross-tolerance to meperidine but not to pentobarbital. Ungar and Cohen speculated that the factor responsible for the induction of morphine tolerance is a peptide, soluble in water and insoluble in ethanol or acetone. The mechanism by which this factor induces tolerance was discussed by these workers, and they felt that the "most likely explanation lies in the area of the receptors which could be destroyed by the transfer factor or whose formation could be inhibited." Another speculation involved the transfer by the dialysate of something analogous to memory that might contain a chemical code.

Several years later, Huidobro and Miranda[44] also published a paper in which they described the presence in tolerant mice of a substance that modifies morphine response. They described a principle that was obtained from extracts of whole mouse and from urine of mice implanted with morphine pellets that shortens and reduces the intensity of abstinence in white mice. Additionally, they found a substance that prolonged morphine analgesia in mice implanted with morphine pellets. Extracts were prepared by a long and elaborate procedure from both whole mice and mouse urine. Both extracts were injected intraperitoneally into groups of recipient mice, and the recipients were tested for their analgesic response on the hot plate or for the intensity and duration of the abstinence syndrome. The results showed that the injection of whole-mouse or urine extracts from implanted animals did not modify the behavior of tolerant mice but that the abstinence syndrome elicited in these tolerant animals was always shorter and less intense than that in the untreated control animals. The analgesic effect of morphine was prolonged when nonimplanted test animals were treated with extracts of implanted mice. Huidobro and Miranda also attempted to characterize the active extract and were even able to find spots on thin-layer chromatographic plates that were not present in the extracts of nonimplanted animals. The complexity of the extraction procedure makes any attempt at characterization of the nature of the material extracted hazardous. The thin-layer chromatographic evidence is also difficult to interpret. The pellet itself might account for the additional spots if the implant contained binder of any sort.

It is interesting to note that Ungar and Cohen[43] report attenuation of the morphine effect and Huidobro and Miranda report potentiation of the effect of morphine. The former were unable to find any effect of any tissue except brain, whereas the latter could get the effects with extracts from whole animals and from urine. These differences may be due to differences in the species used or in the techniques of factor preparation or transfer or assay.

That differences in techniques of factor preparation, transfer, and assay are important is borne out by the work of Tirri[45] and of Smits and Takemori.[46] Tirri used brain homogenates from individual animals rather than the pooled homogenates used by Ungar, and, as his measures of tolerance, analgesic effect as measured by the tail-pinch, and the effect of morphine on temperature. He was unable to demonstrate any differences in the reactions of mice receiving brain homogenates from tolerant or nontolerant rats. Smits[46] modified Ungar's method and used heat as a noxious stimulus. He pooled the homogenates and injected them intraperitoneally and, using a random testing procedure in which the experimenter doing the assay did not know the source of the assay, could not confirm Ungar's work. Ungar[47] answered both Tirri's and Smits' criticisms in a paper describing experiments in which he repeated his original protocol and obtained results very much like those he originally reported in his earlier report.[43] He found that the potency of his dialyzed supernatant of rat brain homogenate varied markedly from one preparation to another and that if one measures quantitative changes in response rather than quantal all-or-none responses, then one cannot show differences. He was not able to demonstrate any alteration in drug response using a graded response of any kind. However, the LD_{50} for morphine was shifted significantly to the right in animals receiving the extract. Ungar stressed the importance of following the original protocol exactly and affirmed his belief that any and all attempts by others to simplify the procedure or to change the experimental design in any way foreclosed in advance any possibility of replicating his observations.

At present the matter seems to rest there. No laboratories have been able to replicate the findings of other laboratories insofar as transfer of tissue homogenate or extract is concerned and, as a matter of fact, some laboratories have found it difficult to replicate their own findings under quite similar conditions. Despite this paucity of confirmed observations, the author feels that the possibility of an attenuating or potentiating factor being present in serum or tissues of chronically morphinized animals must be further explored. Much of the work in our laboratories and in Kornetsky's laboratory has been directed toward this end, and the studies that are described below have an important bearing on elucidating possible mechanisms for the phenomenon of tolerance.[48]

IV. THE ROLE OF INTERVAL AND DOSE IN TOLERANCE DEVELOPMENT

If tolerance were truly an immune-like phenomenon, then one might expect to find that there are certain time-related as well as dose-related features in its development. Kornetsky and Bain[49] measured the performance of rats on a voltage-attenuation apparatus after an initial injection of 10 mg/kg morphine sulfate. At various intervals after that initial dose, groups of rats were given a second injection, this time 5 mg/kg morphine sulfate, and tested on the apparatus. Their performance was compared to a control group of animals that was tested on the same experimental days but had not received the initial priming dose. Kornetsky found that even by day 3, and certainly by days 15 and 31, the effect of a second injection of morphine was significantly diminished. This in itself was expected, but what was more interesting was that the degree of tolerance (calculated as the percent of control) was much greater if an interval of 1 month or 6 months intervened between the initial and the second morphine injections. On the day after the initial dose, a second injection had no significant attenuating effect. These findings agree with those observed in mice by Eddy[36] and demonstrated that tolerance is a phenomenon that needs time to develop fully.

In accord with these results is an experiment in our laboratory[37] that demonstrated that the interval between doses is an important factor in the development of tolerance. Rats were given 15 mg/kg injections of morphine sulfate and tested for their response to the drug on the hot plate. They were then divided into three groups: the first was injected with the same dose and tested at weekly intervals; the second group was injected and tested at 14-day intervals; and the third group was injected and tested every 21 days. The results of a typical experiment are shown in Fig. 3. The response of animals receiving morphine at weekly intervals is not significantly different from their initial response until the third injection, whereas the animals receiving morphine at more widely spaced intervals have a significantly attenuated response after the second injection. There are highly significant differences in response between the 1-week and 2- and 3-week groups until the fifth injection of morphine. This experiment again demonstrates the time-related course of tolerance and also shows the importance of the interval between injections as well as the number of injections.

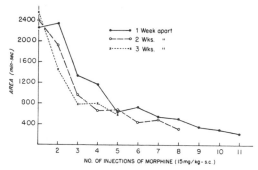

Fig. 3. The effect of dose interval on the development of tolerance in the rat. Rats were given 15 mg/kg morphine sulfate subcutaneously at the designated intervals. Ordinate = area under the time-response curve obtained by the hot-plate assay and abscissa = number of injections of drug. (Adapted from data presented in Cochin and Mushlin.[37])

Another facet of tolerance that needs to be elucidated has to do with the relationship of dose to the intensity and duration of the phenomenon—this needs badly to be known in order to ascertain whether there is a threshold dose that produces measurable drug effect but does not induce tolerance. Miller and Cochin have recently studied the onset, intensity, and persistence of tolerance in mice after several dose levels, and they showed that 6 mg/kg daily did not induce any significant degree of tolerance until day 25 of drug administration despite the fact that this dose did induce a measurable drug effect and that an 8 mg/kg dose induced significant tolerance by the fourth injection.[50] Retesting with the dose given previously after a 37-day drug-free period showed that the response of the group of mice that had received 6 mg/kg returned to the initial day 1 level, whereas the 8 mg/kg group's response was still significantly attenuated, with a mean area of approximately 58% of the inital area. It now seems logical to look for a still lower dose that will be a real threshold dose and induce no measurable tolerance despite measurable drug effect. If such a dose were to be found it would have important theoretical implications.

V. THE EFFECTS OF IMMUNOSUPRESSORS ON TOLERANCE

It has also been of interest to us to look for the effects of immunosuppressive agents and procedures and their effect on tolerance. Friedler[51] studied effects of neonatal thymectomy on the rate and development of tolerance in the rat and could find none. In conjunction with these studies we decided to look for a factor affecting drug action that might be transferred from mother to young across the placental barrier or via the milk. Female rats were given morphine for several weeks and then withdrawn from the drug for 5–6 days and then mated to drug-free males. The sensitivity to drug of the offspring from the previously addicted mothers and their growth pattern were then compared to the drug sensitivity and growth pattern of offspring of nondrug-treated mothers. It was found that there were significant differences in the weights of the offspring of previously morphinized and withdrawn mothers as compared to controls,[52] despite the fact that neither group was exposed to drug either *in utero* or postpartum. The greatest differences were found in the females at 5–6 weeks of age. It was also found that the offspring of the previously treated mothers were significantly less sensitive to the drug than the offspring from untreated mothers.[53]

Although it is not within the province of this chapter to discuss the relationship of tolerance to protein synthesis and the effect of suppressors of protein synthesis on tolerance, the effects of immunosuppressive agents should be mentioned, if only in passing. Most of the investigators using agents such as dactinomycin and cycloheximide have been interested in their effect on protein synthesis per se, but some of these agents are also potent immunosuppressive agents blocking the synthesis of specific proteins and their effect on tolerance may well be through this type of mechanism. Cohen *et al.*[54] reported blocking the development of tolerance with actinomycin D. Cox *et al.*[55,56] reported that dactinomycin blocks the development of acute tolerance to morphine, and it appears unlikely that inhibition of protein synthesis is the mechanism to account for the effect on tolerance. Smith *et al.*[57] found, however, that neither puromycin nor actinomycin D interfered with the short-term tolerance to the lenticular effect of the narcotics but that these agents did prevent the development of long-term tolerance to this effect. Way[58] found that cycloheximide, a potent inhibitor of protein synthesis, profoundly affected the course of development of physical dependence and tolerance in mice. Feinberg and Cochin[59] have also studied the effects of cycloheximide on the development of tolerance in rats under quite different conditions than those in Way's studies. Rats were given either single injections of 10 mg/kg of morphine alone a week apart for a number of weeks or injections of similar amounts of morphine following an injection of 1 mg/kg of cycloheximide given 1 hour previously for the same number of weeks. The results showed that the development of tolerance was blocked by the cycloheximide pretreatment. The rationale for the experimental design in this study was that if morphine does stimulate the formation of antibody-like substances or of a reaction analogous to memory, then cycloheximide should block the synthesis of proteins responsible for such reactions. The results clearly indicated that cycloheximide, a short-acting protein suppressor, did block whatever changes morphine initiates. It is also of interest that the results bear out the observations of Cochin and Mushlin[37] that weekly injections of morphine alone do not induce a significant degree of tolerance, at least for the first few injections.

VI. SUMMARY

There are many aspects of tolerance to the narcotic analgesics that resemble immune reactions. Among these are the remarkable persistence of tolerance for long periods of time after termination of drug administration, the fact that a finite period of time is needed to develop tolerance and that the tolerance induced by a single dose is greater after 6 months than after 3 days, for example, and the fact that there may well be a threshold below which tolerance cannot be detected although the drug effect is clearly discernible. Attempts to transfer tolerance passively to nontolerant animals using serum, tissue homogenates, tissue extracts, or urine extracts from tolerant animals have met with equivocal success and have been difficult to confirm by other investigators or by the very investigators who reported the positive results in the first place. Despite these difficulties, experiments in a number of laboratories have demonstrated that the course of tolerance is affected by a number of immunosuppressive agents. Since such agents are also nonspecific inhibitors of protein synthesis, these observations are capable of being interpreted in many ways, depending on the bias of the interpreter. It is clear that our present state of knowledge does not allow us to draw any conclusions as to the relationship of tolerance and immune reactions.

VII. REFERENCES

1. I. H. Pierce and O. H. Plant, *J. Pharmacol. Exp. Ther. 46*, 201–228 (1932).
2. O. H. Plant and I. H. Pierce, *J. Pharmacol. Exp. Ther. 49*, 432–449 (1933).
3. V. Thompson and E. G. Gross, *J. Pharmacol. Exp. Ther. 72*, 138–145 (1941).
4. F. W. Oberst, *J. Pharmacol. Exp. Ther. 69*, 240–251 (1940).
5. J. Cochin, J. Haggart, and L. A. Woods, *J. Pharmacol. Exp. Ther. 111*, 74–83 (1954).
6. L. A. Woods, J. Cochin, E. J. Fornefeld, F. G. McMahon, and M. H. Seevers, *J. Pharmacol. Exp. Ther. 101*, 188–199 (1951).
7. L. B. Mellet and L. A. Woods, *J. Pharmacol. Exp. Ther. 125*, 97–104 (1959).
8. S. J. Mulé and C. W. Gorodetzky, *J. Pharmacol. Exp. Ther. 154*, 632–645 (1966).
9. S. J. Mulé, C. M. Redman, and J. W. Flesher, *J. Pharmacol. Exp. Ther. 157*, 459–471 (1967).
10. C. F. Schmidt and A. E. Livingston, *J. Pharmacol. Exp. Ther. 47*, 443–472 (1933).
11. J. Cochin and C. Kornetsky, *J. Pharmacol. Exp. Ther. 145*, 1–10 (1964).
12. W. Marmé, *Deut. Med. Wochenschr. 9*, 197–198 (1883).
13. M. H. Seevers, *Fed. Proc. 13*, 672–684 (1954).
14. A. L. Tatum, M. H. Seevers, and K. H. Collins, *J. Pharmacol. Exp. Ther. 36*, 447–475 (1929).
15. M. H. Seevers and G. A. Deneau, *Arch. Int. Pharmacodyn. Ther. 140*, 514–520 (1962).
16. A. Goldstein, L. Aronow, and S. M. Kalman, *in* "Principles of Drug Action," p. 597, Hoeber, New York (1968).
17. L. Shuster, R. V. Hannam, and W. E. Boyle, Jr., *J. Pharmacol. Exp. Ther. 140*, 149–154 (1963).
18. D. B. Goldstein and A. Goldstein, *Biochem. Pharmacol. 8*, 48 (1961).
19. L. Shuster, *Nature 189*, 314–315 (1961).
20. H. O. J. Collier, *Nature 205*, 181–182 (1965).
21. J. Axelrod, *Science 124*, 263–264 (1956).

22. J. Cochin and J. Axelrod, *J. Pharmacol. Exp. Ther. 125*, 105–110 (1959).
23. G. J. Mannering and A. E. Takemori, *J. Pharmacol. Exp. Ther. 127*, 187–190 (1959).
24. M. Vogt, *Amer. J. Physiol. 123*, 451–481 (1954).
25. E. W. Maynert and G. Klingman, *J. Pharmacol. Exp. Ther. 135*, 285–295 (1962).
26. L. M. Gunne, *Acta Physiol. Scand. 58*, Suppl. 204 (1963).
27. W. R. Martin, A. J. Eisenman, J. W. Sloan, D. R. Jasinski, and J. W. Brooks, *J. Psychiat. Res. 7*, 19–28 (1969).
28. T. Akera and T. M. Brody, *Biochem. Pharmacol., 17*, 675–688 (1968).
29. A. Smith, M. Karmin, and J. Gavitt, *J. Pharmacol. Exp. Ther. 156*, 85–91 (1967).
30. M. Weinstock, *Brit. J. Pharmacol. Chemother 17*, 433–441 (1961).
31. E. L. Way, H. H. Loh, and F. Shen, *Science 162*, 1290–1292 (1968).
32. E. L. Way, H. H. Loh, and F. Shen, *Biochem. Pharmacol. 18*, 2711–2721 (1969).
33. C. A. Winter, "Minutes of Twelfth meeting of the Committee on Drug Addiction and Narcotics," App. A, pp. 577–591, NAS-NRC, Washington, D.C. (1953).
34. A. F. Green, P. A. Young, and E. I. Godfrey, *Brit. J. Pharmacol. Chemother. 6*, 572–585 (1951).
35. H. F. Fraser and H. Isbell, *J. Pharmacol. Exp. Ther. 105*, 498–502 (1952).
36. N. B. Eddy, "Minutes of Twelfth meeting of the Committee on Drug Addiction and Narcotics," App. C, pp. 603–618, NAS-NRC, Washington, D.C. (1953).
37. J. Cochin and B. E. Mushlin, *Fed. Proc. 29*, 685 (1970).
38. J. Cochin and C. Kornetsky, *in* "The Addictive States" (A. Wikler, ed.), Proceedings of the Meeting of the Association for Research in Nervous and Mental Disease, Vol. XLVI, pp. 268–279, Williams & Wilkins, Baltimore (1968).
39. H. Krueger, N. B. Eddy, and M. Sumwalt, *Pub. Health Rep.* Part 1, Suppl. 165 (1941).
40. C. Kornetsky and G. F. Kiplinger, *Psychopharmacologia 4*, 66–71 (1963).
41. G. F. Kiplinger and J. W. Clift, *J. Pharmacol. Exp. Ther. 146*, 139–146 (1964).
42. N. B. Eddy and D. Leimbach, *J. Pharmacol. Exp. Ther. 107*, 385–393 (1953).
43. G. Ungar and M. Cohen, *Int. J. Neuropharmacol. 5*, 183–192 (1966).
44. F. Huidobro and H. Miranda, *Biochem. Pharmacol. 17*, 1099–1105 (1968).
45. R. Tirri, *Experientia 23*, 278 (1967).
46. S. E. Smits and A. E. Takemori, *Proc. Soc. Exp. Biol. Med. 127*, 1167–1171 (1968).
47. G. Ungar, *Proc. Soc. Exp. Biol. Med. 130*, 287–291 (1969).
48. J. Cochin, *Fed. Proc. 29*, 19–27 (1970).
49. C. Kornetsky and G. Bain, *Science 162*, 1011–1012 (1968).
50. J. M. Miller and J. Cochin, "Minutes of the Thirtieth Meeting of the Committee on Problems of Drug Dependence," App. 23, pp. 5462–5468, NAS-NRC, Washington, D.C. (1968).
51. G. Friedler, *Diss. Abstr. BV29⁵* (2558-B), (1968).
52. G. Friedler and J. Cochin, *Pharmacologist 9*, 230 (1967).
53. G. Friedler and J. Cochin, *Pharmacologist 10*, 188 (1968).
54. M. Cohen, A. S. Keats, W. Krivoy, and G. Ungar, *Proc. Soc. Exp. Biol. Med. 119*, 381–384 (1965).
55. B. M. Cox, M. Ginsburg, and O. H. Osman, *Brit. J. Pharmacol. Chemother. 33*, 245–256 (1968).
56. B. M. Cox and O. H. Osman, *Brit. J. Pharmacol. Chemother. 38*, 157–170 (1970).
57. A. Smith, M. Karmin, and J. Gavitt, *Biochem. Pharmacol. 15*, 1877–1879 (1966).
58. H. H. Loh, F. H. Shen, and E. L. Way, *Biochem. Pharmacol. 18*, 2711–2721 (1969).
59. M. Feinberg and J. Cochin, *Pharmacologist 11*, 256 (1969).

Chapter 22

SELF-ADMINISTRATION STUDIES IN ANIMALS

James R. Weeks

Experimental Biology Division
The Upjohn Company
Kalamazoo, Michigan

I. INTRODUCTION

Self-administration of drugs by experimental animals is a relatively new experimental approach to the laboratory study of drug abuse. There have been no published biochemical studies on experimental animals who have received drugs by self-administration. Studies have been directed toward refinement of the experimental procedure and evaluation of the responses of animals to different types of drugs. Thus the discussion that follows only outlines briefly the techniques available and points out areas in which biochemical studies could profitably be applied. Self-administration of and behavioral dependence on drugs in animals has been thoroughly reviewed recently.[1]

The great advantage of a self-administration study is that it permits evaluation of behavioral aspects of drug dependence. Indeed, one might consider drug abuse itself as an abnormal or excessive behavioral response in relation to a drug. No one denies that the powerful compulsion to repeat drug administration, whatever its psychological or fundamental biochemical basis, is a major factor in human drug abuse.

II. TECHNIQUES

Self-administration studies of opiates have been limited to intravenous and oral administration to rats and monkeys. For oral administration, drugs are added to the drinking water. Drug effects can then be evaluated by comparing fluid consumption from two or more bottles or by lever-pressing for drug treated water in a variety of situations. Generally animals must be trained by water deprivation to drink the drug solutions before dependence is established.[2] The advantage of oral administration is its simplicity. Large numbers of animals may be studied with a minimum of equipment. On the other hand, oral administration is tied to palatability, thirst, and even the relative position of drinking tubes. In addition, there is a variable delay between ingestion and absorption, a factor that is important in reinforcement and could be of potential importance for correlating dynamic biochemical processes with self-administration of drugs. These objections do not apply when drugs are administered intravenously. This problem area is discussed in detail by Schuster and Thompson[1] and by Myers.[3]

For intravenous self-administration animals are prepared with indwelling venous catheters. The catheter is connected to a motor-driven infusion pump (syringe driver), which the animal can operate by pressing a lever. Animals must wear a suitable saddle or vest but are only slightly restrained by this connection to the pump. Once addiction in such animals is established, regular injections are taken, at a reasonably constant rate, over prolonged periods (up to 3 months in rats and over a year in monkeys). Abrupt withdrawal, or a nalorphine challenge, results in a withdrawal syndrome and increased lever-pressing activity. In addition to the review by Schuster and Thompson[1] mentioned above, studies on opiates in rats have been summarized by Weeks and Collins[4] and in monkeys by Deneau et al.[5] It is of especial interest that morphine-addicted rats will relapse promptly to morphine self-administration after being withdrawn and allowed to recover.[4,6,7]

III. SUGGESTED BIOCHEMICAL STUDIES

In a self-maintained addiction, the total drug intake and dose schedule is more or less balanced to meet the demands of psychic and physical dependence. An overt withdrawal syndrome never develops. Neither are there obvious signs of morphine action following injection, presumably because of the high degree of tolerance that accompanies physical dependence. Other chapters in this book have dealt with the biochemical consequences of chronic morphine administration. Of necessity, arbitrary doses and schedules have been used in these chronic studies. It is very possible that if the doses selected had been too small or given too widely spaced, repeated bouts of early withdrawal might have resulted. Rats start morphine-seeking behavior (lever pressing) before overt abstinence signs are evident.[8] Conversely, too large doses might induce pharmacological changes not consistent with maintenance of a balanced physiological state. Self-administration of the drug would avoid these possible complications.

In the time between morphine injections there is a continuum from full satiation to incipient withdrawal, which culminates in the seeking of another dose. Biochemical studies on the state when this "drive" becomes effective would be possible only by using self-injection techniques. Likewise, the study of satiation without complications of overdosage, and possibly irrelevant actions of the drug, could be managed by requiring the animal to self-administer several small doses.

Physical dependence is certainly a major contributing factor to the maintenance of opiate addiction. The functional and biochemical changes causing this condition are still not understood. Biochemical changes that parallel the development of tolerance and physical dependance merit careful investigation as possible causes. To be fully relevant, changes in the parameter under investigation should also correlate with behavioral aspects (that is, drug seeking) of addiction and withdrawal. Self-administration techniques could be used to establish such correlation. It should be pointed out that physical dependence and a withdrawal syndrome may occur independent of drug-seeking behavior[9] and that the withdrawal signs may become conditioned.[7,10]

Finally, there is the problem of relapse in postaddicts. Wikler[11,12] has suggested that conditioning to withdrawal may play an important role in relapse. Physiological or biochemical changes induced by morphine also may be factors. For example, rats that have received repeated or even single doses of morphine may not return completely to their pre-injection state for several weeks or months.[13-15] Postaddict rats, which had previously maintained their addiction intravenously, relapsed promptly to regular self-injection of morphine. However, when saline was substituted for morphine solution, there followed a period of sustained frequent lever pressing very similar to that observed following substitution of saline for morphine in an actively addicted rat.[4] Such "postaddict" animals would permit assessment of any indirectly mediated biochemical changes attending relapse. Indeed, this technique might provide a method for demonstrating "psychosomatic" changes in an experimental animal.

IV. REFERENCES

1. C. R. Schuster and T. Thompson, *Ann. Rev. Pharmacol. 9*, 483–502 (1969).
2. J. R. Nichols, *Res. Publ. Ass. Res. Nerv. Ment. Dis. 46*, 299–305 (1968).
3. R. Myers, *Psychosom. Med. 28*, 484–497 (1966).
4. J. R. Weeks and R. J. Collins, *Res. Publ. Ass. Res. Nerv. Ment. Dis. 46*, 288–298 (1968).
5. G. A. Deneau, T. Yanagita, and M. H. Seevers, *Psychopharmacologia 16*, 30–48 (1969).
6. J. R. Nichols, *Psychol. Rep. 13*, 895–904 (1963).
7. A. Wikler and F. T. Pescor, *Psychopharmacologia 10*, 255–284 (1967).
8. N. Khazan, J. R. Weeks, and L. A. Schroeder, *J. Pharmacol. Exp. Ther. 155*, 521–531 (1967).
9. W. R. Martin and C. W. Gorodetzky, *J. Pharmacol. Exp. Ther. 150*, 437–442 (1965).
10. S. R. Goldberg and C. R. Schuster, *J. Exp. Anal. Behav. 10*, 235–242 (1967).
11. A. Wikler, *Brit. J. Addict. 57*, 73–79 (1961).
12. A. Wikler, *Res. Publ. Ass. Res. Nerv. Ment. Dis. 46*, 280–287 (1968).
13. W. R. Martin, A. Wikler, C. G. Eades, and F. T. Pescor, *Psychopharmacologia 4*, 247–260 (1963).
14. J. Cochin and C. Kornetsky, *J. Pharmacol. Exp. Ther. 145*, 1–10 (1964).
15. C. Kornetsky and G. Bain, *Science 162*, 1011–1012 (1968).

Chapter 23

OPIATES AND ANTAGONISTS*

Max Fink, Arthur Zaks,
Jan Volavka, and Jiri Roubicek

Division of Biological Psychiatry
Department of Psychiatry
New York Medical College and Metropolitan Hospital Medical Center
New York, New York

> From the first beginning of our knowledge of man, we find him con-
> suming substances of no nutritive value, but taken for the sole purpose
> of producing for a certain time a feeling of contentment, ease, and
> comfort. . . . These substances have formed a bond of union
> between men of opposite hemispheres, the civilized and the un-
> civilized. . . .
>
>   The strongest inducement to a frequent or daily use . . . is to
> be found in the properties they possess; in their capacity to excite the
> functions of the brain-centres which transmit agreeable sensations
> and to maintain for some time the consciousness of experienced emo-
> tions.†

I. INTRODUCTION

In no period of recent world history—surely in no period of American
history—has drug use been so widespread. Tobacco, alcohol, and caffeine
are legally "accepted," and their use is matched by marijuana (cannabis),
barbiturates, "minor tranquillizers," and the opiates. It is the latter drugs
that excite a nationwide anxiety. From a know-nothing view that those who
develop drug dependence are morally weak and deserve punishment, the
nation has gradually begun to pay lip service to a view of the addicted as
medically ill, deserving of treatment. Studies of the opiates, the phenome-
noma of cross tolerance, and of narcotic antagonists in the treatment of the
addicted are now fashionable.

   Except for some direct actions on smooth muscle, the known physio-

* Aided, in part, by NIMH-12567, 13003, and 13358 and the New York State Narcotic
  Addiction Control Commission.
† From L. Lewin, "Phantasies, Narcotic and Stimulating Drugs," Dutton, New York
  (1964).

logical effects of opiates are related almost exclusively to the central nervous system.[1] The prevailing theories of the site of action of opiates and the development of dependence implicate cellular sites of the central nervous system.[2,3] The study of these effects in intact man is difficult; yet one useful way has been the scalp-recorded EEG. The recent development of measurement techniques using digital computers has improved EEG interpretation, facilitating quantitative comparisons and scientific studies.[4,5]

II. PREVIOUS STUDIES

Andrews[6] first reported brain potentials during addiction cycles in "stabilized" addicts, postaddicts, and volunteer civilian personnel. In 50 actively addicted subjects, the EEG was characterized by high-alpha-abundant records with no records with alpha percentages below 30% and most subjects in the 80–100% groups. In postaddicts, low-alpha-abundant records were characteristic. Following drug withdrawal, high alpha indices were maintained in some, but more than half showed a decrease—low levels being maintained for several months.

In another study seven patients were stabilized on morphine, substituted on codeine, and then withdrawn from both drugs. Stabilization on morphine was associated with high alpha activity in occipital leads. Codeine administration also produced increases in alpha abundance, but the levels were lower than morphine.[7] Greater activity from the right hemisphere, both with monopolar and bipolar leads, was found with codeine.

Two postaddicts were described in detail.[8] Regular administration of morphine in doses beyond tolerance level in a subject with a high alpha index slowed the alpha rhythm, with blockade to noxious stimuli. Keeping the dosage constant brought a quick return to preaddiction levels. In another patient with a low alpha index, repeated doses of morphine elicited a high-frequency alpha rhythm, and large doses resulted in a "normal" alpha frequency. Tolerance developed at different rates in these two cases.

Wikler provides the most extensive EEG data in both animals and man.[1,9−12] Wikler and Altschul[12] reported that small doses of methadone or morphine produced a mixture of fast and high-voltage slow activity in dogs. Seizure discharges resulted with large doses of either drug, appearing synchronously in cortical and basal tracings with both spike and wave and sustained spike patterns. Convulsions appeared sooner after subcutaneous injections of methadone than after morphine, and spike and wave activity resembling petit mal was more noticeable after large doses of morphine than after methadone.

In a review, Wikler[9] cites that single subcutaneous doses of morphine (16 mg) in man increased both theta and slow alpha activity, total voltages, and the voltage of beta activity. Single doses of methadone up to 30 mg did not change the EEG, although slow activity appeared in one record; repeated administrations resulted in progressive slowing, with continuous delta activity finally becoming dominant. The repeated administration of ketobemidone elicited progressive slowing with a concomitant increase in low-amplitude fast waves, paroxysmal high-voltage delta activity, and occasional spike and wave activity.

In rats up to 20 mg/kg of morphine produced no EEG changes; 20–50 mg/kg increased voltages and decreased frequencies, and later bursts of 10–12 cps spikelike waves occurred. Curarized or lightly anesthetized cats injected with methadone in increasing doses progressively showed slowing, bursts of diphasic spikes, and then abolition of cortical electrical activity. A single large dose of methadone in curarized or anesthetized cats produced high-voltage spike seizure discharges.

In a later study Wikler[10] reported that 5–150 mg/kg of morphine produced EEG patterns similar to "burst-slow wave" patterns of natural sleep in 12 unanesthetized, uncurarized dogs.

In a detailed study Wikler[11] described the clinical and electrographic effects of morphine and *N*-allylnormorphine (nalline) in postaddicts. Morphine was administered intravenously in ten subjects using a standard 30-mg dose. The clinical effects of morphine were the typical flushing of the skin, a pleasurable "thrill," tingling centered in the "stomach," and itching. Six records were obtained during the "thrill," and in five no change was observed while in one the pattern became more synchronized. In 15 EEG samples taken immediately after the "thrill," the record became synchronized in nine, desynchronized in three, and unchanged in three.

In ten subjects receiving nalline (15–75 mg subcutaneously) the principal effects were anxiety, a dreamy state with vivid fantasies, miosis, pseudo-ptosis, sweating, and diuresis. In ten EEG samples recorded while the subjects reported fantasies or daydreaming, a change in the direction of desynchronization occurred in six, synchronization in two, slowing of alpha by 1 cps in one and unaltered in one.

In similar studies in patients receiving therapeutic doses of morphine for pain, Roubicek[13] reported decreased amplitudes with periods of high-amplitude alpha activity. The frequency of alpha became slower by 1–2 cps, and the changes were described as similar to the first stage of sleep. In chronic morphine addicts with many years of abuse, the alpha activity was usually slower and there was weak and irregular reaction to eye opening. There were frequent beta waves dispersed over the surface of the head with irregular, dispersed theta waves 4–7 cps of low amplitude.

The EEG patterns of the opiate antagonists were reviewed by Martin[14] indicating that these agents were distinguishable from the opiates. Leval-lorphan was shown to antagonize the EEG effects of morphine, and Gold-stein and Aldunate[15] found nalorphine to antagonize the increased integrated amplitudes of morphine, both in the rabbit.

The study by Wikler[11] was less significant for the description of the EEG changes of morphine and nalline than for the behavioral–EEG cor-relations suggested. Wikler observed the following for the three drugs studied (mescaline, morphine, and nalline):

In general, and regardless of the nature of the drug administered, shifts in the pattern of the electroencephalogram in the direction of desynchronization occurred in association with anxiety, hallucinations, fantasies, illusions or tremors, and in the direction of synchronization in association with euphoria, relaxation or drowsiness. However, the converse was not true, since such experiences often occurred without any apparent change in electroencephalographic pattern.

This view, of a behavioral–EEG association, has been the basis for our extensive studies of the EEG effects of psychoactive drugs.[16–18] It has been possible to relate the induced EEG changes to the clinical applications of various drugs and to the identification of new psychoactive drugs. It was within this framework that EEG studies were undertaken as part of our studies of the treatment of opiate dependence. We wished to classify the EEG patterns of opiates and their antagonists within an EEG–behavior association hypothesis, to examine the processes of antagonism and tolerance, and to use the EEG as an objective index of the duration or degree of opiate activity. The studies are ongoing, and we are summarizing the available data, including reports previously published or in press.

III. SUBJECTS AND METHODS

The subjects are male narcotic addicts, voluntarily admitted to the Metropolitan Hospital Mental Health Center in New York. The median age of the subjects is 26 years, ranging from 17 to 54 years. Subjects are first detoxified with decreasing doses of methadone and then maintained drug-free for at least 1 week before participating in experimental studies. Each subject has at least one EEG examination before the first experimental session. Experiments are carried out in air-conditioned but not sound-proofed EEG laboratories. Each study consists of 15–30 min of resting EEG, followed by an intravenous injection of an experimental agent, usually in distilled water, within 2 min.

Subjects are interviewed before the injection, at the end of the experimental session, and again 24 hr after the session. With the exception of interviewing time (approximately 10 min), the EEG is recorded throughout the experimental session. The subjects are supine, with their eyes closed, and at irregular intervals are asked to open their eyes for 20–40 sec. Grass VII (eight-channel) electroencephalographs and needle electrodes are used. The EEG and EKG are recorded on paper, and two EEG derivations are also FM tape-recorded for digital computer analysis.[4] While different electrode combinations have been recorded in different studies, the F4–OZ and O1–Cz derivations are analyzed most often. We have noted differences in the degree to which different electrode pairs respond to the drugs given, but the patterns are more similar in different parts of the scalp and the similarities are emphasized here.

Subjects have received diacetylmorphine (heroin; 15–25 mg/2 ml/2 min), cyclazocine (0.5–3.0 mg/2 ml/2 min), and naloxone (0.5–2.0 mg/2 ml/2 min). Where the intravenous route was unavailable, each drug has given subcutaneously. The subjects have been treated by oral methadone (20–150 mg/day), cyclazocine (3.0–5.5 mg/day), naloxone (0.2–3.0 g/day), and tybamate (4.0–12.0 g/day) for periods of a few weeks to 4 years. When subjects have been on chronic oral medication, acute administration of heroin (25–75 mg/2 ml/2 min, "heroin challenges") have been given to assess the duration of antagonism and cross tolerance.[19]

IV. OBSERVATIONS

A. Heroin

In a recent summary Volavka *et al.*[20] reviewed the data in 63 subjects, including data published earlier by Fink *et al.*[21] and Zaks *et al.*[19]

The initial (resting) EEG was found to be regular, usually with dominant alpha activity and free of visually appreciable slow waves in 34 subjects.

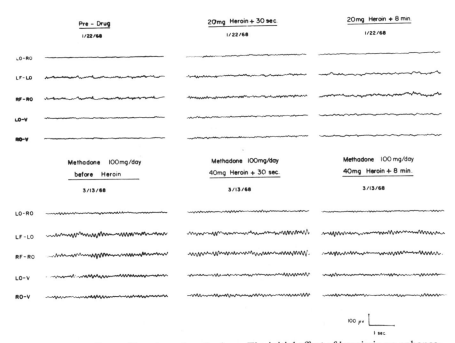

Fig. 1. EEG effects of heroin and methadone. The initial effect of heroin is an enhancement of alpha activity, followed by a decrease in the amount of alpha and an increase in slow waves. Chronic methadone elicits enhanced alpha activity and increased amplitudes. The effects of injected methadone at twice the predrug dose is blocked.

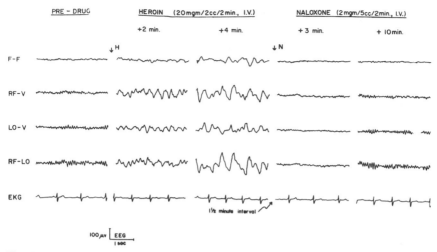

Fig. 2. Heroin-induced seizures blocked by naloxone. An alpha predrug record, replaced by high-voltage theta activity within 2 min after 20 mg of heroin, is followed by high-voltage EEG seizure activity accompanied by a clinical motor seizure. Intravenous naloxone blocked the seizure and reestablished the preheroin record.

Low-voltage irregular recordings displaying a mixture of beta and theta activity were seen in 16 subjects, and 8 subjects showed occasional theta activity in their waking records. Five subjects had a large amount of slow activity and/or paroxysmal activity in their tracings.

The EEG response to heroin usually occurred in two stages; The *early response*, during the injection or within 2 min after its termination, was characterized by an increase in alpha amplitudes, a decrease in mean alpha frequency and, occasionally, pronounced alpha spindling (Fig. 1). The increase in alpha *amplitude* occurred in 42 subjects (67%) and was more prominent in the subjects with intial low-voltage, irregular EEG activity. A decrease in mean alpha *frequency* was observed in 33 subjects (52%).

The *late response* started usually 3–5 min after the end of the injection of heroin and was characterized by a decrease in alpha amplitudes, fragmentation of alpha rhythms, and an increase in theta abundance (Figs. 1, 2, 4). Alpha activity reappeared in short runs, but eventually it was replaced by slower activities. Theta activity was prominent in 61 of 63 subjects (97%), and these gradually increased in amplitude and decreased in frequency, followed in 25 subjects (40%) by the appearance of delta activity.

Paroxysmal activity occurred in 17 subjects. Two subjects developed clinical attacks, one with a clonic seizure pattern (Zaks *et al.*).[19] Both had normal preheroin recordings and were without prior seizure history (Fig. 2). In one the recording immediately before and during the clonic seizure was dominated by high-voltage delta waves and sharp waves. The remaining 15 subjects showed paroxysmal activity in the EEG only, which developed from a background of slower (theta and delta) activity (ten cases) or from an alpha background (seven cases). (Two subjects developed paroxysmal activity on both predrug backgrounds.)

The occurrence of a paroxysmal response was not dependent on the dose of heroin but was related to the type of the predrug EEG with the paroxysmal response developing less frequently in EEG records showing low-voltage irregular patterns than in any other EEG type.

B. Methadone

In 12 subjects who received chronic methadone to 100 mg/day in whom we have 5–7 records each, we have observed a progressive decrease

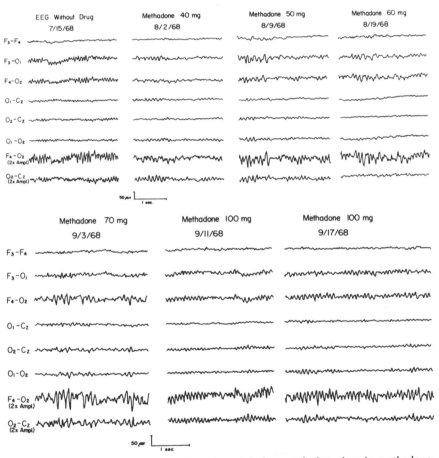

Fig. 3. Chronic oral methadone. The sequential changes during chronic methadone administration include slowing and decreased voltages, similar to those with heroin. During treatment, slow wave activity is prominent and is associated with drowsiness. By the time 100 mg is established, the pretreatment alpha-dominant record recurs.

in mean alpha frequency and an increase in percent time alpha with increasing doses of methadone (Roubicek *et al.*).[22] Beta activity decreases, and theta activity, both random and in bursts, increases in amount. Voltages increase, and the variability of frequencies and amplitudes decrease (Figs. 3 and 5). In eight patients these changes, which were seen after the early doses, vanished in later recordings as the EEG returned to the pre-methadone patterns. Of eight EEG-adapted subjects, two had slightly abnormal and six had normal pre-methadone EEG records. Of the four patients without EEG adaptation, one had an abnormal and the other three had normal pre-methadone EEG records.

Following the acute administration of oral methadone (20–40 mg), the EEG in postaddicts free of other drugs exhibited a slowing of the dominant alpha activity by 2–3 cps. After 4–7 min there are widely dispersed theta waves (4–7 cps), occasionally with much higher amplitudes. These changes are occasionally followed by periods of desynchronized activity with dissolution of alpha, flattening of the graph mixed with long runs of higher and hypersynchronized alpha and theta waves.

In patients receiving 100 mg of methadone daily, we have given varying doses of heroin (25, 50, and 75 mg) at different periods after their daily oral methadone (4–72 hr) to assess the duration of heroin "blockade." When EEG changes were produced by heroin in these methadone-tolerant subjects, the changes are increased theta frequencies (4–7 cps) dispersed or in runs of several seconds duration (Figs. 1 and 5). Sometimes delta activity was accompanied by short or long periods of desynchronization with flattening of the record and alpha fragmentation. Administration of placebo did not produce these effects.

In clinical assessments in subjects receiving increasing single daily doses of methadone to 100 mg during a 2-month induction period, blockade to heroin was complete for 25-mg heroin to 60 hr, for 50 mg heroin to 48 hr, and for 75 mg to 24 hr. (Withdrawal symptoms first became evident 50–60 hr after a 100-mg dose of methadone had been replaced by placebo; Zaks *et al.*[23]) While there was agreement between the verbal reports of subjects (that is a "high" or "no effect") and the EEG effects, there were disagreements in the reports in one-third of the examinations—particularly at the higher heroin dosages (Roubicek *et al.*[22]).

C. Naloxone

Naloxone (*N*-allylnoroxymorphone) is a narcotic antagonist that effectively blocks the clinical effects of heroin.[21,24,26] In our initial studies we were unable to elicit any EEG effects of intravenous naloxone in "clean" subjects.[21] On chronic oral administration of 0.2–3.0 g/day we have also not elicited any EEG patterns. (Indeed, except for one subject at 2.4g, we have noted no behavioral effects of naloxone.)

In subjects who receive intravenous heroin (15–25 mg), intravenous naloxone given after the end of the postheroin interview results in three patterns. Immediately after the naloxone injection, the EEG is dominated

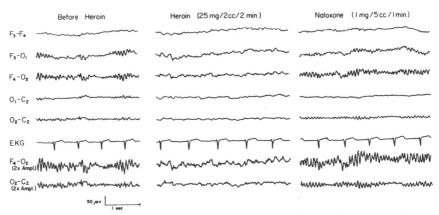

Fig. 4. Heroin and naloxone. Another example of the effects of heroin and the re-establishment of the pretreatment record by naloxone.

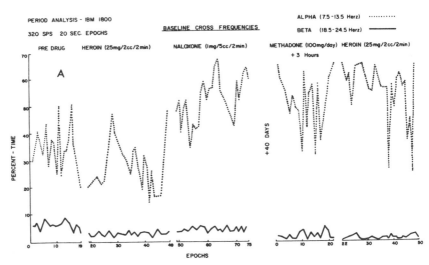

Fig. 5A. Heroin–naloxone: computer outputs. The three parts of Fig. 5 are graphic examples of the quantitative EEG changes with heroin in a drug-free subject, blockade by naloxone, changes on chronic methadone administration, and the effects of heroin challenges. ("320 sps" = a digitizing rate of 320 samples/sec and "epoch" = size of each sample analyzed.) Part A shows baseline cross frequencies. The amount of alpha is reduced by heroin and enhanced by naloxone. Chronic methadone enhances alpha activity, and the heroin is still effective. Beta frequencies are reduced by heroin and are reestablished with naloxone. Chronic methadone reduces fast frequencies, and this reduction is further enhanced by heroin.

PERIOD ANALYSIS⁻ IBM 1800

320 SPS   20 SEC EPOCHS   AMPLITUDE

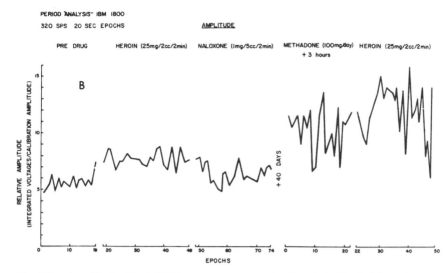

Fig. 5B. Amplitude. The initial increase accompanying heroin is blocked by naloxone. The effect of chronic methadone increases amplitudes. The heroin effect is still present.

by high voltage and very regular alpha activity, usually at a frequency slightly slower than the preheroin recordings. After the injection is completed, the mean alpha frequency gradually increases and the amplitude decreases, the EEG approximating the appearance of the preheroin initial recording. These changes occurred in 33/63 records.

A second effect is the continuation of slow activity during the injection and a gradual disappearance within several minutes. This pattern was seen in 19 subjects. In two cases slow activity was very suddenly erased during the injection.

In nine cases no appreciable changes occurred during or after the administration of naloxone. Each of the four subjects who received naloxone intramuscularly or subcutaneously qualified for this group. In other subjects in this group, slow activity produced by heroin was already waning before the naloxone was given.

The dramatic ability of naloxone to immediately antagonize the central effects of heroin was most clearly seen in two subjects who experienced clinical seizures. Intravenous naloxone abolished the clinical seizure and the EEG activity as rapidly as it was administered (within 40 sec).

In a chronic experiment in which 5 postaddicts received orally various daily doses (200–1500 mg) of naloxone, intravenous heroin (25–75 mg) was given from 4 to 24 hr later. For several minutes after the administration of 50–75 mg heroin, 6–24 hr after naloxone, slow activity (4–7 cps) appeared, sometimes in hypersynchronous bursts, followed by longer periods of desynchronization and some slow activity of low amplitude. These EEG changes were connected with behavioral symptoms of drowsiness and

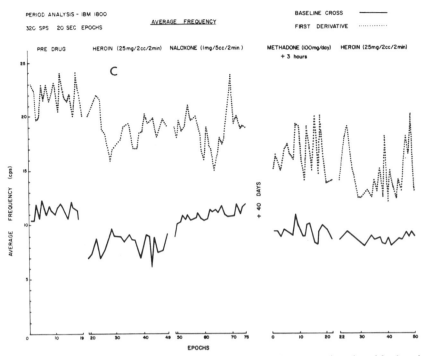

Fig. 5C. Average frequency. The baseline cross-average frequency is reduced by heroin and reinstated by naloxone. Methadone reduces the average frequency, and heroin has little effect. The first derivative average frequency follows a parallel course.

reports of a "high." In subjects tested 4 hr after oral naloxone with 50 mg of heroin or up to 12 hr for 25 mg of heroin, there were neither EEG nor behavioral effects of heroin.*

D. Cyclazocine

Cyclazocine, a narcotic antagonist chemically similar to nalorphine, was synthesized in the search for a nonaddicting analgesic and has been useful in the treatment of opiate dependence.[27–29] In addicts withdrawn from opiates, single oral daily administration of cyclazocine produces an effective blockade to the clinical and physiologic effects of heroin for periods up to 28 hr.

We observed the acute and chronic effects of cyclazocine in psychotic subjects and in postnarcotic addicts.[21,30] In 14 acute administrations of 0.5–1.5 mg, seven subjects became restless, dysphoric, and withdrawn. The EEG showed a progressive decrease in amplitudes and in the amount of

---

* In one subject receiving 3.0 g naloxone daily in single doses, a 25–mg heroin challenge 24 hr after his last dose was "blocked."

alpha (Fig. 6). Fast activity increased and desynchronized, poorly organized low-voltage activity became prominent. When chlorpromazine (0.1 mg/kg) was given on three occasions, sleep and stupor occurred, and the EEG showed high voltage slow waves with inhibition of responses to acoustic stimuli and eye opening.

In postaddicts intravenous cyclazocine exhibited similar EEG patterns and was compared to LSD, marijuana, or a narcotic. Heroin given after cyclazocine was appreciated as a blank or a mild sedative. The EEG after heroin showed a small increase in amplitudes and in the amounts of alpha, and the periods of desynchronization were shorter and less frequent.

In postaddicts receiving chronic oral cyclazocine, (1.5–4 mg) we observed a great variability of electrical activity. The records were predominantly desynchronized, the abundance of beta waves (20–25 cps) increased, and occasional lower amplitude theta activity was seen. After intravenous heroin challenges (25–75 mg) we again found EEG signs of heroin influence (slow waves, 4–7 cps, and longer periods of desynchronization with some burstlike phenomena of high amplitude of alpha and theta activity), which was usually, though not always, associated with the behavioral reports of the patients. The degree of blockade appeared less than that seen in the acute studies.

We undertook comparisons of cyclazocine with known psychoactive compounds in rabbits[31] Cyclazocine antagonized the EEG effects of heroin and was distinguishable in its patterns from morphine and pent-

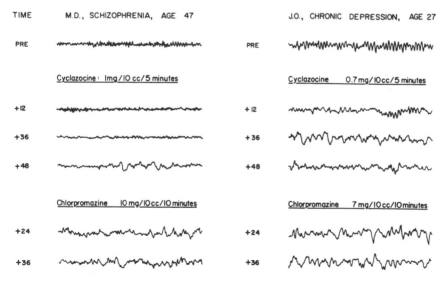

Fig. 6. EEG effects of cyclazocine and chlorpromazine. Two subjects received intravenous cyclazocine followed by intravenous chlorpromazine. The cyclazocine induces decreased amplitudes, desynchronization, and increased beta and theta frequencies. With chlorpromazine, slowing is enhanced and fast frequencies are reduced.

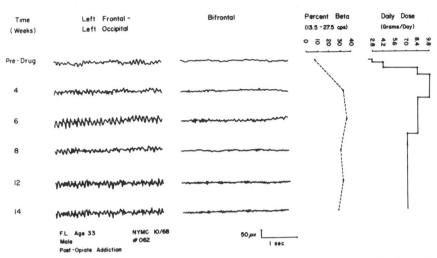

Fig. 7. EEG effects of tybamate. Chronic tybamate induces increased amplitudes and increased fast frequencies. Adaptation of the fast activity does not occur as drug levels are maintained.

azocine. Its profile of activity was found similar to low-dose imipramine, and when the two compounds were tested together they acted synergistically to antagonize the EEG actions of physostigmine and methamphetamine.*

E. Tybamate

The EEG effects of tybamate were observed during a clinical study in the treatment of addiction.[33] EEG records were obtained in 16 subjects before and during chronic tybamate administration. Twelve records before tybamate were predominantly alpha or alpha/theta records, and four were beta dominant records. With tybamate, beta frequencies from 18 to 35 cps increased as oral dosages rapidly rose to 7.0 g/day (Fig. 7). At higher dosages beta activity was accompanied or replaced by theta activity of high voltage. Beta percent time increased from less than 10% to 30–40% and persisted during the period of high drug intake. Adaptation in EEG fast wave activity was not observed, except in two subjects who exhibited sharp increases in theta activity replacing the beta preponderance. Spindles of 24–28 cps were common in the bifrontal leads.

V. DISCUSSION

In this review four distinct EEG patterns are consistently reported in

---

* The EEG profile of cyclazocine was found to be so similar to imipramine that following our view of an association of EEG and behavior we recommended a clinical trial of cyclazocine as an antidepressant. This has been completed, and antidepressant activity was indeed observed.[32]

studies of opiates and their antagonists: Morphine, methadone, heroin, codeine and other opiate substances produce a pattern of synchronization and slowing, with increased amplitudes, increased theta (delta) activity or slowing of dominant alpha frequency. Seizure activity is seen at high dosages, and there is blockade of alerting responses. The opiate antagonists nalorphine and cyclazocine elicit increased beta and theta activities, with decreased amplitudes and without seizure activity. Naloxone appears to elicit neither EEG nor observable behavioral effects on acute or chronic administration. Tybamate is of the class of beta producers, eliciting fast beta activity of high amplitude and reduced variability of amplitudes and frequencies.

In these profiles the EEG effects generally parallel the clinical effects—the most unusual and specific being that of naloxone, which elicits neither EEG nor behavioral changes except with regard to the blockade of heroin. The EEG patterns of the four varieties of drugs studied are similar to those reported by others and characterize compounds useful in treating opiate dependence. From this vantage point methadone has a profile like opiates. In its early administration methadone produces EEG slowing—and this central effect is reflected in the associated changes in clinical behavior. With clinical adaptation and the establishment of tolerance during the eighth to twelfth weeks of treatment, EEG adaptation also occurs. In some patients, in whom the EEG fails to return to a premethadone pattern, the clinical behavior is marked by altered vigilance, drowsiness, and continued abuse of other substances. While we have not studied chronic heroin use, we would anticipate a similar adaptation to be seen in the EEG as clinical dependence occurs.

Our studies have been dedicated to the alert, waking subject. Disturbances in sleep behavior are common during addiction and withdrawal. In studies of the EEG patterns during all-night sleep, Lewis et al.[34] reported the immediate effects of heroin administered on three successive nights to be a reduction in percent time REM sleep, with an increase on successive nights. On withdrawal there was a moderate but immediate percent time REM increase.

VI. SUMMARY

Electrographic studies of opiates and opiate antagonists in man show consistent patterns that reflect dosage and the development of tolerance. The EEG patterns of morphine, heroin, and methadone are characterized by increased slowing and amplitudes and decreased variability and fast frequencies. Seizure activity may be observed when dosages are rapidly administered.

The antagonist cyclazocine exhibits a different pattern, with decreased amplitudes, increased beta and theta activities, and variability. Naloxone appears unique in not elaborating consistent EEG changes.

Tolerance to methadone is reflected in a return of EEG patterns to pretreatment levels after 3–6 months of treatment.

These observations are discussed in relation to the EEG–behavioral association hypothesis.

VII. REFERENCES

1. A. Wikler, "Opiate Addiction," C. C Thomas, Springfield, Illinois (1953).
2. W. R. Martin, *Res. Publ. Ass. Nerv. Ment. Dis. 46*, 206–225 (1968).
3. J. H. Jaffe, and S. K. Sharpless, *Res. Publ. Ass. Nerv. Ment. Dis. 46*, 226–246, (1968).
4. M. Fink, T. Itil, and D. M. Shapiro, *Comp. Psychiat. 8*, 521–538 (1967).
5. M. Fink, D. Shapiro, C. Hickman, and T. Itil, *in* "Computers and Electronic Devices in Psychiatry" (N. S. Kline, and E. Laska, eds.), Grune & Stratton, New York, pp. 109–123 (1968).
6. H. L. Andrews, *Psychosom. Med. 3*, 399–409 (1941).
7. H. L. Andrews, *in* "National Research Council, Report of Committee on Drug Addiction 1929–1941, and Collected Reprints 1930–1941," pp. 1479–1490, Washington, D.C. (1941).
8. H. L. Andrews, *Psychosom. Med. 5*, 143–147 (1943).
9. A. Wikler, *J. Pharmacol. Exp. Ther. 2*, 435–506 (1950).
10. A. Wikler, *Proc. Soc. Exp. Biol. Med. 79*, 261–265 (1952).
11. A. Wikler, *J. Nerv. Ment. Dis. 120*, 157–175 (1954).
12. A. Wikler and S. Altschul, *J. Pharmacol. Exp. Ther. 98*, 437–446 (1950).
13. J. Roubicek, "Clinical Electroencephalography," *Cesk. Zdrav. Nakl.*, Prague (1959).
14. W. R. Martin, *Pharmacol. Rev. 19*, 463–521 (1967).
15. L. Goldstein and J. Aldunate, *J. Pharmacol. Exp. Ther. 130*, 204–211 (1960).
16. M. Fink, *in* "Psychopharmacology—A Review of Progress, 1957–1967" (D. Efron, J. Cole, J. Levine, and J. R. Wittenborn, eds.), pp. 497–507, U.S. Government Printing Office, Washington, D.C. (1969).
17. M. Fink, *Ann. Rev. Pharmacol., 9*, 241–268 (1969).
18. M. Fink and T. Itil, *in* "Psychopharmacology—A Review of Progress, 1957–1967" (D. Efron, J. Cole, J. Levine, and J. R. Wittenborn, eds.), pp. 671–682, U.S. Government Printing Office, Washington, D.C. (1969).
19. A. Zaks, A. Bruner, M. Fink, and A. M. Freedman, *Dis. Nerv. Syst. (Suppl.) 30*, 89–92 (1969).
20. J. Volavka, A. Zaks, J. Roubicek, and M. Fink, "Electrographic Effects of Diacetylmorphine (Heroin) and Naloxone in Man" *Neuropharm.—Psychopharm.* (in press).
21. M. Fink, A. Zaks, R. Sharoff, A. Mora, A. Bruner, S. Levit, and A. M. Freedman, *Clin. Pharmacol. Ther. 9*, 568–577 (1968).
22. J. Roubicek, J. Volavka, A. Zaks, and M. Fink, "Electrographic Effects of Chronic Administration of Methadone" *Neuropharm.—Psychopharm.* (in press).
23. A. Zaks, M. Fink, and A. Freedman, "Naloxone Treatment of Opiate Dependence," *J. Am. Med. Assoc.* (in press).
24. F. F. Foldes, J. Lunn, J. Moore, and I. Brown, *Amer. J. Med. Sci. 245*, 23–30 (1963).
25. F. F. Foldes, M. Schapira, T. A. Torda, D. Duncalf, and H. P. Shiffman, *Anesthesiology 26*, 320–328 (1965).
26. D. R. Jasinski, W. R. Martin, and C. A. Haertzen, *J. Pharmacol. Exp. Ther. 157*, 420–426 (1967).
27. A. Freedman, M. Fink, R. Sharoff, and A. Zaks, *J. Am. Med. Assoc. 201*, 191–194 (1967).
28. A. Freedman, M. Fink, R. Sharoff, and A. Zaks, *Amer. J. Psychiat. 124*, 1499–1504 (1968).
29. W. R. Martin, C. W. Gorodetsky, and T. K. McClane, *Clin. Pharmacol. Ther. 7*, 455–465 (1966).
30. M. Fink, T. Itil, A. Zaks, and A. M. Freedman, *in* "Neurophysiological and Be-

havioral Aspects of Psychotropic Drugs" (A. G. Karczmar and W. P. Koella, eds.), C. C Thomas, Springfield, Illinois pp. 62–71 (1969).

31. R. P. White, W. G. Drew, and M. Fink, *Biol. Psychiat. 1*, 317–330 (1969).

32. M. Fink, J. Simeon, T. Itil, and A. M. Freedman, *Clin. Pharmacol. Ther. 11*, 41–48 (1970).

33. F. Veress, V. Major, M. Fink, and A. M. Freedman, *J. Clin. Pharmacol. 9*, 232–238 (1969).

34. S. A. Lewis, I. Oswald, J. I. Evans, and M. O. Akindele, *Electroencephalog. Clin. Neurophysiol.* (in press).

Chapter 24

OPIATE ANTAGONISTS IN THE TREATMENT OF HEROIN DEPENDENCE*

Max Fink, Arthur Zaks,
Richard Resnick, and Alfred M. Freedman

Department of Psychiatry
New York Medical College and Metropolitan Hospital Medical Center
New York, New York

I. INTRODUCTION

Opiate addiction is a complex social and psychological disorder of diverse origins that defies conventional therapeutic efforts, both social and medical. Treatment is discouraged by the high rate of recidivism. Rehabilitation programs frequently reduce recidivism either by the careful selection of suitable patients (accepting only those with the best motivation for recovery) or by imprisonment and primitive systems of parole or by methadone substitution, which satisfies opiate craving by cross tolerance. Each treatment program for opiate dependence—whether forced imprisonment, group therapy, reeducation, psychotherapy, or social rehabilitation— requires an adjuvant to "engage" the subject in the treatment schedule. Methadone substitution fulfills such a role, although the addicting properties of methadone is a continuing hazard.

Another use of drugs in therapy is derived from a theory of opiate dependence. Patients—addicts withdrawn from heroin—report a craving for narcotics as soon as they return to the community in which their "habit" flourished. Reports of their experiences led to a theory of relapse based on a conditioning model, with emphasis on the role of subtle environmental cues and physiological components.[1] In this view drug dependence proceeds in several phases. In the first phase the subject is aware of tensions

---

* Aided, in part, by grants from the USPHS MH-12567, 13003, and 13358, and from the New York State Narcotic Addiction Control Commission. Supplies of cyclazocine were made available by Sterling–Winthrop Research Laboratories, Rensselaer, New York, and naloxone by Endo Laboratories, Garden City, New York.

or discomforts and seeks and obtains relief in narcotic use. Reinforcement of the habit occurs with each relief by drug use in a period termed "episodic intoxication." If the drug has an addicting potential, it reduces tension or discomfort less and less efficiently, and a second state, that of physical dependence, develops. Dependence is further reinforced as the addict fails to maintain an adequate drug intake, and periods of tension and discomfort increase in frequency and duration.

Sooner or later the addict must be withdrawn because the supply of opiates runs out. He is admitted to a hospital or is arrested and thus enters the phases termed "physiological disequilibrium" and "conditioned abstinence." These states may exceed the period of simple withdrawal and probably persist for many months.

This theory of dependence led to the suggestion that the blockade of the relief afforded by narcotics during the period of conditioned abstinence may lead to extinction of physical dependence and conditioned drug-seeking behavior. The opiate antagonist nalorphine was suggested as a suitable drug for blockade, but its short duration of action and the high incidence of hallucinogenic effects limited its clinical usefulness. The development of a long-acting related drug, cyclazocine, led to its trial and introduction in clinical trials.[2,3] Cyclazocine use is associated with agonistic effects of opiates, sometimes considered unpleasant, and clinical programs have been few. The development of a pure opiate antagonist, naloxone, has stimulated renewed interest in clinical trials.[4] This report describes our clinical trials with cyclazocine and naloxone in an effort to provide "engagement" to the rehabilitation program and to test the conditioning theory or opiate dependence.

In these studies of antagonists, randomly assigned subjects also have received other agents, such as methadone[5-8] and tybamate.[9] The subjects of these studies are male opiate addicts who volunteered for admission to an in-patient treatment center in a municipal hospital in the East Harlem section of New York City. The population of this community is of very low income, predominantly black and Puerto Rican, with low educational levels and large numbers receiving extensive welfare assistance. The ages for those reported here ranged from 17 to 54, with a mean of 26 years, and the duration of addiction was from 2 to 30 years. All patients applying for treatment were accepted for cyclazocine induction except those who were actively psychotic, dependent on nonnarcotic drugs, or physically ill.

Male patients applying for treatment are asked to sign a permit for an "experimental" treatment of their addiction. Some volunteered and were placed on a waiting list but never appeared for admission. Others left during the detoxification period or immediately thereafter.

On admission to the study ward, the narcotic usage of each patient is estimated by inquiry. Methadone is given in reducing amounts for detoxification, usually within 4–7 days, and a drug-free period of observation allows medical and laboratory examinations before induction with an antagonist.

II. CYCLAZOCINE

Cyclazocine is an *N*-substituted benzomorphan derivative, chemically similar to nalorphine hydrochloride. It is an active analgesic with some subjects reporting dysphoria, vivid imagery, and anxiety with its use.

The first clinical trials of cyclazocine in opiate addiction by Martin[2] and Jaffe and Brill[3] defined 4 mg/day as an effective dose for clinical antagonism to opiates. They observed disturbing subjective effects when cyclazocine was administered rapidly with the development of tolerance to these changes when the daily amount of cyclazocine was slowly increased. Tolerance failed to develop to the antagonism for opiates.

Cyclazocine is supplied in liquid vehicle (0.1 mg/ml), in capsules of 0.25 mg, in tablets of 0.1 and 1.0 mg, and for intravenous use (0.5 mg/ml).

A. Induction

We first attempted to develop an induction schedule that would minimize dysphoria. In our initial study (1966–1967) 60 patients completed detoxification and began a cyclazocine induction.[10] Fifty-eight completed the study period, with only two patients beginning the active cyclazocine treatment and signing out against medical advice.

Induction was defined as achieving a daily therapeutic dose of 4 mg. In our first subjects the dosage of cyclazocine had been increased gradually over 40 days. To reduce this time span, groups of four patients each were tested in induction periods of 10, 15, 20, and 30 days. A daily increment of 0.4 mg (10-day schedule) brought forth complaints of somnolence, headache, irritability, "fuzzy thinking," illusions, and in two instances visual hallucinatory experiences. These symptoms were not reported on the 15-day schedule with dosages divided in two increments daily except that somnolence, weakness, and irritability appeared during the first 4 days. The 20- and 30-day induction periods were well tolerated without secondary symptoms.

Based on this experience we defined a 15-day schedule as optimal. After a drug-free period of 7–10 days to permit laboratory tests, cyclazocine is started at 0.2 mg/day and increased 0.2 mg/day for 10 days, using the liquid vehicle form. If this is well tolerated, the dose is increased by 0.4 mg/day for 5 days. Medication is given twice daily, and a maintenance dose of 4 mg/day is usually reached in 15 days. The medication is gradually changed to a single 4-mg dose on day 20 by giving progressively larger amounts each morning.

Reports that naloxone might antagonize the agonistic effects of cyclazocine prompted us to reassess a more rapid induction schedule. Patients are currently inducted to 4 mg cyclazocine within 4 days, at daily increments of 1 mg. Initially each patient was given 0.5–1 g oral naloxone with each daily dose of cyclazocine. Recently naloxone has been necessary in only a few patients, and it is now given in 0.3–0.6 g at the patient's request.

Fifteen patients inducted on this rapid schedule requested naloxone only occasionally, usually during day 3–6.[11]

B. Heroin Challenge

During the induction phase the value of cyclazocine as an effective narcotic blocking agent is discussed both in individual and group therapy sessions with all the resident addicts. They are told that the injection of a narcotic in their usual amounts would have neither euphoric nor systemic effects. To reinforce this suggestion, they are offered an opportunity for an intravenous injection of heroin in the laboratory when they are receiving a full daily dose of cyclazocine.[5,7,10] We have determined that 25 mg of heroin/2 ml saline/2 min elicits both the behavioral effects and EEG changes in postaddicts 7–10 days after a final dose of methadone. (In our earlier studies 15–20 mg of heroin was used.)

In patients receiving 4 mg of cyclazocine in single daily doses we have observed complete blockade to 25 mg of heroin in 6 hr.

In 18 patients challenged 24 hr after 4 mg of cyclazocine, 12 experienced no clinical effects and 6 became euphoric. These described an initial "rush" or "buzz" in their abdomen. This pleasant sensation was intense, moved to the throat, and then spread "all over," persisting for approximately 30 min. The pupils were constricted. Following this challenge, the dose of one patient was raised to 5.5 mg, and there was no response to a second heroin challenge 24 hr after a dose of cyclazocine. In five patients cyclazocine was increased to 5 mg, and in subsequent challenges they still responded with a slight and short-lasting euphoria.

Tolerance to the antinarcotic activity of cyclazocine was not observed. Patients receiving a heroin challenge up to 5 months after their first daily 4-mg dose of cyclazocine showed as much blockade to heroin as when the challenge was given immediately at the end of the induction period. From these experiences it is probable that the effective duration of antinarcotic activity of a single 4-mg dose on chronic administration is at least 20 hr, waning rapidly after 24 hr, with a peak 6–8 hr after a dose of cyclazocine.[12,13]

C. Secondary Effects

The secondary effects of chronic cyclazocine use are similar in patients with long and short induction periods. Increased libido, constipation, elation, anxiety, dizziness, headaches, restlessness, and insomnia are most frequent during the initial 4–6 days of treatment but generally abate during the second week. Increased libido and constipation have persisted for longer periods, being occasionally noted in the third and fourth weeks of treatment. In our early studies two patients reported visual hallucinations, but these have not recurred as our experience increased.[5,10]

The secondary effects do not persist in the aftercare program, where patients have received cyclazocine for more than 24 months. Some reported episodic drowsiness and constipation, but the effects noted during the early weeks of treatment are not reported, even on specific inquiry.

D. Maintenance

Following discharge, patients are asked to return three times weekly. At each visit they are given their daily dose of cyclazocine and a supply for the intervening day or days.

The patients receiving cyclazocine report a variety of changes in behavior. Sexual activity is increased and more satisfactory. Social activities such as dating, going to parties and dances, and leisure time spent with nonaddicts increase. Anxiety is less, and drug-seeking behavior is curtailed. But while cyclazocine serves to inhibit interest in narcotics, the use of marijuana, amphetamines, barbiturates, and alcohol has not been affected. Criminality appears to be less, and interest in vocational activities is enhanced.

Patients report an ability to interrupt cyclazocine intake for a few days without withdrawal. A day after cyclazocine they are able to experience the euphoric and sedative effects of heroin. By reinstituting cyclazocine at the full dosage, they do not become readdicted and can continue their daily activities.

In a 1968 follow-up of patients discharged on cyclazocine during 1967, 74 patients were traced. Fifteen continued in rehabilitation from 5 to 16 months and were found to be in school or working. Eleven discontinued cyclazocine and entered a methadone maintenance program, and forty-eight returned to heroin use.

In a more recent review of patients in our aftercare facility, nine have continued on cyclazocine for more than 2 years; four for 12–24 months; four 6–12 months; and six less than 6 months.

III. NALOXONE

Naloxone (*N*-allylnoroxymorphone) is an experimental compound 5–8 times as effective as nalorphine in antagonizing opiate effects in animals.[14–17] Tolerance does not develop, nor does withdrawal occur on sudden cessation after 2 weeks of repeated administration. But its duration of action is short—less than 4–6 hr.

A. Single Administration

Two series of acute single intravenous administration studies were undertaken.[4] Nineteen volunteers (patients) received heroin (10–20 mg/2 ml/2 min) followed after 8–32 min by naloxone (0.7–10.0 mg/5 ml/2 min). Ten subjects received two trials for a total of 29 trials. In ten patients naloxone injections were given first, followed by heroin. There studies were done in the laboratory under EEG control.

Following heroin, each patient reported he was "high" (euphoric) or "floating" and exhibited nodding, pupillary constriction, and voice changes.

Naloxone, even at the lowest doses, abolished these effects within 1/2–2 min of the end of the injection. Since these initial studies we have routinely administered 1 mg of naloxone to all "clean" subjects receiving heroin challenges, 10–30 min after heroin. Blockade has been uniformly complete. In two instances we gave 1 mg of naloxone to subjects receiving heroin during a period of chronic methadone use, and in both it precipitated an intense opiate withdrawal syndrome.

When naloxone was given first, there were no effects in five and there was a feeling of coldness for 1–3 min in four subjects.* One patient reported an "amphetamine-like" sense of stimulation. With heroin, eight felt no euphoria and two reported a mild "high"—as if given an injection of weak heroin.

B. Chronic Oral Administration

In our initial study seven patients received repeated oral administrations in doses up to 100 mg twice daily.[4] The first doses were 20 mg with increments of 30 mg daily. Heroin challenges were repeated at frequent intervals, and the dosages were increased until complete blockade was achieved. At dosages of 120 mg a full clinical reaction to heroin occurred in one patient, was partial in five, and was absent in one. Complete blockade was evident in five subjects at 200 mg and in one at 160 mg. From these observations we concluded that oral naloxone was effective in blocking 20 mg of heroin up to 10 hr after a divided daily dose.

Our second series of studies are in progress.[18] In these we have set single daily doses of naloxone as a practical and necessary goal for the successful treatment of opiate dependence. Subjects have been given increasing single daily doses of naloxone, and the efficacy and duration of antagonism has been assessed with challenges of 25 and 50 mg of heroin.

One subject was tested 6 hr following daily single 200 mg doses of naloxone. He exhibited no effect from 25 mg of heroin but became euphoric with 50 mg. Naloxone was increased to 400 mg, and this afforded blockade against 25 and 50 mg of heroin but not 75 mg.

The second subject received a daily dose of 400 mg of naloxone and was effectively blocked against the 25, 50, and 75 mg of heroin at 6 hr.

These two subjects were rechallenged 18 hr following drug intake. They did not evidence narcotic antagonism until naloxone was raised to 800 mg daily. At this dose they were protected against both 25 and 50 mg of heroin. They were not given 75-mg challenges.

A third subject was challenged 18 hr postmedication. He did not demonstrate significant narcotic blockade on doses of 600, 800, or 1000 mg of naloxone. On a daily dose of 1250 mg the 25 and 50 mg of heroin challenges were blocked, but the subject became euphoric following the

---

* In subsequent trials with other subjects who had received their latest dose of methadone less than 10 days before, the sensations were accompanied by irritability, gooseflesh, sweating, and other signs of the precipitated abstinence syndrome.

administration of 75 mg of heroin. At 24 hr following 1250 mg of naloxone the subject was protected against a 25-mg challenge but reacted with an intense euphoria to 50 mg of heroin.

The fourth and fifth subjects were brought to a daily single oral dose of 1500 mg. They were first challenged at 18 hr, and both were protected against 25 and 50 mg but experienced a euphoric reaction to heroin following the 75-mg challenge.

At 24 hr the fourth subject demonstrated partial blockade against 25 mg of heroin but became euphoric following 50 mg. The fifth subject demonstrated no narcotic blockade at 24 hr, becoming "high" following 25 mg of heroin.

Two subjects have received 2400 mg of naloxone and one received 3000 in single daily oral doses. In heroin challenges after 24 hr blockade to 25 and 50 mg of heroin was exhibited by the subject receiving 3000 mg.

We have observed few secondary effects of chronic naloxone. One patient at 2.4 g/day reported feelings of depression and tension and requested that the medication be stopped. The subject receiving 3 g/day, who had a heroin challenge without a euphoric response, suffered a grand mal seizure 7 hr after a morning dose and 4 days after a heroin challenge. He had no prior history of seizures and no sequellae.

In no patient receiving any daily dose of naloxone have we observed withdrawal symptoms within the period of observation (up to 3 days).

Laboratory tests, which were performed weekly, were within normal limits except in two instances: One subject had an unexplained elevated white blood count between 12,000 and 18,000 for 7 weeks while he was on naloxone. Repeated medical examinations found no explanation. It returned to normal spontaneously.

All subjects had occasional abnormal liver function tests—primarily SGOT and SGPT. The abnormal findings in these two tests were sporadic and bore no relation to the length of time on naloxone or to dose.[19,20]

Hematological studies consisted of bleeding time, euglobulin lysis time, whole blood coagulation time, clot retraction, Quick prothrombin time, partial thromboplastin time, thrombin time, prothrombin assay, and factors V, VII, X, and XIII determinations. They were performed prior to naloxone treatment, 1 week following drug treatment when the subjects were on a daily dose of 200–500 mg naloxone and 3–8 weeks later on a daily dose of 600–1500 mg. All test values were within normal limits.[18]

IV. DISCUSSION

It has been suggested that narcotic addiction is analogous to a conditioned response—the addict responds to stressful stimuli in his environment with drug-seeking behavior.[1] In this theory the repeated use of heroin without the anticipated subjective effects should lead to extinction of the learned drug-seeking behavior.

This theory was first tested by the use of high doses of methadone as maintenance treatment for addicts.[21] Because methadone is an addicting substance with euphoriant properties, these trials may not be considered an adequate test of the hypothesis. Cyclazocine, which has a length of antagonistic action greater than 24 hr, has been successfully used in the maintenance treatment of opiate dependence.[3,5,6,10] These trials may be considered a more adequate test of the hypothesis. We have observed drug-seeking behavior to continue in these subjects. Use of opiates is markedly reduced, however, and the patients are able to maintain their social and work committments without readdiction. In this experience cyclazocine is less effective in reducing the conditioned drug-seeking response than in providing engagement to a therapy program and preventing the development of tolerance and withdrawal.* Engagement makes possible continued job counseling, reeducation, and social services; prevention of tolerance and withdrawal allows the subject to negotiate our social system without recourse to crime.

Our studies of cyclazocine and naloxone indicate that they are safe and well tolerated and accepted by patients and staff. Engagement to the rehabilitation program takes place rapidly, and the continued contact assists the community adjustment of these patients. A therapeutic optimism is engendered in both patients and staff, and this optimism is useful in the rehabilitation process, supporting the direct effects of the drugs.

In our view an ideal antagonist should have a long duration of action and a minimum of secondary effects. A short length of action requires frequent intake of medication—an impossibility for the poorly motivated postaddict. It is also necessary that secondary effects be minimal, particularly on withdrawal, since patients have shown an intolerance to any discomforts, even for short periods.

Cyclazocine is satisfactory as an antagonist by its duration of action, but its secondary effects preclude wide acceptance. Naloxone, in the oral dosages tested, is effective for more than 24 hr and has no secondary effects— making it the ideal antagonist for therapeutic trial. The large doses required in these trials (3.0 g) must be confirmed, and efforts to provide alternate routes must be tested. We have assessed the possibility of investing naloxone in a silastic-type vehicle and implanting such a tube under the skin to provide continuous sustained naloxone for extended periods. In theory this could be done for months since the parenteral dose necessary for effective blockade of heroin is probably less than 1.0 mg/4–6 hr. Such a development has been proposed and warrants intensive study.

Another theory of opiate dependence is rooted in the potential anti-anxiety effects of opiates. Opiates are used to reduce anxiety and tension occasioned by environmental stimuli, and in this view an antianxiety agent should successfully reduce opiate dependence. Following the favorable

---

* Another aspect of cyclazocine use may be its antidepressant potential.[23] Opiate-dependent subjects frequently exhibit withdrawal, retardation, and loss of drive— symptoms often associated with depressive mood.

reports of Feldman and Mulinos[22] we assessed the clinical efficacy of tybamate.[9] While daily doses of 6–11.2 g/day reduced anxiety symptoms,
opiate hunger remained and subjects rapidly became readdicted on their
return to their communities.

V. SUMMARY

Based on a conditioning theory of opiate dependence, various substances have been proposed to blockade the euphoric effects of opiates and
to reduce dependence by a process of negative reinforcement. Clinical
experiences with cyclazocine and naloxone are described, indicating that
both are effective therapeutic agents. The limitation of the present delivery
systems and the characteristics of an ideal antagonist are described.

VI. REFERENCES

1. A. Wikler, *in* "Narcotics", (D. M. Wilner and G. G. Kassebaum, eds.), McGraw-Hill, New York (1965).
2. W. R. Martin, C. W. Gorodetzky, and T. K. McLane, *Clin. Pharmacol. Ther. 7,* 455–465 (1966).
3. J. H. Jaffe and L. Brill, *Int. J. Addiction 1,* 99–123 (1966).
4. M. Fink, A. Zaks, R. Sharoff, A. Mora, A. Bruner, S. Levit, and A. M. Freedman, *Clin. Pharm. Ther. 9,* 568–577 (1968).
5. A. Freedman, M. Fink, R. Sharoff, and A. Zaks, *J. Amer. Med. Ass. 202,* 191–194 (1967).
6. R. Resnick, M. Fink and A. M. Freedman, *Amer. J. Psychiat. 126,* 1256–1260 (1970).
7. A. Zaks, A. Bruner, M. Fink, and A. M. Freedman, *Dis. Nerv. Syst. (Suppl.) 30,* 89–92 (1969).
8. J. Roubicek, J. Volavka, A. Zaks, and M. Fink, "Electrographic Effects of Chronic Administration of Methadone" (in preparation).
9. F. Veress, V. Major, M. Fink, and A. M. Freedman, *J. Clin. Pharmacol. 9,* 232–238 (1969).
10. A. Freedman, M. Fink, R. Sharoff, and A. Zaks, *Amer. J. Psychiat. 124,* 1499–1504 (1968).
11. R. Resnick, M. Fink, and A. M. Freedman, "Experimental Studies of Cyclazocine, Alone and Combined with Naloxone" (in preparation).
12. M. Fink, A. M. Freedman, A. Zaks, R. L. Sharoff, and R. Resnick, *in* "The Present Status of Psychotropic Drugs" (A. Cerletti and F. J. Bove, eds.), pp. 428–431, Excerpta Medica, Amsterdam (1969).
13. M. Fink and A. M. Freedman, *in* "Modern Trends in Combatting Drug Dependence and Alcoholism" (R. V. Phillipson, ed.), Butterworths, London, pp. 49–59, 1970.
14. M. S. Sadove, R. C. Balagot, S. Hatano, and E. A. Jobgen, *J. Amer. Med. Ass. 183,* 666–668 (1963).
15. H. Blumberg, H. B. Dayton, and P. S. Wolf, *Proc. Soc. Exp. Biol. Med. 123,* 755–758 (1966).
16. F. F. Foldes, J. Lunn, J. Moore, and I. Brown, *Amer. J. Med. Sci. 245,* 23–30, (1963).
17. D. R. Jasinski, W. R. Martin, and C. A. Haertzen, *J. Pharmacol. Exp. Ther. 157,* 420–426 (1967).
18. A. Zaks, M. Fink, and A. Freedman, "Treatment of Opiate Dependence with High Dose Oral Naloxone" *J. Amer. Med. Assn.,* in press.
19. C. E. Cherubin, *Ann. Int. Med. 67,* 23–33 (1967).

20. C. W. Gorodetzky, J. D. Sapira, D. R. Jasinski, and W. R. Martin, *Clin. Pharmacol. Ther.* 9, 720 and 739 (1968).
21. V. P. Dole and M. Nyswander, *J. Amer. Med. Assn. 193*, 646–650 (1965).
22. H. Feldman and M. G. Mulinos, *J. New Drugs 6*, 354–360 (1966).
23. M. Fink, J. Simeon, T. Itil, and A. M. Freedman, *Clin. Pharmacol. Ther. 11*, 41–48 (1970).
24. M. Fink, *Science 169*, 1005–1006 (1970).

NOTE ADDED IN PROOF

In June, 1970, a meeting of clinical and laboratory investigators of opiate agonists was held.[24] A review of more than 450 patients treated with cyclazocine showed a 40% sustained treatment rate. Oral naloxone, to 800 mg/day, was in use in 49 additional patients in two programs with some success.[24]

Chapter 25

BLOCKADE WITH METHADONE*

Vincent P. Dole

Rockefeller University
New York, New York

I. INTRODUCTION

Methadone maintenance, after 6 years of clinical trial, has a demonstrated place in treatment of heroin addicts. Detailed records of all the patients admitted to methadone programs in New York City provide a large amount of supporting data.[1] With the confirmatory findings of programs in other cities,[2] the uses and limitations of this treatment are now fairly well defined.

II. CLINICAL RESULTS

The most recent report of the Methadone Maintenance Evaluation Committee[3] summarized the results of treating 2205 heroin addicts with methadone and rehabilitative services. The overall retention rate of patients in treatment was 82%, with a marked reduction in criminal activity. Before treatment the arrest rate was approximately 20% per year; after the first year of methadone treatment this rate fell to less than 3% per year. A comparison group of untreated addicts, drawn from the detoxification service at the M. J. Bernstein Institute (Beth Israel Medical Center), matched for age and ethnic group, showed continued addiction and criminality. Rehabilitation of the methadone patients was reflected in the high rate (70%) of employment for the program as a whole, and a steady increase in the percentage employed with time on the program.

To test the possibility that these results might have been due to the selection of unusually favorable patients—such as older individuals who might have "matured out" of addiction without any treatment[4]—the data

---

* Supported by grants from the Health Research Council (City of New York Department of Health) and the New York State Narcotic Addiction Control Commission. Conclusions stated here are not necessarily those of the Commission.

on success and failure were analyzed with respect to age. Success was not found to increase with age; contrary to the "maturing out" theory, the patients in the age group 20–30 years had a higher success rate than the older patients. Analysis by ethnic groups showed no significant differences in response.

The limitations of the methadone treatment program in its present form were disclosed by an analysis of the reasons for discharge. Mixed addictions (alcohol, barbiturates, and amphetamines) accounted for about half of the total discharged from the program. This reflects the specificity of the methadone blockade; heroin addiction can be stopped with methadone by a pharmacological action, but there is as yet no equivalent for methadone in the treatment of alcoholism or other non-narcotic abuses. An encouraging result, however, was not shown in the failure statistics: A substantial number of mixed addicts have been rehabilitated with methadone treatment plus good counseling. Most methadone programs treat heroin addicts with complicating addictions because the success rate with these difficult patients is sufficiently high to justify the attempt.

The other major cause for discharge, accounting for one-quarter of the discharges, was psychopathic behavior, persistently uncooperative and disruptive actions, or crimes leading to conviction and jail. For these patients, a simple medical program providing methadone and counseling is clearly inadequate. Although the narcotic blockade succeeds in stopping their addiction to heroin, these disturbed individuals continue antisocial behavior. Controls provided by group pressure, structured environment and, as a last resort, jail are needed for such patients—fortunately the minority.

Death, due to a variety of causes, occurred in 27 patients—a rate of 10 per 1000 patient-years. Each case was studied by the Medical Examiner; none appeared to be related to the pharmacological effects of methadone.[5]

The discharge statistics include 57 persons who left the program voluntarily and were not readmitted. All of these individuals are included in the failure category of the present analysis since the follow-up study of discharged patients has not yet been completed. This exaggerates the failure rate, but in any event does not cause a large error since the voluntary discharges are relatively few (less than 3% of patients admitted to treatment).

III. PHARMACOLOGY

The dose of methadone at the beginning of treatment should be sufficient to control the effects of heroin withdrawal, but not so large as to produce narcotic effects. In general, a dose of 20 mg twice daily is adequate for a patient who has been a regular user of heroin up to the time of admission. This withdrawal from heroin and initiation of methadone treatment can be done on an ambulatory basis if the physician is experienced in the treatment. Thereafter the dose of methadone is increased gradually over a period of 3–6 weeks to the stabilization level of 80–120 mg/day. During the first part of this period, methadone is given in doses divided between morning and evening. Later these doses are consolidated and, finally, the full dose is given in a single administration, usually in the morning.

If the dosage of methadone is increased too rapidly during the early stages of treatment, the patient will be unnecessarily sedated and may develop a transient edema, possibly due to the excessive release of antidiuretic hormone. These pharmacological effects are reversed by reducing the dose of medication, holding it steady at a lower level for a few days, and then resuming the increase at a slower rate.

Tolerance to the constipating effect of methadone is obtained more slowly; all new patients are placed routinely on a colloid laxative. Some male patients also complain of impotence during the first 6 months of treatment. Their difficulty can be relieved in some cases by reducing the dose of medication but, in any event, the impotence tends to disappear after the first year of treatment. Excessive sweating in hot weather has been a symptom noted by some patients, and this phenomenon may be persistent. It has continued to occur in susceptible individuals for as long as 5 years.

Some patients have anxiety symptoms under emotional stress that they interpret as symptoms of abstinence. They fear that the methadone has lost its blocking action, and this fear further increases their anxiety. These symptoms are not relieved by an increase in the dose of methadone but can be controlled by reassurance and by relief of the stressing situation.

Although it is possible to abolish narcotic drug hunger at a lower dosage level (30–60 mg/day) than is usually employed, the blockade is relatively low and the patient can experience euphoriant effects from heroin. Some experimentation with heroin is to be expected in the early stage of treatment, since the behavioral patterns of addicts are not immediately reversed by the pharmacological blockade. It therefore is recommended that the patients be maintained at full blockading doses for at least 1 year.

Physical examination and laboratory testing of patients on admission and at yearly intervals have shown no evidence of somatic damage or of intolerance to the medication. In general the health of patients has improved while in treatment. Patients showing abnormalities in tests of liver function at the time of admission—presumably the residuals of early attacks of hepatitis—have shown improvement or no advance in the abnormalities during periods of more then 3 years of methadone maintenance treatment.[1,6]

IV. CLINICAL SERVICES

In the complex task of rehabilitating an addict, methadone (or any other medication) is only an adjunct. The addict usually comes to treatment in desperation, but without the faith in medical treatment that the physician takes for granted in usual medical practice. The addict–patient is a suspicious and defensive person who has learned to trust no one. It is of fundamental importance in a rehabilitation program that a professional relationship be established between doctor and patient and that it be based on confidence in the physician rather than on punitive controls. Not every addict needs social rehabilitation—some addicts have jobs and families at the time that they come to treatment, and they do well without special rehabilitative services—but in every case the narcotic medication must be prescribed as a medical treatment by a physician under controlled conditions, with assurance that the medicine be taken as prescribed. Well meaning but uniformed physicians have given prescriptions for methadone to addicts without adequate supervision, presumably on the assumption that rehabilitation is a pharmacological effect, and the results have been bad. In all cases, methadone must be used as part of a supervised medical program.

The main services needed by a methadone program for recidivist addicts are helps in housing, school placement, job training, and employment. Without such help the patient is likely to be trapped in his past, even if he stops using heroin. With a jail record, no job skills, and the handicap of minority status, a 30-year-old ex-addict has a long way to go to become a stable resident with a job, a family, and the respect of his community. The achievement of this by a majority of patients in methadone programs in New York City testifies to the potential that exists in apparently hopeless criminal addicts. Equally, this success reflects the dedication of the staff of the methadone program—physicians, nurses, counsellors, and ex-addicts— in bringing out the potential of their patients. Conventional psychiatric treatment is available to patients on the methadone program and has been provided when it has seemed indicated, but these patients have been in the minority. In general, the rehabilitative services employed on the methadone program are of a more immediate, practical nature.

Regular testing of urine for narcotic drugs, barbiturates, amphetamines, and tranquilizers is an essential part of the methadone program, as indeed it should be part of any serious program for the treatment of addicts. The analysis provides an objective measure of the patient's response to treatment and protects the clinician from the psychopathic patients who otherwise would deceive him. Since the urine test is routine, the sample collected at each clinic visit is accepted without resentment by all patients.

V. PROSPECTS FOR THE FUTURE

Perhaps the most important implication of the methadone work has been the demonstration that addicts can be rehabilitated by a simple medical regimen. In the past pharmacological research had been stifled by a theory that addiction was a symptom of character defect that required psychiatric treatment. Under this theory medical treatment would merely mask the symptom and would fail to rehabilitate the patient. The way is now open to an intensive search for better pharmacological agents—drugs with longer periods of action and agents devoid of any danger of misuse.

Further experience with methadone and other drugs will show whether the persistent pharmacological effects of heroin addiction can be reversed. To date the experience with methadone treatment suggests that a total cure will be difficult to achieve. Almost every patient who has been withdrawn from methadone has felt a return of heroin hunger; the hunger had been absent while he was taking methadone, and it disappeared again when methadone was resumed. The traditional assumption that the need for heroin, and therefore the hunger for heroin, will disappear when a patient is socially rehabilitated is not supported by any evidence from other programs and has been repeatedly negated by our experience with withdrawal. The heroin hunger seems to us quite clearly pharmacological in origin and independent of rehabilitation. There is still controversy on this point, however, and further study is in order.

Preliminary studies in our clinic have indicated that methadone can be used effectively as part of a rehabilitation program for adolescents who have failed abstinence-directed treatments. More work is needed to define the special needs of these young patients, such as residential facilities and special school programs to make up their deficiencies and to explore the possibility of low dose, temporary maintenance.

The enormous epidemic problem of heroin addiction in the cities needs an approach that rises above politics and special interest. Methadone treatment obviously cannot provide the total answer. The problem is too complex for a single solution. Abstinence-directed programs and social improvements to eliminate the bad environments that breed addiction obviously are needed if this huge problem is to be brought under control. But the glaring deficiency at the present time is the lack of objective data with which to judge the cost and effectiveness of alternative approaches. The methadone program, so far, is the only treatment in the field with systematic collection of data and independent evaluation. In the interest of the community it is hoped that the other treatment programs will see an obligation to provide data on the successes and failures of their treatment and to permit an independent group of experts to evaluate the results.

VI. REFERENCES

1. Vincent P. Dole, Marie E. Nyswander, and A. Warner, *J. Amer. Med. Ass. 206,* 2708–2711 (1968).
2. "Reports at the Second National Conference on Methadone Treatment," held at Rockefeller University, October 26–27, 1969 (to be published in *Int. J. Addictions*).
3. F. R. Gearing, "Evaluation of Methadone Maintenance Treatment for Heroin Addiction: A Progress Report," presented at the Epidemiology Section of the American Public Health Association, November 13, 1969.
4. C. Winick, *Bull. Narcotics 14,* 1–7 (1962).
5. Reported by Michael Baden, Assistant Medical Examiner, City of New York, at the Second National Methadone Conference, Rockefeller University, October 1969.

6. M. J. Kreek and V. P. Dole, "Medical Status of Patients on the Methadone Program for Three or More Years" (to be published).

Chapter 26

THE NALORPHINE (PUPIL) TEST IN THE DETECTION OF NARCOTIC USE

Henry W. Elliott

Department of Medical Pharmacology and Therapeutics
California College of Medicine
University of California
Irvine, California

I. INTRODUCTION

While there is little doubt that preventive rather than punitive measures will eventually prevail in the control of drug abuse, methods for the detection of narcotic use have simplified the work of parole and probation officers in their attempts to control the illicit use of narcotics in known drug users under their supervision. In California, which has an estimated 18,000 heroin users, the nalorphine test introduced by Terry and Braumoeller in 1955[1] has been used extensively for this purpose, and more recently analysis of urine for narcotics or their metabolites has been used either independently or to supplement the nalorphine test. The tests are not compulsory, but usually willingness to undergo them is made a condition of parole and they are used at the discretion of the parole officer. The tests may be performed at specified times in metropolitan testing centers or, better, parolees are brought in for testing by surprise. The scale of the testing program is impressive. In 1964 the nalorphine test was administered to about 3000 parolees, an average of three times a month. Urinalysis was started in May 1964, and within a few months 1200 specimens a month were being examined; by December 1965, 2000 specimens a month were collected.

The pharmacological basis for the nalorphine test is an antagonism between nalorphine and the morphine-like drugs, which can be detected by examining the pupil. When given to a subject who has ingested enough narcotic to produce a subjective effect, as little as 3 mg of nalorphine will usually produce pupillary dilatation or the pupil diameter will not change. Of the commonly used narcotics, codeine is an exception. Nalorphine will not produce mydriasis in codeine users until they have taken several doses of the drug. On the other hand, in individuals who have not ingested narcotics or have not used them for a period of several hours preceding the test, nalorphine usually produces pupillary constriction or miosis. Therefore, if mydriasis is seen following the administration of nalorphine, the individual being tested is suspected of using a narcotic user. If miosis results, this suggests that the subject being tested is not using a narcotic, at least in the period immediately preceding the test. The nalorphine test is described in detail in a pamphlet issued by the State of California, Department of Public Health,[2] and its place in addiction control has been discussed extensively by Poze.[3]

In addition to the pupil test the analysis of urine for narcotics or their metabolites has become practical with the development of relatively simple and inexpensive analytical procedures that can detect 1–5 μg/ml of narcotic in the urine. Gas chromatographic methods are the most sensitive, but thin-layer chromatography is satisfactory for mass screening of urine samples. In California two tests have been used.[4] One (Test A) detects morphine and its derivatives including heroin, which is metabolized to morphine, and the other (Test B) is used to detect synthetic narcotics and amphetamines. The tests involve extraction of the compounds from urine and separation and identification by thin-layer chromatography as described in a manual issued by the U.S. Public Health Service.[5]

In California the pupil test and urinalysis have been combined. The nalorphine test is administered to parolees, and if the results are positive or equivocal (pupil size does not change following nalorphine), or if the subject exhibits needle marks or there is any other reason for suspicion, a urine specimen is collected on the spot and sent to the laboratory for analysis. In this system the pupil test is used for screening and urinalysis for confirmation.

The nalorphine test was used extensively before it was examined under controlled conditions. However, when it was decided to include this test in a statewide narcotic control program, its advantages, limitations, and reliability were determined experimentally by measuring the effects of the two available narcotic antagonists nalorphine and levallorphan on the pupil diameter of untreated subjects[6] and of subjects who were given known single doses of narcotics.[7] The results of pupil tests done in a metropolitan testing center were compared with the results of analysis of urine for morphine and metabolites using urine voided simultaneously with the pupil test.[8] Orientals heavily addicted to heroin and opium also have been used to evaluate the test,[9] and it also has been studied in subjects given one or multiple doses of codeine.[10] It is apparently (perhaps surprisingly) safe since no reports of serious side effects have been made even at a time when more than 100,000 tests year were being administered in California.

II. MEASUREMENT OF PUPIL SIZE

The amount of change in pupil size resulting from known doses of narcotics or narcotic antagonists has been determined by a photographic method on volunteers with a history of narcotic abuse. In the experimental situation from which most of the following data were derived, pupil diameter was measured in a constantly lighted testing area, 3 mg of nalorphine were administered intramuscularly, and pupil diameter was remeasured 30 min later. Since in the field there is generally time for only one reading, we used the 30-min interval to conform to established practice. The effects of nalorphine alone on pupil diameter have been determined in two different subject populations. In a group of students, 3 mg of nalorphine decreased pupil diameter by 0.3 mm or more. In a group of 200 former drug users given a placebo followed by 3 mg of nalorphine the average decrease in pupil diameter was 0.76 mm. False positive readings in which the pupil diameter increased +0.1 mm or more were obtained in ten subjects for an apparent accuracy of 95%. However, since eight of the ten subjects who gave false positive tests also were given eye drops containing a myrdiatic or miotic agent, the true accuracy was probably 99%. Analysis of data collected at one metropolitan testing center indicates that false positives following nalorphine are probably quite rare. The incidence of positive nalorphine tests (+0.25 mm or greater was considered positive) was 0.39% in more then 60,000 tests and 0.2% in more than 50,000 tests performed the following year. The incidence of false positives was even lower since some of these were true positives. Generally the decrease in pupil size can be easily determined with a simple card pupillometer, but since pupil size may vary spontaneously over a period of time a few false positive tests are probably unavoidable. Their incidence undoubtedly could be reduced by making several readings or by increasing the dose of nalorphine, but the former is not practical because of time considerations and the latter would increase the incidence of unpleasant side effects.

In the field the pupil size is usually determined with a simple card pupillometer by matching the pupil with a black dot of known diameter. Experimentally a hole pupillometer has been evaluated. With this instrument pupil diameter is determined by superimposing a hole of known diameter over the pupil, which is illuminated by light from an ophthalmoscope. The majority of experimental work, however, has been done with a photographic method in which the pupil is photographed with a millimeter scale and later measured at the investigator's leisure. Analysis of data obtained from subjects whose pupil diameters were measured by all three methods simultaneously indicates that the camera is only slightly better than the simple pupillometers, except that the photographic method provides a permanent record. In 12 experiments that involved 136 subjects who received narcotics, 91% of the tests conducted with the camera were positive or equivocal while 83% of the tests made with a card and hole pupillometer were positive. In 35 subjects who were given a placebo instead of a narcotic all three methods yielded negative results in more than 90% of the cases.[11] These results indicate that narcotics or antagonist alone or in combination produce relatively large changes in pupil diameter so that the simple pupillometers are adequate for field work. They should produce few false positive tests since the average decrease in pupil diameter resulting from nalorphine alone is 0.76 mm.

III. NALORPHINE TEST

Since narcotic use is often on a spree basis but may be continuous in certain situations, the nalorphine test has been studied after single doses of various narcotics and during and following continuous administration of morphine[7] and codeine.[10] Results are usually positive for a period of 2 to 4 hr following single intramuscular doses of 15 mg of morphine, 5 mg of heroin, 15 mg of methadone, 150 mg of meperidine, and an oral dose of 25 mg of oxycodone that is the narcotic in Percodan. Negative tests usually resulted after an intramuscular dose of 90 mg of codeine or oral doses of 200 mg of dextropropoxyphene or 90 mg of dextromethorphan. When the tests were administered at intervals up to 36 hr following a single 15-mg dose of morphine, about half the tests were negative at 6 hours and 90% were negative at 12 hr. In subjects given a single dose of a narcotic, 3 mg of nalorphine increases pupil diameter about 0.2 mm. Development of tolerance to a narcotic may be required to increase this value since increasing the dose of nalorphine or giving three 15 mg doses of morphine every 2 hr did not raise it.

Increasing the concentration of narcotic in the body (or brain) by repeated administration results in a more positive nalorphine test for a longer period following the last dose of narcotic. When 30 subjects were given 15-mg doses of morphine every 6 hr for 5 days the nalorphine test was positive 4 hr after the seventh dose of morphine in 29 subjects, and results were positive in nine subjects 20 hr after the last dose.[7] The average increase in pupil diameter was 0.4 mm. Thus, not surprisingly, the nalorphine test is more reliable in the detection of chronic narcotic use than in the detection of sporadic use. The same conclusion was reached in a study of heroin and opium addicts; however, in these subjects either a large dose of heroin intravenously shortly before administration of nalorphine or the onset of severe withdrawal symptoms could prevent the pupillary effects of nalorphine.[9]

The negative response to nalorphine seen after propoxyphene or dextromethorphan was expected since these drugs are much less potent than morphine. The negative results following a single dose of codeine are not well understood, and since narcotic users also will abuse codeine this drug has been studied more thoroughly.[10] In a study in which the dose of codeine was increased over a period of 5 days from 60 to 240 mg orally every 6 hr most subjects gave positive nalorphine tests on the third day. The degree of positivity roughly corresponded to the urinary concentration of morphine, which is a minor metabolite of codeine, suggesting that the nalorphine test becomes positive in codeine users only after body concentrations of morphine achieve a certain level. However, in dogs, which do not metabolize codeine to morphine, given multiple daily injections of codeine the nalorphine test did not become positive until the third day—a finding which suggests that the development of tolerance to codeine is responsible for development of a positive nalorphine test. In another experiment four human subjects were given a continuous intravenous infusion of 30 mg codeine per hour and the nalorphine test was positive during the ninth hour of infusion. Development of a significant body concentration of morphine might explain this since morphine was appearing in the subjects' urine at that time but the development of acute tolerance to codeine cannot be ruled out as an explanation.

In the application of the nalorphine test it has been recognized that miotic or mydriatic drugs might interfere with the actions of the narcotic or antagonist, thus obviating the results of the administration of nalorphine. We have found that the miotic drugs pilocarpine and chlorpromazine decrease pupil size sufficienty to make the test more difficult by forcing the observer to measure small pupils. However, they do not qualitatively change the usual pupillary effects of narcotics and antagonists, which is also true for single doses of the mydriatic drugs atropine, scopolamine, homatropine, amphetamine, and methamphetamine. When sympathomimetic drugs are administered locally their actions are complex and in certain cases may interfere with the pupil test. However, the best antagonists are the narcotics themselves. When morphine or heroin are given in conjunction with nalorphine the usual 3-mg dose of nalorphine is overwhelmed and a false negative response to the test occurs. Codeine also may produce a false negative test when administered intravenously immediately following a dose of morphine.

IV. NALOXONE TEST

In an attempt to increase the sensitivity of the pupil test we have studied in volunteer former drug users the narcotic antagonist naloxone, which has practically no morphine-like actions alone and in conjunction with nalorphine.[12] Subjects were given various narcotics or placebo 3 hr prior to challenge with 0.4 mg of naloxone alone or in conjunction with 3 mg of nalorphine intramuscularly. The narcotics listed in Table I were administered in the doses and by the routes shown. The number of subjects in each experiment is indicated in the columns describing the change in pupil diameter. Pupil diameter was measured at 180 min (all subjects who received the narcotics showed miosis), naloxone 0.4 mg was administered intramuscularly, and pupil diameters were redetermined 30 min later.

Naloxone increased pupil diameter in all subjects who had received methadone, morphine, or oral meperidine and in three of five subjects who received meperidine by the intramuscular route. In a previous study[7] nalorphine did not invariably increase pupil diameter of subjects given these drugs, and the average increase was less ($+0.2$ mm). The short biologic half-life of meperidine is probably responsible for the lack of mydriasis in two of five subjects given meperidine in this study.

TABLE I
Effect of Naloxone 0.4 mg i.m. on Pupil Diameter Following
Administration of Narcotic or Placebo

Narcotic	Dose (g)	Route	Time of Challenge (min)	Change in Pupil Diameter			Mean Change
				In-crease	No Change	De-crease	
Methadone	0.015	i.m.	180	10/10			+0.6
Morphine	0.015	i.m.	180	10/10			+0.5
Morphine	0.015	i.v.	180	5/5			+0.3
Meperidine	0.100	p.o.	180	5/5			+0.4
Meperidine	0.100	i.m.	180	3/5	1/5	1/5	+0.4
Propoxyphene	0.195	p.o.	180	7/10	1/10	2/10	+0.3
Propoxyphene	0.200	i.v.	180	7/7			+0.4
Percodan	0.020	p.o.	180	4/5	1/5		+0.6
Codeine	0.090	p.o.	180	5/10	3/10	2/10	0.0
Codeine	0.060	i.m.	180	7/10		3/10	+0.2
Placebo		i.m.	30	2/5	2/5	1/5	+0.4
Placebo		i.m.	180	3/10	3/10	4/10	0.0
Nalorphine	0.005	i.m.	30	4/10	3/10	3/10	+0.1

All subjects given propoxyphene by the intravenous route and the majority of subjects given propoxyphene or Percodan by the oral route showed an increase in pupil diameter following naloxone. We were previously unable to produce mydriasis with nalorphine in most subjects given propoxyphene by the oral route. Individual differences in drug absorption from the gastrointestinal tract probably account for the lack of mydriasis in four subjects given the drugs by the oral route.

Five of ten subjects given codeine orally and seven of ten given codeine intramuscularly showed mydriasis following naloxone. This is in contrast to our previous finding that nalorphine increased miosis when given following single injections of codeine.[10] In this study after codeine orally the response to naloxone may be random as is true following placebo, but this does not seem likely after codeine given by the intramuscular route. Possibly some individuals can metabolize enough codeine to morphine in 3 hr to account for our results. Naloxone produces only random changes in pupil diameter following placebo, and thus it cannot by itself give a negative narcotic use test. The fact that it also produces random changes following nalorphine (did not antagonize miosis caused by nalorphine) suggests that it might be used in conjunction with nalorphine to increase the sensitivity of the pupil test. In Table II are shown the results of giving nalorphine, nalorphine followed by naloxone, and nalorphine plus naloxone in the same syringe to subjects given methadone or placebo. Pupil diameter decreased in all subjects but one given placebo and antagonists. Nalorphine alone increased pupil diameter in 11 of 14 subjects given methadone. After naloxone, pupil diameter was increased in 13 of 14 subjects and the mean increase in diameter

TABLE II

Effect of Nalorphine and Nalorphine Plus Naloxone on Pupil Diameter Following Administration of Narcotics or Placebo

Narcotic	Dose (g)	Route	Time of Final Challenge (min)	Change in Pupil Diameter			Mean Change
				In-crease	No Change	De-crease	
Methadone	0.015	i.m.					
+ Nalorphine	0.003	i.m.	180	11/14	1/14	2/14	+0.2
Methadone	0.015	i.m.					
+ Nalorphine	0.003	i.m.	210	13/14		1/14	+0.4
+ Naloxone	0.0004						
Methadone	0.015	i.m.					
+ Nalorphine	0.003	i.m.	180	8/10	1/10	1/10	+0.4
− Naloxone	0.0004						
Placebo		i.m.					
+ Nalorphine	0.003	i.m.	180			6/6	−0.9
Placebo		i.m.					
+ Nalorphine	0.003	i.m.	210			6/6	−0.9
+ Naloxone	0.0004						
Placebo		i.m.					
+ Nalorphine	0.003	i.m.			1/5	4/5	−0.7
− Naloxone	0.0004						

was doubled. Nalorphine and naloxone combined increased pupil diameter in eight of ten subjects given methadone. The mean change was again twice that expected for nalorphine alone. This preliminary study suggests that the use of nalorphine plus naloxone provides a more sensitive pupil test for determining the presence or absence of narcotics in humans. The combination should be studied further when naloxone becomes generally available.

V. COMPARISON WITH URINALYSIS

When the standard nalorphine test was compared to the analysis of urine for narcotics and metabolites it was found that sporadic narcotic use could be detected more reliably by urinalysis than by the pupil test. Thus in one study when urine specimens were collected from 30 subjects 36 hr after a single 15-mg dose of morphine 85% of the urines were positive for morphine while the pupil test was positive in one subject.[7] In addition, the occasional use of codeine will not be detected by the pupil test but metabolites of a single dose will appear in the urine for 24–48 hr. However, in a program involving thousands of subjects it might be more economical to reserve urinalysis for suspected cases if there is a high degree of correlation between the pupil and urine test. The question of such a correlation was first studied by analytical methods that would not detect synthetic narcotics.[8] Agreement resulted in 85% of 419 comparisons, partly because 373 of the subjects studied had negative pupil tests, and experimental work indicates that few false positive responses to the pupil test should occur. The incorporation of urinalysis into a large scale statewide testing program provided an opportunity to compare the pupil and urine tests under the condition that urinalysis was requested when the response to the pupil test suggested drug use.[11]

During the last seven months of 1964, about 40,000 pupil tests were performed at one testing center on approximately 1200 subjects, and 1004 urine specimens were collected from these subjects and submitted for analysis. Forty-seven percent of 160 subjects with positive pupil tests and twenty percent of 844 subjects who responded equivocally to the pupil test gave evidence of drug usage by the method of urinalysis (Test A). Approximately the same results were obtained from the analysis of statistics from the same testing center for the period of August 1965 to July 1966, even though urinalysis was done in a different laboratory and Test B for synthetic narcotics was made when requested. Approximately 50,800 pupil tests yielded 101 positive and 1019 equivocal results. Urine specimens obtained from 1075 of these 1120 subjects showed that 53.4% of the 88 subjects with positive pupil tests had narcotics in the urine and 14.7% of 987 subjects who responded equivocally to the nalorphine test had narcotics in the urine. Thus the use of Test B did not improve correlation between the results of the pupil test and the chemical test even though it detected both meperidine and methadone in some urines. The correlation between responses to the pupil test and urine test is low when compared with our original study, but this is largely the result of differences in sample. In the more recent study urine specimens were collected only from those subjects whose pupillary reaction to nalorphine suggested narcotic use, whereas the original study included a much larger group of subjects who reacted negatively to nalorphine. Even so, a roughly 50% incidence of positive response to the nalorphine test in subjects whose urine contained no drug is disturbing and suggests that neither the pupil nor the chemical test may be as reliable in the field as under controlled conditions. In this regard it should be noted that half the subjects who reacted positively to the pupil test in the more recent study and whose urine contained no narcotic either admitted use of heroin within the 48 hr prior to testing or had fresh needle marks on their arms.

There are, of course, numerous possible reasons for the observed discrepancies, including such obvious ones as errors in reading pupil size and in obtaining urine specimens or in the chemical procedures involved in the analysis of urine for narcotics. In addition, emotion or various stimuli may alter pupil size, and experience has shown that perhaps 1% of the population will not consistently show measurable miosis after 3–4 mg of nalorphine.

In conclusion it appears that the pupil test is useful as a screening method for the detection of narcotic abuse and, when combined with urinalysis in suspicious cases, provides a definite deterrent to the continuous use of narcotics.

VI. REFERENCES

1. J. G. Terry and F. L. Braumoeller, *Calif. Med. 85*, 299–301 (1956).
2. State of California Department of Public Health, Berkeley, California. Recommended procedure for narcotic use testing of probationers and parolees (1961).
3. R. S. Poze, *Stanford Med. Bull. 20*, 1–23 (1962).
4. K. D. Parker, C. H. Hine, N. Nomof, and H. W. Elliott, *J. Forensic Sci. 11*, 152–166 (1966).
5. K. D. Parker and C. H. Hine, *Psychopharmacol. Bull. 3*, 18–42 (1966).
6. H. W. Elliott and E. L. Way, *Clin. Pharmacol. Ther. 2*, 713–721 (1961).
7. H. W. Elliott, N. Nomof, K. D. Parker, M. L. Dewey, and E. L. Way, *Clin. Pharmacol. Ther. 5*, 405–413 (1964).
8. E. L. Way, H. W. Elliott, and N. Nomof, *Bull. Narcotics 15*, 29–33 (1963).
9. E. L. Way, B. P. N. Mo, and C. P. Quock, *Clin. Pharmacol. Ther. 7*, 300–311 (1966).
10. H. W. Elliott, N. Nomof, and K. D. Parker, *Clin. Pharmacol. Ther. 8*, 78–95 (1967).
11. H. W. Elliott, N. Nomof, K. D. Parker, and G. R. Turgeon, *Calif. Med. 109*, 121–125 (1968).
12. H. W. Elliott, N. Nomof, and K. D. Parker, Presented at the February 1968 meeting of the NAS-NRC Committee on Problems of Drug Dependence, Indianapolis, Indiana.

SUBJECT INDEX

Calcium (cont.)
role in growth inhibition of *E. coli*, 323
in tolerant animals, 307
transport in brain, 305
Catecholamines and 5-hydroxytryptamine, 229–253
Cerebral cortex slices
effect of morphine on respiration, 304
KCL-stimulated respiration, 177
respiration, 169
uptake of drugs, 131
competitive inhibition, 131
inhibition by metabolic inhibitors, 131
Cerebral enzymes, effect of chronic drug treatment, 180 (*table*)
Cerebral enzymes, effect of morphine, 178–182
Cerebral enzymes, effect of narcotics *in vitro*, 179 (*table*)
Cerebral glycolytic intermediates, effect of acute and chronic morphine, 182–186
Cerebral metabolic effect of morphine *in vitro*, 172 (*table*)
Cerebral metabolism, effect of morphine, 171–178
Chemical reactions of narcotic drugs, 138–139
Choline
effect of morphine on synthesis, 257
uptake in choroid plexus, 125
Choroid plexus, uptake of drugs, 125
Chromogenic reagents for narcotic analgesics, 38 (*table*)
Circular dichroism, of narcotics, 58
Codeine
brain levels of drugs, 111
differential spectrophotometry, 63
drug disposition in the CNS, 111–112
drug metabolism, 145–146

Codeine (cont.)
effect on EEG, 453
effect on growth of tissue in culture, 314
effect on TSH secretion, 290
gas–liquid chromatography, 73
levels in brain, 111
the mass spectrum, 62
polarography, 75
spectrofluorometry, 64
structure, 2
tail-flick test, 62
ultraviolet absorptiometry, 63
uptake in brain slices, 130
uptake in choroid plexus, 125
Color reactions
ammoniacal silver nitrate reagent, 40
cobalt thiocyanate reagent, 40
direct tests on solids, 40
Dragendorff's reagent, 40
Fröhde reagent, 40
iodoplatinate reagent, 40
Marquis reagent, 40
Mecke reagent, 40
microtechnique of Clarke and Williams, 40
official pharmaceutical tests, 41
p-dimethylaminobenzaldehyde reagent, 40
potassium permanganate reagent, 40
sodium tungstate reagent, 40
titanium dioxide reagent, 40
Color reactions for narcotic drugs, 42 (*table*)
Column chromatography
absorption and partition, 19
ion exchange, 20
Concepts of biological transport, 123–124
Corticosteroid hormones, 262–282
CRF, secretion influenced by morphine, 288
Criteria for tolerance and physical dependence, 409–414